Practical
Heating, Ventilation,
Air Conditioning
and Refrigeration

Henry Puzio
Sussex County Vocational-Technical School

Jim Johnson
Pima Community College

Delmar Publishers

1945 - 1995
50 years

I(T)P **An International Thomson Publishing Company**

Albany • Bonn • Boston • Cincinnati • Detroit • London • Madrid • Melbourne
Mexico City • New York • Pacific Grove • Paris • San Francisco • Singapore • Tokyo
Toronto • Washington

NOTICE TO THE READER

Publisher does not warrant or guarantee any of the products described herein or perform any independent analysis in connection with any of the product information contained herein. Publisher does not assume, and expressly disclaims, any obligation to obtain and include information other than that provided to it by the manufacturer.

The reader is expressly warned to consider and adopt all safety precautions that might be indicated by the activities herein and to avoid all potential hazards. By following the instructions contained herein, the reader willingly assumes all risks in connection with such instructions.

The publisher makes no representation or warranties of any kind, including but not limited to, the warranties of fitness for particular purpose or merchantability, nor are any such representations implied with respect to the material set forth herein, and the publisher takes no responsibility with respect to such material. The publisher shall not be liable for any special, consequential, or exemplary damages resulting, in whole or part, from the readers' use of, or reliance upon, this material.

I would like to thank my wife, Sherri and daughter, Jodie, for all of their support.
—Henry Puzio

In the natural order of things, this one is for Jay.
—Jim Johnson

Cover design: Align Design
Cover photos courtesy of Robinair Division-SPX Corporation and Lennox Industries, Inc., celebrating 100 years of comfort.

Delmar Staff
Publisher: Susan Simpfenderfer
Senior Administrative Editor: Vern Anthony
Project Editor: Patricia Konezeski
Art/Design Coordinator: Cheri Plasse

COPYRIGHT © 1996
By Delmar Publishers
a division of International Thomson Publishing Inc.
The ITP logo is a trademark under license

Printed in the United States of America

For more information, contact:

Delmar Publishers
3 Columbia Circle, Box 15015
Albany, New York 12203-5015

International Thomson Publishing Europe
Berkshire House 168-173
High Holborn
London WC1V 7AA
England

Thomas Nelson Australia
102 Dodds Stree
South Melbourne, 320
Victoria, Australia

Nelson Canada
1120 Birchmont Road
Scarborough, Ontario
Canada M1K 5G4

International Thomson Editores
Campos Eliseos 385, Piso 7
Col Palanco
11560 Mexico D F Mexico

International Thomson Publishing GmbH
Königswinterer Strasse 418
53227 Bonn
Germany

International Thomson Publishing Asia
221 Henderson Road
05-10 Henderson Building
Singapore 0315

International Thomson Publishing Japan
Kyowa Building, 3F
2-2-1 Hirakawacho
Chiyoda-ku, Tokyo 102
Japan

All rights reserved. No part of this work covered by the copyright hereon may be reproduced or used in any form or by any means—graphic, electronic, or mechanical, including photocopying, recording, taping, or information storage and retrieval systems—without the written permission of the publisher.

1 2 3 4 5 6 7 8 9 10   XXX   01 00 99 98 97 96 95

Library of Congress Cataloging-in-Publication Data

Puzio, Henry.
    Practical heating, ventilation, air conditioning, and
  refrigeration / by Henry Puzio and Jim Johnson.
       p.   cm.
    Includes index.
    ISBN 0-8273-5591-2 (alk. paper)
    1. Heating.   2. Ventilation.   3. Air conditioning.
  4. Refrigeration and refrigerating machinery.   I. Johnson, Jim,
  1950-   . II. Title.
  TH7012.P89   1996
  697—dc20   94-30878
                                                CIP

CONTENTS

FIGURES

PREFACE

The authors of this book wrote it because they believed that there was a need for a new, comprehensive HVAC/R book that took a truly practical approach. Practical to us means that which is best suited to a specific task; our intent has been to create the book we felt would be most efficient, useable, and useful to the students we both have encountered over many years of teaching and working in the field.

To this end, this book is comprehensive and current, yet we have striven not to go beyond the level of a technician, especially a technician new to the field. Many of the books currently on the market are encyclopedic, more reference than textbook. We are hoping that we have created a textbook that most students will actually be willing to read. We have worked hard to use language most students will be able to understand. We have tried to make technical terms understandable and have taken care to explain new concepts clearly and succinctly.

This book is, most of all, focused on the major HVAC/R systems from a service orientation—"how to get the job done" is what we wanted to answer for the new student. We fully discuss the how's and why's of HVAC/R systems, component parts, theory of operation, applications, service and repair procedures, and diagnostic procedures. We include troubleshooting techniques throughout the book.

ORGANIZATION

This book is organized into what we as authors feel are logical sections and units. As no one book can fit all needs, we have tried to make the sections and units discrete enough to be re-organized to your individual tastes or needs.

We have started this book with the real basics every student should have prior to working on HVAC/R equipment; units one through three cover safety, tools, and practical HVAC/R math.

The second unit follows up with another essential topic, electricity. We found that many other books do not cover this topic until much later. We felt that earlier coverage gave students information that is essential at the beginning of the course.

The next four sections cover refrigeration, domestic refrigeration (refrigerators, freezers and room air conditioners), residential heating and cooling, and commercial. The important topic of CFC recovery, recycle and reclaim was covered in its own complete section within the third unit (refrigeration). The final unit of the book covers the HVAC/R business, including regulating and licensing agencies, insurance policies, factory authorized service centers, computers, and customer relations.

FEATURES

- Clear behavioral OBJECTIVES open every chapter.
- Carefully chosen ILLUSTRATIONS are used throughout. The emphasis is on the right picture, versus cluttering the book with a picture of every possible item.
- SUMMARIES conclude every chapter.
- KEY TERMS are bold faced on first use and collected into a chapter-ending KEY TERM list.

- End-of-chapter REVIEWS are designed to emphasize critical thinking skills, to integrate academic topics (especially math and English), to promote independent work and classroom discussion, and to provide basic review questions like those found in most other books. The questions at the end of the chapter in this book are both plentiful and varied and are meant to be a real addition to your class, not just an afterthought or something the publisher told us we had to do.
- SAFETY is emphasized throughout and safety guidelines are included in all units as applicable.
- GLOSSARY of terms is extensive, and is provided in both English and Spanish.

ANCILLARIES

This textbook is accompanied by an excellent Workbook. This workbook contains laboratory exercise job sheets structured in a competency-based format; corresponding progress charts are included in the Instructor's Guide. The Workbook also includes other exercises which encourage learning through problem solving exercises that require critical thinking skills, group projects, math and English exercises, and practice tests. There is also a unique exercise section on resume writing/application form completion that is tailored for the HVAC/R technician. The Instructor's Guide contains answers to all of the end-of-chapter review questions as well as answers to the Workbook exercises.

We hope you find this book to be a useful addition to your classroom and an excellent reference for your students when they complete the course. We welcome your comments and suggestions, and ask that you address them to:

> Delmar Publishers Inc.
> Technical Careers Team
> 3 Columbia Circle
> Box 15015
> Albany, NY 12212-5015

The Authors

SECTION ONE

Safety, Tools and Math

Students who decide to enter a trade and technical field are sometimes surprised to find that their study in the skill-related areas of their career choice must be preceded by a review of common sense issues and the study of fundamentals such as math basics and the proper use of tools. The tendency of some to rush into learning about the things for which they have a strong interest can create situations that result in ineffective job performance or, worse, an unsafe situation for technicians and those working with them.

In this session, we'll discuss safety issues that you must always be aware of, as well as tools used in the HVAC/R trade. In addition, basic math skills, without which a technician cannot fully function, will be covered. Investing time in these fundamental topics before you begin the studies of electricity and refrigeration will, in the long run, pay off in substantial dividends.

UNIT ONE

Safety

OBJECTIVES

After completing this unit, the student will be able to:

1. Understand effective safety practices.

2. Relate electrical and compressed gas hazards to bodily injury.

3. Relate on-site and in-shop safety to accident and injury prevention.

As a heating, ventilation, air conditioning, and refrigeration (HVAC/R) technician, the reader must develop a safety awareness that will lead to an attitude and consciousness resulting in safe work practices and accident prevention. An effective technician is able to overcome the "It can't happen to me" attitude.

Safety consciousness requires an education in safety, both generally and specifically. As you go about your work, take the time to ask yourself: "Is what I am about to do unsafe in any way to myself, to others, or to property and equipment?" Safety consciousness includes understanding the necessity for a clean and orderly work environment, being aware of possible accident situations (where respect replaces fear), and knowing the importance of rules and regulations. It's also necessary to learn the correct way to perform a task and to properly use and maintain tools and machines.

ELECTRICAL SAFETY

Because approximately 85 percent of the service problems you'll encounter in air conditioning and refrigeration service are electrical in nature, you'll probably be exposed to electrical hazards on a daily basis so you will need to be familiar with safe and proper procedures related to equipment usage.

The human body consists mostly of water so the resistance the human body offers to electrical energy may not prove adequate to prevent serious injury. The three factors that determine the extent of harm from electrical shock are the amount of current flow in the circuit, the path of current flow, and the length of time the current flow travels through the body. To gain a full understanding of how electrical shock affects the body, consider the following information.

If the current in the circuit is:

A. less than 5 milliamperes ...no sensation

B. 2 to 10 milliamperes ....muscle contraction

C. 2 to 25 milliamperes ....painful shock, unable to let go

D. over 25 milliamperes ....violent muscular contraction

E. over 100 milliamperes ......paralysis of breathing, burns

F. 50 to 200 milliamperes .....heart convulsions, death

We've identified the current in the above examples in **milliamperes**. One **ampere** is equal to 1,000 milliamperes. To put this in perspective, consider an average air conditioning unit. It's common for a unit used on the average residence to have a current draw of 15 amperes. Pause for a moment to consider how many milliamperes are in 15 amperes and review the information above to consider how many milliamperes it takes to deliver a fatal shock.

It is important to realize that the *current* is the shock factor, not the amount of voltage in a circuit. The body doesn't even have enough resistance to prevent a shock from voltage as low as 115 **volts alternating current (VAC)**. In many cases, voltages less than 115 volts can be fatal.

ELECTRICAL SAFETY RULES

1. NEVER assume a circuit is off. ALWAYS check with **voltmeter**.

2. Remove all rings and bracelets when working with electrical circuits.

3. NEVER bypass electrical safety devices.

4. Tag and lock all disconnect boxes when working on equipment.

5. Stand on dry, nonconductive surfaces when working on live circuits.

6. Make sure all circuits are properly fused.

7. Don't run electrical wires directly over refrigerant lines.

8. Use properly grounded power tools.

9. Work with insulated ladders.

10. Don't work when you're tired or taking medication that will make you drowsy.

11. Discharge all capacitors before handling.

12. Wear safety shoes with nonconductive soles.

13. Wear safety glasses.

14. Avoid using volatile liquids such as alcohol, turpentine, or lacquers around electrical equipment. Fumes may ignite by a spark.

15. NEVER use tools with damaged cords.

SAFETY AND COMPRESSED GASES

As an HVAC/R technician you will work with a variety of compressed gases on a daily basis, so proper handling of tanks and equipment is vital to your safety and to the environment.

Oxygen/Acetylene

Never use oxygen or acetylene to pressurize a refrigeration system. Oxygen will explode on contact with oil. Acetylene will explode under pressure if not properly dissolved in acetone, as used in commercial acetylene cylinders.

Nitrogen/Carbon Dioxide

Dry nitrogen or dry carbon dioxide can be used to pressurize HVAC/R systems and perform leak tests, provided specific safety precautions are followed. Nitrogen (N_2) contains pressure in excess of 2,000 pounds per square inch (PSI) at normal room temperature. Carbon dioxide (CO_2) contains pressure in excess of 800 PSI. Working with pressures this extreme can be dangerous.

NEVER attempt to pressurize a system by installing a pressure regulator on the cylinder. This regulating valve assembly, illustrated in Figure 1–1, should be equipped with two functioning gauges—one that indicates cylinder pressure and one that indicates discharge or **downstream pressure**.

Always install a relief valve or a frangible, disk-type safety relief in the supply line set to release at 175 pounds per square inch gauge (PSIG). When leak testing a refrigeration system, set the regulator to deliver a maximum pressure of 150 PSIG.

CYLINDER SAFETY RULES

1. Don't drop or bump cylinders.

2. Keep cylinder in a vertical position.

3. Fasten securely, when transporting, to prevent tipping. (refer to Figure 1–2)

4. NEVER heat a cylinder with a torch.

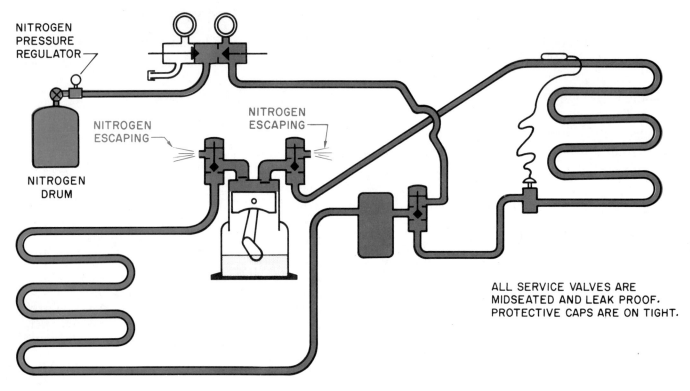

Fig. 1–1A   Regulating valve assemblies.

Fig. 1–1B   Regulating valve assemblies.   *Photo by Bill Johnson.*

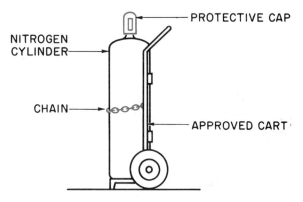

Fig. 1–2   Pressurized cylinders should be moved safely on an approved cart and chained. The protective cap must be secured.

5. Heat by immersing the lower portion of the cylinder in warm water if heat is necessary to withdraw gas from a cylinder.

6. Never heat a cylinder over 110°F.

7. ALWAYS move cylinders with the protective caps in place.

8. Wear safety glasses. (see Figure 1–3)

REFRIGERANTS

Group one refrigerants, such as those most commonly used in the HVAC/R systems (R-12,

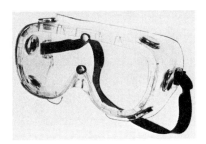

Fig. 1–3   Safety glasses. *Photo by Bill Johnson.*

R-22, and R-502), are generally considered chemically nontoxic and nonflammable. However, because any gas under pressure can be hazardous, you should adhere to all rules regarding cylinder handling.

Never fill a refrigerant cylinder completely full with liquid. Fill only to 80 percent capacity to allow for normal expansion.

A sudden release of liquid refrigerant contacting the skin or eyes can cause serious frostbite damage to the skin tissue. If a burn occurs, flush immediately with cold water, apply ice packs, and seek medical attention.

TORCH AND SOLDER SAFETY

When servicing and installing HVAC/R equipment, you will be required to use a torch for **soldering** and **brazing**. Always keep a fire extinguisher close at hand when working in a construction area or in any situation where combustible materials are nearby. When working with a torch in tight spots where combustible materials are nearby or where a finished surface will be damaged by excessive heat, a heat shield such as the one shown in Figure 1–4 should be used.

SOLDERING SAFETY RULES

1. Always solder in a well-ventilated area.
2. Always make sure tanks are properly secured.
3. Keep tanks and hoses out of aisle ways and work areas where someone may trip over them.
4. Use a proper flint lighter when lighting a torch.

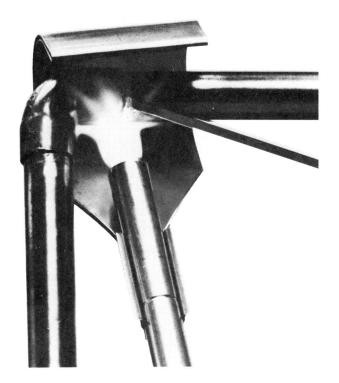

Fig. 1–4 A shield used when soldering. *Courtesy Turbotorch Division, Victor Eq. Co., Denton, Texas.*

5. Make sure a refrigeration system is vented to prevent a build-up of gas pressure during soldering.
6. Watch your flame pattern at all times to prevent personal injury or equipment damage.
7. Treat any burn injuries immediately with a crushed ice pack.
8. Use care in handling work that has been heated.
9. Wear safety glasses.
10. Wear protective clothing when required.

LIFTING AND MOVING HEAVY OBJECTS

Muscle strains and sprains are often caused by improper methods of lifting. Learning to lift the safe way will prevent back injury. Before lifting, bend the knees, keep the back erect and, then, lift gradually with the leg muscles as shown in Figure 1–5.

Heavy equipment should be handled with hand trucks, dollies, and lifts. Cranes are used to lift equipment that is to be placed on the roof.

Fig. 1–5   Use legs, not back, to lift objects. Keep back straight.

Helicopters are also used in places where cranes may not reach.

GENERAL ON-SITE AND IN-SHOP SAFETY

As an HVAC/R technician, you will work in a variety of in-shop and on-site locations. In all situations, basic rules of safety should be observed. Your state of mind and attitude can cause—or prevent—serious accidents, injury, or even death. Safety means freedom from accidents caused by falling, inhaling, temperature exposure, electrical contact, or carelessness.

Unit One Summary

- The air conditioning, refrigeration, and heating service technician must be educated in safety both generally and specifically.

- It is essential that work habits lend themselves to developing a positive safety attitude.
- Electrical hazards are common and the HVAC/R technician should use care at all times when working with electricity.
- Current, rather than voltage, is the shock factor in electricity.
- Never pressurize a refrigeration system with oxygen or acetylene.
- Always use a regulator when pressurizing a system or using compressed gases.
- Always work in a well-ventilated area when working with gases.
- Keep a fire extinguisher close at hand when soldering or brazing.
- Learn to lift properly to avoid injury.
- When exposed to liquid refrigerant burns, immediately flush affected area with cold water and apply ice packs.

Unit One Key Terms

Milliamperes

Amperes

Voltmeter

Volts alternating current

Downstream pressure

Soldering

Brazing

UNIT ONE REVIEW

1. Complete the following sentence on safety consciousness. "I am unsafe when I
   a. _____, b. _____, c. _____,
   d. _____."

2. What are the three most important factors that determine the extent of harm from electrical shock?

3. Of the 15 electrical safety rules, give the number(s) of the rule(s) that concerns a technician's dress or appearance.

4. Of the 15 electrical safety rules, give the number(s) of the rule(s) that could affect a technician's physical condition.

5. Of the 15 electrical safety rules, give the number(s) of the rule(s) that carelessly leaves a technician's safety to the decisions of others.

6. What is the first fire protective device needed when soldering or brazing?

7. Of the 10 soldering safety rules, give the number(s) of the rule(s) that, if ignored, can result in burns.

8. Why must a regulator be used when working with gas cylinders?

9. What could cause oxygen under pressure to explode?

10. Why must care be taken when working with nitrogen and carbon dioxide cylinders?

11. Of the eight cylinder safety rules, give the number(s) of the rule(s) that involves the transport and positioning of the cylinder.

12. What will happen if a refrigeration system is pressurized with oxygen or acetylene?

13. What is the maximum recommended pressure used when testing a refrigeration system for leaks?

14. Why must rings and other jewelry be removed when working with electricity?

15. How many milliamperes can kill a person?

16. A refrigerant cylinder should never be filled to more than _____ percent capacity.

17. What is the correct method for treating a refrigerant burn?

18. List three things a technician can do to avoid electrical shock.

19. Describe three rules to follow when handling gas cylinders.

20. List four attitudes that could cause accidents.

21. List four accident sources and the methods for preventing them from occurring.

22. Why should a refrigeration system be vented during soldering?

23. List the three safety tips to help prevent muscle strains and sprains when lifting.

24. Write down the procedure to be followed in case of serious frostbite.

25. Write a paragraph in complete sentences using the following terms: flint lighter, safety glasses, soldering, burns, flame, ventilated, and fire extinguisher.

26. Write a paragraph in complete sentences using the following terms: muscle strain, technician, platform, nitrogen cylinder, hand truck, and hospital.

UNIT TWO

Hand Tools for HVAC/R Technicians

OBJECTIVES

After completing this unit, the student will be able to:

1. Identify basic hand tools used in the HVAC/R trade.

2. Relate the use of tools to their specific application.

3. Describe the methods of proper use and care of hand tools.

Most of the work you'll be doing as a refrigeration and air conditioning service technician will be accomplished through the use of hand tools. To work effectively in the industry, you need to be familiar with the different types of wrenches, screwdrivers, hammers, punches, pliers, and nut drivers you'll be using throughout your workday. In some cases, you may even need to use a tap and die set in order to perform a repair or to accomplish an installation.

WRENCHES

Wrenches are tools that are designed to tighten or loosen a hex head bolt or nut. Good quality wrenches are made of alloy steels for strength (usually chromevanadium) and are accurately machined and hardened for toughness. The shape of the opening fits over the nut or bolt snugly to prevent slipping or damage to the nut or the head of the bolt.

Open End Wrenches

Open end wrenches, such as the one shown in Figure 2–1, fit easily over the flat, parallel sur-

faces of a nut or bolt. It's important that the proper size wrench be used to prevent slipping or distortion of the nut, as illustrated.

Box Wrenches

Box wrenches are used for tightening or loosening hex head nuts, bolts, and screws. They generally have two heads and may be flat in design or offset, making them easier to use in tight places. A standard box wrench is shown in Figure 2–2.

A *ratchet-type box wrench* is also frequently used by HVAC/R technicians. A tool with the ratcheting feature such as the one illustrated in Figure 2–3 saves time because it allows you to rapidly tighten or loosen a bolt or nut.

Fig. 2–1 Open end wrench. *Photo by Bill Johnson.*

Fig. 2–2 Standard box wrench. *Photo by Bill Johnson.*

Fig. 2–3   Ratchet box.   *Courtesy Klein Tools, Inc.*

Fig. 2–4   Combination wrench.   *Photo by Bill Johnson.*

Fig. 2–5   Adjustable open end wrench.   *Photo by Bill Johnson.*

Fig. 2–6   Flare nut wrench.   *Photo by Bill Johnson.*

Fig. 2–7A   Refrigeration service valve wrench.   *Photo courtesy Bill Johnson.*

Fig. 2–7B   Air conditioning and refrigeration reversible ratchet box wrench.   *Courtesy Klein Tools, Inc.*

Combination Wrenches

A *combination wrench*, as the name implies, has both an open and a box end. Using combination wrenches cuts down on the amount of tools you have to carry. A box wrench also comes in handy when you're working in a very tight spot. You may have to use the open end of the tool and then switch to the box end in order to accomplish the removal of the nut or bolt. A combination wrench is shown in Figure 2–4.

Adjustable Open End Wrenches

Adjustable wrenches are commonly referred to as *crescent wrenches* because the name of the company (Crescent) is so closely associated with the design of the product. These wrenches are available in a variety of sizes. When using this type of wrench, it's important that you make sure it is adjusted to fit properly on the parallel surfaces of the nut or bolt. Failure to do so could result in damage to the hardware or personal injury if the wrench slips. Figure 2–5 shows an adjustable wrench.

Flare Nut Wrenches

A *flare nut wrench* is similar in appearance to a standard open end wrench with the exception of the ends, which are formed closer to that of a hex-shaped flare nut. (see Figure 2–6) This type of

wrench is commonly used by HVAC/R technicians when tightening the flare nuts on the tubing of refrigeration systems. Because of the design, you have to slide the wrench over the nut from behind rather than slip it straight onto the nut. The 12-point contact that the wrench supplies ensures maximum contact on the flat hexagonal surfaces of the flare nut or valve body. This provides maximum torque without slippage.

Service Valve Wrench

A *service valve wrench* is a specialty tool used on the stems of refrigeration system valves. There are several styles of service valve wrenches, which have a 3/16, 1/4, and 5/6 square opening. The square openings, designed to fit over the square stems of valves and fittings used in HVAC/R equipment, will be located either on the ratchet end of the tool or on the fixed end.

In addition to being used on refrigeration systems, a service valve wrench is often used to open the valve on acetylene tanks. Figure 2–7 shows two different styles of service wrenches—one equipped with a 6-point socket at the fixed end.

Fig. 2–8   Torque wrench.   *Courtesy Snap-on Tools Corporation.*

Fig. 2–9   Pipe wrench.   *Photo by Bill Johnson.*

Torque Wrenches

Torque wrenches, such as the one shown in Figure 2–8, are designed to tighten nuts, bolts, and fittings to a measured degree of tightness. They are designed so that the degree of torque may be read on a scale or measurement indicator located on the handle. In many cases, compressor manufacturers specify that certain bolts be tightened to specifications.

Pipe Wrenches

A *pipe wrench* can be designed to get a grip from the inside diameter (internal pipe wrench) or the outside diameter of a pipe. The proper place to use a pipe wrench is on metal piping used on water supply or drain lines (galvanized) or on black iron pipe used on gas supply lines. The internal pipe wrench is generally made to fit national pipe sizes from 3/8" to 1". External pipe wrenches (see Figure 2–9) are the most commonly used of the two.

WRENCH SAFETY RULES

1. Never use wrenches as hammers. This may render the tool unsafe and will reduce the service life of the tool as well as the quality of work you can do with it.

2. Always pull on a wrench rather than pushing on it. Using this method will help prevent injury if the wrench should slip or if the nut or bolt should suddenly break loose or simply break.

3. Keep tools clean and free of oil and grease.

4. Don't use an extra length of pipe as a "cheater bar" to provide additional turning force. If you don't have enough leverage with the wrench you're using, get a bigger one.

5. Use wrenches only on the nuts, bolts, or pipes for which they were intended.

6. Before applying force with an adjustable wrench, make sure you have it properly adjusted.

SCREWDRIVERS

Figure 2–10 shows some of the **screwdrivers** and screwdriver bits you may use in the HVAC/R trade. A good grade of screwdriver is of quality design with steel shafts and tempered tips that allow it to withstand twisting torque.

Some screwdriver tips are magnetized in order to hold small screws until you can get them started. Using a starting screwdriver, one equipped with two blade clips to hold the screws, is another method of holding a screw in place until it can be started.

Offset screwdrivers are formed with the tips at a 90° angle and are used in tight spots where you can't use a regular screwdriver with a full handle.

SCREWDRIVER SAFETY RULES

1. Never use a screwdriver in place of a chisel.

2. Never use a screwdriver as a pry bar.

3. Always be sure that the screwdriver tip is correctly formed to prevent slipping that could result in injury or damage to the tool or fastener.

HAMMERS

Some of the more common types of **hammers** used in the HVAC/R trade are shown in Figure

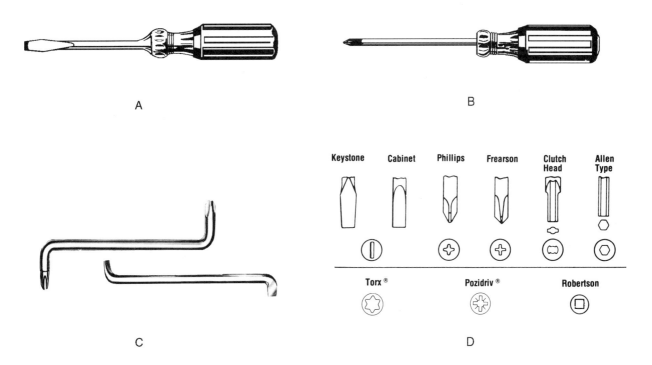

Fig. 2–10   Screwdrivers (a) straight or slot blade, (b) Phillips tip, (c) offset, (d) other standard screwdriver bit types. *Courtesy Klein Tools, Inc.*

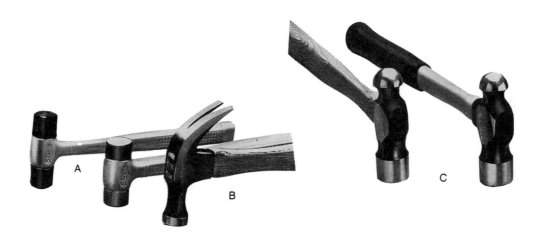

Fig. 2–11   Common types of hammers (a) soft face, (b) claw, and (c) ball peen. *Courtesy Snap-on Tools Corporation.*

2–11. The *ball peen hammer* is used for hammering on metal tools such as punches, chisels, or to drive a swaging tool (a tapered tool inserted into soft copper tubing to expand the inner diameter of the tube, allowing you to fit two pieces of tubing together in preparation for soldering).

The *soft-faced hammer* is designed for hammering on materials that a steel-faced hammer would mar or break. The soft-faced hammer has a face that is made of either plastic, rubber, wood, or another specialty material. The *claw hammer* is used primarily for driving and pulling nails.

HAMMER SAFETY RULES

1. Use the correct type and size hammer for the job.

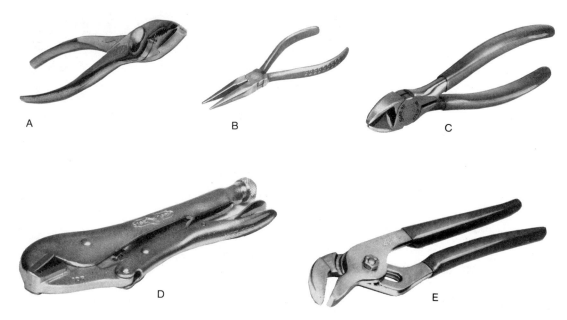

Fig. 2–12   **Basic types of pliers (a) general purpose (combination pliers), (b) needle nose pliers, (c) side (diagonal cutting pliers), (d) vise grip, and (e) adjustable (slip-joint pliers).**   *Courtesy Snap-on Tools Corporation.*

2. Wear safety glasses.

3. Don't use a hammer that has a split handle or loose head.

4. Discard a hammer that is chipped or otherwise damaged.

5. Strike the hammer solidly and squarely to avoid glancing blows.

6. Don't strike a hammer against another hammer or tool not made for that purpose.

PLIERS

Pliers are used to grip, turn, bend, and, in some cases, to strip wire. The basic types of pliers are shown in Figure 2–12. General purpose pliers have two positions, enabling you to use them on various size items. *Slip-joint pliers*, sometimes referred to as channelocks due to the close association of the brand name, open wider than standard *combination pliers* and have handles that give you better leverage.

Needle nose pliers are used for holding or reaching small items in a cramped location and can also be used to cut wire. *Diagonals* (side cutters) are used for cutting wire. *Vise grip pliers* can

be adjusted to fit a nut or bolt, then locked into place by squeezing the handles together.

FILES

Files are classified according to their length, shape, and cross-sectional form. They are used to remove burrs from metals, to shape and form material (so mating parts fit), and to clean damaged threads on the ends of bolts.

Flat files are for smoothing metal surfaces and have a double cut. *Half-round files* are generally used to enlarge holes in metal, as are *round files* (sometimes referred to as rat-tail files).

Fig. 2–13   **General-purpose files and file shapes.**   *Courtesy Snap-on Tools Corporation.*

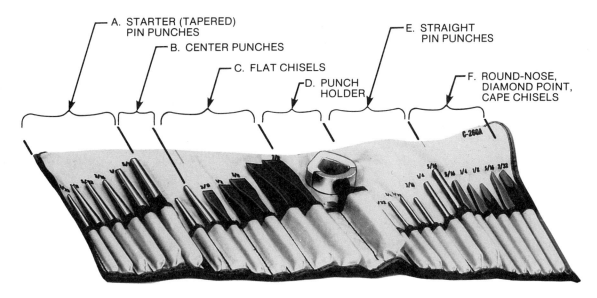

A. STARTER (TAPERED) PIN PUNCHES
B. CENTER PUNCHES
C. FLAT CHISELS
D. PUNCH HOLDER
E. STRAIGHT PIN PUNCHES
F. ROUND-NOSE, DIAMOND POINT, CAPE CHISELS

Fig. 2–14   Chisel and punch set (a) starter (tapered) pin punches, (b) center punches, (c) flat chisels, (d) punch holder, (e) straight pin punches, and (f) round-nose, diamond point, cape chisels. *Courtesy Snap-on Tools Corporation.*

CHISELS AND PUNCHES

Chisels and **punches** are made of hardened, tempered alloy steels that are designed to withstand dulling and becoming deformed by heavy blows. The *pin punch* is designed for driving straight and tapered pins into and out of hubs and shafts, and *center punches* are used to make an indentation in metal to start a drill.

The *flat* or *cold chisel* is designed for cutting off bolt heads or splitting nuts. Chisels and punches range in size and shape, as shown in Figure 2–14, to suit a variety of service requirements.

HACKSAWS

Hand hacksaws are used to cut heavy gauge metals. The thickness of the metal being cut determines the length of the saw blade and the number of teeth per inch. Fine pitch blades with 24 or 32 teeth per inch are used for lighter gauge metals while more coarse blades with 8, 10, 12, or 14 teeth per inch are used for thicker metals. When using a hacksaw, be sure you have the blade properly installed with the teeth pointing toward the front of the saw. This ensures that you'll be able to use an effective cutting stroke by

Fig. 2–15   Hacksaw. *Photo by Bill Johnson.*

applying pressure while you're pushing away from your body. Figure 2–15 shows a hacksaw.

Unit Two Summary

- Much of the work performed by refrigeration and air conditioning technicians is done with the use of hand tools. To perform tasks effectively, you must be familiar with a variety of wrenches, screwdrivers, hammers, and specialty tools used in the HVAC/R trade.

- Wrenches, whether they are open end, box, combination, or adjustable, must be used correctly to ensure safety and prevent damage to equipment.

- A service valve wrench is a specialty tool used by HVAC/R technicians to open and close valve stems on refrigeration equipment.

- Flare nut wrenches are used by technicians installing refrigerant lines with flare nuts. The 12-point contact surface of the flare nut wrench ensures maximum contact on the flare nut.

- A torque wrench is used to perform some specialty repairs or to properly assemble compressors.

- Good quality screwdrivers have steel shafts and tempered tips.

- In addition to standard steel-face hammers, some hammers have plastic, rubber, or wood faces to protect surfaces from damage.

- A technician may use various types of pliers such as slip joint, vise grip, side cutters, or needle nose pliers to cut wire and make repairs on HVAC/R equipment.

- Files are used to remove burrs, shape parts to fit, and to enlarge holes in metal. Chisels are used to cut off bolt heads or split nuts. Punches are used to remove or install shaft pins or mark metal prior to drilling.

- Hacksaw blades range from fine-pitch blades to more coarse blades, depending on the material being cut. Always be sure to install a hacksaw blade with the teeth pointing toward the front of the saw.

Unit Two Key Terms

Wrenches	Files
Screwdrivers	Chisels
Hammers	Punches
Pliers	Hacksaws

UNIT TWO REVIEW

1. What is a combination wrench?
2. Why are some adjustable wrenches referred to as "crescent" wrenches?
3. A loose fit can result in slipping or distortion of the nut. True or False?
4. A tool that is flat in design or offset works easier in tight places. True or False?
5. Flare nut wrenches have 12 points and supply maximum contact on flat hexagonal surfaces for maximum torque without slipping. True or False?
6. What is the purpose of a torque wrench?
7. What are the specialty uses of a service valve wrench?
8. What are two types of pipe wrenches?
9. When working with a wrench, should you pull or push when applying pressure?
10. Can wrenches be used as hammers in an emergency? Explain.
11. Why are screwdrivers manufactured with tempered tips?
12. What are the two things that will shorten the life of a screwdriver?
13. What are some of the hammer faces that are used on specialty hammers other than steel?
14. Give the uses for the ball peen hammer.
15. List the purposes of pliers.
16. When would a technician use a flat file? a half-round file? a round file?

17. What is the purpose of a pin punch? a center punch?

18. When using a hacksaw, how would you determine the length and the number of teeth you need?

19. Write a paragraph in complete sentences explaining hammer safety using the following terms: safety glasses, hammer, split handle, chipped, and squarely.

20. Explain procedures you would follow if you observed a fellow technician engaging in unsafe work practices.

UNIT THREE

Practical HVAC/R Math

OBJECTIVES

After completing this unit, the student will be able to:

1. Identify whole numbers, fractions, decimals, and percents.

2. Add, subtract, multiply, and divide whole numbers.

3. Describe ratio and proportion.

4. Identify units of measure.

Working as an HVAC/R technician, it's inevitable that you will find yourself in a work situation where you'll need some math skills. It may be that you'll have to measure a distance, figure a ratio, or add up a customer's bill. It's true that pocket calculators are making it easier for technicians, but that doesn't absolve you of the responsibility of having an understanding of math concepts. In the same way that you have to develop a feel for the technical work you do as a service tech, you also have to develop a feel for the math work—if for no other reason than to be able to recognize that an answer from a pocket calculator is incorrect due to a wrong number or a wrong sign having been entered.

In this unit we'll discuss fundamentals of math and how they relate to the job of an HVAC/R technician, as well as a practical approach to addition, multiplication, subtraction, and division.

WHOLE NUMBERS

Whole numbers are the easiest of all to work with because our (American) numbering system is simply based on 10 individual figures known as *digits* (0, 1, 2, 3, 4, 5, 6, 7, 8, 9) and anything we refer to as a number contains one or more digits.

Each place a digit can occupy in a whole number has what is known as *place value*. Knowing the place value of each digit means we can read the number and know how much it means. Each place value increases as we move from right to left and that increase is 10 times the value of the place to the right. For example:

5 = Five

55 = Fifty-five

555 = Five hundred fifty-five

5555 = Five thousand five hundred fifty-five

Place values are arranged in groups of three to make them easier to read. Numbers of four digits or more often have a comma, many times considered to be optional, as in the case with the numbers 4,575 and 4575. Both are correct in their expression.

Adding Whole Numbers

Addition is the simplest task to perform. The terms that apply here are *addends* (the numbers being added together), and *sum* (the answer after adding). When adding a group of simple numbers, the order of their listing doesn't matter.

$$7 + 4 + 6 = 17 \quad 4 + 7 + 6 = 17 \quad 6 + 7 + 4 = 17$$

Another thing to consider when adding whole numbers is that they can be grouped any way we choose, which leads to solving problems more quickly when a group can be recognized. $7 + 4 + 6$, for example, could be shown as $7 + (4 + 6) \rightarrow 7 + 10 = 17$. Numbers that add up to 10 are relatively easy to create within a given problem and doing so can increase your speed in solving problems and reaching answers. As an example, take a look at the numbers below and determine the sum as quickly as possible.

$$7 + 3 + 6 + 8 + 2 + 4$$

The answer is 30. One way to get the answer would be to add the numbers in sequence—$7 + 3$ is 10, plus 6 is now 16, then add 8 to reach 24, add 2 for 26, then add 4 to reach 30. Another way to approach the problem would be to scan the group of numbers to determine the sets of 10 available—$7 + 3 = 10$, $6 + 4 = 10$, and $8 + 2 = 10$. $10 + 10 = 20$, then 10 more makes 30.

Adding single digit numbers through this process is simple. However, in the event you find yourself working with a variety of numbers, a column may be the way to a quicker solution.

$$
\begin{array}{r}
9 \\
21 \\
144 \\
6 \\
83 \\
\underline{67}
\end{array}
$$

When using the column method, add the units column (the one farthest to the right) and if the number is greater than 10, carry the second number to the next column to the left (the tens column) and add it to the total in the tens column. The final answer of 330 is calculated by carrying to the next column if necessary.

Subtraction of Whole Numbers

Subtraction is the inverse of addition. The terms to consider in subtraction are *minuend* (the initial number we want to remove another number from), *subtrahend* (the number we want to take away), and *remainder* (the answer).

There are some differences in approaching subtraction. One of them being that order and grouping are important, meaning we can't rearrange the numbers in the way we did with addition. The other factor is that, unlike addition, subtraction problems are solved two numbers at a time.

Borrowing

When working with subtraction problems, borrowing is necessary when any digit in any place of the subtrahend is larger than the same place digit in the minuend. A simple way to look at it is that you "borrow 10" from the number to the left. For example:

$$
\begin{array}{r}
68 \\
\underline{-\,29}
\end{array}
$$

In the example shown above, we can't subtract 9 from 8, so we have to borrow from the next column to the left and change the 8 to an 18. At this point we can subtract 9 from the new number, 18. Going back to the left column to complete the problem means that we're now working with a 5 instead of a 6. Taking this simple approach, we can determine the answer to the above problem to be a remainder of 39.

Multiplication of Whole Numbers

One simple way to describe **multiplication** is that it's a shortcut to addition. If, for example, you have six bags of wire nuts on your service truck and each bag contained eight wire nuts, you could determine the total number of wire nuts on hand by adding:

$$6 + 6 + 6 + 6 + 6 + 6 + 6 + 6$$

Or, you could simplify the problem by multiplying:

$$6 \times 8 = 48$$

Terms to be aware of regarding multiplication are *factors* (the numbers multiplied), *multiplicand* (the first of two numbers multiplied), *multiplier* (the second of two numbers multiplied), and *product* (the answer). In most cases, you'll find it easier to accomplish multiplication problems by setting up the smaller number as the multiplier and the larger number as the multiplicand.

Carrying Partial Product

In some cases, multiplication problems require that we carry partial product to the next column. For example:

$$\begin{array}{r} 43 \\ \times\ 5 \\ \hline \end{array}$$

In the example above, when we multiply 3 by 5, we reach a product of 15. The proper way to record this is to list the 5, then carry the one to the next column, which is to be added after the multiplication process has been accomplished. Multiplying 4 times 5 offers a product of 20 and then by adding the 1 that was carried allows us to record the answer as 215.

Dividing Whole Numbers

Division is the inverse of multiplication. Rather than asking what the product is when 6 is multiplied by 7 (42), we're asking how many times does 7 divide into 42.

Terms to deal with in division are *dividend* (the number to divide into), *divisor* (the number to divide by), and *quotient* (the answer). In division the order is important, for example, 20 divided by 10 is not the same as 10 divided by 20.

Averaging

Finding the **average** of a set of numbers is accomplished in three steps:

1. Add all of the numbers together to find a total.
2. Count the numbers in the set.
3. Divide the total of step one by the count from step two.

As an example, we'll calculate the average of this set of numbers:

$$20, 30, 40, 50$$

STEP ONE: Adding $20 + 30 + 40 + 50 = 140$

STEP TWO: $140 \div 4$ (the total amount of numbers in the set) $= 35$

Fractions

One way in which HVAC/R technicians have to deal with **fractions** is in relation to measurement. Accomplishing piping in an installation or modification of a refrigeration system, for example, would require the understanding of fractions as shown on a measuring tape. In most cases, you would be cutting tubing within a tolerance of ⅛ of an inch. On most measuring tapes, each inch is divided into 16 parts, each identified as 1/16 of an inch. (see Figure 3–1)

A fundamental rule to keep in mind about fractions it that in the event you find the need to add, a common denominator must be found. As an example, let's assume that you find it necessary to replace two pieces of tubing while working on a refrigeration system. When measuring the old tubing, you find that the first piece measures 4⅜" and the second piece measures 5½". Before you could determine the total length of tubing you would need to replace the two pieces, you would first need to find a common denominator— in this case, the fraction ½ could be converted into eighths:

$$\tfrac{1}{2} = \tfrac{4}{8}$$

$$4\tfrac{3}{8} + 5\tfrac{4}{8} = 9\tfrac{7}{8}$$

Fig. 3–1 Rule.

Decimals

When you think of **decimals**, think in terms of hundredths. Most of us have a head start in understanding decimals because we're used to working with money. If an item costs $2.37, we know that the real cost is two whole dollars and 37 hundredths of a third dollar.

Decimals, which are fundamentally a kind of fraction, and a period (the decimal point), separate a whole number from a fraction of a number. The whole number is to the left of the decimal point and the fraction of the number is to the right of the decimal point.

Adding Decimals

The addition of decimals can be accomplished by using the same care as we discussed when adding columns of whole numbers. As an example of adding decimals, we'll use money. Suppose that of the service calls you completed in a day, two of the customers preferred to write checks rather than be billed. To be able to report the total amount of money collected, you would need to add decimals. An example of the checks would be as follows:

$$\begin{array}{r} \$265.38 \\ \underline{149.76} \\ \$415.14 \end{array}$$

Notice that in order to arrive at the correct amount of money collected, we had to carry from the right column to the left column until one column didn't equal more than 10.

Percent and Percentages

The word **percent** means "per hundred." In the example of decimal addition the total was $415.14. That could be explained as 415 whole dollars and 14 hundredths (14%) of another dollar.

One area that you may have to work with percentages is in calculating a discount. Referring again to the checks collected, suppose that one of the customers was entitled to a 10% discount on the service work. One simple way to calculate the

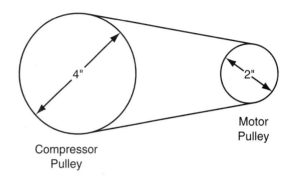

Fig. 3–2   Comparison of 4" compressor pulley and 2" motor pulley, ratio of 2 to 1.

discount would be to move the decimal point one place to the left.

As an example, if the customer who was billed $149.76 was entitled to the discount, moving the decimal point one place to the left would offer the solution that 10% of the bill would be $14.97.

Ratio and Proportion

A **ratio** compares two numbers. One of the most common applications for ratios is in the comparison of two pulleys as shown in Figure 3–2. If, for example, an air handler had a motor pulley that was two inches in diameter and a blower pulley that was four inches in diameter, the pulley assembly would be described as having a ratio of 2 to 1 (sometimes expressed as 2:1).

What that means in a practical sense is that in the event that the motor turned at a rate of 500 revolutions per minute (RPM), the blower pulley would turn at a rate of 250 RPM. In other words, the motor pulley would have to make two revolutions before the blower pulley would make one full revolution.

Unit Three Summary

- Our system of numbers uses 10 individual figures called digits. A number contains one or more digits. Each place a digit can occupy in a number has a value known as place value. The place value increases ten times from right to left.

- Adding whole numbers is the simplest of math concepts. The terms to recognize in adding whole numbers are addends and sum. Subtracting whole numbers is the inverse of addition and the terms to recognize in relation to subtraction are minuend, subtrahend, and remainder. Multiplication is a shortcut for addition and the terms to recognize are factors, multiplicand, multiplier, and product. Division is the inverse of multiplication and the terms to recognize are dividend, divisor, and quotient.

- Fractions must be recognized by the HVAC/R technician because working with a measuring tape relates to fractions. Measuring tapes usually divide an inch into 16 parts, each identified as ⅟₁₆ of an inch.

- A decimal number is a kind of fraction. In decimals, the decimal point separates the whole number from the fraction of a number. The whole number is at the left of the decimal point and the fraction is at the right of the decimal point.

- The term ratio is often used when referring to a motor pulley and a blower pulley. Ratio means a way of comparing two numbers.

Unit Three Key Terms

Whole Numbers	Average
Addition	Fractions
Subtraction	Decimals
Multiplication	Percent
Division	Ratio

UNIT THREE REVIEW

Please do not use a calculator.

1. Add the following numbers:

   a. 29   b. 30   c. 149   d. 960   e. 1025
     + 48   + 87   + 851   + 97   + 101

2. Subtract the following numbers:

   a. 50   b. 108   c. 153   d. 733   e. 1230
     − 42   − 59   − 54   − 79   − 329

3. Multiply the following numbers:

   a. 12   b. 40   c. 44   d. 79   e. 789
     × 9   × 8   × 7   × 40   × 29

4. What is the value each place a digit can occupy in a number known as?

5. What are the terms related to addition?

6. What are the terms related to subtraction?

7. What are the terms related to multiplication?

8. What are the terms related to division?

9. Find the averge temperature from the five readings:

   79°, 84°, 80°, 69°, 65°

10. Find the average of the following weights:

    110#, 145#, 126#, 133#

11. When adding fractions, what is it often necessary to convert one of the fractions to?

12. Change the following fractions into another fraction of equal value:

    a. $\frac{7}{8} = \frac{}{16}$ b. $\frac{10}{16} = \frac{}{8}$ c. $\frac{3}{8} = \frac{}{16}$ d. $\frac{4}{16} = \frac{}{8} = \frac{}{4}$

13. Add the following fractions:

    a. $7\frac{1}{8}$ b. $3\frac{1}{2}$ c. $5\frac{1}{4}$ d. $6\frac{1}{2}$
       $+9\frac{3}{16}$ $+4\frac{3}{8}$ $+8\frac{5}{16}$ $+3\frac{7}{16}$

14. Change the following fractions to decimals: (Hint: 3 divided by 16 = .1875)

    a. $\frac{1}{16}$

    b. $\frac{1}{8}$

    c. $\frac{3}{8}$

    d. $\frac{7}{16}$

15. Add the following fractions: (give answers in decimals).

    a. $\frac{3}{16}$ b. $\frac{7}{16}$
      $+\frac{1}{4}$ $+\frac{3}{8}$

16. Add and subtract the following decimals:

    a.   271.34   b.   456.07
       $+\underline{61.36}$ $-\underline{45.08}$

17. Define the term percent.

18. Give the totals after subtracting the discount:

    a. Discount 10% from $43.20

    b. Discount 5% from $70.49

19. Define the term ratio.

20. The ratio between two pulleys is 3 to 1.

    a. If the larger pulley turns at 450 RPM, find the RPM of the smaller pulley.

    b. If the smaller pulley has a 5" diameter, find the diameter of the larger pulley.

SECTION TWO

Electricity

In recent years, the demands for electrical energy have increased dramatically, resulting in greater public awareness of the use of electrical resources. This awareness, along with spiraling energy costs, has paved the way for the development of improved HVAC/R systems. They're more compact, higher-capacity, energy-efficient systems that require precise control of temperature, humidity, and air movement. As a result, electronic control components have become an integral part of these more efficient systems and the electrical circuitry is more complex. This means that now, more than ever, you need a good working knowledge of the electrical theory, circuitry, and component devices associated with HVAC/R equipment.

In this section, we'll cover the background information necessary for you to understand electricity, the components used in HVAC/R systems, and lay the foundation for you to develop the skills that will enable you to take a practical approach to troubleshooting and servicing heating, air conditioning, and refrigeration equipment.

UNIT FOUR

Electricity

OBJECTIVES

After completing this unit, the student will be able to:

1. Define matter, compounds, elements, and atoms.

2. List the subatomic components of an atom.

3. Describe six methods through which electricity is produced.

4. Define the terms volt, ohm, and ampere.

5. Differentiate between a series and a parallel circuit.

WHAT IS ELECTRICITY?

Asking the questions "What is electricity?" "Where does it come from?" and "How does it perform useful work?" is a good beginning to understanding the principles behind electricity. In Ben Franklin's day, electricity was thought to have been a fluid with both positive and negative charges. Today we understand that the basics of electricity revolve around what is known as *electron theory* (also referred to as atomic theory). Electrons are tiny particles that are too small to be seen yet exist in all matter.

COMPOUNDS

All matter is made up of molecules. When a molecule is divided, a **compound** results. A compound is two or more elements that, when combined chemically, produce a distinct substance that has the same proportion of each element in its composition. For example, water is a compound made from two elements, hydrogen and oxygen.

ELEMENTS

An **element** is a substance that contains only one type of atom. Elements are the basic materials that make up all matter. There are 102 known elements, 92 of which are natural and 10 of which are synthetic (man-made through atomic research).

ATOMS

The smallest part of an element that still maintains the properties of that element is known as an **atom**. Consider the elements of hydrogen and oxygen in water. They are in turn made up of atoms of hydrogen and oxygen. An illustration of this is shown in Figure 4–1.

When the atom is divided, three subatomic particles are identified: the neutron, the proton, and the electron. Neutrons and protons are located in the center of the atom in what is known as the *nucleus* and electrons are in orbit around the nucleus. Figure 4–2 shows an elementary atom.

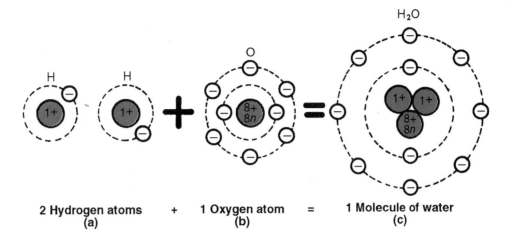

Fig. 4–1   Atomic structure of a water molecule: one atom of oxygen and two atoms of hydrogen.

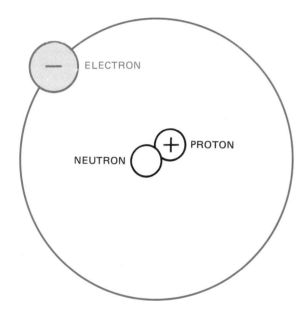

Fig. 4–2   Principal parts of an atom.

Neutron

The **neutron** is located in the nucleus of the atom and as its name implies, carries a neutral electrical charge. Although it is a separate particle, it is not important in the electrical behavior of the atom.

Proton

The **proton** is a permanent part of the nucleus. Although it's only about one-third the

diameter of the electron, it is 1,840 times heavier, making it extremely difficult to dislodge from its position. The proton carries a positive charge and, like the neutron, is not the most important component in the electrical behavior of the atom.

Electron

Although much larger than a proton, the **electron** is lighter and, as a result, is easier to dislodge from its orbital path around the nucleus. All atoms have a certain number of electrons in various orbits around the nucleus. The farther the electron's orbit is from the nucleus, the easier it is to dislodge. Electrons in orbits far from the nucleus are referred to as *free electrons* (sometimes you'll see the term valence electrons used) and electrons in orbits close to the nucleus are called *planetary electrons*. Electrons carry a negative charge.

The attraction of planetary electrons to the nucleus is very strong so they are very difficult to dislodge from their orbital path. The free electron is the most important one in the electrical behavior of the atom, because in order to understand electrical theory, you have to buy into the concept that the movement of an electron from the orbital path of one atom to the orbital path of another atom is the basis behind current flow.

THE LAW OF ELECTRICAL CHARGES

A fundamental theory to understand regarding the behavior of an atom is the law of electrical charges, which states, "Like charges repel, unlike charges attract." This basic law of physics applies to the discussion of the atom and the electron being "attracted" to the nucleus of the atom. The proton holds the positive charge and the electron holds the negative charge.

The electrical charges on electrons and protons produce lines of force called *electrostatic fields*. When these fields interact, they can either attract or repel one another. One way to look at it is:

Proton (+) repels Proton (+)

Electron (−) repels Electron (−)

Proton (+) attracts Electron (−)

When an atom contains the same number of electrons and protons, the opposite charges cancel one another, giving the atom a neutral charge. When an atom contains fewer electrons, it has a positive charge; when it contains more electrons than protons, it has a negative charge. Every electron contains energy. When electrons move in random directions, the effects of the movements are canceled. When electrons move in the same direction, a current is formed and the resulting energy can be used to perform useful work. Electrical current is the impulse of energy that one electron transmits to another as it moves through its orbit. Figure 4–3 shows an electron being knocked out of its orbit.

How fast does this process of one electron knocking another out of its orbit and resulting in what we've referred to as current flow happen? Instantaneously! An electron's movement from orbit to orbit takes place at a rate of 186,000 miles per second.

CONDUCTORS AND INSULATORS

Some materials, because of their atomic structure, are good **conductors** of electricity, while

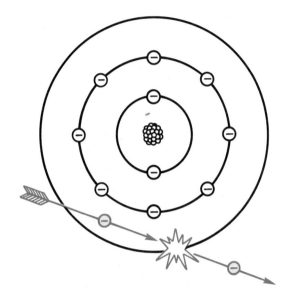

Fig. 4–3   An electron knocked out of orbit by another.

others are not because their atomic structure is different. A good conductor of electricity has one free electron in an outer orbit of the nucleus. Silver is the best conductor of electricity because an atom of silver is larger than an atom of other materials commonly used as conductors, such as copper and aluminum. A silver atom contains five orbits and a copper atom contains only four orbits. Figure 4–4 shows a comparison between a copper and a silver atom.

Because silver is so expensive, its use is limited to contact points in relays, contactors, and other switching devices used in HVAC/R equipment. Aluminum is subject to a buildup of corrosion on its surface so its use is also limited. Copper is the most widely used because of its low cost and because it is almost as good a conductor as silver.

The atomic structure of an **insulator** is different than that of a conductor because there are more valance electrons in the outer orbit of an insulator. Materials such as glass, rubber, or plastic are good insulators because they divide the energy between up to eight valance electrons and, as a result, the energy is not easily transmitted from atom to atom.

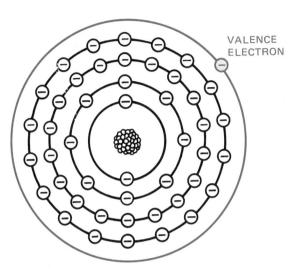

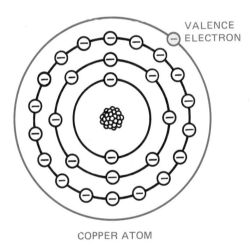

Fig. 4–4   Silver atom and copper atom.

A **semiconductor** is a material that is located somewhere between the conductor and the insulator. It's neither a good conductor of electricity nor a good insulator. Figure 4–5 shows the differences in atomic structure of the conductor atom, the semiconductor atom, and the insulator atom.

Figure A is a typical conductor with only one electron in orbit. Figure B shows a semiconductor that has four valence electrons, and the insulator shown in Figure C has eight valence electrons. In reality, a conductor of electricity can have one or two free electrons, a semiconductor can have three or four, and an insulator has seven or eight valence electrons.

PRODUCING ELECTRICITY

Electricity is produced in various ways. The general idea in all applications is to apply a form or force of energy that will cause the free electrons to leave their orbital path. This can be accomplished through friction, chemicals, pressure, heat, light, and the method that is most important for us to understand when working with the power applied to HVAC/R equipment, magnetism.

Friction

Electricity produced by friction is known as **static electricity**. This occurs when two materials are rubbed together and electrons are transferred from one material to another. You may have

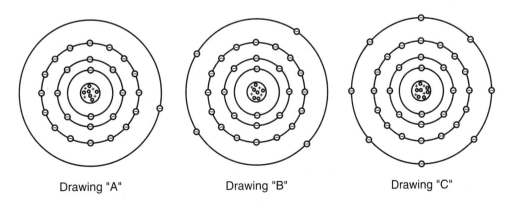

Drawing "A"   Drawing "B"   Drawing "C"

Fig. 4–5   Atomic structure (a) conductor atom, (b) semiconductor atom, (c) insulator atom.

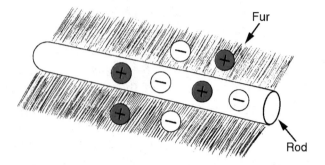

Fig. 4–6   Rod comes into contact with fur, no movement, charges are equalized.

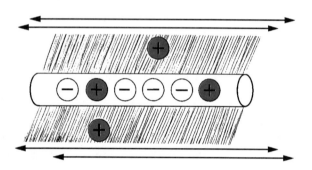

Fig. 4–7   Rod and fur are rubbed together, rod becomes negatively charged.

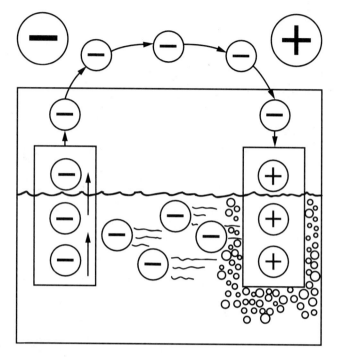

Fig. 4–8   Electricity produced by chemical. Basic wet cell battery.

demonstrated static electricity at a very young age when you discovered you could shuffle across the carpet in your socks, then touch your little sister with the tip of your finger and create a discharge of electricity.

The action of discharge creates an *arc*, a sustained discharge of electricity across a gap. Figures 4–6 and 4–7 show an illustration of static electricity. A rod is in contact with fur but because there is no movement of the rod or the fur, positive and negative charges are present in equal quantities in both materials. In Figure 4–7 the rod and fur are rubbed together and the fur loses electrons to the rod. The rod is negatively charged and the fur becomes positively charged.

Chemicals

Electricity can be produced through chemical means by immersing certain metals, such as copper and zinc bars, in an electrolyte solution. The two bar-shaped metals are called *electrodes*. The two electrodes and the electrolyte solution comprise a single unit known as a **cell**. A wet cell is illustrated in Figure 4–8.

In this process, sulfuric acid is mixed with water to form the electrolyte solution. When mixed in a solution, the sulfuric acid breaks down into two chemicals, hydrogen and sulfate. The positive atoms of the hydrogen are equal to the negative atoms of the sulfate, which means that the solution has no net electrical charge. When the copper and zinc bars are added to the solution, a chemical reaction takes place. The zinc bar will combine with the sulfate atoms and assume a negative charge. Positive charges of the zinc will be given off and the zinc will hold a surplus of electrons.

The positive zinc atoms will combine with the sulfate atoms and neutralize them so that the solution now has more positive atoms and can assume a positive charge. The positive hydrogen atoms in the solution will now attract electrons from the copper bar to again neutralize the solution while the copper bar, with a lack of electrons, will assume a positive electrical charge.

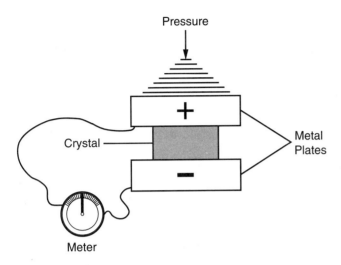

Fig. 4–9 Electricity produced by pressure or piezoelectricity.

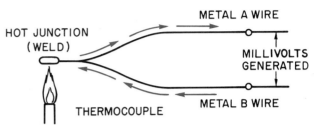

WHEN HEATED, ELECTRONS FLOW IN ONE DIRECTION IN ONE TYPE OF METAL (A) AND IN THE OPPOSITE DIRECTION IN THE OTHER TYPE (B).

Fig. 4–10 Thermocouple. *Courtesy Robertshaw Controls.*

Pressure

Electricity is produced by pressure, that is, when a force is applied to certain materials, it causes electrons to leave one side of the material and accumulate on the other side. This results in a buildup of positive and negative charges on opposite sides of the material. When the pressure is released, the electrons will return to their orbits. This type of electrical action is used in microphone and photographic applications and is known as **piezoelectricity**. Figure 4–9 illustrates the concept of electricity produced by pressure.

Heat

Electricity produced by heat is referred to as **thermoelectricity**. This occurs when two dissimilar metals are joined together and heat is applied to their point of contact. Electrons will be driven from one metal and forced to accumulate on the other. The more heat that is applied, the greater the charge buildup between the two metals. Figure 4–10 shows a typical application of this concept—a thermocouple used in heating equipment.

When the heat is removed, the two metals will cool and the electrical charges will neutralize. The cooling process for most thermocouples usually takes less than a minute. Another variation of this process is used in a device known as a ther-

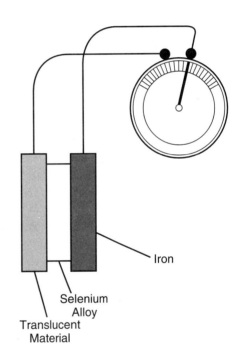

Fig. 4–11 Photocell electricity produced by light.

mopile in some HVAC equipment. A *thermopile* is a group of thermocouples and is used to create a higher voltage.

Light

Light itself is a form of energy made up of particles called *photons*. When photons in a light source strike an object, an energy is released causing the atoms in the object to release electrons. The **photoelectric cell** used in photovoltaic systems that collect light energy from the sun and produce electricity is a good example of this process. A photoelectric cell is shown in Figure 4–11.

Magnetism

Magnetism is the most popular method of producing electrical energy. The principle, known as **electromagnetism**, centers around the process of producing electricity by cutting the lines of force of a magnetic field. All magnets have a north and south pole (a positive and negative) and a magnetic field. Figure 4–12 shows a magnet and the lines of force that exist between the north and the south poles.

The process of producing electricity through magnetism is simple. When the lines of force of a magnetic field are cut by a conductor, an energy is induced into the conductor. A simplified illustration of this concept is shown in Figure 4–13. A

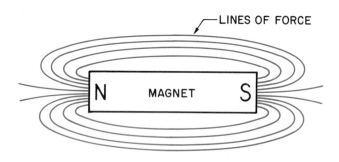

Fig. 4–12   Magnet with lines of force between north and south poles.

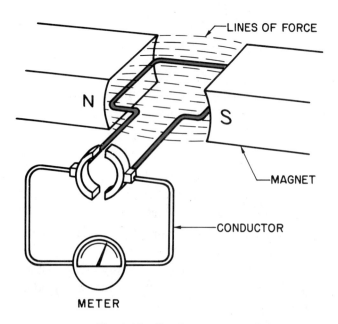

Fig. 4–13   Simple generator.

simple generator would be a wire, which is rotated through the lines of force of the magnetic field.

Electricity produced in this manner is called *alternating current* because as the conductor cuts the lines of force and goes from positive to negative and back again, the charge alternates from negative to positive. This means that the current is continually reversing itself a given number of times within a given time frame. For example, in the United States, alternating current changes direction 60 times per second. When discussing the number of cycles per second, we use the term *hertz*. HVAC/R equipment, lighting, office equipment, and appliances in the home are all designed to operate on alternating current.

VOLTAGE

In order for electrons to flow in an electrical circuit, pressure must be applied. This pressure, referred to as *electromotive force (EMF)* is actually the difference of potential energy that exists between two charged objects. A force field exists when two unlike charges attract, whereas the negative charge is considered to have a high potential. Current always flows from a low potential to a high potential, so in other words, from a positive charge to a negative charge. Figure 4–14 is an illustration of voltage.

By definition, a **volt** is the amount of pressure required to force one ampere of current through a resistance of one ohm. In certain applications, the term *millivolt* (one-thousandth of a volt) is used and in other applications the term *kilovolts* (one thousand volts) is used.

1 millivolt (MV) = 0.001 volts

1 kilovolt (KV) = 1,000 volts

RESISTANCE

Resistance opposes the flow of current in an electrical circuit. All devices installed in an electrical circuit to perform work (called loads) will have a resistance. An **ohm** is the unit of measurement used to define the level of resistance of a given material or component. One ohm is the

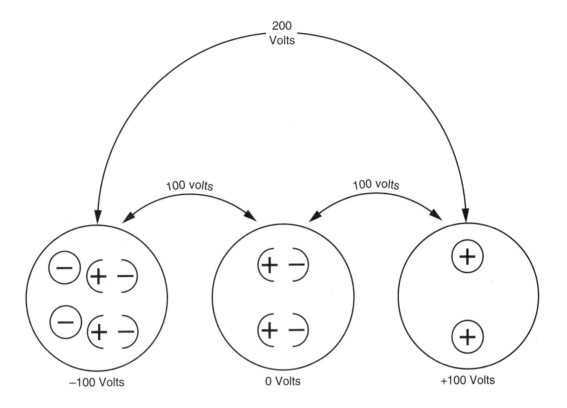

200
Volts

100 volts

100 volts

−100 Volts

0 Volts

+100 Volts

Fig. 4–14   Voltage is the difference in potential.

amount of resistance that will still allow the current flow of one ampere under a pressure of one volt.

All devices and conductors, regardless of their atomic structure, offer some resistance to current flow. The wire connecting the generating station to the electrical grid that serves the homes and businesses with electricity has resistance. This resistance creates voltage drop. Voltage drop is the reason that electrical generating stations send electrical energy out at such a high level (many thousands of volts), when in fact the home only needs power delivered at a rate of 240 volts and a business or industrial user may only need power delivered at 440 volts.

CURRENT

The amount of current flowing through a conductor is determined by the number of electrons that will pass through a given point in one second. The unit of measurement used for current flow is **ampere**. One ampere is the amount of current that will flow through a resistance of one ohm under a pressure of one volt. One ampere is equal to 6,280,000,000,000,000,000 electrons per second. A smaller scale used to measure current flow is the milliamp (MA), which is one-thousandth of an ampere.

For an illustration of the method of measuring voltage and amperage in a circuit, refer to Figure 4–15. It shows that voltage is measured by touching the probes of a test meter across the terminals connected to a load, while the amperage flowing in a circuit is measured by allowing the current to flow through a test meter. Another method of measuring current flow in a circuit is with a meter known as a clamp-on meter, which we'll be discussing in later units.

GENERATOR

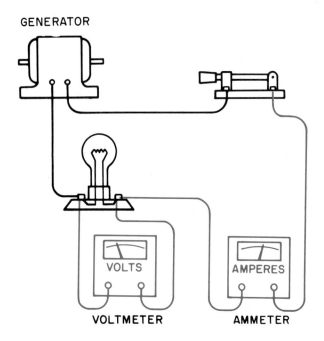

Fig. 4–15   Voltage is measured across the resistance.
Amperage is measured in series.

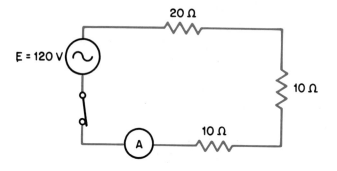

Fig. 4–16   A series circuit.

TYPES OF ELECTRICAL CIRCUITS

In dealing with electricity, the three types of
electrical circuits you have to identify and understand are the series circuit, the parallel circuit, and
the series-parallel circuit.

In a series circuit, the current has only one
path in which to flow. The current starts at the
source of supply and will flow through each component of the circuit, returning to the source of
supply, as shown in Figure 4–16.

Series circuits are most commonly used in
system control circuits that contain protective

Fig. 4–17   Current flow starts at source of supply, divides
following different paths through circuit resistances, then
returns to the source of supply.

devices and control switches. When there is an
open circuit to any part of the series circuit, the
current will cease to flow. A good example of this
is a string of Christmas lights wired in series.
When one bulb burns out, none of the remaining
bulbs will light.

A parallel circuit has two or more paths for
current to flow. As illustrated in Figure 4–17, the
current flow will start at the source of supply and
will divide to follow different paths through circuit resistances before returning to the source of
supply.

Parallel circuits are most commonly used in
load circuits to supply full line voltage to
equipment such as motors, accessory devices,
and household appliances. If one part of a parallel circuit is broken, the voltage supply will
still be delivered to the remaining loads in the
circuit.

The series-parallel circuit is, as the name
implies, a combination of components wired in
series, such as switches, with full line voltage
delivered to several components, such as motors.
Figure 4–18 shows a practical application of a
series-parallel circuit—a wiring diagram that
shows the switches wired in series with several
loads wired in parallel.

As you can see in the diagram, the compressor, the condenser fan motor 1, the condenser fan
motor 2, and the contactor coil all receive the
voltage applied in the circuit at L1 and L2, while
the low pressure switch, high pressure switch, and
thermostats are all wired in series to allow the
system to work.

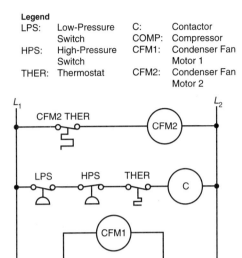

Legend

LPS:	Low-Pressure Switch	C:	Contactor
HPS:	High-Pressure Switch	COMP:	Compressor
		CFM1:	Condenser Fan Motor 1
THER:	Thermostat	CFM2:	Condenser Fan Motor 2

Fig. 4–18   Switches wired in series with several loads wired in parallel.

FUSES AND CIRCUIT BREAKERS

A short circuit can be defined as an unauthorized path for current to flow. It is a condition of very low or no circuit resistance accompanied by a very high current flow. A short circuit causes excessive temperatures that could damage power sources and circuit load devices, burn the insulation on conductors, and create potential fire hazards.

Because this potential exists in all circuits, protection is required to prevent conductors and load devices from being subjected to such current overload conditions. Fuses and circuit breakers provide protection to current overloads.

Fuses

A **fuse** has the following characteristics:

1. A fuse senses when current overload exists.

2. A fuse will open the circuit before any damage is done to the circuit components.

3. A fuse has no effect on the circuit during normal operation.

Many fuses employ a low-melting-point alloy, which is formed into a link contained in a housing and inserted into the circuit to be protected. Figure 4–19 illustrates a fuse in a circuit. When a current overload, accompanied by an increase in circuit temperature, is experienced, the alloy will melt, breaking the circuit. Once a fuse melts, it is no longer operable and must be replaced.

Figure 4–20 illustrates some of the different types of fuses found in HVAC/R equipment. Most fuses used in refrigeration and air conditioning applications are a time delay type. This enables them to tolerate a slight overload condition for a brief period of time, allowing motors to draw locked rotor amperage on initial start. The fuse will open, however, if a high overload condition persists.

Many cartridge fuses are constructed in such a manner that they provide for the replacement of the burned-out element. Fuses or fuse elements should be replaced with the correct ampere rating for the circuit being protected.

Edison base fuses are found in older installations and have since been replaced by the S-type fuse, or the circuit breaker. The S-type fuse with an S-adaptor allows only the prescribed amperage value fuse to fit into the base. Usually, the base has a locking device so that it cannot be removed from the socket and replaced with a base of a larger amperage value.

Circuit Breakers

The circuit breaker has an advantage over the fuse in that after it "trips," it can be used again. The two types of circuit breakers commonly used are the magnetic circuit breaker and the **thermal overload circuit breaker**. A typical circuit breaker is illustrated in Figure 4–21.

Magnetic Circuit Breaker

The **magnetic circuit breaker** allows line current to pass through an electromagnetic coil. During normal operations the circuit breaker switch contacts are held in a closed position by an *armature*. When a current overload is experienced, the current flow through the electromagnetic coil will increase to a magnitude sufficient

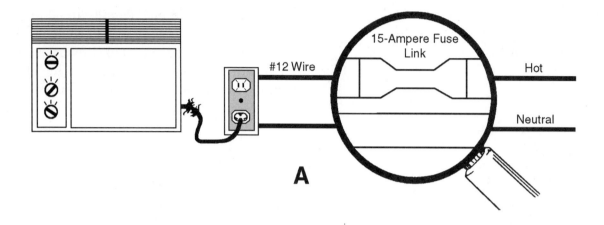

A

Short Circuit

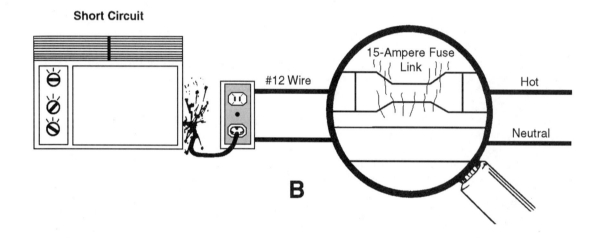

B

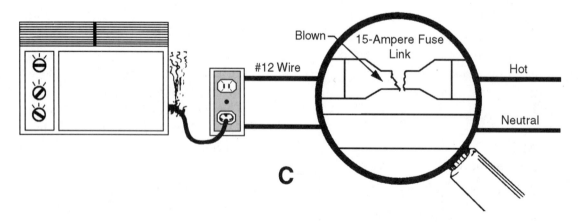

C

Fig. 4–19   How a fuse works.

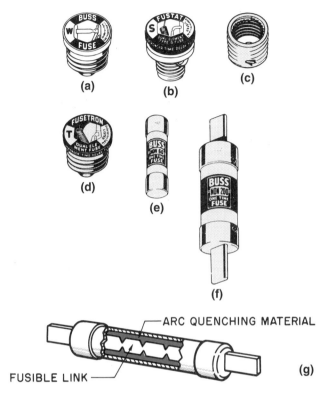

Fig. 4–20   **Types of fuses (a) Edison base fuse, (b) type s base plug fuse, (c) type s adapter, (d) dual-element plug fuse, (e) ferrule-type cartridge fuses, (f) knife-blade cartridge fuse, (g) knife-blade cartridge fuse with arc-quenching material.**
A through F courtesy Bussmann Division, Cooper Industries.

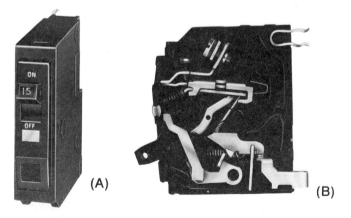

Fig. 4–21   (a) circuit breaker, (b) cutaway.   *Courtesy Square D Company.*

to pull back the armature, opening the circuit breaker contacts. After the overload condition has been corrected the circuit breaker can be reset manually to the On position.

Thermal Overload Circuit Breaker

The **thermal overload circuit breaker**, another common type of circuit breaker, bases its operation on the principle that heat overload will react to a bimetal strip. During normal operation, the current will enter the circuit breaker, pass through the switch contacts and the bimetal strip, and flow to the circuit load device.

In the event of a current overload, the bimetal strip will warp and engage a strip mechanism that will open the circuit breaker contacts, interrupting current flow to the circuit load device. Once the overload condition has been corrected, a sufficient amount of time must be allowed for the bimetal strip to cool before the circuit breaker can be reset.

Electrical Disconnect Switches

One of the most important considerations of the service technician when performing inspections, service, or repair of refrigeration and air conditioning equipment is to ensure that the power to the equipment being serviced is disconnected.

When servicing domestic equipment such as refrigerators and window air conditioners, the technician can safely pull the plug from the receptacle and immediately know that the power has been disconnected. When working with larger or industrial equipment, power is supplied from a panel that is remote from the system being serviced. In this case you would use a disconnect switch located near the equipment to interrupt power while the system is being serviced.

In all motor applications, an electrical disconnect switch is required at or within the sight of the motor control device. The disconnect switch (which may be in the form of a fuse, circuit breaker, or another safety switch) disconnects the controller and the motor from all ungrounded power supplies.

Whenever you interrupt power to a system by means of a power disconnect switch, use a padlock to lock the switch in the open, or Off posi-

Fig. 4–22   Typical three-phase disconnect.

tion. This will protect you from injury during service operations. A disconnect switch is illustrated in Figure 4–22.

In the case of fused disconnect switches, the fuses should be removed. In applications where circuit breakers are used, the circuit breaker should be tagged to identify that the power has been intentionally disconnected and that service is being performed on the system.

These safety precautions are essential and will prevent injury or possible death as a result of power being inadvertently applied.

METERS FOR REFRIGERATION AND AIR CONDITIONING SYSTEMS

There are a wide variety of diagnostic test meters used in the refrigeration and air conditioning industry. These instruments provide you with an efficient and accurate means to measure and determine the condition of an electrical circuit. You can use them to diagnose the condition and operation of components, monitor the operation of the refrigeration system, and evaluate its condition.

With the rapidly changing and advancing technology of the industry, it is extremely impor-

tant that you be aware of the diagnostic tools available and also become proficient in their use.

Instrument Selection

The selection of a particular test meter can become quite a complex matter. One instrument may have multiple functions, which makes it easier to handle but, in reality, may make it more difficult to use. For example, in the event of meter failure, all functions may be lost. Single function instruments have only one function and are easier to use, but using them will require you to carry several types of meters.

The selection of a particular meter will depend largely upon the type of equipment being serviced. In some applications a multi-function meter would best fit your needs, while in other applications a single-function meter would be more appropriate. Your preference is also a consideration in regard to instrument selection.

CARE AND SERVICE OF TEST METERS

Test meters are very sensitive devices and should be handled with care. Extreme temperatures, humidity, dust, dirt, and rough handling will affect the accuracy of the instrument and reduce its service life. Most test meters are equipped with protective cases that should be used during transportation and storage.

Outside of minor repairs, such as broken leads or calibration, no attempt should be made to make internal repairs to a meter. Most manufacturers or service centers have both a reasonable repair time and qualified service technicians, should the need for a repair arise.

Always follow the manufacturer's recommendations for calibrating and using a specific test meter. The following suggestions serve as a general guide:

1. Use a meter in the same position in which it was calibrated.

2. Use batteries of good quality and carry spare batteries at all times.

3. Remove the batteries from the meter if the meter is not used frequently.

4. Check the meter's battery compartment periodically for possible corrosion.

5. Some meters are equipped with protective fuses, so carry a spare.

6. Check the conditions of the test leads periodically. It may be wise to carry a spare set of leads.

7. Keep the selector switch in the Off position and the leads disconnected when storing a meter.

8. Read the meter face from an angle that prevents a deceptive reading, a condition referred to as parallax.

Because you will use a variety of test meters under a variety of working conditions, extreme care should be taken to prevent damage to the meter and to prevent injury to yourself. Competent technicians adhere to the following procedures:

1. Read the manufacturer's instructions carefully to become familiar with the proper use of the test meter before using it.

2. Make sure all test leads are clear of fan blades, belts, heat, and any other hazardous condition that could cause damage to the leads or the instrument, or could cause injury to yourself.

3. Always start a diagnostic voltage check with the highest scale indicated on the instrument and work downward until the reading you desire is reached.

4. Protect all meters from excessive heat, cold, or moisture.

5. Make sure the power and wiring is disconnected from a component being tested with an ohmmeter.

6. Remove loose jewelry and rings when working in a live electrical circuit.

7. Handle all meters with care during use and storage.

8. Use only the fuse designated for an instrument.

9. Always check the scale and rating of a test meter before applying it to a circuit.

10. Use the correct meter for the required application.

11. Do not attempt to field repair test meters.

Ammeters

The most common and convenient **ammeter** used by HVAC/R service technicians is the snap-around ammeter, shown in Figure 4–23. A current flow through a conductor produces a magnetic field around the conductor and a snap-around ammeter will function when clamped around one circuit conductor.

The meter jaws contain a split steel core that carries the magnetism through the meter coil and induces a current flow causing the pointer to move. A rectifier installed in the circuit ensures one-way current flow.

It is important when using the snap-around ammeter to clamp the meter jaws around one circuit conductor only. This is because the current

Fig. 4–23 Snap-around ammeter. *Courtesy Amprobe Instrument, Div. of Core Industries.*

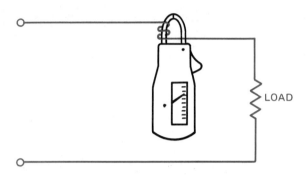

Fig. 4–24   Two turns of wire in the ammeter jaw will double the ampere reading.

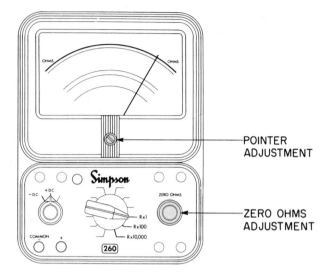

Fig. 4–25   Ohmmeter.   *Courtesy Simpson Electric Co., Elgin, Illinois.*

flow in two conductors is usually opposite and their magnetic fields will tend to cancel each other, causing the meter to read "zero."

In some cases the current flow of the circuit being tested may be too low for the ammeter pointer to move, even on the low range scale. Most ammeter manufacturers have a multiplier accessory available to overcome this problem. Another solution is to wrap the single conductor around the meter jaws several times to magnify the reading. To obtain the correct current reading, divide the meter reading by the number of times the conductor coil is wrapped around the meter jaws. (see Figure 4–24)

When the current flow of the circuit being tested is unknown, set the meter scale to the highest point and then work downward. The most desirable scale to use is when the meter reading is about 60 percent of the full meter scale.

Never connect an ammeter across a voltage source where a direct path may exist from the source of the voltage to ground, such as in the terminals of a battery. To do so will result in damage to the meter.

VOLT/OHMMETERS

The purpose of the **ohmmeter**, illustrated in Figure 4–25, is to test the condition of a circuit or an electrical component of the circuit. The ohmmeter works essentially the same as the ammeter except that it utilizes its own power supply in the form of a dry cell battery.

A rheostat, a resistor, and test probes are wired in series with the meter coil and battery. When the two test probes are shorted together or placed across the circuit or device being tested, a very small amount of current produced by the battery will flow through the circuit or device. The amount of resistance in the circuit will determine the amount of needle movement registered on the meter scale.

The ohmmeter scale is calibrated opposite of the way a voltmeter and ammeter are calibrated. The ohmmeter scale increases in value from right to left. Full deflection of the needle to the right, or zero on the scale, indicates no resistance. No needle movement on the scale indicates an infinite resistance.

The term *continuity* refers to a complete circuit. An open circuit will indicate that there is an infinite resistance. A measurable resistance refers to the actual resistance of a component of the circuit or the circuit itself.

Ohmmeter scales are marked in ohms and megohms (MEG), or 1,000,000 ohms. The letter K is used to identify 1,000 ohms. Most ohmmeters have more than one ohm scale, usually R × 1, R × 100, and R · × 1,000. For general field servicing of refrigeration and air conditioning equipment you should have an ohmmeter with readings in multiple scales.

When using an ohmmeter the test probes are first crossed or shorted together. The meter calibration knob, or rheostat, is then adjusted to bring the meter pointer to zero on the scale. The test probes are then placed to read the resistance. There are several guidelines that should be followed when using an ohmmeter.

1. Make sure all power is disconnected from the circuit.

2. Select the ohm scale to be used prior to calibrating the meter to zero ohms.

3. Isolate that part of the circuit being tested to prevent feedback readings from parallel circuits. Components of the circuit should also be isolated for the same reason.

4. Make sure the meter battery is fresh.

5. Look directly at the meter face to prevent false readings resulting from parallax.

Using the Voltage Function

A **voltmeter** measures the amount of electromotive force in a circuit and operates in a manner similar to the ammeter. A multiplier resistor with a known resistance value is added to prevent a direct short in the meter circuitry. The greater the current flow through the meter circuitry reflected by an increase of the circuit voltage, the greater the resulting magnetic field. This is indicated by a greater needle movement of the meter pointer on the meter scale.

The operation of the meter is best explained by saying that the rate of current flow in a circuit will vary with the voltage difference between electrical charges. If the rate of current flow through a given material increases, this will indicate a greater voltage across that material. If the rate of current flow decreases, this will indicate a lower voltage present in the circuit.

In the operation of the voltmeter, the current flow through a given material, referred to as a *multiplier resistor*, is measured. As illustrated in Figure 4–26, the multiplier resistor determines the scale range of the voltmeter. If the voltage for

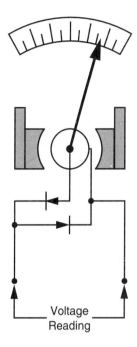

Fig. 4–26   Multiplier resistor circuit determines scale range of voltmeter.

a given value is high, the current flow will also be high, as indicated on the meter scale in volts. If the voltage for a given value is low, the amount of current flow will also be low, as indicated on the meter scale.

In the operation of multi-range voltmeters, the multiplier resistors are located within the meter and either a switching device or terminal connections are used to select the desired scale shown externally on the meter.

When using a voltmeter, start with the highest scale on the meter and work downward until a voltage reading is indicated at about 60 percent of the meter scale. As illustrated in Figure 4–27, the voltmeter is placed parallel in the circuit being tested.

Another type of voltmeter is illustrated in Figure 4–28. With this voltage detector no connection to the circuit is necessary, nor does any current have to flow to detect the presence of voltage. These instruments emit a loud computer-like sound, the rate of which increases as the voltage source is approached.

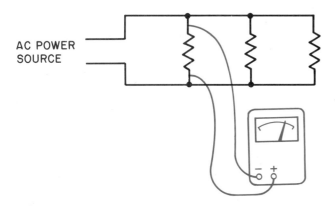

Fig. 4–27 Voltmeter is placed parallel in a circuit to check voltage condition.

Fig. 4–28 Circuit tester. *Courtesy TIF Instruments, Inc.*

Wattmeters

As you have learned, the power consumption, or wattage, can be determined in a circuit by multiplying the circuit voltage by the amperage draw of the circuit, or:

$$\text{wattage} = \text{voltage} \times \text{amperage}$$

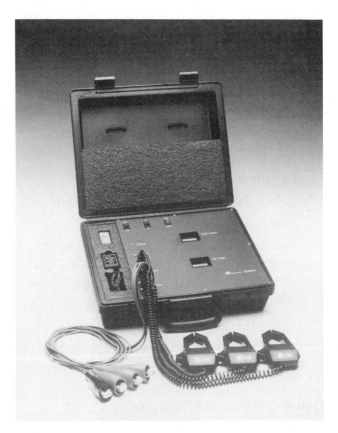

Fig. 4–29 Wattmeter. *Courtesy TIF Instruments, Inc.*

This formula can be applied to an alternating current circuit to determine power when only the voltage and current are in phase. That is, strictly when there is a purely resistive load, such as in the case of resistance heaters.

In most applications of alternating current circuits, the circuit load is reactive due to the presence of inductance and capacitance. Thus, the formula will not measure the true power consumed in the circuit.

The **wattmeter**, which is illustrated in Figure 4–29, measures the true power consumed in a circuit by all of the electrical devices, regardless of the load. When using a wattmeter, set it to the highest range and adjust it downward after an initial reading is observed.

Megohmmeter

The **megohmmeter** insulation tester, illustrated in Figure 4–30, is used to test the electrical

Fig. 4–30 Megohmmeter tests electrical insulation and resistance. *Courtesy A.W. Sperry Instruments Inc.*

insulation and resistance in motors, transformers, cables, switches, electrical wiring, relays, and appliances. With the electrical demands being placed on today's equipment and systems, the megohmmeter is becoming more popular in the service industry as a part of preventive maintenance programs.

Unit Four Summary

- In recent years, energy-efficient HVAC/R equipment has been developed because of rising energy costs and public awareness of the use of natural resources. Newer units have more complex electrical systems so that now, more than ever, a technician has to have a firm understanding of the fundamentals of electricity.

- The fundamental concept behind electricity is based on atomic theory. All matter is made up of molecules, elements make up molecules and the smallest part of an element that still maintains the properties of the element is the atom.

- The three principle parts of the atom are the electron, proton and neutron. The electron carries a negative charge, the proton carries a positive charge, and the neutron is neutral. The neutron and proton are at the nucleus of the atom and the electron orbits the nucleus.

- Electricity can be produced by chemical means, from pressure, heat, light, friction, or through magnetism. Alternating current is produced through the concept of cutting the lines of force of a magnetic field.

- Voltage refers to the pressure in a circuit. Resistance is measured in ohms and current flow in a circuit is measured in amperes.

- The three types of circuits found in HVAC/R equipment are the series circuit, parallel circuit, and series-parallel circuit.

- A fuse senses when an overload exists in a circuit and opens the circuit before any damage is done to the components in the equipment. A wide range of fuses are used in the various applications in which HVAC/R equipment can be found. Time delay fuses are used in motor applications so that a fuse won't open during the startup of the motor.

- Circuit breakers work in the same capacity as a fuse in an electrical circuit but have the advantage of being reset after they trip. Two types of circuit breakers are the magnetic circuit breaker and the thermal overload circuit breaker.

- Disconnect switches are located for the convenience and safety of the service technician. Following proper safety precautions related to disconnect switches prevents serious injury.

- An HVAC/R technician uses a variety of meters when servicing and troubleshooting refrigeration and air conditioning systems. The selection of a meter is based on the type of equipment being serviced and the technician's personal preference.

- Test meters are sensitive devices and beyond changing batteries or performing minor repairs such as broken leads, all other servicing should be left to manufacturers and service centers. Always be familiar with meters being used and follow manufacturer's instructions in order to prevent meter damage or incorrect diagnosis of equipment and components.

Unit Four Key Terms

Compound	Proton
Element	Electron
Atom	Conductor
Neutron	Insulator

Semiconductor	Electrical Circuits
Static Electricity	Fuse
Cell	Thermal Overload Circuit Breaker
Piezoelectricity	
Thermoelectricity	Magnetic Circuit Breaker
Photoelectric Cell	Ammeter
Electromagnetism	Ohmmeter
Volt	Voltmeter
Ohm	Wattmeter
Ampere	Megohmmeter

UNIT FOUR REVIEW

1. What is the basis of the theory of electricity?

2. What is a compound?

3. How many known elements are there?

4. What is the smallest part of an element that still maintains the properties of that element?

5. In H_2O (water), how many elements are there? how many atoms?

6. What are the three subatomic particles of the atom?

7. Give the electrical charges for each of the three subatomic particles.

8. Like charges _____, unlike charges _____.

9. What makes a good conductor?

10. Why isn't silver more widely used as a conductor in the HVAC/R industry?

11. List three commonly used conductors of electricity.

12. How is an insulator different from a conductor?

13. How is electricity produced?

14. What are six ways of producing electricity?

15. Match the method of producing electricity with its description:

    a. Certain metals, such as copper and zinc bars, are immersed in an electrolyte solution.

    b. Force is applied to certain materials, causing electrons to leave one side of the material and accumulate on the other side.

c. Two materials are rubbed together and electrons are transferred from one material to another.

d. Occurs when two dissimilar metals are joined together and heat is applied to their point of contact.

e. Photons strike an object and energy is released.

f. Electricity is produced by cutting the lines of force.

16. What is the current called when the lines of a magnetic field are cut by a conductor?

17. What is the measurement of resistance in an electrical circuit?

18. What is the measurement of current flow in an electrical circuit?

19. Fill in the blanks using volt, ampere, ohm.

a. A _____ is the amount of pressure required to force one _____ of current through a resistance of one _____ .

b. One _____ is the amount of resistance that will allow the current flow of one _____ under a pressure of one _____.

c. One _____ is the amount of current that will flow through a resistance of one _____ under the pressure of one _____ .

20. Why does voltage drop occur in an electrical circuit?

21. What is the difference between a series circuit and a parallel circuit?

22. What is the purpose of a test meter?

23. Name several conditions to which test meters are sensitive.

24. What are some suggestions for handling batteries used in test meters?

25. How does an ammeter work?

26. How is an ammeter set when the circuit is unknown?

27. How is an ohmmeter different from an ammeter?

28. What does a voltmeter measure?

29. What does a wattmeter measure?

30. What is a megohmmeter used for?

wattage = voltage × amperage

31. What is the wattage of a compressor with a motor operating at 230 volts drawing about 27.5 amperes?

32. What is the wattage of a compressor with a motor operating at 460 volts drawing about 12.1 amperes?

33. Write a paragraph in complete sentences using the following terms: atom, electrons, negative charge, good conductor, protons, copper, glass, and insulator.

34. Write a paragraph explaining the procedure to be followed by a competent technician in using a test meter.

UNIT FIVE

Direct Current, Alternating Current, Electrical Power

OBJECTIVES

After completing this unit, the student will be able to:

1. Describe the difference between alternating current and direct current.

2. Understand the relationship between voltage, amperage, and resistance.

3. Understand how the electric company charges its customers for electricity.

Air conditioning equipment operates on alternating current (AC). As we described in Unit Four, alternating current is generated through the process of cutting the lines of force of a magnetic field with a conductor. From a practical standpoint, this means that alternating current fundamentally reverses its direction from negative to positive every time the conductor cuts the lines of force of both poles of the magnet.

DIRECT CURRENT

Unlike alternating current, **direct current (DC)** electricity does not reverse direction. DC flows in one direction only and is a type of current produced by a battery rather than an alternator. A good example of producing direct current is a dry cell battery you would find in a flashlight. An illustration of the difference between alternating current and direct current is shown in Figure 5–1.

As you can see, the *waveform* (commonly referred to as a sine wave) for alternating current shows that during the time frame of ⅟₆₀ of a second, the energy reaches both a positive and a neg-

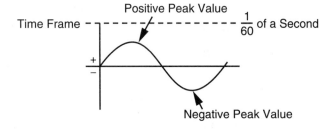

Alternating Current

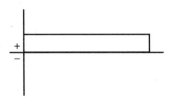

Direct Current

Fig. 5–1   Alternating current, direct current.

ative peak value. The time frame of ⅟₆₀ of a second relates to a two-pole generator that is spinning at 3,600 RPM, designed to achieve a frequency of 60 cycles per second (60 hertz).

The waveform for direct current is different. The time frame shown is the same, but there is no alternation or reversal of current flow. Direct current flows in one direction.

While the current supplied to homes and offices to operate HVAC/R equipment is alternating current, newer equipment containing printed circuit boards utilizes direct current to power the solid state components within the electronic control components. In this case, a component within the printed circuit board known as a *diode* acts to change alternating current to direct current.

OHM'S LAW

Ohm's Law illustrates the relationship between the three units of electrical measurement: volts, amperes, and resistance. To work with Ohm's Law, an identifying letter is used for each of the three electrical units.

The letter E, which stands for electromotive force, is used to represent voltage. The letter I, which stands for induction, is used to represent current flow in amperes because it is understood that a current is induced in a circuit. The letter R is used to represent the resistance in a circuit.

One way to illustrate Ohm's Law is as follows:

$$E = I \times R$$

What this equation says is that in order to find the voltage in a circuit, you multiply the current by the resistance. Two other equations used when working with Ohm's Law are:

$$I = E \div R \quad R = E \div I$$

Dividing the voltage (E) by the resistance (R) in a circuit will tell you what the current draw is in the circuit. Dividing the voltage (E) by the current (I) will yield an answer to the question "What is the resistance of the circuit?"

The thing to keep in mind when working with equations like these is that the factor you want to find is listed first. It's not like reading 2 + 2 = 4, the way we all learned basic arithmetic in elementary school. Instead, it's 4 = 2 + 2.

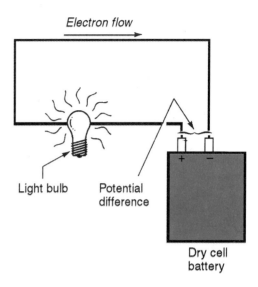

Fig. 5–2   **A dry cell battery supplying electric potential (voltage to an electric circuit).**

If any two factors of a circuit are known, Ohm's Law can be used to determine the third factor. As an example, let's assume that we have a battery that is powering a lightbulb. An example of this circuit is shown in Figure 5–2.

The two factors we know about this circuit are that the voltage applied is 12 volts and the resistance of the lightbulb is 6 ohms. The question to answer is "What is the current flow in the circuit?" In order to answer it, we can use Ohm's Law.

STEP ONE: Select the appropriate equation.

$$I = E \div R$$

STEP TWO: Apply the known factors to the equation.

$$I = 12 \div 6$$

Ohm's Law tells us that the current flow in this circuit is 2 amperes. The voltage (12 volts) divided by the resistance (6 ohms) yields the answer.

Another method used when working with Ohm's Law is the *memory wheel*. As shown in Figure 5–3, the memory wheel can be used as an aid in calculating either the resistance, voltage, or amperage in a circuit.

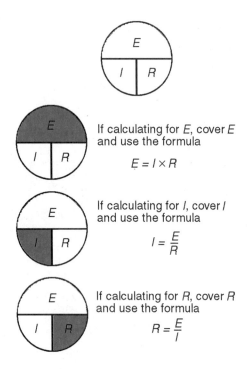

Fig. 5–3 Using Ohm's Law (not applicable on AC inductive circuits).

If calculating for *E*, cover *E* and use the formula

$$E = I \times R$$

If calculating for *I*, cover *I* and use the formula

$$I = \frac{E}{R}$$

If calculating for *R*, cover *R* and use the formula

$$R = \frac{E}{I}$$

Ohm's Law, in the strictest sense, applies to DC circuits—but the fundamental idea behind the process should give you an understanding of electrical circuits. Using Ohm's Law in what is known as an inductive circuit on alternating current (such as an electric motor) would not yield a totally correct answer because of the effects produced by an electrical winding in a motor when electrical energy is applied. You can come closer, though still not perfectly correct, with what is known as a resistive load, such as a heating element in an electric furnace.

For our example, and to give you an understanding of how the electric company charges for electricity, we'll use an electric furnace that operates on 240 volts AC and has one element with a resistance of 12 ohms. The first step in determining an approximate cost of operation is to use Ohm's Law to find the current draw in the circuit. If you refer again to Figure 5–3, you can see that the memory aid, when used to calculate the current in a circuit, directs you to cover the I. This refers to the equation I = E ÷ R. Current is found by dividing the voltage by the resistance.

In our electric furnace circuit, this would mean dividing 240 volts by 12 ohms, shown as:

$$240 \div 12 = 20 \text{ amps}$$

Keep in mind that our example makes use of even numbers and ideal conditions—something you'll rarely encounter in the field when working with HVAC equipment. Another factor to consider is that when the heating element heats up, the resistance of the element changes. This causes the calculated current to differ from the actual reading of a meter. Also, a slight fluctuation of the applied voltage (either up or down) will create a condition in which the meter reading would not agree exactly with the calculated amperage.

ELECTRICAL POWER/WATTS

A fundamental definition of electricity is that it is a form of energy that performs useful work. When electrical energy is converted to heat, light, or mechanical energy, it is referred to as power. **Electrical power** is measured in **watts**.

The term watt is familiar to most people because they've been exposed to it through common items such as lightbulbs and hair dryers. It's common knowledge that the higher the wattage rating, the more work you'll get out of a component. A 100-watt lightbulb burns brighter than a 40-watt lightbulb.

A simple formula used to calculate the power in a circuit is:

$$P = I \times E$$

Power is equal to the amperage in a circuit multiplied by the voltage.

Many technicians simply state "Volts times amps equals watts."

To carry it further in our explanation, we'll refer again to the electric furnace heating element. We calculated that the current draw of that circuit was 20 amps. Using the formula for calculating power (P = I × E), 20 amps multiplied by 240 volts equals 4,800 watts.

This information, along with other factors about the utility company and the amount of time the heating element is energized, will lead us to the approximate cost of operating the element.

The utility company that supplies the electrical energy to the customer measures the rate of power consumption in kilowatts. One kilowatt is equal to 1,000 watts. Charges for electrical energy used are listed in kilowatt/hour (KW/H).

The formula used by the utility company to determine the cost of power consumption is:

$$E = \frac{P \times T}{1,000}$$

E (energy in this case, not electromotive force as we identified it when working with Ohm's Law) is equal to P (power in watts) multiplied by T (time), which is then divided by 1,000, a figure representative of the kilowatt.

Using this equation and adding the kilowatt/hour charge by the utility company, we can determine the cost of operating the element for a given length of time. For our example, we'll assign a cost per kilowatt/hour of 9 cents and determine that the element is in use four hours a day for 30 days.

From a practical standpoint, therefore, if you were servicing an electric furnace with one heating element and your customer asked what it was costing them to operate the furnace for a month, you could give an answer with the following information:

Amp draw of the element (listed on equipment tag)—20 amps

Applied voltage (on equipment tag)—240 VAC

Hours of use per day (information from customer)—4 hours

Charge per kilowatt/hour (from utility company)—9 cents

To determine the operating cost for a month, take the following steps:

STEP 1: Determine the power consumption in watts.

$$P = E \times I$$

240 volts × 20 amps = 4,800 watts

STEP 2: Determine the power consumption in kilowatt/hours for one 24-hour period.

$$E = Power \times Time \div 1,000$$

4,800 watts × 4 hours = 19,200

19,200 ÷ 1,000 = 19.2 kilowatt/hour

STEP 3: Determine the operating cost in a 24-hour period at 9 cents per kilowatt/hour.

19.2 kilowatt/hour × 0.09 = $1.728

STEP 4: Determine the operating cost for a month.

$1.728 × 30 days = $51.84

Our conclusion, then, is that an electric furnace with one 12-ohm heating element being energized at 240 volts for four hours per day would cost a total of $51.84 to operate for one month.

Unit Five Summary

- HVAC/R equipment operates on alternating current. Direct current differs from alternating current in that it only flows in one direction. A good example of direct current is a dry cell battery found in a flashlight. Direct current may be put to use in some HVAC/R equipment that utilizes printed circuit boards. In electronic controls, alternating current is supplied to the printed circuit boards where a component known as a diode converts AC to DC.

- Ohm's Law illustrates the relationship between the three basic electrical units: voltage, amperes, and resistance, and if two factors about a circuit are known, the third can be found by using an Ohm's Law equation. The Ohm's Law memory wheel can be used as an aid to solve electrical math problems. Ohm's Law, in its strictest sense, applies to DC circuits.

- To calculate the power consumption of a circuit, multiply the voltage by the current. Electrical power is measured in watts. The higher the wattage draw of a component, the more work that component can perform.

- The utility company that supplies electrical energy to the consumer charges for it by the kilowatt/hour. One kilowatt is equal to 1,000 watts. If the rate per kilowatt/hour is known, the cost of operating a piece of equipment can be calculated.

Unit Five Key Terms

Direct Current   Electrical Power

Ohm's Law   Watts

UNIT FIVE REVIEW

1. What is direct current?

2. How is alternating current different from direct current?

3. What is voltage? amperage? resistance? (from Unit Four)

4. Give the symbols for volt (voltage), current flow in amperes (amperage), and resistance (ohm).

5. _____ _____ states "To find voltage in a circuit, multiply the current by the resistance."

6. Put the symbols from Question 4 into a simple equation.

7. What is power?

8. Power is measured in _____.

9. If power is equal to the amperage in a circuit multiplied by the voltage, write down its simple equation.

10. Give the equation for volts times amps equals watts.

11. What is a kilowatt?

12. If energy is power in watts multiplied by time (in hours) divided by 1,000, write down its simple equation.

   Example: Simple Algebra

   $$A = B \times C$$

   $$A/C = B \qquad A/B = C$$

   Remember: In algebra, to move a symbol to the other side of the equal sign in a multiplication equation, one has to divide.

   Ohm's Law: $E = I \times R$

13. Write the equation to solve for I.

14. Write the equation to solve for R.

   Remember: If any two factors of a circuit are known, Ohm's Law can solve for the third factor.

15. a. What is the resistance given 240 volts, 20 amperes?

  b. What is the current flow given 100 volts, 50,000 ohms?

  c. What is the voltage given 4 ohms, 3 amperes?

16. Solve the following problems:

  a. The freezer uses 3 amperes at 115 volts. How many watts?

  b. The refrigerator uses 5 amperes at 115 volts. How many watts?

  c. The water heater uses 39 amperes at 115 volts. How many watts?

17. The utility company supplies electrical energy at 13 cents per kilowatt hour. Using the answers from Question 16, what is the cost to power these appliances for 30 days (in use three hours per day)? (Hint: Use the energy equation to find the kilowatts per hours used; P was solved in Question 16.)

  a.

  b.

  c.

18. Why is the letter E used to identify voltage in Ohm's Law equations?

19. Why is the letter I used to identify current in Ohm's Law equations?

20. What is the formula for calculating the cost of operating a piece of equipment?

UNIT SIX

HVAC/R Electrical System Components

OBJECTIVES

After completing this unit, the student will be able to:

1. Describe the difference between a switch and a load.

2. Identify the different types of switching devices used in HVAC/R equipment.

3. Describe the method of operation of an electric motor.

4. Identify the different types of electric motors and relate them to their specific applications.

As a technician working with HVAC/R equipment, you must have an understanding of the switches used to control various types of systems, their classification, and an understanding of basic wiring. Three types of switches, manually actuated switches, switches controlled by temperature and pressure, and flow control switches, are all used to operate the various motors and other loads that make up the equipment used for heating, comfort cooling, and refrigeration.

SWITCHES

Switches are used to make or break an electrical circuit in order to control the operation of a load device. In refrigeration and air conditioning applications, a switch may be controlled either manually or mechanically.

A switch is classified by the number of poles, the number of contacts the switch contains, and the throw (how the switch is operated). For example, a single-pole, single-throw switch has one set of contacts and two positions—open and closed—while a double-pole, single-throw switch has two sets of contacts but the same two positions. Another factor that describes a switch is whether or not it is a single break or double break switch. An illustration of the various types of switch contact points is shown in Figure 6–1.

Push-Button Switches

One type of switch used in the industry is the push-button switch. Push-button switches, which open and close a set of contacts by manually pressing a button, are used in motor applications to stop-start, forward-reverse, or jog-run motor operation. The most common application of the push-button switch in refrigeration and air conditioning applications is in conjunction with a magnetic line starter or contactor to stop-start a motor operation from a remote location.

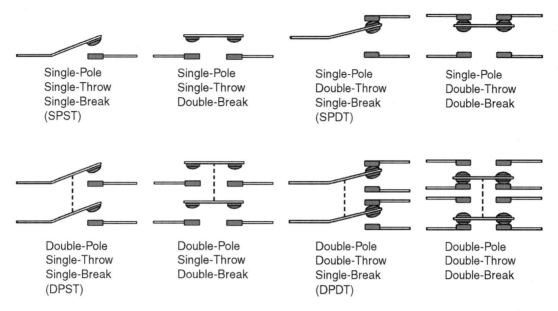

Fig. 6–1   Types of switch contact points.   *Courtesy Honeywell Micro Switch.*

Disconnect Switches

Another type of switch that is considered to be manual is the service disconnect switch. You'll find a switch such as this either mounted on the cabinet of the equipment or in close proximity to the unit. A fused disconnect switch, in accordance with electrical codes, is used in conjunction with a circuit breaker in most installations. For example, a 40-amp circuit breaker in the main panel would supply power to the A/C unit, but 40-amp fuses would also be used to protect the equipment. A disconnect switch that is within sight of the unit also provides for your safety while working on the equipment. Figure 6–2 shows a fused disconnect switch as it is used with HVAC/R equipment.

A variation of the fused disconnect is the non-fused disconnect switch. A non-fused switch can be used when the electrical code and the equipment manufacturer allow for a specially designed circuit breaker in the main panel. The switch still allows you to disconnect the power to the unit, it just doesn't contain any fuses. A non-fused disconnect is shown in Figure 6–3.

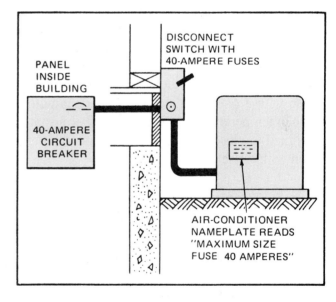

Fig. 6–2   Fused disconnect switch is within sight of the unit and contains the 40-ampere fuse called for on the air conditioner nameplate as the branch-circuit protection.

Mechanical Switches

Mechanical switches can be controlled by temperature, pressure, humidity, fluid flow, or manual operation. A mechanical switch in the form of a thermostat, for example, is used in

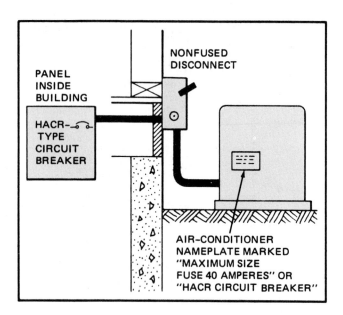

Fig. 6–3   Non-fused disconnect. Permitted only when the label on the breaker and the label on the air conditioner bear the letters "HACR."

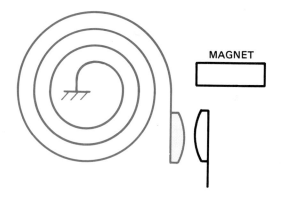

Fig. 6–4   Snap action thermostat.

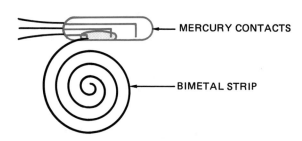

Fig. 6–5   Mercury-type thermostat.

almost all heating and cooling systems. In these applications, a temperature-sensing element moves a set of contacts by means of a mechanical linkage. The cooling thermostat is designed to close with a rise in temperature and open with a drop in temperature, while a heating thermostat is designed to close with a drop in temperature and open with a rise in temperature.

Bimetal Thermostats

A bimetal thermostat, also referred to as a *snap action thermostat,* is the type commonly found in homes and light commercial applications. In this type of thermostat, a bimetal strip made of two dissimilar metals, wrapped in a spiral shape, reacts to temperature change and controls the operation of the heating or cooling equipment. A magnet is used with this type of thermostat to make sure the contacts come together quickly and make a definite connection. The magnet itself is encased in glass and does not actually make contact with the switching assembly; however, its magnetic field still provides the attraction that pulls the contact arm toward the magnet once the bimetal has reacted and caused it

to come close enough to be attracted. A simplified drawing of a snap action thermostat is shown in Figure 6–4.

Mercury Bulb Thermostat

Another common type of thermostat used in HVAC equipment is the *mercury bulb thermostat,* in which a bimetal strip has a mercury bulb fastened to it. When the strip reacts to the temperature change by turning, the mercury inside the bulb shifts and provides contact between the wires routed inside the bulb. It's common for this type of thermostat (shown in Figure 6–5) to be used in a heating/cooling application, which means that the three wires routed inside the bulb and the mercury, depending on the position of the bulb, will make contact from the center wire to either the wire on the left of the bulb or the wire on the right.

Both the snap action and the mercury bulb thermostat will be found in applications where a low-voltage control system is in use. The full-line operating current in an HVAC system cannot be passed through the contacts of these thermostats. A transformer within the unit steps the voltage down to 24 volts. A low-voltage control system allows smaller wiring to be used for the control circuitry and ensures the safety of the consumer should they decide to attempt to work with the thermostat without turning the power off.

HEATING/COOLING ANTICIPATORS

Within the thermostat assembly is a component referred to as an **anticipator**. A heating anticipator is designed to provide a small amount of heat to the temperature-sensing section of the thermostat while the thermostat itself is calling for heat. This little extra heat "fools" the thermostat and causes it to break the circuit just before the thermostat set point is reached. This system prevents a condition known as "overshooting" or "overrunning" the thermostat.

A heating anticipator is a small wrap of resistive wire and in some low-priced thermostats it may be a fixed anticipator. In most cases, the heating anticipator in a thermostat can be adjusted to match the current draw of the gas valve of the furnace. A heat anticipator that is not properly adjusted can cause short cycling of the burners, resulting in a complaint by the customer that the furnace isn't providing enough heat.

In the case of a cooling anticipator, it provides a small amount of heat between the cycles of the A/C. Unlike the heating anticipator, the cooling anticipator is energized during the off cycle, not while the unit is performing its assigned task of modifying the temperature of the conditioned space.

A cooling anticipator is not adjustable. It consists of a resistor mounted to the subbase of the thermostat assembly; during the off cycle of the unit, current passes through the resistor causing a small amount of heat to be generated.

PRESSURE SWITCHES

Pressure switches are not usually found in HVAC equipment unless the system is of an older vintage. It is common, however, to find pressure switches used to control and protect the operation of refrigeration systems commonly found in grocery stores and in restaurant equipment. A pressure switch is a switch that senses pressure, then controls an electrical circuit to the components within a refrigeration system. A pressure switch has terminals that allow for connection to the electrical system and a flare fitting that allows for connection to the refrigeration system. Their cut-in and cut-out points are adjusted by the technician installing and servicing the equipment.

In commercial refrigeration systems, a low-pressure switch such as the one shown in Figure 6–6 can be used as a device to cycle the unit off and on and can also be used solely as a protective device. A high-pressure switch is similar in appearance, but is connected to the high-pressure side of the refrigeration system and is used as a protective device only. Figure 6–7 shows a high-pressure switch. In some cases you may find a low-pressure/high-pressure switch assembly combined.

Another type of pressure switch used on commercial refrigeration equipment is the oil pressure safety switch. The purpose of the oil pressure safety switch is to monitor the level of oil pressure being delivered by the oil pump that is built into the types of compressors used in restaurant and grocery store equipment. Like a combination high/low pressure switch, it has two tubes with flare connections that connect to the refrigeration system and compressor, but it's easy to tell the

Fig. 6–6 Low-pressure switch. *Courtesy Ranco Controls.*

Fig. 6–7   High pressure switch.   *Courtesy Ranco Controls.*

difference between the two devices. The high/low pressure switch tubes both connect to the bottom of the switch assembly. With an oil pressure safety switch, one tube exits the top of the switch assembly and the other tube exits the bottom.

CONTACTORS

Basically, a **contactor** is a switching relay used to establish or interrupt electrical power to a load device, usually a motor. In a single-phase, low-horsepower motor, the current flow is relatively low and motor operation can be controlled by the system cycling and safety controls. As the motor size and current flow increase, the motor current flow must be handled through a contactor. Contactors are also used to provide remote control of the load device being served as well as low-voltage control circuit contactor operation.

When the system-controlling device closes, it will allow current in the control circuit to flow through the contactor holding coil. A magnetic field will be created within the core of the coil. This magnetic field is strong enough to attract and pull an armature, which has a set of movable contacts. The movable contacts will close against a set of stationary contacts, allowing the current to flow to the device being served. When the system-controlling device is satisfied, it will open, interrupting the current flow to the contactor coil.

Fig. 6–8   Contactor.

When the coil is de-energized, the armature will drop and the current flow to the load device will be interrupted. A simplified drawing of a contactor is shown in Figure 6–8.

There are many types of contactors available to suit a variety of applications on specific types of equipment. General purpose contactors are designed to fit a wide variety of situations while definite purpose contactors are designed to fit a specific application. Definite purpose contactors are rated for the specific load device on which they'll be used and they are also smaller in size and more economical to use than general purpose contactors.

Contactor holding coils are available in a variety of voltage ratings, usually 24 volts, 115 volts, and 208/230 volts. Other voltage ratings may be available as required. Voltage rating is usually identified on the coil and should be double-checked to ensure it's the proper rating for the application.

When the coil is energized and the armature closes, the contact points will allow a full current flow to the load device. For this reason, it's important that the contact points are properly rated for a given application. The current flow rating of the contactor should be greater than the full load and locked rotor amperage of the motor (or other load

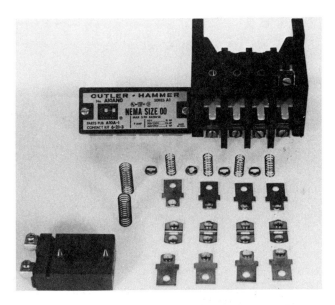

Fig. 6–9 Parts that may be replaced in a contactor: the contacts, both movable and stationary; the springs that hold the contacts; and the holding coil. *Photo by Bill Johnson.*

Fig. 6–10 Magnetic line starter with overload relays.

device) being used, plus any other devices that may be operated through the contactor.

Contactors are available in one, two, three, or four sets of contact points, which are referred to as *poles*. For example, a single-phase motor would normally require two sets of contacts, or a two-pole contactor. A three-phase motor would require a three-pole contactor unless the manufacturer's design allowed for the making and breaking of only two legs of power to start and stop the motor.

The contact points of a contactor are usually manufactured of silver and cadmium and are large enough to rapidly dissipate heat. In most cases, a visual inspection of the contact surfaces is all that is necessary to determine a faulty contact.

In some cases, a contactor must be replaced as a complete unit; however, at times, replacement parts are available. Figure 6–9 shows the parts that can be replaced in some contactors.

MAGNETIC LINE STARTERS

Essentially, **magnetic line starters** are contactors equipped with accessory devices such as motor overload protectors, stop-start switches,

and auxiliary contacts. Figure 6–10 shows a typical starter, which is usually used with higher horsepower, three-phase motors because they incorporate a means of overload protection as well as the ability to start and stop the current flow to a motor. The auxiliary contacts are used in more sophisticated circuits for the remote control and interlocking of starters controlling various load devices. An example of this would be the application of water-cooled equipment in which a water-circulating pump in a water tower would have to be energized before the compressor of the refrigeration system is allowed to start.

RELAYS

Many **relays** used to operate fan motors in HVAC equipment operate in the same manner as a contactor in that a coil, when energized through the control circuit, controls another circuit to a load. While the contact points of a contactor are normally open until the holding coil is energized, a relay can contain both normally open contacts and normally closed contacts. The contacts will change position when the coil of the relay is energized. A simplified drawing of a relay with both normally open and normally closed contacts is shown in Figure 6–11.

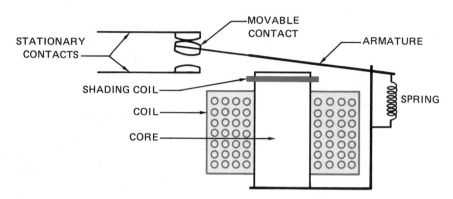

Fig. 6–11   Simple relay with normally open and normally closed contacts.

Fig. 6–12   Several different types of relays used in the industry.

Potential Relay

Another type of relay commonly used in HVAC/R equipment is the *potential relay*. This type of relay is used to start a compressor, primarily in conjunction with a start capacitor, a device that provides an extra boost to start a motor. Figure 6–12 shows the potential relay, along with other types of relays used in HVAC/R equipment.

Current Relay

A current relay is a compressor starting relay that is used on fractional horsepower compressors found in refrigerators, freezers, and some restaurant and bar equipment. A current relay is also referred to as a magnetic relay or amperage relay. It plugs directly into the compressor terminal pins in most cases and may be used in conjunction with a start capacitor.

CAPACITORS

The two types of **capacitors** used in HVAC/R equipment are start capacitors and run capacitors. A start capacitor is used to provide extra starting torque on refrigeration system compressors and other motors. A run capacitor is used to increase the operating efficiency of a motor.

Capacitors are rated in microfarads. Run capacitors can be as small as 1.5 microfarads,

Fig. 6–13   Two capacitors used in the industry (a) start capacitor, (b) run capacitor.

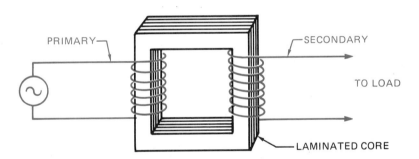

Fig. 6–14   A step-down transformer.

used on small motors, and rated as large as 60 microfarads, used on larger motors and refrigeration system compressors. Start capacitors are not available in ratings lower than 75 microfarads or higher than 600 microfarads. A start capacitor and run capacitor are shown in Figure 6–13.

TRANSFORMERS

Transformers are commonly found in HVAC/R equipment and used to step down voltage for the control system of the unit. A step-down transformer is a device that accepts a given voltage at its primary winding and delivers a lower voltage from its secondary winding. This step down in voltage is accomplished by wrapping an iron core with a given number of windings on the primary side, then wrapping the same iron core with a proportionately smaller number of windings on the secondary side. An illustration of a simple step-down transformer is shown in Figure 6–14.

TIME DELAY RELAYS

Time delay relays are used in a wide variety of applications in HVAC/R equipment. In gas furnaces, a time delay relay allows the heat exchanger to heat up before the blower motor is energized. In comfort cooling systems, they are used to prevent a compressor from restarting too soon in the event of a power failure or if the customer turns the thermostat off and then quickly turns it on again. On larger commercial systems, time delay relays may be used to accomplish what is known as a part winding start on large compressors.

Time delay relays, like a standard relay, contain a set of contact points that carry the equipment system operating voltage to a load. They also have two terminals that are connected to the control voltage circuit. The control section of the time delay relay is usually made up of some kind of heating or bimetal device that, after being energized for a period of time, reacts and causes the switching contacts of the relay assembly to close.

SEQUENCERS

A **sequencer** is commonly found in an electric furnace that has two or more heating elements. As a call for heat comes from the thermostat, one heating element in an electric furnace is energized, then those that follow are energized in sequence. From a practical standpoint, the function of the sequencer is to prevent too heavy a load on the electrical system at the initial start-up of the furnace. Electric heating elements draw a lot of current and energizing two or more at the same time would cause a momentary overload on the electrical system.

TIMERS

Timers are used in various types of HVAC/R equipment in both the domestic and commercial areas. Heat pumps have timers to initiate and terminate necessary defrost cycles. Low-temperature grocery store and restaurant equipment also use timers to periodically shut the refrigeration cycle down and start a defrost cycle to keep the unit free of frost, which allows maximum operating efficiency. Frost-free refrigerators and freezers commonly use a defrost time to break the circuit to the system compressor and energize a heating element that accomplishes the defrost cycle.

ELECTRIC MOTORS

In the application of **electric motors**, electrical energy is converted into mechanical energy in order to create a rotating motion that drives compressors, fans, timers, pumps, and other devices in refrigeration, air conditioning, and heating systems.

The operating principle is based on the laws of magnetism in that like poles will repel each other and unlike poles will attract each other. A simple electric motor is illustrated in Figure 6–15 where a magnet mounted on a pivot is placed between the poles of a permanent magnet.

The pivoting magnet is referred to as the rotor (the rotating part of the motor) and the stator (the stationary part of the motor, thus named because it remains stationary). The rotor will move with

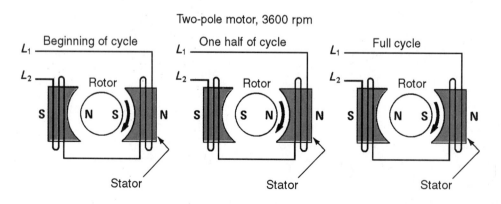

Fig. 6–15   Complete cycle of operation of an electric motor.

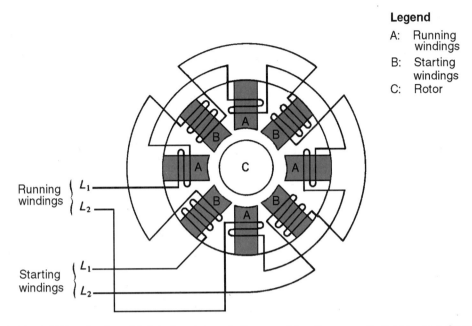

Legend
A: Running windings
B: Starting windings
C: Rotor

Running windings { L_1
{ L_2

Starting windings { L_1
{ L_2

Fig. 6–16   Additional poles added to the stator winding providing the magnetic field to start the motor.

the repulsion and attraction of the poles of the two permanent magnets.

In order for an electric motor to run continuously, a rotating magnetic field must be produced by reversing the poles of the motor or by reversing the polarity of the rotor or stator. Because an alternating current of 60 hertz (60 cycles per second) changes direction 120 times per second, the resulting current flow changes the polarity of the stator poles each time the current reverses. The alternating polarity of the stator windings in a motor allow it to run continuously.

One problem to overcome in motor operation is starting the rotation of the rotor. If the rotor is stopped in line with the stator poles, the north pole will attract the south pole and the rotor will not move. Even when the polarity of the stator poles is reversed, the repelling forces of the stator and rotor may be at such an angle that the rotor is not able to turn.

If additional poles are added to the stator winding as shown in Figure 6–16, the opposite polarity of the stator poles and the rotor will provide the necessary magnetic field strong enough to start the motor.

It's true that if the rotor were to stop midway between these poles, the magnetic attraction would again be equal in both directions and the rotor would not move. This problem is easily solved if the current flow to the different poles in the stator is controlled so that it does not reach all of the poles at the same time. As a result, the current flow to one pole will be out of phase with the current flow to another pole, creating a two-phase current that will start the motor. In all single-phase applications, this second phase of electrical current is required to start the motor.

SYNCHRONOUS SPEED

The natural state of rotation of the magnetic field in the stator of the motor is called **synchronous speed**. The rotor will follow the rotation of the stator's magnetic field and the speed of the motor is directly related to the speed at which the stator's magnetic field rotates. The synchronous speed of a motor is dependent upon three things:

1. The frequency of the applied alternating current

2. The number of poles in the stator

3. The amount of slip

During one AC cycle in a motor with a two-pole stator, the main plane is reversed once, then restored to its original position. In one AC cycle, the magnetic field of the stator will make one complete revolution. For example, if an alternating current of 60 cycles per second is applied, the synchronous speed of the motor will be 60 revolutions per second. Multiplying 60 revolutions per second by 60 seconds per minute would result in 3,600 revolutions per minute. The second factor that determines the synchronous speed of a motor (the number of poles in the stator) would change the RPM from 3,600 to a different number.

The synchronous speed of a motor can be calculated by multiplying the frequency of the applied current by the factor 120, then dividing that result by the number of poles in the motor. (The reason the number 120 is used in working with this equation is because the magnetic field builds up and collapses twice each second—from a positive peak to a negative peak—in a 60-cycle system.)

We could, for example, calculate the speed of a four-pole motor by using the equation and the following steps:

STEP ONE: Multiply the AC frequency (60) × 120 = 7,200

STEP TWO: Divide the factor arrived at in Step One by the number of poles in the motor. 7,200 ÷ 4 = 1,800

Our conclusion, then, is that a four-pole motor operating on a 60 hertz system has a synchronous speed of 1,800 RPM.

In actual application, some slip is experienced between the speed of the rotor and the magnetic field produced in the stator in order for the field to cut the rotor conductors and induce a current into them. *Slip* is the difference between the synchronous speed and the actual motor speed. The slip of an induction motor is approximately 3 to 4 percent of the synchronous speed. For example, a motor with a speed of 1,800 RPM minus a 3 percent slip will experience a speed of about 1,750 RPM.

In the application of induction motors, the amount of slip will be less and the speed of the motor will be slightly higher with a low load. The speed of a motor is known as the full-load speed. The difference between a no-load and a full-load

condition is so small that the induction motor is considered to be a constant speed motor.

SHADED POLE MOTORS

The **shaded pole motor** has a very low starting and running torque (the motor's force of turning action) and is primarily used in small horsepower fans and timer applications. When comparing it to a single-phase induction motor in which the start winding is used to create the torque required to start the motor, the shaded pole motor is different in that each of the stator winding poles has a slot or groove cut into the surface, banded by a solid copper band or wire. Figure 6–17 shows the construction method of a shaded pole motor.

Upon the initial start of the motor, an induced current is created in the shaded pole from the main winding. The resulting magnetic field produced in the shaded pole is out of phase with the magnetic field of the main winding. The main pole of the motor has a magnetic field that is building up and collapsing at a rate of speed and sequence, while the shaded pole has a magnetic field that is building up and collapsing at a different rate of speed and sequence. This results in the torque required to start the motor.

The rotation of a shaded pole motor is the direction of the rotating magnetic field, or from the unshaded pole toward the shaded pole. To reverse the direction of a shaded pole motor it must be disassembled and the position of the stator must be reversed. Figure 6–18 shows the rotation of the motor in the direction toward the shaded pole.

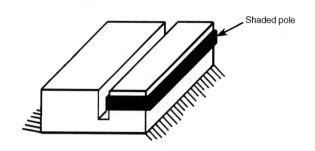

Shaded pole

Fig. 6–17   Shaded pole motor construction.

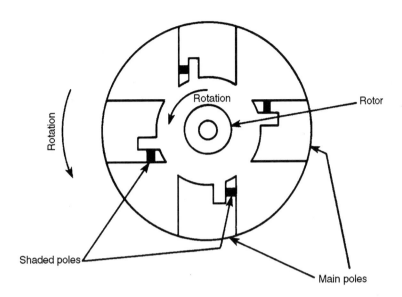

Fig. 6–18   Layout of a shaded pole motor, with counterclockwise rotation in the direction of the shaded poles.

PERMANENT SPLIT CAPACITOR (PSC) MOTORS

The **permanent split capacitor (PSC) motor** is used in refrigeration and air conditioning systems to drive compressors, fans, pumps, and blowers. It has a low to moderate starting torque and excellent running characteristics. A single speed PSC motor has two windings, a run (main) winding and a start (phase) winding. Both windings are almost the same in size and length and both are energized during the starting and running of the motor. A run capacitor is wired in series with the start winding to provide the split-phase electrical characteristic necessary to start the motor and maintain the split phase between the start and run windings during motor operation. This provides good running torque and operating efficiency. A simplified schematic diagram of a PSC motor is shown in Figure 6–19.

A PSC motor is commonly used in multi-speed fan and compressor applications. Connected in parallel with the start winding, additional main or auxiliary windings are wired in series with each other. Through a switching device, the desired motor speed may be achieved by placing one or more of the windings in series with each other as shown in Figure 6–20.

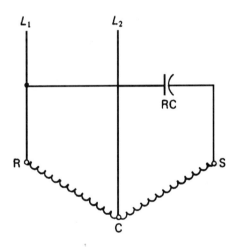

Fig. 6–19   Schematic of a permanent split capacitor motor.

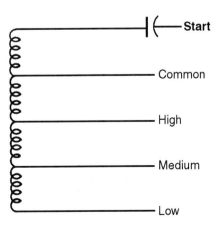

Fig. 6–20   Schematic diagram of a three-speed PSC motor.

For high-speed operation, only the start and high speed windings are energized. For medium-speed operation, the start winding, high speed, and medium speed windings are energized. To achieve a low-speed operation, all windings are energized. As the resistance in the main windings increases, the speed decreases. Similarly, as the resistance in the main windings decreases, the speed of the motor increases.

CAPACITOR START/INDUCTION RUN MOTORS

Capacitor start/induction run motors are used in applications that require high starting torque. In these applications, a start capacitor is wired in series with the start winding to produce a very high value of current in the start winding (which is out of phase with the current in the run winding). Once the motor has started and reached about 75 percent of its running speed, a centrifugal switch or a relay will remove the start winding from the circuit. Figure 6–21 shows a capacitor start motor with the start capacitor mounted on the top of the motor and Figure 6–22 shows a simplified schematic diagram of this type of motor.

CAPACITOR START/CAPACITOR RUN MOTORS

A **capacitor start/capacitor run motor** has a very high starting torque and excellent running characteristics. The motor is essentially a combination of the capacitor start/induction run motor and the permanent split capacitor motor. As Figure 6–23 illustrates, a start capacitor is wired in series with the start winding and in parallel with the run capacitor.

On the initial start of the motor, the start capacitor will provide the necessary phase displacement between the start and run windings to accomplish the high starting torque. Once the motor has started and reached approximately 75 percent of its running speed, a switching relay (in this case, a potential relay) will remove the start capacitor from the circuit, allowing the motor to continue to operate on the run and start windings. The run capacitor provides the slight split in the phase between the two windings, resulting in improved running torque, efficiency, and power.

SPLIT PHASE MOTORS

Starting problems are resolved in **split phase** or **single-phase motors** through two sets of windings wired to the stator. A start winding is made of small, high resistance wire and is located near the top of the stator core. A run winding is made of a relatively large, lower resistance wire and is located in the bottom part of the stator core.

Because the run winding is larger and located near the bottom of the stator core, it is more inductive than the start winding. Due to the greater number of turns in the start winding and because of its higher resistance, the current flow

Fig. 6–21   Capacitor start motor with the start capacitor mounted on the top of the motor.

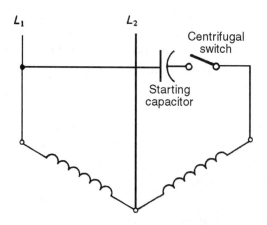

Fig. 6–22   Schematic diagram of a capacitor start motor switch.

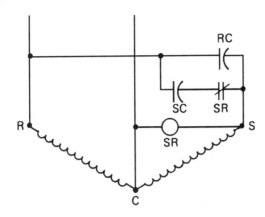

Fig. 6–23 Schematic diagram of a start capacitor wired in series with the start winding and in parallel with the run capacitor.

in the start winding will lag the current flow in the run winding, creating the two-phase effect required to start the motor.

Because of the start winding's high resistance, it will draw a large amount of current. If the start winding remains in the past during the brief time frame required to start the motor, the motor will trip on overload protection or the winding will burn out.

THREE-PHASE MOTORS

Three-phase or **poly-phase motors** are widely used in commercial and industrial air conditioning and refrigeration systems. The operation of the motor is similar to that of a single-phase motor except that the rotating magnetic field in the stator is generated in three phases instead of two. The motor is simple in construction and does not require start windings, capacitors, or starting relays. The operation of a three-phase motor is usually controlled by a magnetic line starter or contactor in response to a manual or automatic control device.

In the application of the three-phase motor, the stator winding is comprised of three single-phase windings, which are displaced 120° out of phase from each other. The applied voltage across the three windings will also be 120° out of phase from each winding. During motor operation, each phase will experience a change in peak current flow at different times, and at any given time, one of the motor windings will be

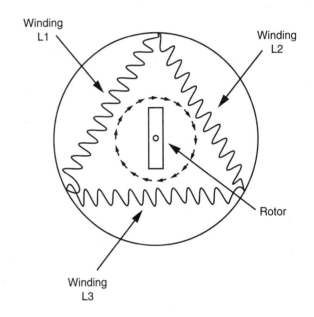

Fig. 6–24 Rotation of rotor in three-phase rotating magnetic field.

positioned to provide maximum torque as shown in Figure 6–24.

The two basic types of windings in a three-phase motor are the *wye winding* and the *delta winding*. While there is no difference in the operating characteristics of the two connections, the two types of windings give the design engineer more flexibility in designing motor windings for a given application.

Most three-phase motors have windings that are equipped to operate on multiple voltages. To accomplish this, the circuits of each phase are divided equally into two sections and are

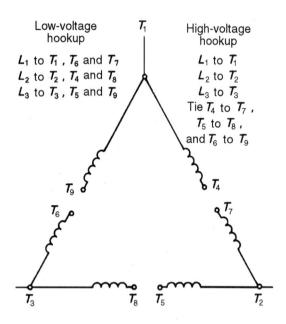

Low-voltage hookup

L_1 to T_1, T_6 and T_7
L_2 to T_2, T_4 and T_8
L_3 to T_3, T_5 and T_9

High-voltage hookup

L_1 to T_1
L_2 to T_2
L_3 to T_3
Tie T_4 to T_7,
T_5 to T_8,
and T_6 to T_9

Fig. 6–25   Three phase motor schematic showing the wiring connection for low voltage and high voltage.

arranged to be externally connected in series for high-voltage applications and in parallel for low-voltage applications. Figure 6–25 gives you a look at a three-phase motor schematic and how wiring is connected to a high-voltage and low-voltage system.

To discuss our example, we'll assume that the motor identified as the one connected to the start winding system is able to operate on either a 230 volt, three-phase system, or a 460 volt, three-phase system. The schematic indicates that the motor will have nine wire leads labeled T1 through T9.

To accomplish the low-voltage hookup, we would proceed as follows:

STEP ONE: Tie the wires labeled T1 and T7 together, then connect them to L1 of the power supply.

STEP TWO: Tie the wires labeled T2 and T8 together, then connect them to L2 of the power supply.

STEP THREE: Tie the wires labeled T3 and T9 together, then connect them to L3 of the power supply.

STEP FOUR: Tie the wires labeled T4, T5, and T6 together.

To wire the same motor to a high-voltage power supply (460 VAC) we would proceed as follows:

STEP ONE: Tie the wire labeled T1 to L1 of the power supply.

STEP TWO: Tie the wire labeled T2 to L2 of the power supply.

STEP THREE: Tie the wire labeled T3 to L3 of the power supply.

STEP FOUR: Tie the wires labeled T4 and T7 together.

STEP FIVE: Tie the wires labeled T5 and T8 together.

STEP SIX: Tie the wires labeled T6 and T9 together.

In the case of the schematic diagram identified as the delta winding system, we would proceed as follows for the low-voltage hookup:

STEP ONE: Tie the wires labeled T1, T6, and T7 together, then connect them to L1 of the power supply.

STEP TWO: Tie the wires labeled T2, T4, and T8 together, then connect them to L2 of the power supply.

STEP THREE: Tie the wires labeled T3, T5, and T9 together, then connect them to L3 of the power supply.

In the case of a high-voltage hookup of a delta winding system, we would take the following steps:

STEP ONE: Tie the wire labeled T1 to L1 of the power supply.

STEP TWO: Tie the wire labeled T2 to L2 of the power supply.

STEP THREE: Tie the wire labeled T3 to L3 of the power supply.

STEP FOUR: Tie the wires labeled T4 and T7 together.

STEP FIVE: Tie the wires labeled T5 and T8 together.

STEP SIX: Tie the wires labeled T6 and T9 together.

When working with three-phase motors, you'll find that a guide for wiring the motor is usually printed on the inside of the cover, which you remove to expose the motor leads.

While the nameplate of a three-phase motor specifies the current draw of the motor by line or by phase, the total motor amperage draw is calculated by multiplying the nameplate amperage by 1.73. For example, if the nameplate on a given motor indicates that the current draw is 10 amperes, the total motor amperage draw would be 17.3 amperes.

Reversing the direction of a three-phase motor is a simple task. It's accomplished by interchanging any two of the three leads that provide voltage to the motor.

Unit Six Summary

- Switches are used in HVAC/R equipment to make or break an electrical circuit to a load device and are classified by the amount of poles, the number of contacts, and the throw. A single-pole, single-throw switch has one set of contacts and its position is either opened or closed. Push-button switches are often used with magnetic line starters to start and stop a motor from a remote location.

- Mechanical switches can be controlled by temperature, pressure, humidity, fluid flow, or manually by a technician. In HVAC systems, a cooling thermostat is designed to close with a rise in temperature and a heating thermostat is designed to close with a drop in temperature. Heating thermostats may have an adjustable heat anticipator, which is a component of the thermostat that provides a small amount of heat during the run cycle of the furnace, and a cooling thermostat has an anticipator that is in the form of a resistor mounted to the thermostat subbase.

- Pressure switches are used in commercial refrigeration systems and are found in the form of low-pressure, high-pressure, and oil-pressure monitoring switches.

- Contactors are switching relays used to establish or interrupt a circuit to a load device. Contactors have holding coils that, when energized, create a magnetic field that will attract and pull an armature, causing a set of movable contacts to connect with a set of stationary contacts.

- General purpose contactors are designed for a wide variety of applications and definite purpose contactors are designed to fit a specific application. Both types of contactors are available with one, two, three, or four sets of contact points, which are referred to as poles. A two-pole contactor would have two sets of contacts. Contactors differ from magnetic starters in that they have no overload protection while starters have overload protection built in.

- Magnetic line starters are commonly used with higher horsepower, three-phase motors. Starters may also contain a set of auxiliary contacts.

- In the application of electric motors, electrical energy is converted into mechanical energy in order to create a rotating motion that drives compressors, fans, timers, and pumps. The operating principle of the electric motor is based upon the laws of magnetism: Like poles will repel each other and unlike poles will attract.

- The natural rate of rotation of the magnetic field in the stator of a motor is called synchronous speed. The rotor will follow the rotation of the stator's magnetic field and the speed of the motor is directly related to the speed at which the stator's magnetic field rotates. Multiplying the AC frequency by the factor 120, then dividing by the number of poles of the motor will give the RPM of the motor.

- Some slip is experienced between the speed of the rotor and the magnetic field produced in the stator. Slip is the difference between the synchronous speed and the actual speed of a motor. The slip of an induction motor is approximately 3 to 4 percent of the synchronous speed. A motor with a synchronous speed of 1,800 RPM, minus a 3 percent slip, will have a speed of 1,750 RPM.

- Shaded pole motors have a very low starting and running torque and are primarily used in small horsepower fan, pump, and timer applications. To reverse the rotation of a shaded pole motor, it must be disassembled.

- Permanent split capacitor (PSC) motors are widely used in the HVAC/R industry. The start and run windings in a PSC motor are very similar in size and length and there is very little difference in the resistance of the motor windings. A run capacitor is wired in series with the start winding and both windings are energized during the starting and running of the motor.

- Capacitor start/induction run motors are used in applications that require high starting torque. A start capacitor is wired in series with the start winding and once the motor has reached about 75 percent of its running speed, a centrifugal switch or relay will break the circuit to the start capacitor and start winding.

- Capacitor start/capacitor run motors have a very high starting torque and excellent running characteristics. It is a combination of a capacitor start/induction run motor and a PSC motor.

- Split phase motors have two sets of windings, one for starting and one for running. A start winding in a split phase motor is made up of wire that is smaller in diameter than a run winding and also has a greater number of turns. Because of these factors, a start winding will have a greater resistance than a run winding.

- In three-phase motors, the stator winding is made up of three single-phase windings displaced 120° out of phase from each other. The two basic types of windings in a three-phase motor are the wye winding and the delta winding. Most three-phase motors are equipped to run on high or low voltage and they require no capacitors or relays.

Unit Six Key Terms

Switches

Anticipator

Pressure Switches

Contactors

Magnetic Line Starters

Relays

Capacitors

Transformers

Time Delay Relays

Sequencers

Timers

Electric Motors

Synchronous Speed

Shaded Pole Motor

Permanent Split Capacitor Motor (PSC)

Capacitor Start/Induction Run Motors

Capacitor Start/Capacitor Run Motors

Split Phase or Single-Phase Motors

Three-Phase or Poly-Phase Motors

UNIT SIX REVIEW

1. What is the purpose of a switch in an electric circuit?

2. Give the difference between a mechanical switch and a manual switch.

3. In what applications are stop-start push-button switches found?

4. Identify the type of switch:

   a. Open and close a set of contacts by pressing a button.

   b. Can be fused or non-fused but still allows the technician to disconnect power to the unit.

   c. Made up of two dissimilar metals wrapped in a spiral shape that reacts to temperature change.

   d. Senses pressure, then controls an electrical circuit to the components within a refrigeration system.

5. What is the purpose of a contactor?

6. What is the major difference between a starter and a contactor?

7. Name the device found in the electric furnace with two or more heat elements to prevent too heavy a load on the electrical system startup.

8. How does a start capacitor differ from a run capacitor?

9. What is the function of the heat anticipator in a thermostat?

10. What is the purpose of a step-down transformer in an HVAC/R system?

11. What is the basic operating principle of an electric motor?

12. Identify the type of motor:

   a. A start capacitor is wired in series with the start winding to produce high starting torque. The start winding is out of phase with the run winding.

   b. Induced current is produced in each pole at a different rate of speed and sequence for low torque to start the motor.

   c. Does not have a start capacitor. A run capacitor is wired in series with the start winding to provide the split phase necessary to start the motor and maintain the split phase between start and run windings.

   d. A start capacitor is wired in series with the start winding and parallel with the run capacitor. There is a slight split in the phase between the start winding and run winding.

   e. Two sets of windings are wired to the stator. Start winding is located near the top of the stator core and the run winding is located in the bottom of the stator core.

13. The speed of a motor is determined from its synchronous speed. This speed, which is measured in revolutions per minute (RPM), is dependent upon:

   • frequency (applied alternating current)

   • number of poles (in the stator)

   • amount of slip

   In the United States the standard frequency is 60 cycles (hertz)

   $$\frac{60 \times 120 \text{ (factor)}}{\text{number of poles}} = RPM$$

   a. If two poles, find the RPM.

   b. If four poles, find the RPM.

Slip is the difference between the synchronous speed (formula above) and the actual motor speed. The actual motor speed will be less. Three percent to 4 percent of slip occurs in an induction motor.

   c. Determine the actual motor speed in (a) if 3 percent slip.

   d. Determine the actual motor speed in (b) if 4 percent slip.

14. What are the three factors that affect the synchronous speed of an electric motor?

15. What are the characteristics of a shaded pole motor?

16. How is a shaded pole motor's direction reversed?

17. What are the characteristics of a permanent split capacitor motor?

18. Which winding has the higher resistance in a split phase motor?

19. What is the purpose of a centrifugal switch or relay in a split phase motor?

20. What are the characteristics of a three-phase motor?

21. Are start capacitors used with three-phase motors?

22. What two types of windings are used in three-phase motors?

23. How is the direction of a three-phase motor reversed?

24. Write a paragraph, in complete sentences, using the following terms: three-phase motor, magnetic field, start winding, capacitors, starting relays, stator winding, and single-phase windings.

25. Write a paragraph in complete sentences describing this course so far.

SECTION THREE

Refrigeration

In order to function effectively as a technician in the HVAC/R field, you must have a good understanding of the fundamental concepts behind heat transfer, the properties and functions of the chemicals used in refrigeration, and the components that make up a refrigeration system.

In this section, we'll explore and explain the basics behind the operation of a refrigeration system, how it's put together and the components that enable it to do its job of transferring heat. A review of the units that follow is necessary in order to begin your development of troubleshooting skills. A technician can't troubleshoot a system without first knowing what the performance factors of the equipment are supposed to be. Studying an introduction to refrigeration, understanding the refrigerants and oils used in HVAC/R equipment, and understanding specifics of the refrigeration system components will enable you to understand performance factors.

UNIT SEVEN

Introduction to Refrigeration

OBJECTIVES

After completing this unit, the student will be able to:

1. Describe the basic principles of refrigeration.

2. Relate gas and pressure laws to the operation of a refrigeration system.

3. Describe methods of temperature measurement.

4. Accomplish Fahrenheit and Celsius temperature conversions.

5. Determine the amount of heat required to raise the temperature of a given substance.

6. Explain the difference between sensible heat and latent heat.

In order to be able to take a practical approach to troubleshooting and servicing HVAC/R equipment, you have to reconcile some factors regarding the operation of a refrigeration system. Those factors have to do with the fundamental principles that allow a refrigeration system to function. An understanding of heat transfer, the laws of physics, and thermodynamics provides the necessary background for you to develop your skills as a service technician. In this unit we'll discuss gas laws, and the structure of matter, energy, and temperature.

The basic principles of refrigeration and air conditioning are not especially difficult to understand. Because they deal with certain laws of physics, chemistry, and engineering, it will be helpful for you to complete a brief study of those sciences in order to gain a working knowledge of the trade.

Simply put, *refrigeration* is the process of transferring heat by removing it from an area where it's not wanted and shifting it to another area. Whether refrigeration is used for food preservation, comfort cooling, or manufacturing, the principles used and the theory applied are the same.

In mechanical refrigeration, the transfer of heat is accomplished through two types of systems: the absorption system, which uses the physics of gas laws and thermal generation to move the refrigerant through the system, and the compression system, which makes use of a mechanical compressor. The compression system is the most widely used in the HVAC/R industry, however, the absorption system is gaining in popularity due to EPA guidelines regarding the venting, recovery, and disposal of refrigerants used in compression systems.

MATTER

Matter is anything that occupies space and has weight. All matter is composed of molecules. A molecule is the smallest particle of a substance that still retains the properties of that substance

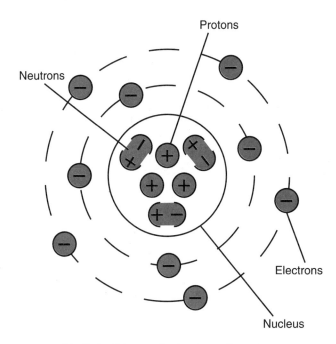

Fig. 7–1 The atomic structure of an atom.

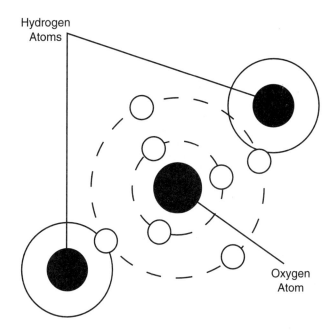

Fig. 7–2 Atomic structure of a water molecule.

and is made up of one or more atoms. Figure 7–1 shows the atomic structure of an atom.

Water is an example of a substance. A molecule of water is made up of two atoms of hydrogen and one atom of oxygen. If a molecule of water was divided further into subatomic parts, it would no longer be water. An illustration of a molecule of water is shown in Figure 7–2.

All molecules of a given substance are alike. However, different materials have different molecules. The characteristics and properties of different materials depend on the agreement of the molecules. The speed, freedom, and number of these molecules determine three things:

1. The state of matter
2. The temperature of matter
3. The effect on other items or mechanisms of which the matter may be a portion

All matter exists in one of the following states:

1. *Solids*, which occupy a definite space and retain a definite shape.
2. *Liquids*, which occupy a definite space but take on the shape of the container in which they are held.

3. *Gases*, which have no form and are composed of molecules that are continually flowing. Gases must be contained in a sealed container and will expand to fill the space in which they are contained.

Figure 7–3 illustrates the three states of matter and related molecular movement.

ENERGY

Energy brings about changes in matter. A law of thermodynamics states, "Energy cannot be created or destroyed, it can only be changed from one form to another." Energy may be classified as thermal, mechanical, electrical, chemical, light, or nuclear. All forms of energy can be converted to heat.

You can identify energy as existing as one of the following types:

1. *Kinetic energy*, which is an energy of motion. An example is the movement of molecules in a material.
2. *Fixed energy*, which is latent or hidden energy. Examples of fixed energy are found in coal, fuel oil, and chemical explosives.

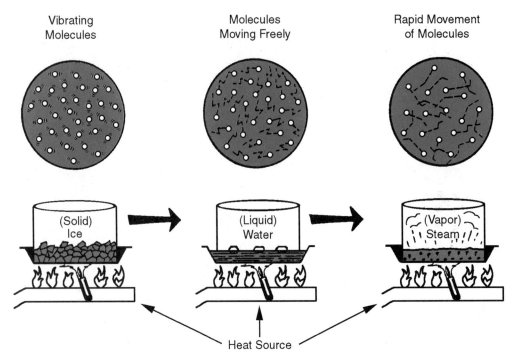

Fig. 7–3   Physical states of matter.

3. *Potential energy*, which is energy at rest. An example of potential energy would be a rock sitting still at the top of a hill.

Regardless of its physical state, the molecules in a material are in a constant state of motion. Introducing kinetic energy to the material will increase the motion of the molecules. The relative motion of these molecules is the focal point in the study of refrigeration.

For example, if we took a close look at a solid such as ice, we would find the water molecules packed tightly together and moving rather slowly. As heat is applied to the ice, the molecules move faster and start to lose their attraction for each other. At 32° Fahrenheit (F), a change occurs—the ice changes to water.

The water molecules are now moving much more rapidly and freely. This is illustrated by the fact that we can pour water. As more heat is applied, the water molecules move even faster. At 212°F, the water changes to steam. In this gaseous state, the molecular particles fly about randomly and have little attraction for each other.

Throughout our example, the application of heat caused a change in state only. There was no

physical change in the basic composition of the water molecule.

TEMPERATURE

The **temperature** of a substance refers to its heat level, or intensity. Temperature is measured with a thermometer (*thermo* meaning heat, *meter* meaning measure of), which is a glass tube with a sealed top and a bulb at its base. The tube, such as the one shown in Figure 7–4, contains a liquid that expands and contracts with the change in temperature. Mercury is commonly used in thermometers.

A thermometer is calibrated to the desired system of measurement. The two types of temperature-measuring systems in international use are the Fahrenheit scale, which is used in the United States, and the Centigrade or Celsius scale, which is used in Europe and in all scientific work. The Celsius system is also used in technical publications and for specifications of refrigeration systems. For that reason, you need to be familiar with and have a working knowledge of the Celsius system.

The Fahrenheit and Celsius scales are shown in Figure 7–5. The freezing and boiling points of

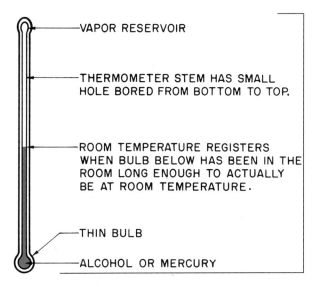

Fig. 7–4 Simple thermometer.

VAPOR RESERVOIR

THERMOMETER STEM HAS SMALL HOLE BORED FROM BOTTOM TO TOP.

ROOM TEMPERATURE REGISTERS WHEN BULB BELOW HAS BEEN IN THE ROOM LONG ENOUGH TO ACTUALLY BE AT ROOM TEMPERATURE.

THIN BULB

ALCOHOL OR MERCURY

water are used as reference points because water is the most common element and has relatively constant freezing and boiling points. In the Fahrenheit scale, there are 180 divisions between the freezing and boiling points of water (32 to 212). In the Celsius scale, there are 100 divisions between the freezing and boiling points of water (0 to 100).

ABSOLUTE TEMPERATURE SCALES

The Celsius and Fahrenheit scales shown in Figure 7–5 indicate **relative temperature**. Also indicated are scales referred to as **absolute temperature scales**, which are used in some calculations and specifications. The Rankine scale is the absolute for the Fahrenheit scale and the Kelvin scale is the absolute for the Celsius scale. Absolute temperatures may be measured only where the reference scale indicates a true zero temperature. *Absolute zero*, the temperature at which theoretically all molecular motion would cease, is the basis for the two absolute temperature scales.

To illustrate the concept of absolute temperature, we'll use this example: The pressure of a gas will decrease by 1/273 for each degree Celsius that it is cooled. This condition exists for all gases. Therefore, the temperature at which no pressure is exerted is the same point as the temperature at which there is no heat.

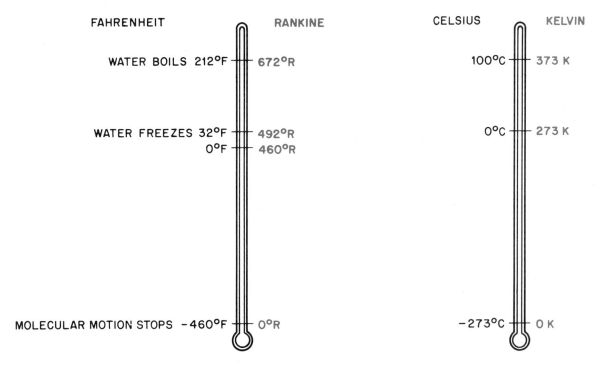

FAHRENHEIT   RANKINE   CELSIUS   KELVIN

WATER BOILS 212°F — 672°R   100°C — 373 K

WATER FREEZES 32°F — 492°R   0°C — 273 K
0°F — 460°R

MOLECULAR MOTION STOPS −460°F — 0°R   −273°C — 0 K

Fig. 7–5 Comparisons of Fahrenheit, Rankine, Celsius and Kelvin temperature scales.

FAHRENHEIT AND CELSIUS TEMPERATURE CONVERSIONS

At times you'll find it necessary to convert a given temperature to a different scale. To convert degrees Fahrenheit to degrees Celsius, you can use a pocket calculator and the following formula:

Degrees Celsius = (degrees Fahrenheit – 32) ÷ 1.8

As an example, we'll convert 86° Fahrenheit to the Celsius scale by going through the following steps to make use of the formula.

STEP ONE: Subtract 32 from the Fahrenheit temperature.

$$86 - 32 = 56$$

STEP TWO: Divide by 1.8

$$56 \div 1.8 = 30$$

Our conclusion then, is that 86° Fahrenheit is equal to 30° Celsius.

To convert Celsius temperature to the Fahrenheit scale, you can use this formula:

Degrees Fahrenheit = (degrees Celsius × 1.8) + 32

As an example, we'll convert 15° Celsius to the Fahrenheit scale using the following steps:

STEP ONE: Multiply the given Celsius temperature by 1.8

$$15 \times 1.8 = 27$$

STEP TWO: Add the figure 32.

$$27 + 32 = 59$$

The conclusion is that 15° Celsius is equal to 59° Fahrenheit.

Temperature Measuring Devices

A wide variety of instruments that give accurate temperature readings are available for almost any application you might encounter. One of the most common and useful instruments is the pocket thermometer, such as the one shown in Figure 7–6.

The thermometer shown is a glass tube type. The bulb usually contains mercury or another type of easy-to-read red fluid. Be careful when

Fig. 7–6   Glass tube pocket thermometer.

Fig. 7–7   Pocket type dial thermometer.

using this type of thermometer. The glass tube is fragile and breaks easily. On occasion, the fluid column in the tube may separate, which means you will have to carefully apply heat to the thermometer tube, causing the fluid to expand and reconnect with the fluid in the bulb. Glass tube thermometers are available in a variety of temperature ranges, the most popular being one that reads temperatures from –40°F to 140°F.

Another type of pocket thermometer is the dial thermometer, shown in Figure 7–7. This design uses a bimetallic coil that expands and contracts with temperature changes. The movement of the coil, which is connected to a pointer, is converted to reflect temperature changes on a calibrated scale.

Most dial thermometers have a calibration nut located under the thermometer head that is used to recalibrate the thermometer occasionally. To accomplish the calibration of a thermometer of this type, place the sensing step in a glass containing a mixture of ice and water, then calibrate the thermometer to 32°F according to manufacturer's directions.

When you find it necessary to take a remote temperature reading, you would use a remote bulb unit such as the one shown in Figure 7–8. This

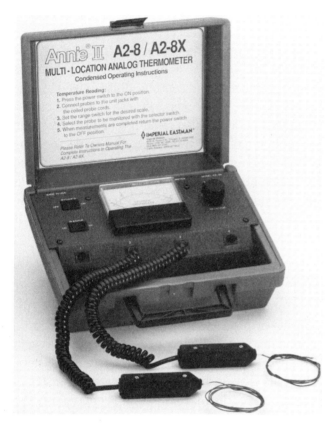

Fig. 7–8   Analog type electronic thermometer. *Courtesy Imperial Eastman, Imperial Tools & Accessories.*

Fig. 7–9   Electronic thermometer. *Courtesy Thermal Engineering.*

Fig. 7–10   Recording thermometer. *Courtesy Bacharach Inc.*

type of thermometer can record more accurately than the pocket thermometer and is available with a variety of probe lengths and tips for taking measurements of remote air and surface temperatures.

The electronic thermometer shown in Figure 7–9 has one or more sensing probes that enable you to take temperature readings from various parts of a refrigeration or air handling system. With an electronic thermometer, a device known as a *thermistor* is located in the sensing probe. When a change in temperature is experienced, there is a change in the resistance of the thermistor. This causes a change in the amount of current flowing through the thermistor (electronic thermometers are battery powered) and meter circuit. The changes in the flow of electrical current are then converted to a temperature scale on the thermometer's dial. The thermistor element makes the electronic thermometer extremely sensitive to the smallest temperature change, giving you accurate readings.

Occasionally you may find it necessary to monitor the temperature of a refrigeration or air conditioning system over a period of time in order to evaluate its performance. A recording thermometer, such as the one shown in Figure 7–10 is used for this purpose.

HEAT TRANSFER

One of the basic laws of thermodynamics states that heat will travel from a level of higher temperature to a level of lower temperature. The greater the temperature difference between the two objects or areas, the greater the rate of **heat transfer**. Without this temperature difference,

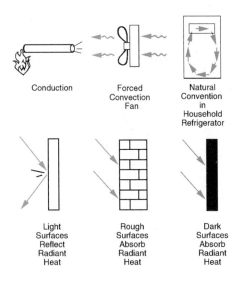

Fig. 7–11   Methods of heat transfer.

there is no transfer of heat. Heat transfer takes place in one of three ways: by conduction, radiation, or convection. The three methods of heat transfer are shown in Figure 7–11.

Conduction

Conduction is the process of transferring heat through physical contact. For example, if a spoon were placed in a cup of hot coffee, the spoon would also become hot after a few minutes. The transfer of heat is caused by molecules that are in direct contact with each other. This is a result of direct molecular collisions. Eventually, the spoon and the coffee would become the same temperature.

Radiation

Radiation is the transfer of heat through the air by heat waves or rays. All substances absorb or emit heat rays. Bodies with higher temperatures emit or radiate heat rays while bodies with lower temperatures absorb heat rays. In radiation, the heat energy can be either filtered, reflected, or radiated, depending upon the surface it strikes or passes through. For example, if a wall is relatively smooth, it will reflect radiant heat. If it is rough, it will absorb radiant heat. A light-colored roof of a building will reflect more heat than a dark-colored roof, which will absorb more heat.

Convection

Convection is the process of transferring heat through fluids using weight. When a gas or liquid is heated, it expands and becomes light in weight. The speed of the molecules increases, which causes them to separate further, resulting in an increase in the volume of the substance and a decrease in weight. The lighter gas or fluid rises and is replaced by colder, heavier gas or fluid. This is referred to as *natural convection*. Using mechanical means to create this movement (such as a fan or pump) is referred to as *forced convection*.

CONDUCTORS AND INSULATORS

Conductors are substances that facilitate the transfer of heat. Good conductors such as copper, aluminum, and steel are an essential part of a well-designed air conditioning or refrigeration system. Materials that do not readily allow the transfer of heat are called **insulators**. Good insulators such as wood, cork, fiberglass, or urethane foam are used in the construction of buildings and spaces conditioned by refrigeration because they reduce the amount of heat that enters or leaves the conditioned space.

HEAT

Heat is the total number of molecules of a substance as they relate to the total amount of energy of the molecules. Often confused with temperature, heat is a direct measurement of quantity, not intensity.

A good example of this would be a cup of water that has been removed from a larger container of warm water. Providing the cup has been immersed in the water long enough for the temperature of all substances to equalize, the temperature of the water in the cup and the container are the same and the molecules of water are moving at the same speed. However, the water in the container has a greater number of molecules and a greater quantity of heat because there is a larger amount of water in the container.

As the temperature of a substance decreases, the motion of the molecules also decreases. When all of the heat is removed from a substance, the molecular motion ceases. This is known as absolute zero. As the temperature of a substance increases, so does the motion of the molecules. Heat content depends upon the type of substance, its volume, and the amount of heat that has been added or removed from the substance. The term used to explain the total heat content of a substance is *Btu*, which stands for *British thermal unit*.

One Btu is the amount of heat required to raise the temperature of one pound of water one degree Fahrenheit. When discussing the refrigeration process, we consider this term from the point of view that the refrigeration system has to transfer a given number of Btu's from the conditioned space to a location where the heat isn't objectionable.

Sensible Heat

When we feel a temperature change in a substance, we are experiencing a **sensible heat** change. Sensible heat is the heat that is applied or removed from a substance, which causes a change in the temperature of the substance. The temperature change can be measured with a thermometer or you can feel it with your hand. All substances contain sensible heat.

Latent Heat

Latent heat, sometimes called hidden heat, causes a substance to change from one physical state to another. When latent heat is added to or removed from a substance, the latent heat is not measurable with a thermometer; its effects can only be seen. In this process, the heat units are absorbed and expanded in an intermolecular action, which causes the molecules to either separate or come closer together. The result is a change in state with no change in temperature.

One way to look at latent heat is to focus on ice. Below 32°F, water is in the solid state as ice, above 32°F, water is in its liquid state. *During 32°F, this change in state came about.*

The two types of latent heat are the latent heat of fusion and the latent heat of vaporization. Latent heat of fusion occurs when a substance changes from a solid to a liquid or vice versa. This process could, from a practical standpoint, be referred to as the latent heat of melting or the latent heat of freezing.

Latent heat of vaporization occurs when a substance changes from a liquid to a vapor, or vice versa. This process is also referred to as latent heat of condensation. It's important that you remember that a substance can exist as a solid, liquid, or vapor. In order to bring a substance from a solid state to a vapor state, the following factors must be considered:

1. The initial temperature of the substance as a solid

2. The temperature at which the substance will change from a solid to a liquid

3. The latent heat of fusion of the substance

4. The temperature at which the substance will change from a liquid to a vapor

5. The latent heat of vaporization of a substance

The process of changing a substance from a solid to a vapor, taking into consideration the sensible and latent heat changes, is referred to as *total heat content*.

Specific Heat Value

All substances have a **specific heat value** that varies according to the makeup of the substance. In some cases the value changes as the substance changes in state. Metal, for example, has a different specific heat value than wood, and water has a different specific heat value than ice.

The specific heat value of a substance is the quantity of heat required to raise or lower the temperature of one pound of that substance by one degree Fahrenheit. It requires one Btu to raise the temperature of one pound of water one degree Fahrenheit. What this boils down to is that the specific heat of water is one—one Btu.

Although water is used as a basis for specific heat tables, not all substances have the same specific heat value. Most require less heat per pound

SUBSTANCE	SPECIFIC HEAT Btu/lb/°F	SUBSTANCE	SPECIFIC HEAT Btu/lb/°F
Aluminum	0.224	Beets	0.90
Brick	0.22	Cucumbers	0.97
Concrete	0.156	Spinach	0.94
Copper	0.092	Beef, Fresh	
Ice	0.504	Lean	0.77
Iron	0.129	Fish	0.76
Marble	0.21	Pork, Fresh	0.68
Steel	0.116	Shrimp	0.83
Water	1.00	Eggs	0.76
Sea Water	0.94	Flour	0.38
Air	0.24 (Average)		

Fig. 7–12   Specific heat table.

to raise their temperature one degree. Figure 7–12 shows a specific heat table for some common substances.

To calculate the amount of heat to be added or removed from a substance, three factors must be considered:

1. The specific heat value of the substance

2. The weight of the substance

3. The amount of temperature change (TD) desired

We could, for example, determine the quantity of heat required to raise 50 pounds of ice from a temperature of 10°F to a temperature of 300°F. In addition to the factors listed, we would have to consider that with this amount of temperature change, we would cause the material to change in state from a solid to a liquid, then again from a liquid to a vapor. Two changes in state means we would have to consider the factor of latent heat being added twice.

Two constants that are understood in regard to latent heat are:

1. The latent heat of fusion for ice is 144 Btu's per pound. To change one pound of ice at 32°F to one pound of water requires 144 Btu's of latent heat.

2. The latent heat of vaporization for water is 970 Btu's per pound. To change one pound

of water at 212°F to one pound of steam requires 970 Btu's of latent heat.

As an exercise to ensure your understanding of heat, we'll take a step-by-step approach to solving the following problem:

How many Btu's must be added to raise 50 pounds of ice from 10°F to 310°F?

In order to solve the problem, we'll be working with these factors:

1. The total weight of the substance is 50 pounds.

2. The specific heat capacity of ice is .504 Btu.

3. The specific heat capacity of water is 1 Btu.

4. The specific heat capacity of steam is .48 Btu.

5. At the latent heat of fusion, we'll calculate the addition of 144 Btu's per pound.

6. At the latent heat of vaporization, we'll calculate the addition of 970 Btu's per pound.

7. The total temperature change is 300°F.

STEP ONE: Calculate the number of Btu's required to raise the temperature of 50 pounds of ice from 10°F to 32°F. The reason this first calculation covers a temperature rise of only 22° is that we have to stop at the point of the change in state from ice to water to calculate the addition of latent heat.

Step One Calculation: 50 × .504 × 22° = **554.4** Btu's

Before proceeding with Step Two, be certain that you understand where the numbers we've used in the Step One calculation came from. The number 50 represents the total weight of the substance, 50 pounds; the figure .504 represents the specific heat capacity of ice; and the 22° is the temperature difference from 10° to 32°.

Step One Conclusion: It takes 554.4 Btu's to raise the temperature of 50 pounds of ice 22°.

STEP TWO: Calculate the number of Btu's added at the point of change in state. The latent heat of fusion for ice is 144 Btu's per pound.

Step Two Calculation: 50 × 144 = **7,200**

Step Two Conclusion: To bring about the change in state of 50 pounds of ice to water, 7,200 Btu's are required.

STEP THREE: Calculate the number of Btu's required to raise the temperature of water from 32°F to 212°F. We're stopping at 212° because another change in state will occur at that point. Total temperature difference for Step Three is 180°.

Step Three Calculation: 50 (wt) × 1 (sp ht) × 180 (TD) = **9,000**

Step Three Conclusion: To raise the temperature of 50 pounds of water from 32° to 212° requires 9,000 Btu's.

STEP FOUR: Calculate the number of Btu's added for the latent heat of vaporization. The latent heat of vaporization is 970 Btu's per pound.

Step Four Calculation: 50 × 970 = **48,500**

Step Four Conclusion: To change 50 pounds of water to 50 pounds of steam at 212°F, 48,500 Btu's of latent heat are required.

STEP FIVE: Calculate the number of Btu's added to raise the temperature of 50 pounds of steam from 212° to 310°. The specific heat capacity of steam is .48 and the total temperature difference is 98°.

Step Five Calculation: 50 (wt) × .48 (sp ht) × 98 (TD) = **2,352**

Step Five Conclusion: To raise the temperature of 50 pounds of steam from 212° to 310° requires 2,352 Btu's.

STEP SIX: Add the Btu's from each step.

Step One	554.4
Step Two	7,200.0
Step Three	9,000.0
Step Four	48,500.0
Step Five	2,352.0
Total	**67,606.4**

Our conclusion is that in order to raise the temperature of 50 pounds of ice from 10°F to 310°F, 67,606.4 Btu's of heat are required. When considering the process of refrigeration, instead of adding heat to raise the temperature, heat is removed from a substance. You'll recall that we said the definition of refrigeration is the transfer of heat from an area where it's not wanted to a place where it's not objectionable.

A fundamental fact to keep in mind about refrigeration then, is that we don't put the cold in, we take the heat out.

THERMODYNAMIC LAWS OF A SUBSTANCE

In addition to the laws of thermodynamics we discussed regarding the transfer of heat and the three ways in which heat moves—by conduction, radiation, and convection—the operation of a refrigeration system depends on two other factors:

1. When a substance boils, it absorbs heat.

2. When a substance condenses, it rejects heat.

The substance, in the case of a refrigeration system, is known as a *refrigerant*. We'll be discussing refrigerants in detail in later units, so at this point we'll define a refrigerant as a chemical

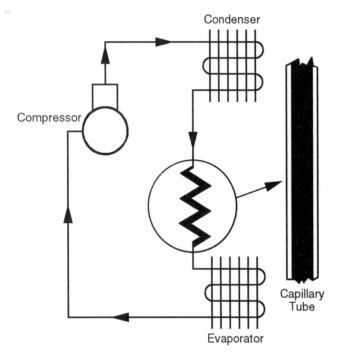

Fig. 7–13 Four components of a refrigeration system. Arrows show the direction of refrigerant flow in a refrigeration system. *Courtesy Ranco Controls.*

that has a very low boiling point. The refrigerant, along with the four basic components that exist in every mechanical refrigeration system, combine to perform the process of refrigeration. Those four components are:

1. compressor
2. condenser
3. evaporator
4. metering device

Along with refrigerants, we'll be discussing each of these components in detail in later units. Figure 7–13 shows the four components of a refrigeration system in their simplest form and the arrows indicate the direction of refrigerant flow in a refrigeration system.

Compressor

A **compressor** in a refrigeration system accepts a low-pressure vapor on its suction side and discharges a high-pressure vapor to the condenser of the system.

Condenser

The **condenser** is the component of the refrigeration system that allows for the rejection of heat. The thermodynamic law, "When a substance condenses, it rejects heat," applies here.

Evaporator

The **evaporator** is the component that allows for the absorption of heat in a refrigeration system. The evaporator would be the component located inside of or in relation to the conditioned space. As the refrigerant changes in state from a liquid to a vapor in this section of the refrigeration system, the thermodynamic law, "When a substance boils, it absorbs heat," explains the function of the evaporator.

Metering Device

A **metering device** creates a controlled restriction in the refrigeration system, creating a drop in pressure, and metering the proper amount of refrigerant into the evaporator. The metering device could be considered as relative to the compressor in that the compressor creates a pressure rise in the system and the metering device creates the pressure drop. This pressure differential is as essential to the operation of a refrigeration system as the thermodynamic laws we've already discussed.

A refrigeration system can do its assigned tasks due to the pressure differential maintained in the system and the changes in state from a vapor to a liquid in the condenser and from a liquid to a vapor in the evaporator.

Pressure

Pressure is defined as a force per unit area. For example, if a force of one pound acts upon an area of one square inch, the pressure exerted is one pound per square inch. Pressure is usually expressed in terms of pounds per square inch or pounds per square foot. When pressure is exerted by a solid, the intensity is concentrated on the base of the solid. When pressure is exerted by a fluid, the pressure is exerted downward and outward. Pressure exerted by

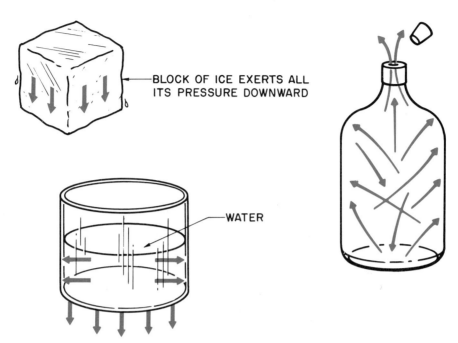

BLOCK OF ICE EXERTS ALL ITS PRESSURE DOWNWARD

WATER

Fig. 7–14   (a) Solids exert all their pressure downward. The molecules of solid water have a great attraction for each other and hold together. (b) The water in the container exerts pressure outward and downward. the outward pressure is what makes water seek its own level. The water molecules still have a small amount of adhesion to each other. (c) Gas molecules travel at random. When a container with a small amount of gas pressure is open, the gas molecules seem to repel each other and fly out.

a gas is in all directions regardless of the size and shape of the container. Figure 7–14 shows pressure being exerted by different substances.

ATMOSPHERIC PRESSURE

Atmospheric pressure is the amount of force transmitted by the air mass surrounding the earth. This air mass, which has weight, is some 50 miles deep and contains air and water vapor. If a column of air one square inch in area was extended from the surface of the earth to the outer limits of the atmosphere, the pressure exerted on the earth's surface (at sea level) would be measured at 14.7 pounds per square inch (14.7 PSI).

Pressure gauges used in refrigeration and air conditioning systems are calibrated to account for atmospheric pressure. An atmospheric pressure of 14.7 pounds per square inch absolute (PSIA) is indicated on pressure gauges as zero pounds per square inch.

Atmospheric pressure is affected by changes in altitude, temperature, and the water vapor con-

tent of the air. The *barometer* is an instrument used to measure the atmospheric pressure and the day-to-day changes in pressure.

A simple barometer, as shown in Figure 7–15, consists of a tube slightly more than 30 inches high. It is sealed at one end and filled with mercury. The open end is inverted and placed in a container also filled with mercury. The level of mercury in the glass tube will drop until the weight of the atmosphere on the surface of the mercury in the pan supports the weight of the mercury in the tube.

At sea level and under average conditions, the height of the mercury in the tube will stabilize at 29.92 inches. Atmospheric pressure is often referred to in inches of mercury or in Hg. One cubic inch of mercury weighs .491 pounds and it takes 2.036 cubic inches of mercury to weigh one pound. A pressure of 14.696 pounds is equivalent to 29.291 inches of mercury. Barometric pressure expressed as inches of mercury can be converted to pounds per square inch absolute by dividing the barometric pressure given by the factor 2.036.

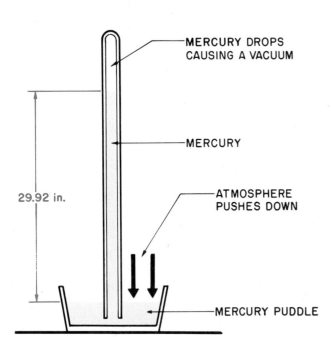

MERCURY DROPS
CAUSING A VACUUM

MERCURY

29.92 in.

ATMOSPHERE
PUSHES DOWN

MERCURY PUDDLE

Fig. 7–15   Simple mercury (Hg) barometer.

Unit Seven Summary

- The basic principles of refrigeration and air conditioning deal with certain laws of physics, chemistry, and engineering. As a service technician you should have an understanding of these principles. Refrigeration is the process of transferring heat from an area where it's not wanted to an area where it's not objectionable. The refrigeration process is the same whether it is being applied to comfort cooling systems, food preservation, or manufacturing. The most common type of refrigeration system is the compression system.

- Matter is anything that occupies space and has weight. All matter is composed of molecules. A molecule is defined as the smallest particle of a substance that still retains all of the properties of that substance. Matter can be found in the physical state of a solid, liquid, or vapor.

- Three types of energy are kinetic energy, fixed energy, and potential energy. Temperature refers to the heat level of a substance and is measured with a thermometer. Two methods of temperature measurement are the Fahrenheit and the Celsius scales. In addition to these relative scales, two absolute tempera-

ture scales known as Rankine and Kelvin are used in engineering.

- Heat always moves from warm to cool in one of three ways: by conduction, radiation, or convection. Materials that facilitate the transfer of heat are known as conductors and materials that do not readily allow the transfer of heat are known as insulators. The specific heat of a substance is the quantity of heat required to raise or lower the temperature of one pound of that substance one degree Fahrenheit. A measure of heat is the Btu (British thermal unit), which is the amount of heat required to raise the temperature of one pound of water one degree Fahrenheit.

- Two kinds of heat are sensible heat and latent heat. Sensible heat brings about a change in temperature and latent heat brings about a change in state but not a change in temperature. The latent heat of fusion occurs when a substance changes in state from a solid to a liquid or from a liquid to a solid. The latent heat of vaporization occurs when a substance changes in state from a liquid to a vapor or from a vapor to a liquid.

- Pressure is defined as a force per square unit area. If the force of one pound acts upon an area of one square inch, the pressure exerted is one pound per square inch (one PSI). Atmospheric pressure is the amount of force transmitted by the air mass surrounding the earth. At sea level it is equal to 14.7 PSI. A barometer is an instrument used to measure atmospheric pressure.

- Atmospheric pressure is often referred to in inches of mercury or Hg. Atmospheric pressure at 14.7 PSI is equal to 29.921 inches of mercury. One cubic inch of mercury weighs .491 pounds and 2.036 inches of mercury weighs one pound. Barometric pressure expressed in inches of mercury can be converted to pounds per square inch by dividing by 2.036.

- In addition to the thermodynamic law of heat transfer, a refrigeration system is dependent upon two other laws of thermodynamics related to substances. When a substance boils it absorbs heat and when a substance condenses it rejects heat. It is also necessary to have a pressure differential system within a refrigeration system in order for it to function.

• The four basic components of any refrigeration system are the compressor, condenser, evaporator, and metering device. The fluid in a refrigeration system is known as a refrigerant, which is a chemical that has a very low boiling point.

• In a refrigeration system, the compressor causes an increase in pressure and the metering device causes a decrease in pressure. In the evaporator the refrigerant changes in state from a liquid to a vapor and in the condenser the refrigerant changes in state from a vapor to a liquid.

Unit Seven Key Terms

Matter	Heat
Energy	Sensible Heat
Temperature	Latent Heat
Relative Temperature	Specific Heat Value
Absolute Temperature Scales	Compressor
Heat Transfer	Condenser
Conduction	Evaporator
Radiation	Metering Device
Convection	Pressure
Conductor	Atmospheric Pressure
Insulator	

UNIT SEVEN REVIEW

1. What is the definition of refrigeration?
2. What are the two types of refrigeration systems? Which system is the most common?
3. Define matter.
4. Define molecule.
5. Why is the most widely used compression system of removing heat being replaced by the absorption system?
6. Of the three states of matter, how do solids differ from liquids?
7. Label each of the following substances as demonstrating either kinetic energy, fixed energy, or potential energy:
    a. wood
    b. baseball in flight
    c. brick on top of a chimney
    d. car brakes engaged
    e. pop can on the edge of a table
    f. natural gas
8. Define temperature.
9. What are the two absolute temperature scales?
10. Heat travels from a level of _____ temperature to a level of _____ temperature.
11. What is a British thermal unit (Btu)?
12. How is temperature affected by sensible heat?

13. How is temperature affected by latent heat?

14. What is specific heat?

15. When a substance boils, it _____ heat, and when a substance condenses, it _____ heat.

16. What is the procedure for calibrating a dial type pocket thermometer?

17. What is the fundamental law of thermodynamics in regard to heat transfer?

18. List the three ways in which heat moves.

19. What are the two thermodynamic laws in relation to the change in state of a substance?

20. Define atmospheric pressure.

21. C = Degrees Celsius   F = Degrees Fahrenheit

    Equations

    To solve for:

    $$\text{Degrees Celsius} = \frac{F - 32}{1.8}$$

    To solve for:

    $$\text{Degrees Fahrenheit} = (C \times 1.8) + 32$$

    a. Convert 25° Celsius to Fahrenheit.

    b. Convert 41° Fahrenheit to Celsius.

    c. Convert 14° Fahrenheit to Celsius.

22. (Q)uantity of heat (in Btu's) = (W)eight × (S)pecific heat × (T)emperature change.

    A swimming pool holds 1,200 cubic feet of water at 61° Fahrenheit. How many Btu's are required to bring the pool up to 90° Fahrenheit? (Hint: Specific heat of water = 1, and one cubic foot of water weighs 62.4 pounds.)

23. It was 20° Fahrenheit when a person caught a salmon in Alaska and put the fish under six pounds of ice. Taking a boat ride to town, a temperature rise to 65°F occurred. Since there was one pound of ice left, how many Btu's were necessary to melt the ice?

    Substitute the numbers into the following equations:

    $$Q = W \times S \times T \text{ (S = specific heat for ice)}$$
    $$+$$
    $$144 \times W \text{ (latent heat)}$$
    $$+$$
    $$W \times S \times T \text{ (S = specific heat for water)}$$

24. Explain the connection between the following pairs of terms:

    a. absolute zero / heat

    b. conductor / physical contact

    c. radiation / heat waves or rays

    d. convector / weight

25. Write a paragraph, in complete sentences, to give some idea of how the four components of refrigeration (compressor, condenser, evaporator, metering device) work together.

UNIT EIGHT

Refrigerants and Refrigeration Oil

OBJECTIVES

After completing this unit, the student will be able to:

1. Describe the function of a refrigerant.

2. Relate specific refrigerants to their applications.

3. Differentiate between the three types of refrigerants.

4. Describe the safety procedures to follow when working with refrigerants and refrigeration oil.

As a service technician working with refrigeration and air-conditioning systems, you will have to work with chemicals such as refrigerants and refrigeration oils. Understanding how a refrigerant performs its task of absorbing heat on the low-pressure side of the system and dispensing heat on the high-pressure side of the system will help you when you're troubleshooting a malfunctioning refrigeration system. Knowing the properties of a refrigeration oil and how it works to keep a compressor lubricated will give you the background necessary to determine the cause of a failure and prevent it from occurring again.

In this unit, we'll discuss the types of refrigerants commonly used in HVAC/R systems, the importance of proper lubrication in a refrigeration system, and the refrigerants that are being developed to replace the chlorofluorocarbons (CFCs), which, under the Clean Air Act of 1990, must be phased out of use in the refrigeration industry.

REFRIGERANTS

The purpose of a **refrigerant** is to absorb heat and transfer it from an area where it is objectionable to an area where it is not objectionable. We have learned that a certain amount of heat can be absorbed by a liquid or a vapor when a temperature differential exists. The amount of heat absorbed is small, however, compared to the amount of heat absorbed by a liquid when it is changing to a vapor through the latent heat process.

Almost any liquid can be used to absorb heat by evaporation. However, only a certain number of liquids have the suitable physical and thermodynamic properties necessary to be used as a refrigerant. For example, water has many desirable characteristics, but it boils at too high a temperature to be used in most cooling applications, and it freezes at too low a temperature for other applications. There is no perfect refrigerant. A refrigerant is chosen for a specific application because it has the most desirable characteristics for that application; any negative characteristics of the refrigerant must be accepted and dealt with. As a rule, the characteristics of a good refrigerant are as follows:

1. Possesses a high latent heat of vaporization
2. Detects leaks easily
3. Is nontoxic
4. Is nonflammable
5. Mixes readily with oil and carries it in the solution as a vapor and a liquid
6. Has no harmful reaction with oil and moisture
7. Has a high resistance to electricity
8. Has a low cost
9. Is readily available

PROPERTIES OF REFRIGERANTS

Vaporization Pressure

Vaporization pressure, the pressure at which a refrigerant vaporizes, is important because the refrigerant must vaporize without requiring too low of a suction pressure. Ideally, the evaporator pressure in the low side of the system should be as close as possible to atmospheric pressure (14.7 PSIA). This allows the system to operate under a positive pressure at all times. A refrigerant that requires a vacuum to evaporate has a tendency to draw air into the system. The pressure in the low side of the system and in the evaporator will be the same. The temperature of the evaporating refrigerant will be the same as the temperature indicated on the temperature/pressure chart, along with the corresponding evaporator or system low-side pressures.

Latent Heat of Vaporization

The latent heat of a refrigerant is the amount of Btu's of heat required to change a liquid into a vapor when the temperature remains constant.

The **latent heat of vaporization** of a refrigerant is the amount of Btu's of heat per hour required to vaporize one pound of liquid refrigerant at atmospheric pressure. For example, as we learned in Unit Seven, to convert one pound of water at atmospheric pressure with a temperature of 212°F into steam at the same pressure and temperature requires 970 Btu's. This quantity of heat is the total latent heat value of one pound of water at atmospheric pressure.

When the latent heat value of a refrigerant is high, less refrigerant needs to be circulated for each ton of refrigeration effect produced. Therefore, a lower capacity compressor can be used.

Condensing Pressure

The pressure at which a refrigerant vapor will liquify depends upon the temperature of the cooling medium. In refrigeration and air conditioning applications, it is desirable to keep the **condensing pressure** as low as possible. Generally, the condensing temperature should be approximately 30° to 35° above the prevailing ambient temperature. This additional temperature is a result of the heat of compression and friction experienced by the refrigerant vapor in the compressor during the compression cycle.

Critical Pressure and Critical Temperature

Every refrigerant has a pressure above which it will remain a liquid even when more heat is added. This is due to the molecular movement of the refrigerant becoming so intense that the molecules cannot come into contact with each other in order to liquify. This pressure is referred to as the **critical pressure** of a refrigerant.

Every refrigerant also has a temperature above which it will not remain a liquid regardless of the pressure applied to it. This is referred to as the **critical temperature** of a refrigerant.

Density of a Refrigerant

Density is defined as weight per unit volume. The **density of a refrigerant** is expressed in weight per cubic foot (lb/cu ft). High-density refrigerants are desirable because they keep refrigerant lines small and reduce installation and construction costs. In applications where the refrigerant piping has long or high vertical runs, a low-density refrigerant is more desirable because it will reduce the amount of pressure required to circulate the refrigerant.

Specific Volume

The **specific volume** of a refrigerant is the number of cubic feet of vapor formed when one pound of the refrigerant is vaporized. The specific volume of a refrigerant is expressed in cubic feet per pound (cu ft/lb). In applications that require a high-density refrigerant, a refrigerant with a low specific volume should be used. In applications that require a low-density refrigerant, a refrigerant with a high specific volume should be used.

Enthalpy

Enthalpy is the total heat content of a refrigerant. It is the amount of sensible and latent heat required to change one pound of refrigerant liquid to a vapor. Enthalpy is expressed in Btu's per pound (Btu/lb). Refrigerants with a high enthalpy will absorb more heat per pound from the conditioned area than those with a low enthalpy.

TYPES OF REFRIGERANTS

There are a variety of refrigerants available for specific and general use. The National Refrigeration Safety Code classifies refrigerants into three major groups:

1. Group One refrigerants, which are the safest.

2. Group Two refrigerants, which are toxic and somewhat flammable.

3. Group Three refrigerants, which are highly flammable.

Group One Refrigerants

Group One refrigerants are the safest and most commonly used refrigerants.

Halogenated Refrigerants

Halogenated refrigerants were originally developed by the DuPont Company as Freon refrigerants. They are most commonly used in normal refrigeration and air conditioning applications. These refrigerants are made of compounds containing one or more halogens, such as chlorine, bromine, iodine, or fluorine. All Freon refrigerants use a numerical identification code now standard with all manufacturers.

Refrigerant-12, Dichlorodifluoromethane (CCl_2F_2), is most commonly used in household and commercial refrigeration applications. As a liquid, it is clear, virtually odorless, and nontoxic. Its primary use is in reciprocating and rotary compressors. R-12 is stored in a white container. Although considered nonflammable, if exposed to an open flame, the compounds that make up the refrigerant will break down and produce a toxic phosgene gas.

Refrigerant-22, Chlorodifluoromethane ($CHClF_2$), has many of the same physical characteristics as Refrigerant-12. However, R-22 has a much higher latent heat of vaporization as well as a much higher saturation temperature/pressure relationship and a lower specific volume. R-22, therefore, has a greater heat absorption characteristic than R-12. This allows for the use of a smaller displacement compressor. It is widely used in refrigeration and air conditioning applications that utilize a reciprocating compressor. R-22 is stored in a green container.

Refrigerant-11, Trichlorofluoromethane (CCl_3F), is a low-pressure refrigerant commonly used in large-capacity centrifugal compressor applications. Liquid R-11 is also commonly used to flush contaminated refrigeration and air-conditioning systems that have experienced a hermetic compressor motor burnout or a heavy concentration of moisture. R-11 is stored in an orange container.

An azeotropic refrigerant is the result of a mixture of refrigerants that, after being combined, exhibit a new maximum and minimum boiling point. They are primarily used in special applications that utilize reciprocating compressors.

Refrigerant-500, R-152a + R-12 (CCl_2F_2/ CH_3/CHF_2), is used in commercial and industrial refrigeration systems that utilize reciprocation compressors. R-500 has an excellent latent heat of vaporization and about 20 percent greater refrigeration capacity than R-12. It mixes well with oil, but its ability to go into a solution with water is critical. In this case, the use of system driers is important as the system must be thoroughly dehydrated. R-500 is stored in a yellow container.

Refrigerant-502, R-22 + R-115 ($CHClF_2$/ $CClF_2CF_3$), is confined to reciprocating compressors where its relatively low condensing pressure and temperature help increase the service life of compressor valves and internal parts. R-502 is used in medium- and low-temperature applications that range from 0°F to –60°F. When operating in these low temperature ranges, oil will tend to separate from the refrigerant and stratify in various parts of the system's low side. For this reason, R-502 systems are equipped with special devices to ensure that the oil returns to the compressor. R-502 is stored in an orchid-colored container.

Group Two Refrigerants

Group Two refrigerants are considered toxic, irritating to breathe, and are somewhat flammable. With the exception of ammonia, R-717, which is used in large-capacity industrial absorption systems, Group Two refrigerants are no longer used. Group Two refrigerants include:

1. R-1130, dichloroethylene ($C_2H_2Cl_2$)
2. R-160, ethyl chloride (C_2H_5Cl)
3. R-40, methyl chloride (CH_3Cl)
4. R-61 1, methyl formate ($C_2H_4O_2$)
5. R-764, sulfur dioxide (SO_2)

Group Three Refrigerants

Group Three refrigerants are not in common use except in specific and controlled applications. When mixed with air, they form a highly combustible mixture and are extremely dangerous. Refrigerants in this group include:

1. R-600, butane (C_4H_{10})
2. R-170, ethane (C_2H_2)
3. R-290, propane (C_3H_8)

REFRIGERANT NAME	NUMBER	COLOR
Trichloromonofluoromethane	R–11	Orange
Dichlorodifluoromethane	R–12	White
Monochlorotrifluoromethane	R–13	Pale Blue
Monochlorodifluoromethane	R–22	Green
Trichlorotrifluoroethane	R–113	Purple
Dichlorotetrafluoroethane	R–114	Dark Blue
Dichlorotrifluoroethane	R–123	Light (silver) Gray
Tetrafluoroethane	R–134a	Light (sky) Blue
Refrigerant 152A/12	R–500	Yellow
Refrigerants 22/115 (48.8% R-22/51.2% R-115)	R–502	Orchid
Refrigerants 23/13 (40.1% R-23/59.9% R-13)	R–503	Aquamarine
Ammonia	R–717	Silver

Fig. 8–1 Cylinder color code for some common refrigerants.

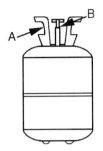

Fig. 8–2 Disposable refrigerant cylinder (a) handle and safety guard for refrigerant valve, (b) refrigerant valve. *Courtesy The DuPont Company.*

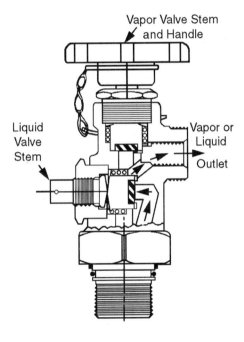

Fig. 8–3   Refrigerant cylinder valve.

Refrigerant Cylinders

Refrigerant storage cylinders may be found in a variety of sizes to best suit the service application. The disposable service cylinder is the most popular cylinder in use today because, being made of aluminum, it is lightweight and easy to handle. It is available in quantities ranging in weight from a few ounces to 50 pounds. (see Figure 8–2) The cylinder service valve is equipped with a ¼" male flare fitting. When the cylinder is placed in an upright position, vapor can be drawn from the service valve. When the cylinder is inverted, liquid can be drawn from the cylinder. (see Figure 8–3) Disposable aluminum service cylinders in the one-pound capacity require a special adaptor valve that is clamped to the top of the cylinder. When installed, the valve punctures the cylinder, allowing the refrigerant to be drawn.

A 145-pound-capacity steel storage cylinder is used when the refrigeration system requires a substantial amount of refrigerant; it may also be used to transfer refrigerant to a smaller capacity service cylinder. (see Figure 8–4)

REPLACEMENT REFRIGERANTS

The Group One refrigerants we've discussed up to this point (those most popularly used in refrigeration systems) are being phased out of pro-

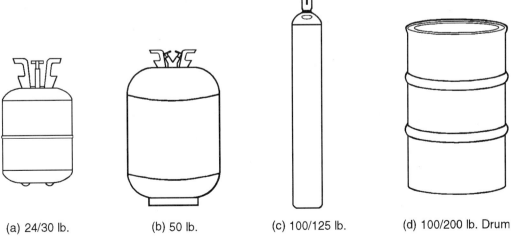

| (a) 24/30 lb. | (b) 50 lb. | (c) 100/125 lb. | (d) 100/200 lb. Drum |

Fig. 8–4   Common refrigerant storage cylinders a, b, c-disposable cans: (a) 24/30-lb, (b) 50-lb, (c) 100/125-lb, (d) drum.   *Courtesy The DuPont Company.*

duction under the Clean Air Act of 1990. The two refrigerants most affected by the EPA regulations concerning recovery of chlorine-based refrigerants are R-11 and R-12. All technicians need to be aware that it is illegal to vent refrigerants into the atmosphere and that all persons working with refrigerants must be certified through an EPA-sanctioned exam.

The reasoning behind the Clean Air Act is that chlorine-based refrigerants are causing a depletion of the ozone layer, a shield that protects us from the ultraviolet rays of the sun. Scientists contend that refrigerants vented into the atmosphere cause damage to the ozone, which will result in increased skin cancer and other health problems.

The most common replacement refrigerants for chlorofluorocarbons (CFCs) are known as hydrofluorocarbons, which do not contain any chlorine. In some cases, an HCFC (hydrochlorofluorocarbon) is being used in place of a CFC. HCFCs are said to be less harmful to the atmosphere because they don't remain as stable as CFCs due to the hydrogen added to the chemical compound. R-22 is classified as an HCFC.

An alternative to R-11 is HFC-123 and an alternate for R-12 is HFC-134a. The reason these refrigerants were developed as replacements is that they have properties similar to the refrigerants they are designed to replace. HFC-123 was developed for use mainly in chiller systems and HFC-134a for use in automobile air-conditioning systems, refrigerators and freezers, and in any commercial refrigeration systems that have operating conditions in which R-12 would be used.

New systems are designed to use the newly developed refrigerants and, in some cases, systems can be converted to use a replacement refrigerant. Most of the problems in replacing refrigerants with the new alternatives stem from compatibility problems with the oil. Diminished operating capacities are also a consideration. In any case, replacement refrigerants cannot be considered a drop-in replacement for those refrigerants that are being phased out of production. Figure 8–5 shows some of the alternate refrigerants that are available as replacements for CFCs.

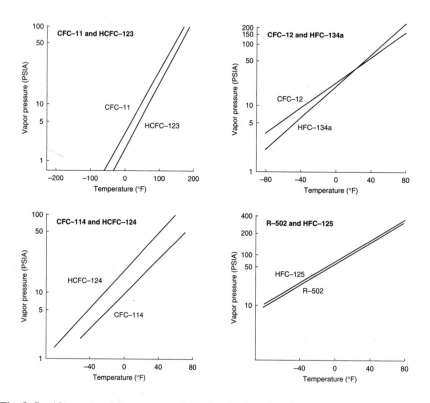

Fig. 8–5   Alternate refrigerants available for CFC replacement.   *Courtesy The DuPont Company.*

THE THREE R'S

The three R's of refrigerants are recovery, recycling, and reclaiming. As a technician, your area of responsibility lies primarily with recovering refrigerant according to EPA guidelines. When working with an HVAC/R system, you'll need to recover the refrigerant, either for reuse in the system once the repair is completed or for return to a refrigerant reclaiming company.

Recovery is the removal of the refrigerant from a system into a container. **Recycling** is defined as cleaning the refrigerant for reuse, and **reclaiming** is the process of remanufacturing the refrigerant for use. When refrigerant is recovered from a refrigeration system, oil is also bound to be recovered, along with contaminants. For this reason, refrigerant-recovery equipment is designed with filtering systems and oil traps to prevent as many problems as possible.

When working with residential and light commercial refrigeration systems, refrigerant-recovery equipment such as that shown in Figures 8–6 and 8–7 is used. When working with large commercial refrigeration systems, a liquid pump (shown in Figure 8–8) is used.

Regardless of the type of recovery equipment used, it must be approved and registered with the EPA. With each manufacturer, the method of operating the refrigerant-recovery equipment is somewhat different, requiring the technician to become familiar with the specific equipment being used.

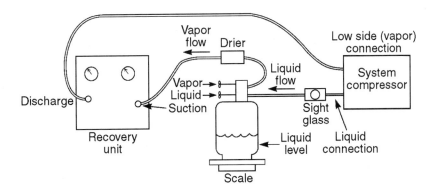

Fig. 8–6   Liquid recovery equipment.

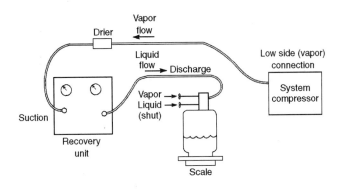

Fig. 8–7   Vapor recovery equipment.

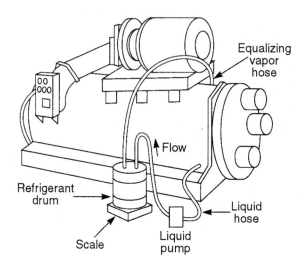

Fig. 8–8   Liquid pump for large amounts of refrigerant.

REFRIGERANT OILS

Refrigerant oil in a compressor serves two basic purposes:

1. To lubricate areas of friction in the compressor.
2. To act as a cooling medium for the dissipation of heat caused by the friction of the moving parts inside the compressor.

A good-quality refrigerant oil has several characteristics that should be considered when selecting an oil for a specific application. These characteristics are discussed in the following paragraphs.

A refrigerant oil must remain fluid or have a low pour point at low system operating temperatures. The pour point of an oil is the lowest temperature at which an oil will still flow. Although the temperature in the compressor may never reach such low levels, some oil will circulate with the refrigerant and will find its way to the evaporator. The low temperatures that may be encountered in the evaporator could cause a poor grade of oil to congeal and stratify in the evaporator, preventing heat transfer and the oil's return to the compressor.

A refrigerant oil must remain stable and resist decomposition under high temperatures, especially at the discharge valve of the compressor where the highest temperatures in the system are reached. This could cause a poor grade of oil to carbonize.

The **viscosity** of a refrigerant oil is also important. Viscosity is the resistance to fluid flow and is measured by the number of seconds it takes a specified quantity of fluid or oil to flow through a calibrated orifice at a specific temperature. The temperature application of the system is the determining factor as to what viscosity of oil should be used. A heavier, or high-viscosity oil that takes longer to flow is more applicable in a system operating under high temperatures, while a low-viscosity oil is more applicable in a system operating under low temperatures. Always consult the manufacturer's specifications for the proper viscosity of oil recommended for a given application. (see Figure 8–9)

Refrigerant oil must not react chemically with the refrigerant used in the system, metal components of the system, motor windings in a hermetic compressor, or system contaminants, such as air. In the case of oil used in a hermetic compressor, the **dielectric strength** of the oil, which is a measurement of the oil's resistance to the flow of electrical current, is important. The dielectric strength is expressed in terms of the voltage required to cause an electrical current to arc

REFRIGERANT	VISCOSITY	SERVICE CONDITION
		Compressor Temperature
All	150 150/additives	Normal
Halogen Ammonia	150/additives 300 300/additives	High
		Evaporator Temperature Above 0°F (–18°C)
Halogen	150 150/additives	
Ammonia	300	
Halogen	150 150/additives	0°F to –40°F (–18°C to –40°C)
Ammonia	150 150/additives	
Halogen	150 150/additives	Below –40°F (–40°C)
Ammonia	150 150/additives	
Halogen	500	Automotive compressors

Fig. 8–9   Refrigerant oils should be chosen according to compressor temperature, evaporator temperature, and kind of refrigerant used.

across a gap one-tenth of an inch wide between two poles immersed in the oil.

The refrigerant oil must be dry and as free of moisture as possible. Refrigerant oil has an affinity for moisture and should be stored in a sealed container until ready for use. To prevent possible contamination, purchase only the quantity of oil needed at any one particular time.

The **floc point** of a refrigerant oil is another important consideration when choosing an oil for a specific application. All refrigerant oils contain some degree of wax. When the oil temperature is low enough, the wax will separate from the oil. These suspended clusters of wax particles will collect in the lower temperature parts of the system, such as the metering device and evaporator. This will result in a restriction of the metering device and a poor transfer of heat in the evaporator coil. For this reason, it is important that the floc point of the oil used is

satisfactory for the temperature application of the system.

The oil used in refrigeration and air-conditioning systems should be highly refined. The color of a well-refined refrigerant oil is light gold; any variation from that color is an indication of possible system contamination.

General Precautions

Every safety precaution should be taken when working on refrigeration and air-conditioning systems and when handling refrigerants and refrigerant oil. The boiling point of the commonly used refrigerants is extremely low—for example, the boiling point of R-22 is –41°F. Thus, exposure to these liquid refrigerants could cause freezing, frostbite, and other serious injuries. For this reason, it is important to wear the proper clothing and safety glasses to protect the body and eyes.

Always follow the prescribed procedures for either discharging refrigerant from a system or charging a system.

If liquid refrigerant comes into contact with the eyes, flush them gently with water and seek medical attention immediately. If liquid refrigerant comes into contact with the skin, follow these first aid procedures:

1. Immediately soak the exposed area in lukewarm water.

2. Apply mineral oil, petroleum jelly, or another similar ointment to the affected area.

3. Keep the exposed area clean, and prevent irritation from clothing. If possible, do not cover the area with a bandage.

4. Immediately seek medical attention.

Although halocarbon refrigerants are classified as nontoxic and nonirritating, one should not be exposed to the gas for an extended length of time. If exposed to an open flame or electrical heating element, halocarbon refrigerants will decompose and form acids and a poisonous gas called phosgene. If this should occur, immediately ventilate and evacuate the contaminated area until the fumes have dispersed.

In general, follow these precautions when handling refrigerants to prevent equipment damage and personal injury:

1. Always work in a well-ventilated area.

2. Wear safety glasses and protective clothing when servicing a system.

3. Make sure that there are no open flames when exposing refrigerant to the environment.

4. Do not breathe refrigerant vapors, especially those emanating from a system with a hermetic motor burnout.

Extreme caution should also be taken when handling refrigerant cylinders. To avoid the buildup of high hydrostatic pressures, cylinders should not be full of liquid at temperatures exceeding 130°F. The resulting pressure from the temperature could rupture the cylinder. Refrigerant cylinders are equipped with a pressure relief device designed to relieve the cylinder's pressure when the temperature reaches 125°F.

Refrigerant cylinders should not be filled to more than 80 percent of their capacity. This allows room for gas expansion as the temperature of the cylinder increases. Never use a torch or other heat source to warm a cylinder to increase refrigerant pressure for transferring or charging purposes. Disposable cylinders should not be refilled when empty.

In general, follow these safety procedures when handling refrigerant cylinders:

1. Always open the cylinder valves slowly.

2. Avoid exposure to high ambient temperatures.

3. Replace the cylinder valve's protective cap after use.

4. Never force connections.

5. Secure cylinders properly when in use or in storage.

6. Do not dent cylinders or handle them roughly.

7. Protect cylinders from rusting.

Unit Eight Summary

- The purpose of a refrigerant is to absorb heat and transfer it to a place where its not objectionable. Almost any liquid can be used to absorb heat by evaporation, however, the best refrigerants used in HVAC/R systems have a very low boiling temperature, possess a high latent heat of vaporization, and mix readily with oil.

- The pressure at which a refrigerant vaporizes is important because it must vaporize without requiring a suction pressure that is too low. The latent heat of a refrigerant is the amount of Btu's of heat required to change the chemical from a liquid to a vapor. High-density refrigerants are desirable because they require smaller refrigerant lines and installation and construction costs are reduced. The three types of refrigerants are Group One,

Group Two, and Group Three refrigerants. Group One refrigerants are the safest and most commonly used.

- Refrigerant cylinders are color-coded to ensure that the proper refrigerant is used in a system when servicing. Refrigerants can be purchased in containers of various sizes.

- Some refrigerants are being phased out of production due to the Clean Air Act of 1990. These refrigerants (CFCs) are being replaced by refrigerants (HFCs) that are described as being environmentally safe. It is against the law to vent refrigerants into the atmosphere and refrigerant-recovery equipment must be used when working with HVAC/R equipment.

- A refrigerant oil serves to lubricate the compressor components and act as a cooling medium caused by the friction of components within the compressor. A good refrigerant oil must have a low pour point, be of a high viscosity, not react chemically with the metal components in a refrigeration system, and be resistant to electrical flow.

Unit Eight Key Terms

Refrigerant

Vaporization Pressure

Latent Heat of Vaporization

Condensing Pressure

Critical Pressure

Critical Temperature

Density of a Refrigerant

Specific Volume

Enthalpy

Group One Refrigerants

Group Two Refrigerants

Group Three Refrigerants

Recovery

Recycling

Reclaiming

Refrigerant Oil

Viscosity

Dielectric Strength

Floc Point

UNIT EIGHT REVIEW

1. What is the purpose of a refrigerant?

2. Why is water undesirable as a refrigerant?

3. What are the nine characteristics of a good refrigerant?

4. Given as one of nine characteristics for a good refrigerant, what makes high latent heat of vaporization so good?

5. While condensing pressure at which a refrigerant vapor will liquify depends on temperature, what causes condensing temperature to be 30° to 35° above ambient temperature?

6. How is the density of a refrigerant expressed?

7. Explain the difference between the critical pressure of a refrigerant and its critical temperature.

8. Why is the specific volume of a refrigerant, measured in cubic feet of vapor, important to know?

9. What are the three types of refrigerants?

10. What are the applications for R-12?

11. Why are refrigerant drums color-coded?

12. Define refrigerant recovery, recycling, and reclaiming.

13. Describe enthalpy.

14. What should technicians know about the EPA (Environmental Protection Agency) concerning the recovery of chlorine-based refrigerants?

15. What does it mean to say that an oil has a high viscosity?

16. Quality refrigerant oil has some important characteristics. Connect these following characteristics with their description: stable, high viscosity, dielectric strength, satisfactory floc point.

    a. Resistance to decomposition under high temperatures.

    b. Resistance to the flow of electrical current.

    c. Resistance to flow under high temperatures.

    d. Resistance to formation in clusters of wax particles at low temperatures.

17. Area is expressed in square measurements.

    Volume is expressed in cubic measurements.

    12 inches × 12 inches = 144 square inches

    144 inches × 12 inches = 1,728 cubic inches

    144 square inches = 1 square foot

    1,728 cubic inches = 1 cubic foot

    a. 1 foot × 1 foot =

    b. 1 foot × 1 foot × 1 foot =

18. A cylinder of nitrogen holds 224 cubic feet with a pressure of 2,160 pounds per square inch.

    a. How many pounds per square foot are being exerted against the cylinder wall?

    b. Using a gas flow of 14 cubic feet an hour, how long will the cylinder last?

19. Write a brief paragraph to explain the three Rs of refrigerants.

20. Are Group One halogenated refrigerants the safest? Explain.

21. Write a paragraph to explain the procedure to be followed if a liquid refrigerant comes into contact with the skin.

UNIT NINE

Techniques for Handling Tubing

OBJECTIVES

After completing this unit, the student will be able to:

1. Identify the types of tubing used in HVAC/R systems.

2. Describe the methods of cutting, reaming, swaging, flaring, and bending tubing.

As a service technician, part of your job entails performing installation of new systems and replacing sealed system components on existing systems. In each application it's necessary for you to be able to properly perform the tasks related to working with tubing as well as to identify the different types of tubing and the requirements for handling each type. In this unit, we'll discuss the fundamentals of refrigeration and air conditioning tubing and how to work effectively with it.

TUBING

Most of the **tubing** used in refrigeration and air-conditioning systems is made of copper. Aluminum is sometimes used in the fabrication of evaporator and condenser coils. The tubing in an HVAC/R system serves two major functions:

1. To provide a passageway for the refrigerant to circulate through the system.

2. To provide for oil return to the compressor.

The terms tubing and piping can be used interchangeably. **Piping** usually refers to heavy, thick-walled material such as iron or steel. The walls of this type of material are thick enough to cut threads into and the threaded fittings are then used to make connections. Thick-walled piping can also be connected by welding. Tubing has a comparatively thin wall and is joined together by means other than threading or welding.

ACR Tubing

Air conditioning and refrigeration (ACR) tubing is manufactured specifically for use in the sealed system. A controlled manufacturing process ensures that the tubing is free of air, dirt, and moisture. The tubing may then be sealed and a holding charge of nitrogen added to prevent contaminants from entering.

When using ACR tubing, the end plugs should be left in place until you're ready to use the tubing. Once a length of tubing has been cut, it should be immediately capped to prevent contamination. All tubing sizes in refrigeration and air-conditioning applications are measured in terms of the outside diameter (OD).

ACR tubing that is classified as type K, or heavy-walled tubing, is approved for HVAC/R systems and special application uses. Type L tubing, or medium-walled ACR tubing, is most common and is also approved for HVAC/R system use. Type K and type L ACR tubing are available in soft or in hard-drawn form.

Nominal Size Tubing

Tubing used in plumbing and heating applications is called **nominal size tubing**. When nominal size tubing is measured, the outside diameter is actually ⅛" larger than the inside diameter (ID). Tubing of this type is not meant for use in HVAC/R systems.

Nominal size tubing is available in hard-drawn and soft form and is available in three types, identified as types L, M, and K. Type L is a medium-walled tubing and is the most common. Type M is a thin-walled tubing, while type K is a heavy-walled tubing used in special applications.

Soft Tubing

Soft tubing is very flexible and easy to form, bend, and flare. It's an annealed copper, which means it has been through a process that heats the material, then allows it to cool. It's available in rolls of 25, 50, and 100 feet, in sizes of ³⁄₁₆" to ¾".

When using soft tubing for installations that have long runs, it's important that you provide proper support for the tubing. Soft tubing is very flexible and will bend if heavy strains are placed at points between the supporting claims. Soft tubing can be hardened by bending it repeatedly, a process known as work-hardening.

Hard-Drawn Tubing

Hard-drawn, **or rigid tubing**, is primarily used in commercial applications. It's available in OD sizes of ³⁄₁₆" to ¾" and in lengths of 20 feet. Because hard-drawn tubing shouldn't be bent, soldered fittings are used to make tubing connections.

FITTINGS

There are a wide variety of fittings available for ACR tubing. The type of fitting used depends upon the installation requirements, the material used, and the permanency of the joint required. Flared fittings and soldered fittings are the most popular types of fittings used.

Flared Fittings

Flared fittings are used in conjunction with soft tubing. These types of fittings are forged from brass, which eliminates the porosity that could result in a refrigerant leak. In applications where the connection is not subject to extreme temperature changes, a long flare nut is recommended. In applications where the connection is subject to extreme temperatures, a short flare nut is recommended. The short flare nut prevents a capillary action that draws moisture into the space between the tubing and the flare nut.

When joining flared fittings, a drop of refrigerant oil should be placed on the surfaces of the connection to prevent the tubing from twisting when the connection is made. As a general rule, a flare connection should be hand-tightened, then wrench-tightened one full turn with a flare nut wrench. Over tightening will cause the tubing to split while under tightening will result in a refrigerant leak.

Soldered Fittings

Soldered fittings are used in field installations where type L or M hard-drawn tubing is used. Soldered fittings are available in the form of tees, 90° elbows and 45° elbows, and couplers for straight connections.

TECHNIQUES FOR HANDLING TUBING

The techniques for handling tubing are among the basic skills you must master as an HVAC/R technician. Cutting, flaring, swaging, and forming are tasks you must be able to perform on a regular basis.

Cutting and Reaming

Soft copper tubing and hard-drawn tubing may be first cut with a handheld tubing **cutter** and then **reamed** to remove any burrs. When using a tubing cutter on copper tubing, follow these steps:

1. Place the cutter on the tubing without over tightening.

Fig. 9–1   Reaming the tubing.   *Photo by Bill Johnson.*

2. Rotate the cutter a few revolutions, tightening the cutter a little more with each rotation until the final cut is made.

3. Ream the tubing as shown in Figure 9–1. Don't over ream; just remove any burrs that may be on the inner surface of the tubing. Take care to prevent chips from falling into the tubing.

Swaging

Tubing is **swaged** in order to join two pieces of soft copper tubing of the same size without using fittings. The advantage of swaging is the elimination of a possible leak from two mechanically joined pieces of tubing. The two types of swaging tools, shown in Figure 9–2, that are commonly used are the punch and the lever.

Both types of tools come in various sizes so that different size tubing can be swaged. The procedure for swaging tubing is as follows:

1. Place the tubing in the flaring block until it extends above the block as illustrated in Figure 9–3. Place the swaging tool adjacent to the tubing to check your distance.

2. Hold the block and tubing in one hand. The last two fingers should be under the block. Hold the tubing by the second and third fingers and hold the swaging tool between the thumb and index finger. This technique is shown in Figure 9–4.

3. Tap the swaging tool lightly to start the swage. Continue to hammer the tool into the tubing. Remove the swaging tool when it has gone as far as it can into the tubing. You can develop a feel for swaging tubing by listen-

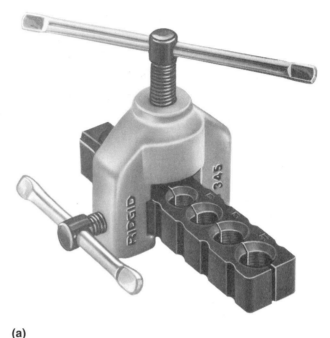

(a)

(b)

Fig. 9–2   Two types of flaring tools used for making standard single flares on HVAC/R tubing. (a) is a split block type and (b) is designed to adjust to the tubing. Both types have a 45 degree chamfer to accomplish the flare shape.   *(a) 345 Flaring Tool published by courtesy Ridge Tool Company, and (b) Imperial Eastman, Imperial Tools and Accessories.*

Fig. 9–3   Aligning proper tubing depth for swaging operation.   *Photo by Bill Johnson.*

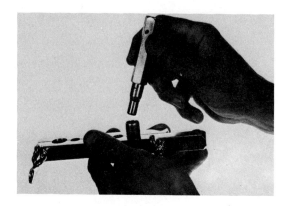

Fig. 9–4 Holding the swage for swaging operation. *Photo by Bill Johnson.*

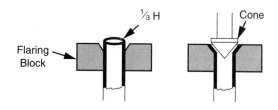

Fig. 9–5 Adjustment of tubing height for flaring operation.

Fig. 9–6 Mechanical bender. *Courtesy Imperial Eastman, Imperial Tools & Accessories.*

ing closely as you're hammering. When the tool has gone in as far as it can, there will be a change in sound. If the swaging tool gets stuck in the tubing, you may have to twist it out.

Flaring

Soft copper tubing is **flared** in order to fabricate a mechanical joint using flared fittings. Flare connections are of a single or double thickness. The single flare is most commonly used and the procedure for flaring tubing is as follows:

1. Place the tubing in the flaring block as shown in Figure 9–5. A rule of thumb is to extend the tubing above the block by a distance equal to the diameter of the tubing being flared. Remember to slip the nut onto the tubing before beginning the flaring process.

2. Slip the flaring tool onto the block and twist it to lock it into place. Put a drop of refrigeration oil on the flaring cone to act as a lubricant.

3. Tighten the flaring cone into the tubing using a few turns, then loosen it to check your progress. Repeat the procedure until the flare is complete.

The tightening and loosening procedure will make a clean, polished flare and reduce the process of work-hardening of the tubing. Don't over tighten the flaring tool. This will cause the flare to split or wash out. Check the fit of the flare in the seat of the flare nut or by placing the flared tubing on the surface of a male connection and inspecting the fit closely.

The critical part of the flaring process is in placing the tubing in the flaring block. If the tubing extends too far above the block, the flare will be too large. If the tubing doesn't extend far enough above the block, the flare will be too small. Tubing must be properly cut and reamed to ensure a good flare.

Bending

Tubing can be bent with a mechanical **bender**, which is shown in Figure 9–6, or with a spring bender, such as the one shown in Figure 9–7. Soft tubing can also be formed by hand.

Precise bending of tubing takes a considerable amount of time and practice and must be done carefully to prevent collapsing or kinking. Bends should be made gradually as the tubing is formed into the desired shape. As a general rule, the min-

Fig. 9–7 Spring bender. *Courtesy Imperial Eastman, Imperial Tools & Accessories.*

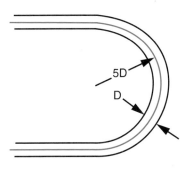

Fig. 9–8 A minimum safe bending radius for HVAC/R tubing is 7 to 10 times the radius of the tubing.

imum radius for tubing to bend is about 7 to 10 times the radius of the tubing, as shown in Figure 9–8.

Constricting

In certain applications, it may be necessary to slip-joint two pieces of tubing and **constrict** a joint. This is accomplished by inserting a piece of small tubing into one of a larger diameter. The space is then constricted by the use of a constrictor wheel installed on the tubing cutter. As the wheel is tightened, the outer tubing is constricted

against the inner tubing. After constricting, the joint can be soldered.

Unit Nine Summary

- Most of the tubing used in HVAC/R equipment is made of copper. Aluminum may be used in the manufacture of evaporator and condenser coils. ACR tubing performs the two functions of allowing a path for refrigerant flow and the return of oil to the compressor.

- Piping is a term that refers to heavy, thick-walled material while tubing has a comparatively thin wall. Piping is joined by threading, and tubing is joined by soldered fittings or by swaging.

- ACR tubing is manufactured in a controlled environment to prevent contamination and the tubing is capped when it is shipped. Nitrogen may be used to prevent the migration of contaminants into the tubing. The two types of tubing used in HVAC/R systems are type K and type L tubing.

- HVAC/R system tubing may be bought in the form of soft or hard-drawn. Soft copper comes in rolls of 25, 50, and 100 feet while hard-drawn copper comes in 20-foot lengths. HVAC/R tubing is measured by its outside diameter.

- Soldered fittings in the form of tees, 90s, or 45s may be used. Couplers are also available for joining two pieces of tubing for a long, straight run. Flared fittings are used on soft copper while soldered fittings are used on hard-drawn copper.

- Swaging is a process used to join two pieces of soft tubing together without the use of a fitting. Swaging tubing offers the advantage of only one solder joint instead of two. Bending of tubing may be accomplished with a mechanical bender, a bending spring, or by hand. In some applications a constricted joint known as a slip-joint may be used.

Unit Nine Key Terms

Tubing

Piping

Air Conditioning and Refrigeration Tubing (ACR)

Nominal Size Tubing

Soft Tubing

Hard-Drawn Tubing

Flared Fittings

Soldered Fittings

Cutter

Reamed

Swaged

Flared

Bender

Constrict

UNIT NINE REVIEW

1. What is the difference between tubing and piping?

2. What two purposes does ACR tubing serve?

3. How is air conditioning and refrigeration tubing different from other kinds of tubing?

4. What is nominal size tubing?

5. Nominal size tubing is identified as types L, M, and K. Match each type with the following descriptions:

   a. thin-walled tubing

   b. heavy-walled tubing

   c. medium-walled tubing and most common

6. How is soft tubing different from hard-drawn tubing?

7. What can be done to harden soft tubing?

8. What are the classifications of tubing?

9. Describe the process of annealing tubing.

10. Why should tubing be reamed after it has been cut?

11. What is swaging?

12. What is the advantage of swaging tubing rather than using a fitting?

13. Describe the common types of fittings used in connecting ACR tubing.

14. What is the minimum radius required for tubing to bend?

15. What is the relationship between flared fittings and soft tubing?

16. What is the relationship between soldered fittings and hard-drawn tubing?

17. Match the method with its description: reaming, swaging, and flaring.

    a. Remove burrs on the inner surface of the tubing.

    b. Used to join tubing of the same size without using fittings.

    c. Used to fabricate a mechanical joint using special fittings.

18. To prevent collapsing or kinking, give the general rule for bending tubing.

19. Diameter = 2 radii

    a. 5 diameters is equal to how many radii?

    b. 8 radii is equal to how many diameters?

20. Wall thickness for a length of tubing is .030", and the inside diameter is .375".

    a. What is the outside diameter of the tubing?

    b. Round off this number to the closest fraction.

21. Write an explanation to show how to make a slip-joint.

UNIT TEN

Soldering Alloys and Soldering Techniques

OBJECTIVES

After completing this unit, the student will be able to:

1. Identify the various types of solder used in the HVAC/R industry.

2. Describe the equipment used in soldering.

3. Describe the proper procedures to follow in preparing tubing and soldering joints.

When employers consider hiring technicians, one of their major considerations is their ability to solder. Understanding the techniques used in soldering, the types of solders used (and their applications), along with developing your coordinated skills when making joints, are extremely important. In this unit we'll discuss the types of alloys used in soldering and the equipment you'll use when accomplishing installations or repairs on HVAC/R systems.

BRAZING AND SOLDERING ALLOYS

A wide variety of alloys are available to suit just about any brazing or soldering operation that may be required in the installation and repair of refrigeration and air-conditioning systems. It's advisable to refer to the manufacturer's product specifications, which will assist you in selecting the proper alloy for a specific application. As a refrigeration and air-conditioning technician, you will need to be familiar with the terms associated with soldering and brazing.

Solidus The temperature at which an alloy begins to melt.

Liquidus The temperature at which an alloy is completely melted.

Melting Range The difference between the solidus and liquidus temperatures.

Liquidation The partial melting of an alloy, which causes a separation of two or more of its components.

Ferrous An alloy that contains iron as the major component.

Brazing The process of joining together metals by means of heat and a filler alloy with a liquidus temperature above 800°F.

Brazing Alloy A nonferrous filler metal used in brazing, generally with liquidus temperature of above 800°F.

Soft Solder An alloy with liquidus temperature below 1,000°F.

For the most part, brazing and soldering alloys can be categorized into two groups, general purpose and specific purpose alloys.

General Purpose Alloys

General purpose alloys will braze together dissimilar metals such as steel, copper, bronze, and brass. They may be sold by the avoirdupois weight system in which a pound contains 16 ounces or by the troy system in which a pound contains 12 ounces.

Silver solder is a general purpose filler metal used for brazing all metals except aluminum and magnesium. Although the process is commonly referred to as silver soldering, it's actually a brazing process.

Various types of silver solder are available for use in a wide range of brazing applications. Some silver solders contain a toxic metal called cadium, which will emit a toxic cadium oxide fume when melted. These fumes can cause serious illness or even death. When using an alloy that contains cadium, it's important for you to remember to work in a well-ventilated area. Cadium-free alloys are also available and should be your first choice when selecting a silver solder. It's common to use silver solder when working with domestic refrigeration systems.

Specific Purpose Alloys

These filler alloys will braze together only like metals: copper to copper joints, brass to brass, or bronze to bronze. They may be sold by either weight system.

Phos/Copper (Sil-Phos)

This family of specific purpose brazing alloys, **phos/copper (sil-phos)**, is used primarily for brazing nonferrous metals. The phosphorous content in the alloy acts as a flux to deoxidize copper surfaces. As a result, you can often forego the use of flux when soldering copper tubing. The temperature range of these alloys is from around 1,190°F solidus to 1,500°F liquidus.

Soft Solders

Tin-lead solders are the most common soft solders in use. These solders are for a specific use only and are not recommended where tubing or a joint will be subjected to stress or vibration.

Tin-Antimony

The **tin-antimony solder**, sometimes referred to as 95/5, contains 95 percent tin and 5 percent antimony. It has a temperature range of 452°F solidus to 464°F liquidus. This isn't much of a melting range (some techs refer to it as the "plastic range") so it's important that you work carefully with 95/5—do not apply too much heat too fast. If you get the work too hot, the solder will run off and not stick to the tubing properly.

It's not a good idea to use 95/5 on a brass joint. When melted, the antimony will absorb zinc from the brass, resulting in a brittle joint. A flux must be used with 95/5, commonly used on solar hot-water systems.

50/50 Solder

50/50 solder used some years back was 50 percent tin and 50 percent lead. Commonly used on water piping, federal regulations have since dictated that a lead-free solder be used. 50/50 solder has a melting range from 360°F to 420°F. Somewhat easier to work with than 95/5, 50/50 solder also requires a flux to condition the tubing and ensure a good joint.

AIR ACETYLENE UNITS

An **air acetylene torch** is one of the most common pieces of equipment used for brazing and soldering in the installation and repair of refrigeration and air-conditioning systems. Figure 10–1 shows a type of air acetylene unit sold under the brand name of Turbo-Torch. An air acetylene unit consists of the following components:

1. acetylene tank
2. pressure regulator
3. hose
4. torch handle
5. torch tip (may be available in various sizes)

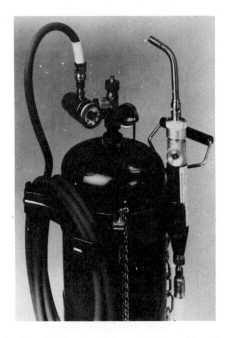

Fig. 10–1 40-cubic-ft. acetylene torch (B tank) with regulator handle and torch tip. Note that the tank is strapped to its holder. *Courtesy Turbotorch Division, Victor Eq. Co., Denton, Texas.*

Acetylene tanks used in field-service applications come in two sizes—the MC tank, which contains 10 cubic feet of acetylene and the B tank, which contains 40 cubic feet of fuel. Each tank has a different size pressure regulator connection, and if the two are to be interchanged, an adaptor must be used.

An acetylene tank under pressure can reach as high as 2,500 PSIG. Being an unstable gas, acetylene can be exploded by impact at a pressure of only 15 PSIG. In order to be able to work with acetylene, a regulator that cuts down the tank pressure to a working pressure of 15 PSIG is used. If you're working with an adjustable regulator, never adjust it above 15 PSIG.

A single-gauge regulator reads the cylinder content only, while a two-gauge regulator reads the working pressure as well as the tank content.

OXYACETYLENE OUTFIT

An **oxyacetylene outfit** consists of two tanks, usually found in a carrier that makes them a portable unit. One tank contains acetylene and the other contains oxygen. With this type of unit,

you'll find that it burns hotter than a standard acetylene torch and will heat up larger tubing faster. Many technicians find that it's comfortable working with an oxyacetylene outfit when both regulators are adjusted to a working pressure of five pounds.

TORCH OPERATING PROCEDURES

(Refer to Figure 10–2)

1. Before making any connections, check the pressure regulators to make sure they are free of dirt, grease, and oil.

2. Before connecting the pressure regulator to a tank, crack the tank valve momentarily to clear the valve of any foreign matter. DON'T DO THIS NEAR AN OPEN FLAME! When working with acetylene tanks, you'll have to use a valve stem wrench or a refrigeration service wrench equipped with a fitting that will fit the stem. The oxygen tank will have a valve knob you can grip with your hand.

3. Use a wrench to secure the pressure regulator to the service cylinder. Pressure regulators themselves are right-hand threads.

4. Use a wrench to tighten the hose connections. Acetylene hose connections are left-hand threads, oxygen hose connections are right-hand threads. Oxygen hoses are green and acetylene hoses are red.

5. Make sure the shutoff on the torch handle is closed. Turn any adjustment knobs on the pressure regulators counterclockwise until there is no pressure on the spring.

6. Open the acetylene tank by turning the stem of the cylinder valve at least one-half turn counterclockwise. If you're working with an oxyacetylene outfit, turn the knob on the oxygen cylinder one full turn counterclockwise. Opening the valves releases the full cylinder pressure to the regulators. It's a good idea to leave the stem wrench on an acetylene tank. In the event of an emergency, you can shut the tank off quickly.

7. Test all connections for leaks by turning the pressure regulator adjustment knob clockwise until the pressure gauge reads about 10 PSIG. Use a soap solution to check for leaks.

8. Open the valve on the handle slightly and use a striker (NOT your pocket lighter) to light the torch. In the case of an oxyacetylene setup, light the acetylene first and then add the oxygen by opening the appropriate handle valve. NEVER open both the oxygen and acetylene valve handles and then try to light the mixture. ALWAYS light the acetylene, then mix the oxygen for the desired flame.

9. Adjust the working pressure of the oxygen or acetylene by turning the pressure regulator adjustment knob clockwise to increase the pressure and counterclockwise to decrease the pressure.

10. To turn off the flame, close the handle valve. When working with oxyacetylene, close the oxygen valve first, then close the acetylene valve. When shutting down a torch assembly for storage, be sure to take the pressure off the regulators by turning the adjustment knobs counterclockwise, then turn the tanks off. Open the handle valves momentarily to bleed off any pressure left in the hoses.

SOLDERING TECHNIQUES
Preparing Tubing

Proper preparation is essential for a good joint. All tubing and fittings need to be cleaned to remove any dirt, oil, or oxidation that may have occurred with tubing exposed to the elements for any length of time. You may clean tubing with a sandcloth, which is specially manufactured for

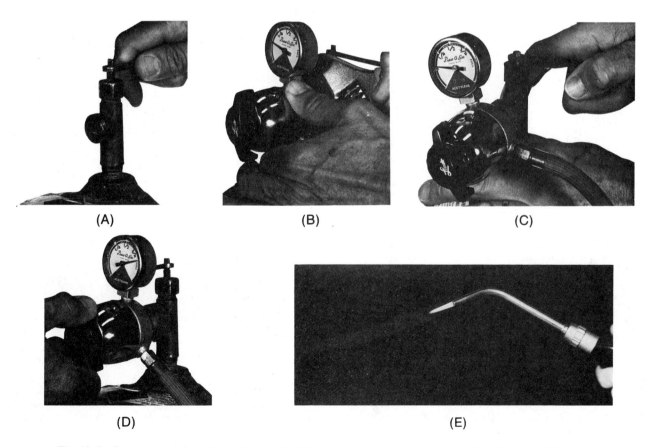

(A)   (B)   (C)

(D)   (E)

Fig. 10–2   Proper procedures for setting up, igniting, and using an air acetylene unit.   *Photos by Bill Johnson.*

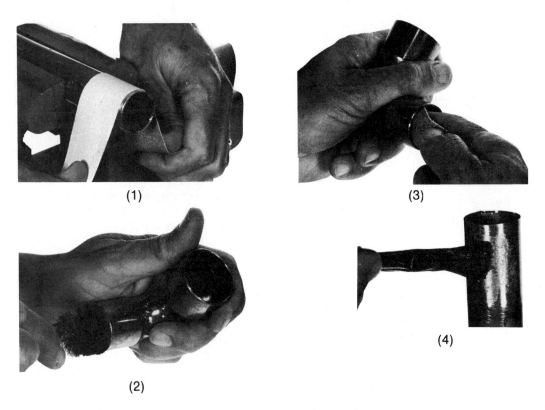

(1)

(3)

(2)

(4)

Fig. 10–3 Preparing the tubing for soldering. *Photos by Bill Johnson.*

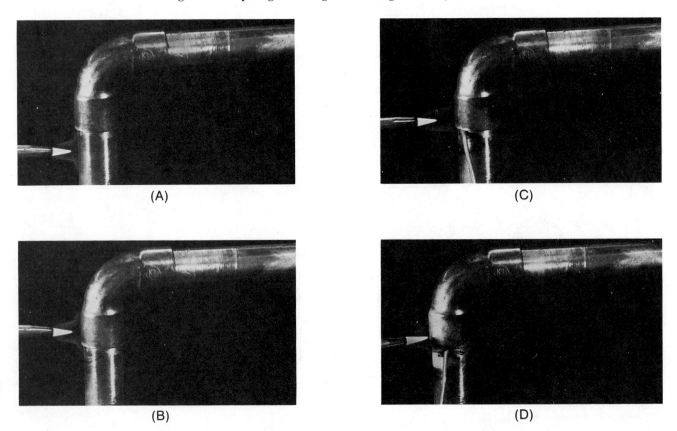

(A)

(C)

(B)

(D)

Fig. 10–4A Soldering a joint with soft solder. *Photo by Bill Johnson.*

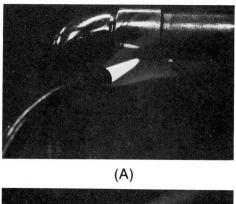

(A)

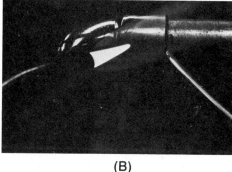

(B)

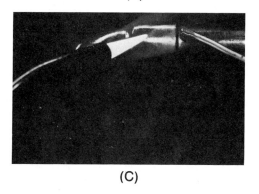

(C)

Fig. 10–4B   Soldering a joint with soft solder.   *Photo by Bill Johnson.*

the purpose, or you can use a wire brush. After you get a piece of tubing or a fitting cleaned, avoid handling it too much on the areas that are to be fitted together. The oils from your body and any dirt, oil, grease, or contaminant you may have on your hands will affect the ability to accomplish a good soldering job.

Using Flux

Flux will prevent oxidation of the metal during the soldering process. If oxidation occurs, it prevents the solder from sticking to the tubing and

fittings. The type of flux you use will depend on the type of alloy you're using for a particular job.

Apply flux sparingly. One of the biggest mistakes inexperienced techs make is using too much flux and, as a result, they don't make a good joint. Take care to prevent flux from getting into the tubing and fittings.

Applying Heat

To make a permanent, leakproof joint, you have to apply heat properly. When the base metal is heated properly, the solder will flow into the joints through what is known as capillary action. Experienced technicians can make solder run "uphill" into a joint by applying the heat in the correct way to the tubing or fitting.

Applying too much heat may prevent the solder from completely sealing the connection. Applying heat unevenly causes a weak or "cold" joint. Proper temperature can be attained by moving the flame around the joint in a waving motion. The solder shouldn't be melted by the flame itself, but rather by the heat conducted through the tubing or fitting.

When joining two different types of metals, such as copper and steel, the heat should first be applied to the tubing that will heat the slowest, then the heat can be conducted to the material that heats up the fastest. On a copper/steel joint, for example, heat the copper first because it is slower to heat up than the steel. By allowing the heat to conduct to the steel, both metals will rise equally in temperature, making the joint easier to accomplish.

SAFETY PROCEDURES FOR SOLDERING AND BRAZING

1. Never drop a cylinder or allow it to fall. Keep them chained or otherwise secured.

2. Keep a fire extinguisher close at hand.

3. Never lay an acetylene tank on its side.

4. Never use oil, grease, or any lubricant on a torch or on any connecting fittings.

5. Don't use matches or a lighter to light a torch. Use a striker.

6. Use eye protection.

7. Work in a well-ventilated area.

8. Never hammer or beat on a valve.

9. Don't use a flame to test for leaks. Instead, use a soap bubble solution.

10. Periodically check all connections for leaks.

11. Keep the valve stem key on an acetylene tank while it's in use. You may have to shut the tank off quickly in an emergency.

12. Never use a torch without a regulator.

Unit Ten Summary

• It's important to develop the ability to solder and be familiar with the types of solder used in the HVAC/R industry. A wide variety of alloys are used in the different applications in the installation and servicing of refrigeration and air-conditioning equipment.

• Some of the terms a technician should be familiar with are solidus, liquidus, and melting range. Soldering alloys are categorized as high-temperature and low-temperature solders. High-temperature solders have a melting range of approximately 1,100°F and soft solders have a melting range of around 360°F. Solder may be sold under the avoirdupois weight system, in which there are 16 ounces in a pound, or the troy system in which there are 12 ounces in a pound.

• Silver solder is a high-temperature solder and some may contain cadium, which emits a toxic fume when melted. Silver solder is commonly used by those working with refrig-erators or similar fractional horsepower systems. Phos/copper is a high-temperature solder commonly used by those working with larger systems.

• 50/50 solder is a low-temperature alloy commonly used on water supply systems. Federal guidelines now require that lead-free solder be used on plumbing installations. 95/5 is a low-temperature solder that is commonly used in the solar industry as well as in the HVAC industry.

• Two kinds of equipment used for soldering are the air acetylene torch and the oxyacetylene torch. An oxyacetylene setup will provide more heat than a single-tank acetylene outfit.

• When preparing tubing for soldering, it must be clean and free of any dirt, oil, or oxidation. Use sandcloth or a wire brush when cleaning. Flux is used to prevent oxidation and ensure a good joint. Solder will flow into a joint through capillary action when heat is properly applied to the base metal.

Unit Ten Key Terms

Solidus	Silver Solder
Liquidus	Phos/copper (Sil-Phos)
Melting Range	Tin-Antimony Solder
Liquidation	50/50 Solder
Ferrous	Air Acetylene Torch
Brazing	Oxyacetylene Outfit
Brazing Alloy	Flux
Soft Solder	

UNIT TEN REVIEW

1. Solidus is the temperature at which an alloy _____ to melt, and liquidus is the temperature at which an alloy is _____ melted.

2. Name the term for the difference between the solidus and liquidus temperatures?

3. Match the solder with its description: silver solder, special purpose alloys, tin-antimony solder, soft solders, 50/50 solder.

   a. Was used commonly on water piping before regulations dictated lead-free solder.

   b. A general purpose filler metal common on refrigeration systems for metals except aluminum and magnesium.

   c. Tin-lead solders are the most common.

   d. Will braze together like metals—for example, copper to copper and brass to brass.

   e. Contains tin with a small melting range of less than 15°F.

4. What are the components of an air acetylene tank?

5. What is a specific purpose alloy?

6. What is the purpose of flux?

7. Describe capillary action.

8. What is a cold joint?

9. For safety, the acetylene regulator should never be adjusted above what pressure?

10. Are oxygen and acetylene hose connections accomplished in the same way? Explain.

11. In the act of soldering, list two causes for a poor joint.

12. Provide the number(s) of the Safety Procedures for Soldering and Brazing that applies once soldering or brazing is underway.

13. A B cylinder contains 40 cubic feet of acetylene and weighs 26 pounds full, or 23.5 pounds empty.

   a. How many cubic feet of acetylene in one pound?

   b. What does a cubic foot of acetylene weigh?

14. A WS cylinder contains 130 cubic feet of acetylene. Some oxyacetylene torches require a working pressure of 5 PSIG on both regulators.

   a. How many cubic inches in 130 cubic feet?

   b. How many hours will the acetylene last at a consumption rate of 2.5 cubic feet an hour?

15. Write a paragraph explaining your thoughts about working with a flame provided by gas or gases coming from pressurized cylinders.

UNIT ELEVEN

Compressors

OBJECTIVES

After completing this unit, the student will be able to:

1. Describe the function of the compressor in a refrigeration system.

2. Differentiate between the types of compressors used in the HVAC/R industry.

3. Explain the necessity for proper lubrication in a compressor.

The compressor has been described as the heart of a refrigeration system. As a technician assigned the task of evaluating the performance of a compressor and troubleshooting possible problems, you can perform more effectively if you have a firm understanding of the design and construction methods used by the industry.

In this unit, we'll discuss the different types of compressors you'll find when working with the wide range of HVAC/R equipment used in homes and businesses.

A compressor serves two purposes in a vapor compression system:

1. To remove the low-temperature, low-pressure refrigerant vapor from the evaporator, reducing evaporator pressure to a point where the desired temperature may be maintained.

2. To raise the pressure of the low-temperature, low-pressure suction vapor to a level where its saturation temperature is higher than the saturation temperature of the condenser cooling medium.

Several types of compressors in common use today include: reciprocating compressors, rotary compressors, screw compressors, centrifugal com-

pressors, and scroll compressors. Reciprocating compressors are the most widely used compressors in all phases of refrigeration and air conditioning applications.

Reciprocating Compressors

Reciprocating compressors are used in the majority of domestic, commercial, and industrial HVAC/R systems and can be classified in several ways:

1. *By the number of cylinders.* The number of cylinders may range from a single cylinder with its operating piston to as many as 16 cylinders.

2. *By cylinder arrangement.* Cylinder arrangement may be vertical, horizontal, or radial. Cylinders may be arranged to form a V, a W, or a double V.

3. *By type of crankshaft.* The design of the crankshaft may be of the crank-throw type or the eccentric shaft.

4. *By construction.* Compressor construction may be either open, accessible hermetic, or nonaccessible hermetic.

Reciprocating compressors are similar in construction to the reciprocating engine in an automobile. Both have cylinders, a cylinder block, pistons, connecting rods, valves, piston rings, cylinder heads, crankshafts, a lubricating system, bearing seals, and gaskets. A refrigeration compressor, however, is driven by an outside force, such as an electric motor.

The Compression Cycle

In order to evaluate and discuss the reciprocating compressor further, a basic knowledge of compression theory is essential. Figures 11–1, 11–2, and 11–3 illustrate the operation of a simple reciprocating compressor.

Figure 11–1 shows the piston at the top of its stroke, which is referred to as *top dead center (TDC)*. In this position, the discharge valve is closed due to the spring and vapor pressure acting on top of it. The suction valve is also closed due to the vapor pressure contained in the clearance pocket above the head of the piston and the underside of the cylinder head.

The piston starts its downward travel in Figure 11–2. An area of low pressure is created

above the head of the piston due to the increase of cylinder volume. When the pressure in the cylinder falls below the suction pressure, the suction

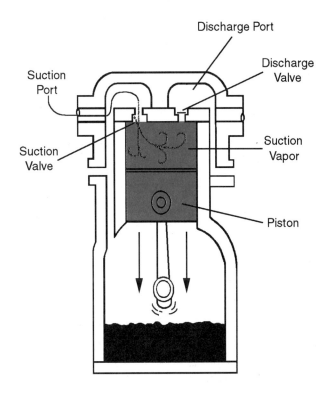

Fig. 11–2   Piston starts downward travel.

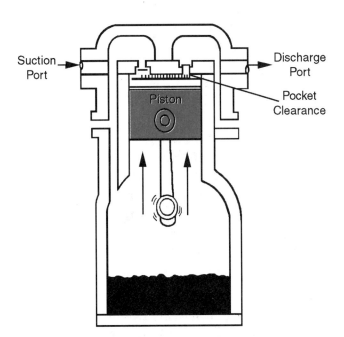

Fig. 11–1   Piston at top dead center.

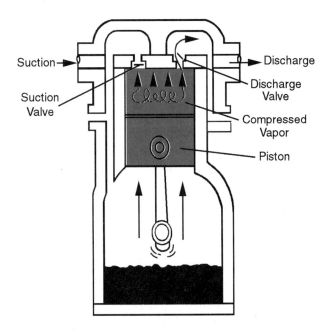

Fig. 11–3   Piston starts upward travel.

valve will open, allowing suction vapor to flow into the cylinder. The cylinder will continue to fill with vapor until the piston reaches the bottom of its stroke, or *bottom dead center (BDC)*. When the vapor pressure of the refrigerant in the cylinder equals the pressure of the vapor in the suction line, the intake valve will close as a result of spring tension.

The piston starts its upward stroke in Figure 11–3, compressing the refrigerant vapor in the cylinder. As the volume of the cylinder decreases, the pressure of the vapor within the cylinder will increase. When the pressure in the cylinder becomes greater than the pressure holding the discharge valve closed, the discharge valve will open. This allows the compressed vapor to enter the discharge line and system condenser. This cycle is repeated every two strokes of the piston, or every revolution of the crankshaft.

COMPRESSOR CONSTRUCTION

The main parts of a reciprocating compressor include a crankcase, cylinder, piston, crankshaft, connecting rod, cylinder head, and valves. While compressor design may vary from compressor to compressor, all are constructed in basically the same way.

Crankcase

The body of the compressor is referred to as the **crankcase**. It contains the bearing surfaces for the crankshaft and stores the oil required to lubricate the internal moving parts.

The compressor crankcase may be constructed as a single piece or as two pieces. The single-piece design is cast with the crankcase and cylinder as one block. A two-piece design consists of a crankcase and a cylinder bolted together with a gasket between them.

Compressor crankcases are usually made of cast iron with some nickel added to increase density and prevent the seepage of refrigerant through the metal. Fractional horsepower compressors usually have fins cast with the cylinders to dissipate heat and improve cooling. Water-

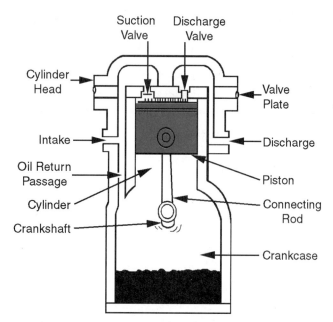

Fig. 11–4   Compressor crankcase assembly.

cooled condensing units may be equipped with a water jacket around the compressor cylinders to provide cooling. Other models of compressors employ an external oil cooler as a means of dissipating heat. Some compressors are manufactured with cylinder liners that can be replaced in the event of wear. Figure 11–4 illustrates the assembly of a typical compressor crankcase.

Pistons

A **piston** is usually constructed of cast iron, although in some applications it may be made of a die-cast aluminum. The piston head is attached to the connecting rod by a piston pin. The piston pin is usually made of case-hardened steel and is hollow to reduce weight. The piston pin is of a free-floating design, which allows the freedom of pin rotation in the connecting rod and piston bushing.

Most pistons are equipped with two or three compression rings. In larger compressors, piston heads are fitted with two types of rings. Upper, or compression rings, are designed to prevent refrigerant vapor from blowing past the piston during the compression stroke. Lower, or oil-control rings, are designed to prevent oil from flowing

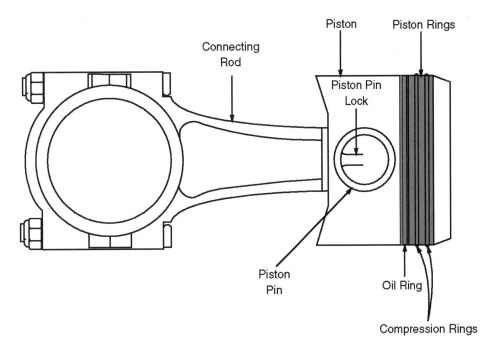

Fig. 11–5   Plug piston.

past the piston head. Piston rings are usually constructed of cast iron, although in some applications bronze rings are used.

The two designs of pistons used in reciprocating compressors are plug pistons and trunk pistons. The plug piston, which is illustrated in Figure 11–5, is sometimes referred to as the automotive piston. A plug piston is used when the suction and discharge valves are located in the head of the compressor. This type of piston is grooved in the upper section. It's usually fitted with two or three compression rings at the top of the piston and an oil-control ring at the bottom of the piston. In some of the smaller compressors, oil grooves in the piston take the place of the oil-control ring.

The trunk piston, or valve in a head piston, has one or more openings in the piston head for the passage of refrigerant vapor. The suction valves are located in the head of the piston. Figure 11–6 illustrates a trunk piston.

Crankshaft

The purpose of the **crankshaft** is to convert rotary motion into reciprocating motion. The crank-throw crankshaft, which is illustrated in

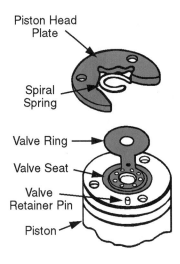

Fig. 11–6   Trunk piston.

Figure 11–7, is the most common design in current use. Its design is very similar to the crankshaft used in automotive engines.

Some compressors, especially smaller ones, use an eccentric-type mechanism, such as the one illustrated in Figure 11–8. The advantage of the eccentric crankshaft is the elimination of the connecting rod caps and bolts as well as a larger bearing surface for the connecting rod. This results in

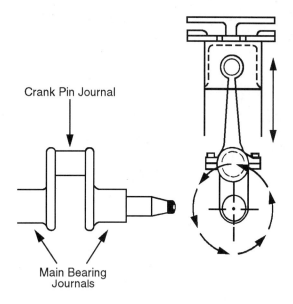

Fig. 11–7   Crank-throw crankshaft.

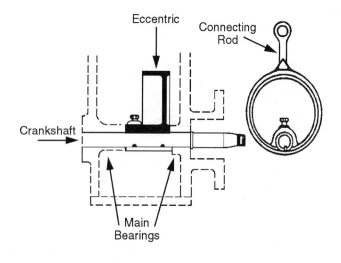

Fig. 11–8   Eccentric crankshaft.

less wear and smoother operation. Crankshafts are usually manufactured from a dropped-forged or a cast-steel alloy.

Crankshaft Seals

The **crankshaft seal** provides a means of sealing the part of the crankshaft that protrudes out of the crankcase and connects to an external motor or another drive means. The crankshaft seal keeps the refrigerant pressure and oil inside the compressor crankcase. The crankshaft must be able to project through the seal and be free to rotate within it at a rapid speed.

Crankshaft seals operate by using two surfaces that rub together. One surface is attached to the crankshaft and sealed with an O-ring. It is designed to rotate with the crankshaft. The second surface is stationary and is mounted on the compressor housing with a leakproof gasket. The two surfaces are made of different materials such as hardened steel and bronze, or ceramic and carbon. One surface is usually soft; the other is hard. The two surfaces must be thoroughly lubricated or excessive wear and leakage will result. Figure 11–9 illustrates some of the crankshaft seals commonly used.

Connecting Rod

A **connecting rod** is used to connect the piston to the crankshaft. It is usually made from a dropped-forged steel. A connecting rod rides on the crankshaft. Its movement comes from the throw journal of the crankshaft, or in the case of an eccentric crankshaft, from a disk of cast iron mounted off-center on the turning shaft.

A connecting rod used with a crankshaft that has a throw design has split lower ends that clamp around the crankshaft journal. In the eccentric crankshaft, the connecting rod usually has a cast-iron bearing surface, and the crank-throw end is a solid ring that slips over the crankshaft.

Cylinder Head

The **cylinder head** holds the compressor valves and valve plate in position and also provides a passage for suction vapor to enter the compressor and discharge vapor to exit the compressor. A cylinder head is usually made from cast iron and is attached to the cylinder with bolts. A sealing gasket is located between the cylinder head and the cylinder.

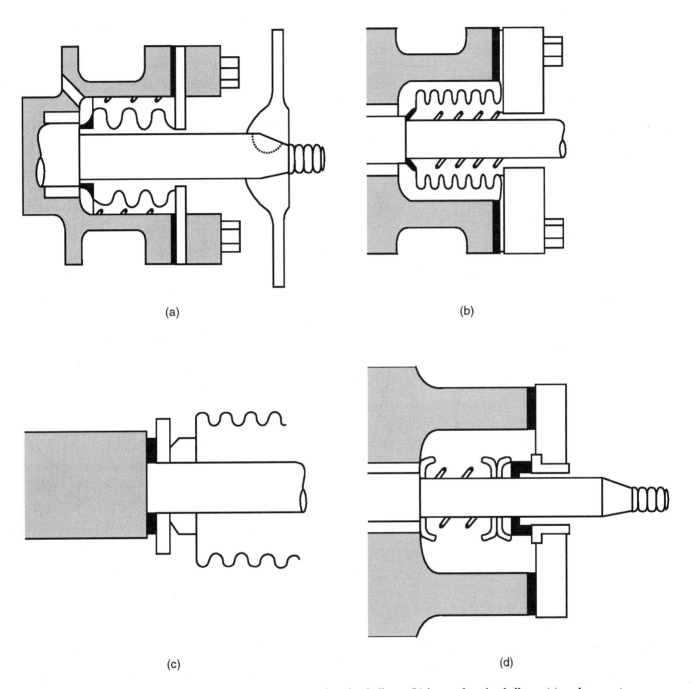

(a)

(b)

(c)

(d)

Fig. 11–9 **Four common crankshaft seal designs (a) external spring bellows, (b) internal spring bellows, (c) replacement crankshaft seal shoulder, and (d) synthetic rubber nonbellows.**

Valves and Valve Plates

A **valve plate** contains the compressor valves, valve seats, retainers, and springs. It is located between the head of the compressor and the top of the cylinder. Valve plates are constructed of cast iron or hardened steel. The three basic types of compressor valves are the poppet valve, the reed or flapper valve, and the ring valve.

The **poppet valve** is not used extensively in modern refrigeration and air-conditioning compressors. Poppet valves were popular in old, large, slow-speed compressors that used ammonia as the refrigerant. A poppet valve is shaped like a

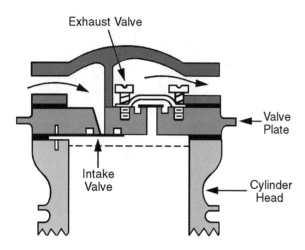

Fig. 11–10   Reed valve.

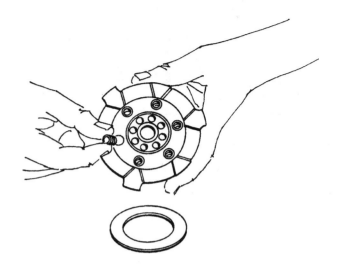

Fig. 11–11   Typical ring valve.

mushroom with a stem and flat cap, and is more commonly used in automotive engines. It is an extremely heavy and sturdy valve requiring the use of an actuating spring as well as other hardware.

The **reed valve**, or **flapper valve**, is quiet, simple in operation and construction, and long-lasting. It is the most widely used valve in modern compressors. The reed valve is constructed of a high-carbon alloy steel, anchored at one end and free to move or flap at the other end. Beneath the free end is a port, or opening, that the valve alternately covers or uncovers. Reed valves bend open as a result of the force of refrigerant vapor pressure and snap closed as a result of the tension of the spring steel from which the valves are made. Retainers are usually placed above the valve to limit the amount of travel when opening. A reed valve is shown in Figure 11–10.

The **ring valve** is heavier and sturdier than the reed valve. Ring valves are suitable for use over a wide range of operating pressures and compressor speeds. The ring valve does not bend but rises and falls in order to open and close its ports. Figure 11–11 illustrates a typical ring valve.

Compressor Service Valves

Compressor service valves provide a means of access to the system for testing and servicing. In conventional compressors, the suction and discharge service valves are bolted to the compres-

sor's head or body. In some partially sealed compressors, the service valves are welded to the compressor dome.

Compressor service valves are designed with a backseated or frontseated construction to control refrigerant flow between the compressor, refrigerant lines, and the gauge manifold ports. Figure 11–12 illustrates a typical compressor service valve.

In the backseated design, shown in Figure 11–13, the service valve stem is turned all the way in a counterclockwise direction. When the service valves are in this position, refrigerant can enter the line port of the valve and flow into the compressor suction port on the low side of the system or leave the compressor discharge port and flow into the discharge line of the system. The service port for the gauge manifold will be in the closed position, which is the position of the service valve during normal operations.

In the backseated and cracked design, shown in Figure 11–14, the service valve stem is turned about one-fourth of a turn clockwise back from the backseated position. The refrigerant flow is the same as when the valve was in the backseated position except that the service port to the gauge manifold port is now open.

In the frontseated design, shown in Figure 11–15, the service valve stem is turned all the way

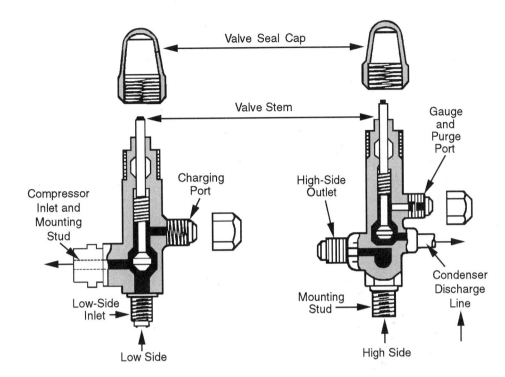

Fig. 11–12   Compressor service valve.

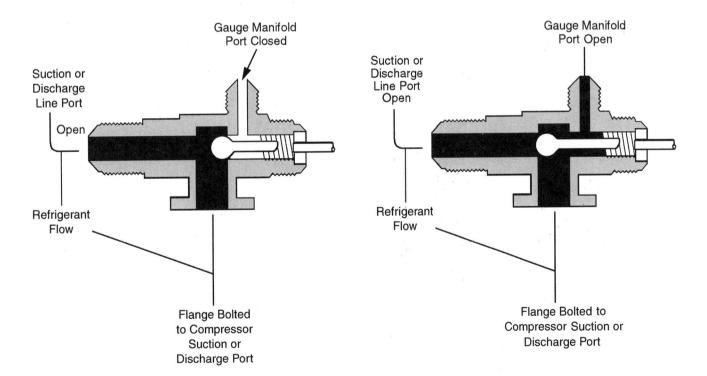

Fig. 11–13   Compressor service valve in backseated position.

Fig. 11–14   Compressor service valve in backseated position and cracked.

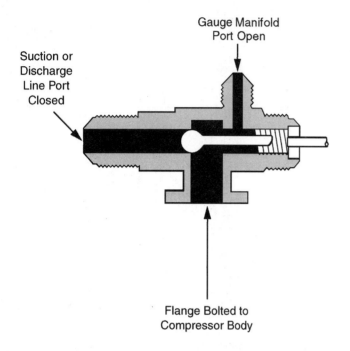

Fig. 11–15   Compressor service valve frontseated.

in a clockwise direction. When the suction service valve is placed in this position, refrigerant flow from the suction line to the compressor is restricted and the gauge manifold port is open. When the discharge service valve is placed in this position, refrigerant flow from the compressor to the discharge line is restricted and the manifold service gauge port is open. If the compressor is operated when the discharge service valve is in the frontseated position, excessive pressure buildup in the head of the compressor could cause system damage or personal injury.

COMPRESSOR LUBRICATION

The two types of **compressor lubricating systems** currently used are the pressure system and the splash system.

Larger-horsepower compressors usually have oil pumps and a pressure lubrication system through which oil is circulated to the bearings and other moving parts. Other compressors utilize a splash lubrication system through which oil is thrown from the crankcase to various projections

and grooves that catch the oil and direct it to different parts of the compressor.

COMPRESSOR DISPLACEMENT

The displacement of a reciprocating compressor is the volume displaced by the piston. **Compressor displacement**, which is expressed in cubic feet per minute, may be calculated by the following equation:

$$\text{Displacement} = \frac{3.1416 \times D \times L \times N \times n}{4 \times 1.728}$$

D = cylinder bore in inches, L = length of stroke in inches, N = RPM of the compressor, and n = number of cylinders.

Clearance Volume

When the piston of a reciprocating compressor is at top dead center, there must always be a distance between the top of the piston and the bottom of the valve plate. This space prevents the piston from striking the bottom of the valve plate and is referred to as the *mechanical clearance*; the volume therein is referred to as the **clearance volume**. Clearance volume is illustrated in Figure 11–16.

Because the discharge valves are on top of the valve plate, there is also more space in the discharge valve ports of the valve plate. This residual space is also part of the clearance volume. At the end of the compression stroke of the piston, this area remains filled with hot compressed discharge vapor at the prevailing discharge pressure. When the piston starts its downward stroke, the residual high-pressure refrigerant vapor expands and its pressure is reduced. This pressure must fall below the pressure in the system suction line before any suction vapor can flow into the cylinder. As a result, a loss of capacity is experienced during the first part of the suction stroke. As the compression ratio increases, a greater loss of capacity will be experienced. Therefore, it is evident that the clearance volume must be held to a minimum. In other words, the clearance volume must be reduced to a minimum on low-temperature applications in

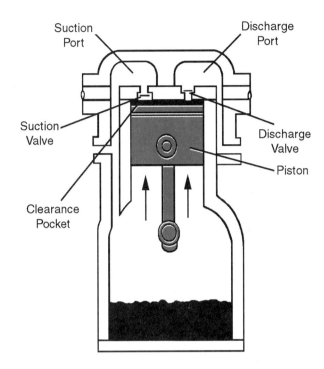

Fig. 11–16   Clearance volume of a piston at top dead center position.

order to obtain the desired capacity. When the suction pressure is high, the compression ratio is low. Consequently, system capacity is not affected, and the clearance volume is not critical.

Compression Ratio

The **compression ratio** of a compressor is the ratio of the absolute discharge pressure to the absolute suction pressure, and is expressed as:

$$\text{Compression ratio} = \frac{\text{absolute discharge pressure}}{\text{absolute suction pressure}}$$

As an HVAC/R technician, you should be concerned when a compression ratio is excessive. A high compression ratio indicates that the system is operating at a high discharge pressure and/or a low suction pressure. High compression ratios result in a loss of efficiency and excessive superheating of the discharge gas. Both of these conditions could result in compressor damage. Compression ratios of 10 to 1 are industry-accepted values.

Compressor Capacity

Compressor capacity is the volume of the actual amount of refrigerant pumped by the compressor. For all practical purposes, actual compressor capacity cannot be measured under normal field conditions. However, the volummetric efficiency, which can be measured, is the ratio of the capacity of the compressor to its displacement and is expressed as:

$$\text{Volummetric efficiency} = \frac{\text{capacity}}{\text{displacement}}$$

Many variables affect the volummetric efficiency of a compressor, such as:

1. Amount of heat exchange between the vapor and the cylinder wall
2. Type of refrigerant
3. Piston fit
4. Valve leakage
5. Compression ratio
6. Amount of vapor restriction as it passes through the valves

ROTARY COMPRESSORS

Rotary compressors are popular in fractional horsepower applications such as domestic refrigerators and window air conditioners. Compared to the construction and operation of the reciprocating compressor, the rotary compressor offers the following advantages:

1. Quietness of operation
2. Smoothness of operation
3. Ability to move large quantities of refrigerant at low compression ratios
4. Fewer moving parts

The two basic types of rotary compressors commonly used are the stationary blade compressor and the rotating blade compressor.

Stationary Blade Rotary Compressor

A simple **stationary blade compressor** is illustrated in Figure 11–17. The compressor con-

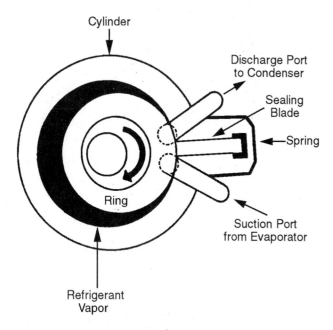

Fig. 11–17   Stationary blade rotary compressor.

sists of a stationary cylinder, roller, and crankshaft assembly. The crankshaft has an eccentric on which the roller is mounted. The blade is held against the roller by a spring and serves to separate the suction and discharge ports of the compressor. The roller does not come into contact with the cylinder; the small clearance is sealed by an oil film. Figure 11–18 illustrates the action of the compressor at the various points of its operating cycle.

Rotating Blade Rotary Compressor

In the **rotating blade compressor**, a series of slots that fit sliding vanes are cut into the rotor. These vanes slide easily in and out of the slots. When the rotor is spun, the vanes tend to fly out by centrifugal force or, in some cases, with the aid of springs. They extend only to the limit of the cylinder wall. The cylinder head is off center so that there is a crescent-shaped area between the rotor and the cylinder. A rotating blade rotary compressor is illustrated in Figure 11–19.

In the operation of a rotating blade compressor, low-pressure vapor is drawn through the suction port. As the rotor turns, the vapor is compressed because the space between the rotor

and cylinder is continuously reduced, increasing the pressure and the temperature of the refrigerant. When the compressed refrigerant vapor reaches the discharge port, it passes into the high-pressure area of the compressor dome. During operation, both a fine film of oil and the design of the compressor will prevent any discharge refrigerant from leaking into the suction side of the compressor. During the system's off cycle, a check valve installed in the suction line will prevent any discharge pressure from leaking into the suction side of the system.

One of the advantages of the rotary compressor is that during the off cycle, high-side and low-side pressures of the system will equalize, allowing for the use of a low-starting-torque compressor motor.

SCREW COMPRESSORS

Screw compressors are used in large-capacity systems ranging from 20 to 300 tons or more. The compressor consists of two rotors with helical meshing lobes that trap the refrigerant vapor as they revolve in the compressor cylinder.

Figure 11–20 illustrates a cross section of a screw compressor. The rotor consists of one male and one female assembly. The motor drives the male rotor and, in turn, drives the female rotor, which is meshed with the male rotor. The operation of a screw compressor is as follows:

1. Refrigerant vapor is drawn in from an inlet port and enters the space between the lobes.

2. As the rotors turn, the lobes move over the inlet point, closing it. Screw action forces the vapor toward the discharge port, compressing it against a discharge plate.

3. At a given point, the lobes will move and open the discharge plate through which the compressed refrigerant is forced out.

A capacity-control slide in the housing wall is used to vary the capacity of the screw compressor. Because the rotors are helixes, the pumping action of the compressor is continuous, resulting in very little vibration during operation.

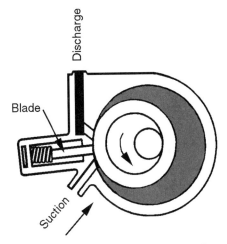

1. Completion of intake stroke; beginning of compression.

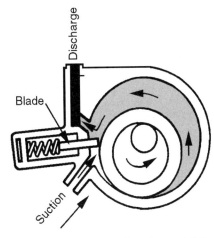

2. Compression stroke continued; new intake stroke started.

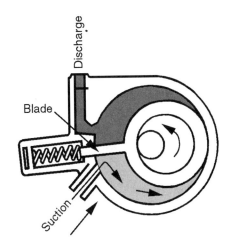

3. Compression continued; new intake stroke continued.

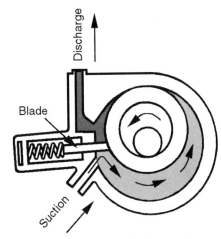

4. Compression vapor discharged to the condenser; new intake stroke continued.

Fig. 11–18   Operation of rotary blade compressor at various points of its operating cycle.

The compressor oil system injects oil into the compressor to seal clearances and absorb heat of compression. The higher operating speeds of the compressor require a good bearing and lubricating system. For this reason, the oil lubricating system utilizes an oil pump, an oil cooler, and an efficient oil-separation system. Because of the efficient oil sealing and cooling systems, there is no reexpansion of discharge refrigerant vapor to reduce compressor volummetric efficiency. Thus, a single-stage screw compressor can be used in many applications that would otherwise require a two-stage reciprocating compressor.

Screw compressors are available as open, externally driven compressors or hermetic, internally driven compressors. The open compressors are usually designed for ammonia systems, while the hermetic compressors are designed for use with halocarbon refrigerants.

CENTRIFUGAL COMPRESSORS

Centrifugal compressors are used in large-capacity systems ranging in size from 50 to 5,000 tons. The centrifugal compressor compresses the refrigerant vapor by whirling it at a high speed,

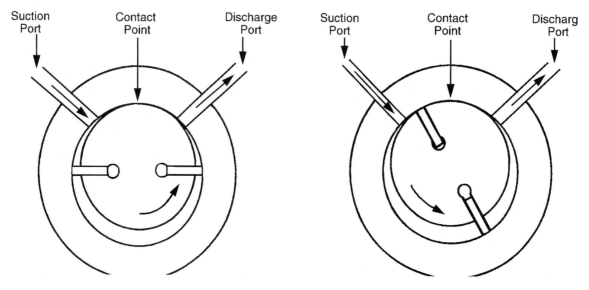

Fig. 11–19   Rotating blade rotary compressor.

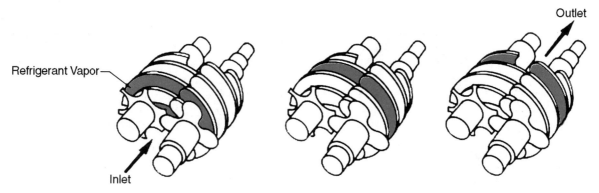

One Stage of Compression as Refrigerant Moves Through Screw Compressor

Fig. 11–20   Tapered machined gears in a screw compressor.

causing it to be thrown by centrifugal force where it is then caught in a channel. An impeller, which is shown in Figure 11–21, is designed to draw vapor from the center of the crankshaft and discharge it at high velocity to the outside edge of the impeller. Because the pressure gained is relatively small, several impellers are placed in series or stages to create a greater pressure differential and to handle larger volumes of refrigerant vapor. The discharge vapor of one stage will enter the intake of the next stage. Each impeller is progressively smaller in size because the pressure of the refrigerant vapor increases progressively even

though the amount of refrigerant vapor passing from one stage to the next remains the same. If each impeller were the same size, the compressed vapor would reexpand to fill the opening and all compression would be lost.

Lubrication is only needed at the end bearings of the crankshaft. This reduces the circulation of oil in the system, allowing the refrigerant to be compressed and improving overall system heat transfer.

Centrifugal compressors operate efficiently over a wide temperature range, from 50°F to 120°F. Because the compressors have a positive

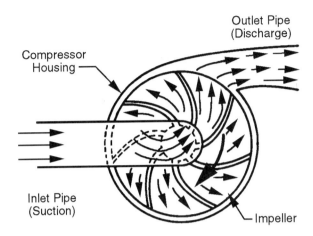

The turning impeller imparts centrifugal force on the refrigerant forcing the refrigerant to the outside of the impeller. The compressor housing traps the refrigerant and forces it to exit into the discharge line. The refrigerant moving to the outside creates a low pressure in the center of the impeller where the inlet is connected.

Fig. 11–21   Impeller.

displacement, they are efficient over a wide range of operating loads and remain reasonably efficient to as low as 20 percent of their normal design capacity.

Centrifugal compressors are available in hermetic, semihermetic, and open designs. They may be direct-driven or belt-driven. Small-capacity units may be driven by electric motors while high-capacity units may be driven by steam turbines.

SCROLL COMPRESSORS

For many years the piston-type compressor has been the primary design used in residential air-conditioning and heat pump applications and has offered good efficiency and reliability. The design and operating parameters of the piston-type compressor are well-developed and understood. However, high energy costs and pending regulations have compelled manufacturers to develop a compressor of higher efficiency than that of the piston-type compressor. The scroll compression system was designed to meet this need.

The operation of the **scroll compressor** centers on two identical scrolls mated to form con-

Scroll Gas Flow

Compression in the scroll is created by the interaction of an orbiting spiral and a stationary spiral. Gas enters an outer opening as one of the spirals orbits.

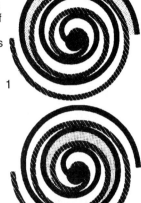

1

2
The open passage is sealed off as gas is drawn into the spiral.

3
As the spiral continues to orbit, the gas is compressed into an increasingly smaller pocket.

4
By the time the gas arrives at the center port, discharge pressure has been reached.

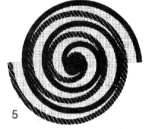

5
Actually, during operation, all six gas passages are in various stages of compression at all times, resulting in nearly continuous suction and discharge.

Fig. 11–22   Operation of scroll compressor. *Courtesy Copeland Corp.*

centric spiral shapes, such as in Figure 11–22. During the compression cycle, one scroll, the fixed scroll, remains stationary while the other scroll, the orbiting scroll, orbits around it. The orbiting scroll does not rotate or turn, but merely orbits the stationary scroll. The orbiting scroll draws suction vapor into the outer crescent-shaped pocket formed by the two scrolls. The centrifugal action of the orbiting scroll seals the flanks of the scrolls.

As the orbiting motion continues, vapor is forced toward the center of the scroll, and the vapor pockets are compressed. When the compressed vapor reaches the center, it is discharged into a chamber and discharge port at the top of the compressor. The discharge pressure pushing down on the top of the scroll helps seal off the upper- and lower-edge tips of the scroll.

During a single orbit, several pockets of vapor are compressed simultaneously, providing smooth and continuous compression. The suction and discharge processes are continuous.

In the application of the scroll compressor, only two components—the fixed and the orbiting scrolls—are required to compress the refrigerant vapor. These two components replace the 15 required in a piston-type compressor (see Figure 11–23).

The scroll compressor has several distinct advantages over the piston-type compressor:

1. The suction and discharge processes of the scroll compressor are physically separated, reducing heat transfer between the suction and discharge gases (Figure 11–24). In the piston-type compressor, the cylinder is exposed to suction and discharge gases, causing high heat transfer and the reduction of compressor efficiency.

2. The process of compression and vapor discharge is very smooth. The scroll compressor compresses vapor in approximately one and one-half revolutions, compared to less than one-half of a revolution for a piston-type compressor. The discharge process occurs for a full 360° of rotation, versus 30° to 60° of rotation for a piston-type compressor (Figure 11–25).

3. The scroll compressor has no valves, while the piston-type compressor requires both suction and discharge valves. The scroll does not require a dynamic valve, which eliminates all valve loss (Figure 11–26).

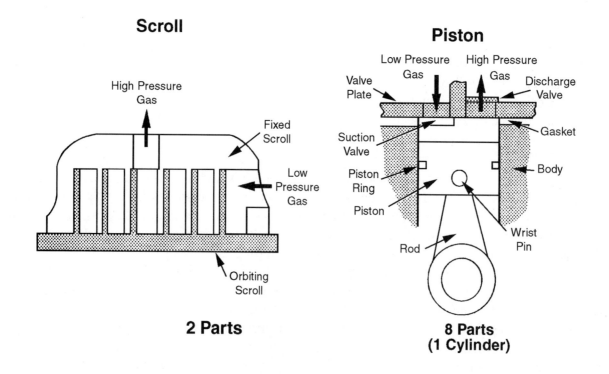

Fig. 11–23   Comparison of scroll and piston compressor parts.   *Courtesy Copeland Corp.*

Scroll

High Pressure
Gas
Discharge

Low
Pressure
Gas
Suction
and Discharge
Separated

Low Heat
Transfer

Piston

Low Pressure
Gas

High Pressure
Gas

Discharge
Valve

Single
Suction
and
Discharge
Chamber

Suction
Valve

High Heat Transfer
at This Point

Fig. 11–24   Suction and discharge are separated in the scroll compressor.   *Courtesy Copeland Corp.*

Scroll

Piston

360°........Length of Discharge.......30° to 60°
>540°.......Length of Compression.......<180°

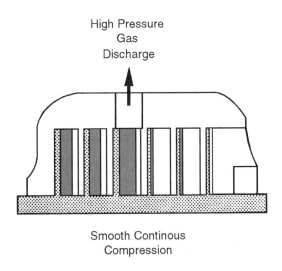

High Pressure
Gas
Discharge

Smooth Continous
Compression

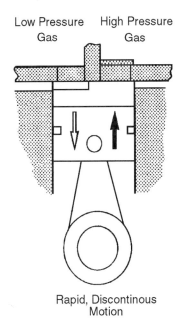

Low Pressure
Gas

High Pressure
Gas

Rapid, Discontinous
Motion

Fig. 11–25   Discharge process occurs full 360 degrees of rotation.   *Courtesy Copeland Corp.*

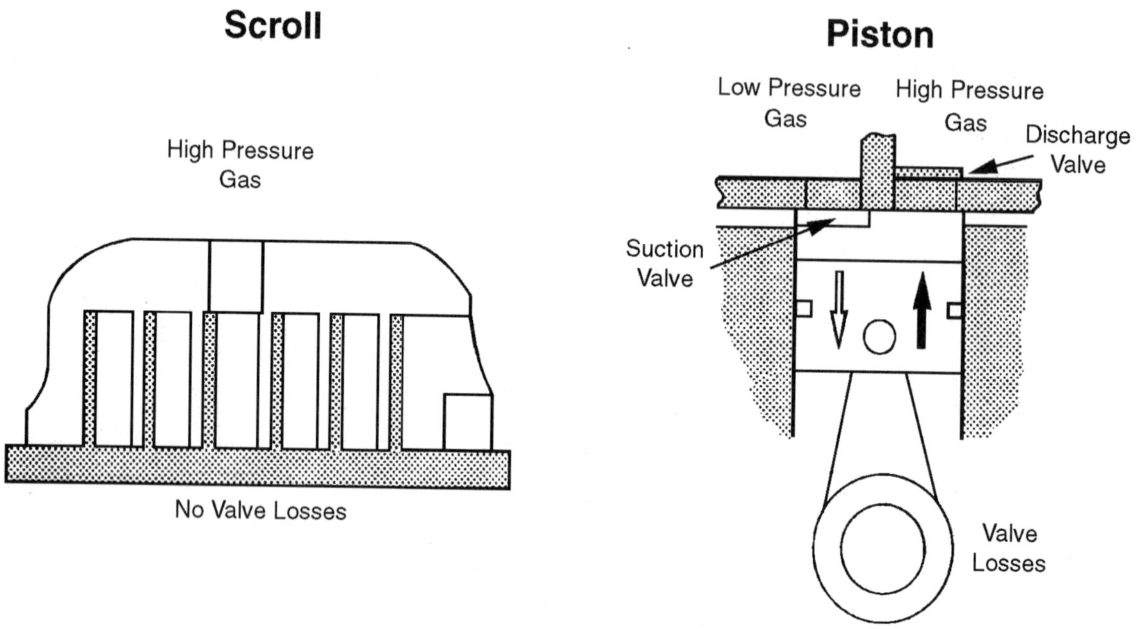

Scroll

High Pressure
Gas

No Valve Losses

Piston

Low Pressure   High Pressure
Gas   Gas
                              Discharge
                              Valve
Suction
Valve

Valve
Losses

Fig. 11–26   Scroll does not require dynamic valves.   *Courtesy Copeland Corp.*

Unit Eleven Summary

- The two primary functions of a compressor in a refrigeration system are to remove the low-temperature, low-pressure vapor from the evaporator and to raise the vapor pressure to a point where the refrigeration system can function as designed.

- Compressors are classified by cylinder arrangement and crankshaft design. The crankcase is the body of the compressor and many compressors are equipped with service valves to provide access to the sealed system. Compressors are lubricated through either a splash system or a forced lubrication pump system.

- Clearance volume is the space between the top of the piston at top dead center (TDC) and the bottom of the valve plate. The capacity of a compressor is the actual volume of the refrigerant pumped.

- Rotary compressors are quiet, smooth in operation, have fewer moving parts than a piston-type compressor, and can be of a stationary blade or rotating blade design.

- Screw compressors and centrifugal compressors are used in high- and very-high-capacity systems while reciprocating compressors are most commonly used in low-capacity systems.

- Scroll compressors are constructed without valves and have fewer moving parts than piston compressors.

Unit Eleven Key Terms

Reciprocating Compressor

Crankcase

Piston

Crankshaft

Crankshaft Seal

Connecting Rod

Cylinder Head

Valve Plate

Poppet Valve

Reed Valve

Flapper Valve

Ring Valve

Compressor Service Valve

Compressor Lubricating Systems

Compressor Displacement

Clearance Volume

Compression Ratio

Compressor Capacity

Rotary Compressor

Stationary Blade Compressor

Rotating Blade Compressor

Screw Compressor

Centrifugal Compressor

Scroll Compressor

UNIT ELEVEN REVIEW

1. What two purposes does a compressor serve?

2. Name the types of compressors in common use today.

3. How are reciprocating compressors classified?

4. Describe the meaning of top dead center.

5. What is the difference between the trunk piston and the plug piston?

6. What are the functions for each of the following parts?

    a. Crankcase

    b. Upper or compression rings

    c. Lower or oil-control rings

    d. Crankshaft

    e. Connecting rod

    f. Crankshaft seal

7. What is the difference between a crank-throw crankshaft and an eccentric crankshaft?

8. Describe the three types of valves used in compressors.

9. What is the function of the compressor service valve?

10. _____ _____ _____ throws oil from the crankcase to different parts of the compressor.

11. _____ _____ _____ uses pumps to circulate oil to bearings and other moving parts.

12. Describe the refrigerant flow through the suction and discharge service valve when it is in the backseated position, backseated and cracked position, and frontseated position.

13. What methods are utilized to lubricate a compressor?

14. Describe clearance volume.

15. What is the compression ratio of a compressor?

16. How is the compression ratio determined?

17. Describe volummetric efficiency.

18. What variables affect the volummetric efficiency of a compressor?

19. What are the advantages of a rotary compressor?

20. Why do the impellers of centrifugal compressors become smaller in size?

21. Match the following compressors with their descriptions: stationary blade rotary compressor, rotating blade rotary compressor, screw compressor, centrifugal compressor, scroll compressor, reciprocating compressor.

   a. At the start of its downward travel, an area of low pressure is created above the head of the piston.

   b. An impeller is designed to draw refrigerant vapor from the center of the crankshaft and discharge it at high velocity to the outside edge of the impeller.

   c. An orbiting motion forces refrigerant vapor to the center where the vapor is compressed, providing smooth and continuous compression.

   d. Refrigerant vapor is drawn in from the inlet port and as the rotors continue to turn, the lobes move over the inlet point, closing it. The vapor is forced against the discharge port, compressing it.

   e. Slots are cut in the rotor for sliding vanes (blades) to move in and out. Low-pressure vapor is drawn into a crescent-shaped space between the rotor and the cylinder where the refrigerant vapor is compressed as the space between the rotor and the cylinder is reduced.

   f. Spring pressure keeps the blade against the roller (rotor). Low-pressure vapor is drawn into a crescent-shaped space between the roller (rotor) and the cylinder where the refrigerant vapor is compressed as the space between the roller (rotor) and the cylinder is reduced.

$$\text{Volume of a cylinder} = 3.1416 \text{ (pi)} \times D \text{ (diameter)} \times h \text{ (height)}$$

22. Find the volume of the compressor when

   the cylinder diameter is 5 in. and

   the length of stroke is 5 in. (Hint: height)

The volume of a compressor displaced by a piston is measured in feet per minute using the following formula:

$$\text{cubic feet per minute} = \frac{3.1416 \times D \times L(h) \times N(RPM) \times n(\text{cylinder number})}{4 \times 1.728}$$

23. Solve, using the formula, when

   the cylinder diameter is 5 in.,

   the length of stroke is 5 in.,

   RPM is 3,450 with 1 cylinder

24. Knowing the compression ratio is important in preventing compressor damage. Ratios from 10 to 1 are acceptable.

$$\text{Formula for compression ratio} = \frac{\text{absolute discharge pressure}}{\text{absolute suction pressure}}$$

Example: A compressor is operating with a head pressure of 178 PSIG as measured, and a suction pressure of 5 PSIG as measured.

$$\text{Compression ratio} = \frac{178 \text{ PSIG} + 14.7 \text{ atmosphere}}{5 \text{ PSIG} + 14.7 \text{ atmosphere}}$$

$$= \frac{192.7}{19.7} = 9.8 \text{ compression ratio}$$

Compute the compression ratio. Could compressor damage result from the following measurements?

head pressure 170 PSIG

suction pressure 2 PSIG

25. Write a paragraph to explain the advantages the scroll compressor has over the piston-type compressor.

UNIT TWELVE

Condensers

OBJECTIVES

After completing this unit, the student will be able to:

1. Describe the function of a condenser in a refrigeration system.

2. Differentiate between air-cooled, water-cooled, and evaporative condensers.

3. Describe proper service and repair procedures for condensers.

The condenser is located on the high-pressure side of the refrigeration system. In the condenser, the refrigerant changes in state from a vapor to a liquid and the heat absorbed in the evaporator is dissipated. Inexperienced technicians often attribute refrigeration system performance problems to areas other than the condenser because of their lack of understanding about the workings of this section of the sealed system.

In this unit, we'll discuss the different types of condensers you'll encounter when working with the variety of HVAC/R systems in existence and the differences in their methods of operation.

The purpose of a condenser is to transfer the heat absorbed by the refrigerant through the evaporation process to a condensing medium, such as air or water. After leaving the compressor, the high-temperature, high-pressure superheated discharge vapor enters the discharge line and upper tubes of the condenser where sensible heat is removed. This process is commonly referred to as *desuperheating* the hot gas. When the refrigerant is reduced to its saturation temperature, the further removal of latent heat will cause the vapor to start to condense into a liquid in the middle tubes

of the condenser. As the liquid refrigerant flows into the lower tubes of the condenser, its temperature is reduced below saturation and thereby subcooling of the liquid takes place. The total amount of heat dissipated by the condenser is equal to the amount of heat absorbed by the evaporator, plus the amount of heat added to the refrigerant by the compressor due to heat of compression and friction.

The three general classifications of condensers are air-cooled condensers, water-cooled condensers, and evaporative condensers. Condensers can be categorized by airflow design, of which there are two types: forced draft and natural draft (see Figures 12–1A and 12–1B).

Forced draft condensers utilize a fan or blower to mechanically move air across the coil surface to dissipate the heat. Natural draft condensers depend upon the natural movement of warm air currents to rise across the coil surface and dissipate the heat.

Heat flows from the condenser to the cooling medium by means of conduction and convection. Several factors affect the ability of a condenser to transfer heat:

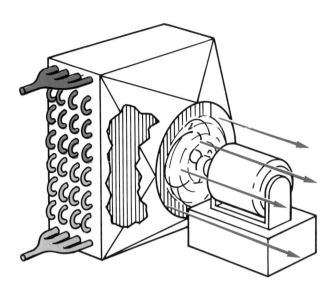

Fig. 12–1A   Forced draft condenser.

1. *Surface area.* The larger the surface area, the more heat that will be dissipated from the condenser.

2. *Temperature difference.* The greater the temperature difference between the refrigerant and the condenser cooling medium, the greater the rate of heat transfer.

3. *Condenser material.* Heat transfer varies with different materials. Efficient conductors of heat, such as aluminum, steel, and copper will increase the capacity of heat transfer.

4. *Condition of the coil.* Cleanliness of the condenser coil and all heat transfer surfaces is important. Dirt and scale that affect airflow or heat transfer surfaces will reduce efficiency.

5. *Flow rate of cooling medium.* As the velocity of the cooling medium over or through the condenser increases, the amount of heat transferred also increases.

AIR-COOLED CONDENSERS

Air-cooled condensers utilize air as the condensing medium. They are used extensively in all types of refrigeration and air-conditioning applications. Air-cooled condensers have the following advantages:

1. Their design is simple.

2. They have a high reliability.

3. They are self-contained.

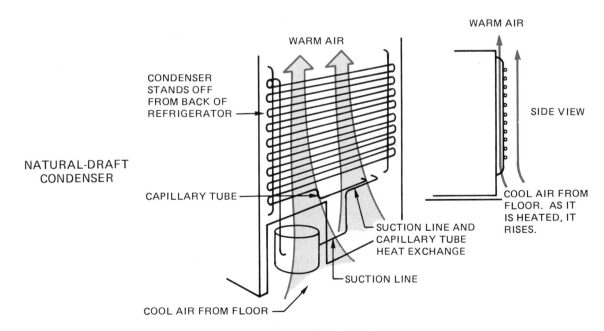

Fig. 12–1B   Natural draft condenser.

4. They are easy and economical to install.

5. They are economical to operate.

Air-cooled condensers also have several disadvantages:

1. The coil tubing and fins must be kept clean. Scales and dirt buildup will greatly reduce heat transfer capabilities.

2. The temperature of the air entering the condenser may vary, resulting in a wide range of operating discharge pressures.

3. There may be an unacceptable noise level from the condenser fans.

4. High ambient temperatures may cause high discharge pressures, making the compressor work harder and reducing system efficiency.

5. Low ambient temperatures may cause low discharge pressures, resulting in a reduction of the pressure difference experienced at the metering device. This will reduce the amount of refrigerant flow into the evaporator and reduce system capacity. Refrigerant velocity is also reduced with a decrease in discharge pressure. This may cause oil logging in the evaporator and reduced heat transfer capabilities. The compressor may also become starved of oil.

WATER-COOLED CONDENSERS

In comparison to air-cooled condensers, **water-cooled condensers** have the following advantages:

1. Lower condensing pressures.

2. Better control of discharge pressures.

3. Smaller and more compact.

4. A lower horsepower compressor.

5. Less power consumption.

Even with the numerous advantages of water-cooled condensers, consideration must be given to the disadvantages:

1. With wastewater systems, the cost of water may be excessive and restrictive. Regulations may also prohibit the use of wastewater systems.

2. The maintenance costs of water-cooled equipment may be three to four times those of air-cooled equipment.

3. If well water is used, the cost of the well, pump, and plumbing must be considered.

4. Water disposal may be a problem. Restrictions may be placed upon pumping wastewater back into a sewer system.

5. If the water source is a river, lake, or pond, pumping costs must be considered. In addition, the possibility of thermal pollution, created by disposing warm water back into the source, may create environmental problems.

6. The installation and maintenance costs will be substantially higher in a recirculating type of system, such as in a water tower system.

Types of Water-Cooled Condensers

The three types of water-cooled condensers in common use are the double tube condenser, the shell and coil condenser, and the shell and tube condenser.

Double Tube Condenser

The **double tube condenser**, shown in Figure 12–2, consists of an S-tube placed within a tube-type arrangement. The water enters the outside tube at the bottom and exits at the top. The refrigerant vapor enters the inside tube of the condenser at the top and exits as a liquid at the bottom.

The refrigerant water counterflow principle provides for excellent heat transfer efficiency in this type of condenser design. Ambient air passing across and between the surfaces of the outside tube removes additional heat, making the double tube condenser a combination air-cooled/water-cooled condenser.

The double tube condenser is small and compact. Because of its high efficiency, it is a popular choice for small commercial applications. Several types of double tube condensers are commonly

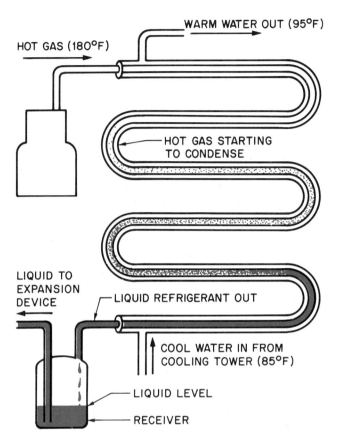

Fig. 12–2 Counterflow principle water-cooled condenser.

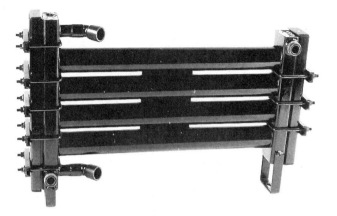

Fig. 12–3 Tube in tube condenser with headers. *Courtesy Standard Refrigeration Company.*

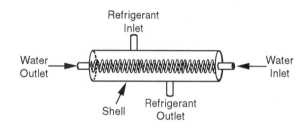

Fig. 12–4 Shell and coil condenser.

used. An efficient design, illustrated in Figure 12–3, has headers installed at each end of the condenser. When the headers are removed, the water tubes are exposed, making them easily accessible for cleaning. The double tube condenser has no liquid storage capacity and therefore must be used with a system receiver when necessary.

Shell and Coil Condenser

The **shell and coil condenser**, illustrated in Figure 12–4, consists of a welded steel shell that contains finned water coils. The refrigerant discharge vapor enters at the top of the condenser, passes around the cold water coils within the shell, and condenses and exits at the bottom. The cold water enters the condenser at the bottom and exits out of the top.

The advantage of the shell and coil condenser is that it is small and compact. It has a liquid storage capability and does not require a separate sys-

tem receiver. Once condensed, the liquid refrigerant is stored at the bottom of the receiver. The shell and coil condenser is popular for small commercial applications.

Although it is inexpensive to manufacture, a disadvantage of the shell and coil condenser is that it cannot be cleaned mechanically. The water tubes must be cleaned by circulating an acid cleaning solution through the coils.

Shell and Tube Condenser

The **shell and tube condenser**, illustrated in Figure 12–5, consists of a cylindrical steel shell that contains finned tubing bundles. The water is circulated through the copper or steel tubing bundles, entering the condenser at the bottom and exiting at the top. The refrigerant discharge vapor enters the condenser at the top, passes around the water tubes, and condenses and exits the condenser at the bottom.

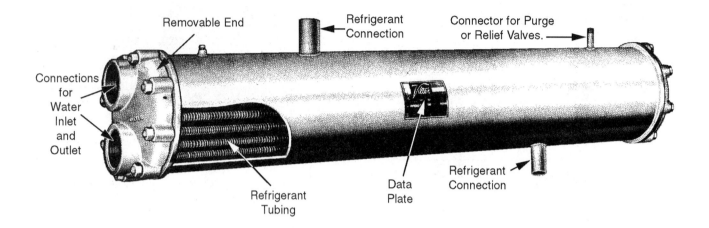

Fig. 12–5 Shell and tube condenser. *Courtesy Vilter Manufacturing Corporation.*

The arrangement of the end plates of the condenser determines how many passes the circulating water will make. This can range from as few as two passes to as many as twenty. The end plates are removable to facilitate the mechanical cleaning of the water tubes.

The shell and tube condenser is used on systems ranging in size from two to several hundred tons of capacity. Because there is sufficient area under the water tubes, the shell and tube condenser can also serve as a liquid receiver.

Condenser Water Flow

The amount of water flowing through a water-cooled condenser varies depending upon the temperature of the supply water and the amount of water flow. The average condensing temperature of the refrigerant in a water-cooled condenser is 105°F. However, condensing temperatures may vary from 90°F to 120°F, depending upon the application. Lower condensing temperatures result in more water usage but higher electrical costs. A compromise must be met, depending on the cost of water and electricity for a given application.

Several variables affect the rate of heat transfer between the refrigerant and water in a water-cooled condenser:

1. The temperature difference (TD) between the circulating water and the refrigerant—the greater the temperature difference, the greater the rate of heat transfer.

2. Condenser surface area—The greater the condenser surface area, the greater the rate of heat transfer.

3. Rate of water flow—The greater the velocity of water flow, the greater the rate of heat transfer.

4. Condenser material—Some condenser metals have a greater heat transfer characteristic than others.

5. Cleanliness of the water heat transfer areas.

In a given application, the design and physical construction of the condenser are fixed. Thus, the primary factors that control the condensing temperature of the refrigerant are the temperature difference between the refrigerant and the circulating water, and the cleanliness of the water circuit.

In wastewater systems, the average water flow through the condenser is approximately 1.5 gallons per minute per ton of refrigeration effect. This will result in an average temperature difference of 15° to 20° between the water entering the condenser and the water leaving it.

In recirculating water tower systems, the water flow rate through the condenser is approximately three to five gallons per minute per ton of refrigeration effect. The average temperature difference between the water entering and the water leaving the condenser is approximately 10° to 15°.

Leak Testing Water-Cooled Condensers

Most of the refrigerant leaks experienced in water-cooled condensers are in the condenser water circuit. Because the pressure of the refrigerant in the system is greater than the pressure of the water circuit, refrigerant will initially leak into the water circuit. Similarly, when the refrigerant pressure is lower than the water circuit pressure, water will leak into the refrigerant circuit. When these conditions occur, severe system contamination and damage to the components will result. For this reason, the possibility of leaks must be considered in servicing water-cooled equipment if the system is found to be low of refrigerant charge.

Water-cooled condensers can be leak-tested by using a halide leak detector or an electronic leak detector. Several methods of leak detection may be employed, depending on the installation and application. In wastewater applications, the air-sensing probe of the leak detector should be placed in an area where the condenser outlet water enters the drain. Any leaking refrigerant will separate from the water and will be detected by the leak detector. In recirculating water tower systems, the air sensing probe of the leak detector should be placed in an area where the condenser outlet water enters the tower.

If the leak cannot be found in this manner, the condenser will have to be isolated from the system and leak-tested. The procedures employed to isolate and leak-test water-cooled condensers vary depending upon the installation and type of condenser.

In wastewater systems, follow these guidelines:

1. If the condenser inlet water supply line is equipped with a shutoff valve, close it.
2. If no shutoff valve is available, shut off the main water supply line valve.

In recirculating water tower systems, follow these guidelines:

1. Shut off the water circulating pump.
2. Isolate the condenser by closing the inlet and outlet water valves.

If no isolation valves are available, the entire system must be drained of water. Follow these guidelines:

1. Drain the condenser of water.
2. Where applicable, such as in the double tube and shell and coil condensers, remove the end manifold housings.
3. Place the air-sensing probe of the leak detector in or at each water tube circuit.

The refrigerant pressure in the system should be high enough to force the leaking refrigerant into the water circuit where it can be detected.

If the condenser design is completely sealed and nonserviceable, such as in the shell and coil condenser and some double tube condensers, a leak cannot be repaired and the condenser will have to be replaced. If the condenser is serviceable, such as the shell and tube condenser and some double tube condensers, a leak can be repaired by plugging the bad tube with a brass plug and brazing it closed. This procedure is not recommended if the leak is large or if several tubes are found to be leaking. In these cases the condenser should be retubed or replaced.

WATER COOLING TOWERS

Water cooling towers are used in recirculating water-cooled condenser applications to cool the circulating water after it has absorbed condenser heat and to allow the water to be recirculated through the system. The underlying principle of the water tower is to take the heat-laden condenser outlet water and introduce it into the tower through a spray header or distribution pan. The water will then cascade through the tower by means of a series of slats or fill. The water will then be collected in a sump. The tower fill slows the travel of the water through the tower, breaking it up into smaller droplets and allowing for more efficient evaporation.

Air introduced by natural prevailing breezes or induced by a fan will evaporate some of the water, in turn cooling the remaining water that has collected in the tower sump. The circulating air will exhaust the warm moist air that has collected in the tower from the evaporation process. A circulating pump will then circulate the cooled water through the condenser to repeat the process. Generally, the water tower system will cool the circulating water to within 5° to 8° of the prevailing ambient wet bulb temperature.

The two types of tower systems in common use today are natural draft cooling towers and forced draft cooling towers.

Natural Draft Cooling Towers

Natural draft or **atmospheric cooling towers** are used in high-capacity commercial and industrial refrigeration and air-conditioning water-cooled condenser applications. Warm condenser outlet water is introduced into the top of the tower by a series of spray headers. As the water cascades through the tower slats, the air inside the tower becomes heated and starts to rise in counterflow to the cascading water. Some of the water will evaporate, cooling the remaining water that has collected in the tower sump to be recirculated. The spray headers also create a pressure drop, which aids in the evaporation and cooling effect of the water. (see Figure 12–6A)

Forced Draft Cooling Towers

Forced draft or **induced cooling towers** are essentially the same as natural draft cooling towers except that a fan is used to circulate air through the tower enclosure. Due to the increased airflow created by the fan, eliminators are installed to reduce the amount of water that can leave the tower. Accessory devices may be installed to cycle the fan in cool weather to maintain the minimum water temperature. In some applications, the tower sump may be installed indoors to prevent the water from freezing during cold weather. A forced draft cooling tower is illustrated in Figure 12–6B.

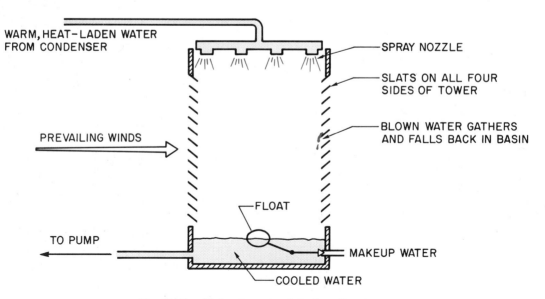

Fig. 12–6A   Piping for natural draft cooling tower.

Fig. 12–6B   **Induced draft evaporative water cooling tower.**
Courtesy Marley Cooling Tower Company.

Water Cooling Tower Components and Terminology

The following items explain some parts and terms commonly used with water cooling towers.

The **water circulating pump**, used in water cooling towers, is not a positive displacement pump and must be properly sized to deliver the proper gallonage of water per minute for the application. Because these pumps do not have a high-capacity lift, they are usually gravity fed from the tower sump.

Makeup water is used to replace water lost through evaporation, drift, or bleed. A ball float and valve assembly is installed to maintain a constant water level in the tower sump. As the water level in the tower sump decreases, the ball float will drop, opening a valve that allows water to enter the sump. As the water level in the tower sump increases, the ball float will rise and, at a predetermined water level, will stop water flow to the tower sump. A makeup water float assembly is illustrated in Figure 12–7.

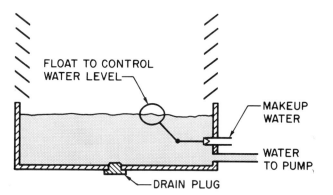

Fig. 12–7   **Makeup water float assembly in a cooling tower.**

As the circulating water in the tower system is evaporated, fresh water will be introduced in order to maintain a constant level of water in the sump. During the process of water evaporation, solidified particles will build up in the circulating water and tower sump. In order to reduce or control this particle buildup, a **bleed line** is installed in the condenser water discharge line feeding the tower. The bleed line is usually installed about six inches above the water level of the tower sump. For each ton of refrigeration capacity, the bleed line will waste approximately two gallons of water per hour.

Drift is water lost from the cooling tower through prevailing breezes and the circulating fan in the tower.

The **tower sump** maintains a constant level and supply of water, which is used to circulate through the condenser water circuit.

The **tower fill** is the material used to slow the travel of the circulating water through the tower, breaking it up into small droplets to allow for more efficient evaporation. Traditionally, the tower fill was manufactured of Florida black cypress or California redwood. Today's technology has allowed the tower fill to be manufactured of man-made, noncorroding materials such as plastic. Figure 12–8 illustrates a cross-sectional view of a tower fill.

Spray headers or a **distribution pan system** allows the circulating water to enter the tower

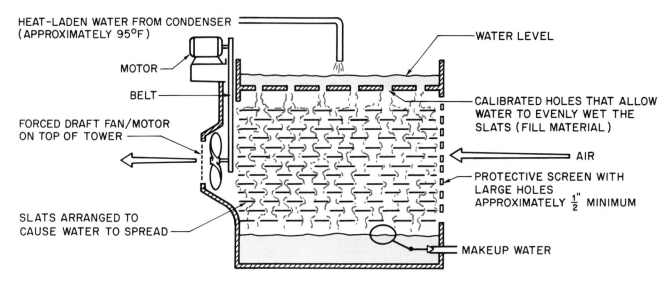

HEAT-LADEN WATER FROM CONDENSER
(APPROXIMATELY 95°F)

MOTOR

BELT

FORCED DRAFT FAN/MOTOR
ON TOP OF TOWER

SLATS ARRANGED TO
CAUSE WATER TO SPREAD

WATER LEVEL

CALIBRATED HOLES THAT ALLOW
WATER TO EVENLY WET THE
SLATS (FILL MATERIAL)

AIR

PROTECTIVE SCREEN WITH
LARGE HOLES
APPROXIMATELY $\frac{1}{2}$" MINIMUM

MAKEUP WATER

Fig. 12–8   Cross section of a tower fill.

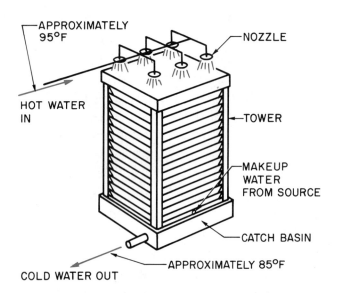

APPROXIMATELY
95°F

NOZZLE

HOT WATER
IN

TOWER

MAKEUP
WATER
FROM SOURCE

CATCH BASIN

APPROXIMATELY 85°F

COLD WATER OUT

Fig. 12–9   Spray header.

and evenly distributes the water over the tower fill. Figure 12–9 illustrates a typical spray header while Figure 12–10 illustrates a distribution pan.

The **tower drain** is installed to collect water that may overflow from the tower sump. The overflow drain should be installed at least two inches from the water level being maintained in the sump.

Eliminators are installed in forced draft water cooling towers to reduce the amount of water loss that may be experienced as a result of the forced convection air movement.

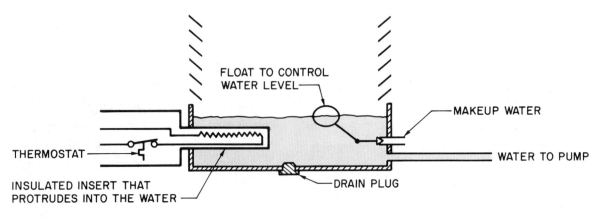

FLOAT TO CONTROL
WATER LEVEL

MAKEUP WATER

WATER TO PUMP

THERMOSTAT

INSULATED INSERT THAT
PROTRUDES INTO THE WATER

DRAIN PLUG

Fig. 12–10   Distribution pan.

WATER REGULATING VALVES

Water regulating valves are used in conjunction with water-cooled condensers to regulate water flow and system discharge pressures. Although primarily used in wastewater applications, water regulating valves may also be found in recirculating water tower applications. The three types of water regulating valves commonly used are the pressure regulating valve, the temperature regulating valve, and the electric regulating valve.

Pressure Regulating Valve

The **pressure regulating valve**, shown in Figure 12–11A, is the most common type of regulating valve in use. It consists of a bellows assembly that is attached to the refrigeration system. Here it can best sense system discharge pressure, preferably at the cylinder head of the compressor or discharge service valve. The bellows assembly works against an adjustable spring pressure. As the discharge pressure increases and overcomes the spring pressure, the valve will open, allowing a greater amount of water flow through the condenser. As the discharge pressure decreases, the valve will close, reducing the amount of water flow through the condenser.

This type of valve will modulate the amount of water flow through the condenser in order to maintain a definite, predetermined discharge pressure. When the compressor cycles off, the valve will close, restricting any water flow to the condenser.

Temperature Regulating Valve

The **temperature regulating valve** works in essentially the same way as the pressure regulating valve except that the temperature regulating valve is operated by a thermostatic element charged with a volatile liquid and connected to the valve bellows assembly. The feeler bulb of the valve is mounted in the condenser water line. As the water temperature rises, the valve will open to allow a greater amount of water flow to the condenser. As the water temperature decreases, the valve will close, reducing the amount of water flow to the condenser. (see Figure 12–11B)

Electric Regulating Valve

The **electric regulating valve** is designed to open when the compressor starts and close when the compressor cycles off. In all applications of this type of valve, the water flow to the condenser is constant. The two types of electric regulating valves, shown in Figures 12–11C and 12–11D, are solenoid valves and motor-operated valves. The solenoid valve is usually used in lower-capacity systems while the motorized-type valve is commonly found in larger-capacity systems.

Valve Selection and Installation

Water regulating valves are selected for a specific application. Consideration must be given to the size of the water line for the application and to the pressure range for the refrigerant being used. Water-in and water-out connections are clearly identified, usually by directional flow arrows. Valve bodies are usually threaded for standard pipe connections. A water filter should be installed on the inlet side of the valve. A water shutoff valve should also be installed before the inlet side of the regulating valve to shut off the water supply to the system in the event that service or valve replacement is required.

Fig. 12–11A   Pressure-type water regulating valve. *Photo by Bill Johnson.*

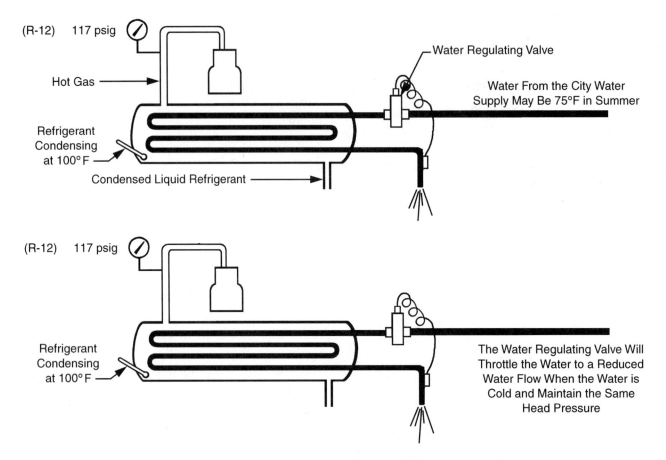

(R-12)   117 psig

Hot Gas

Refrigerant Condensing at 100°F

Condensed Liquid Refrigerant

Water Regulating Valve

Water From the City Water Supply May Be 75°F in Summer

(R-12)   117 psig

Refrigerant Condensing at 100°F

The Water Regulating Valve Will Throttle the Water to a Reduced Water Flow When the Water is Cold and Maintain the Same Head Pressure

Fig. 12–11B   Thermostatic water valve has feeler bulb as shown.

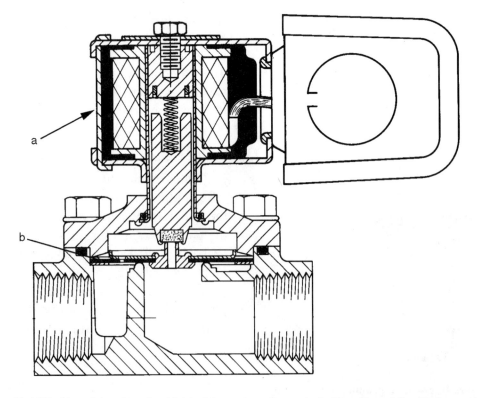

a

b

Fig. 12–11C   Energizing the solenoid (a) of the water valve controls the action of the diaphragm (b).

Fig. 12–11D   Motorized water valve. Motor raises or lowers valve stem. Note gear rack connected to stem. *Courtesy Johnson Controls, Inc.*

Valve Adjustment

There is a direct relationship between the temperature at which the refrigerant will condense and the temperature of the water circulating through the condenser. The normal condensing temperature of a water-cooled system is usually 10° higher than the temperature of the condenser outlet water. If the ideal condensing temperature is 105°F, the temperature of the condenser outlet water should be set for 115°F. In a system utilizing Refrigerant-12 and having a condensing temperature of 105°F, the discharge pressure would be approximately 126.6 PSIG.

Generally, the temperature difference between the water entering and the water leaving a waste water system is 15° to 20°. The water regulating valve can then be adjusted to these specifications.

EVAPORATIVE CONDENSERS

The **evaporative condenser** is a combination of the air- and water-cooled condensers. An evaporative condenser is illustrated in Figure 12–12. In this application, a bare or finned tube condenser coil is located in an enclosure. Water from spray nozzles or a distribution pan is sprayed or dripped over the condenser coils. When the water comes into contact with the hot condenser coils, some of the water will evaporate, cooling the condenser. A fan supplies fresh, cool air in a counterflow direction to the falling water, removing the warm, moist air from the enclosure. The cooled water that has not evaporated collects in the condenser sump where it will be recirculated. A ball float located in the sump maintains a constant water level, replacing water lost through evaporation and drift.

Baffles or eliminators are installed to reduce the amount of water lost through the air leaving the enclosure (drift). Because of the higher rate of water evaporation in the evaporative condenser as opposed to the water cooling tower, the bleed rate is increased to about four gallons per hour per ton of refrigeration effect.

There are several advantages of the evaporative condenser over the water cooling tower:

1. It's more efficient.

2. There is less water loss overall.

3. A smaller horsepower water pump is required due to shorter runs of water lines and less resistance.

4. In cool weather, it can function as an air-cooled condenser.

5. Installation applications are more flexible.

The evaporative condenser is used in numerous refrigeration and air-conditioning applications. The design of the installation and the control system vary with each application. In some applications, the condenser may be installed indoors using ductwork to deliver the exhaust to the circulating air. In cold weather, an indoor sump may be employed to prevent the water from freezing. Thermostatically controlled louver fans

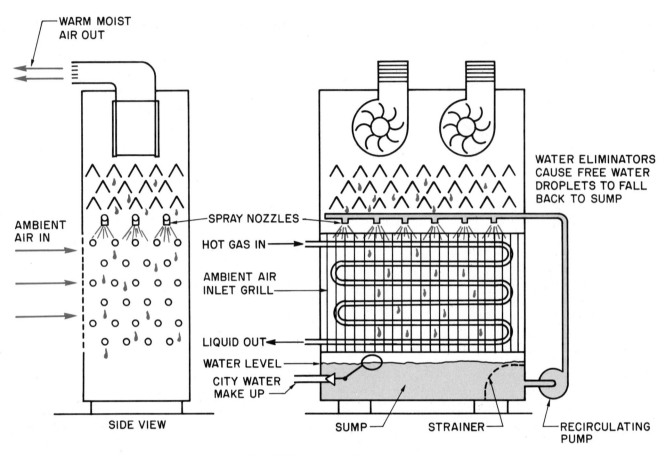

Fig. 12-12 Evaporative condenser.

may also be used to regulate airflow during cold weather operation in order to maintain minimum system discharge pressures. In other applications, electric heaters may be installed to maintain minimum water temperatures.

Unit Twelve Summary

- A fan is used to force air across the surface area of a fan cooled condenser while a forced draft condenser depends only on natural air circulation. Heat flows from a condenser through conduction and radiation and the ambient temperature affects the condenser's ability to dissipate heat.

- Water-cooled condensers are more efficient than air-cooled condensers. Recirculating water-cooled condensers use a water tower to cool the condenser water while wastewater systems circulate water in a heat exchange system to cool the refrigerant, then the water is pumped down a drainage system. A wastewater system circulates 1.5 gallons of water per minute for each ton of refrigeration capacity.

- Recirculating systems use three to five gallons of water per minute for each ton of refrigeration capacity and a system of this type uses both forced air and natural draft water tower systems. Makeup water replaces water lost in a recirculating system due to drift, evaporation, and bleed.

- Drift is the water lost through wind in a water tower and tower water bleed reduces the amount of solidified buildup in a tower recirculating system.

- Water regulating valves control water flow to the condenser in order to maintain the desired discharge temperature in a refrigeration system.

- Evaporative condensers are considered to be a combination of a water-cooled and an air-cooled condenser because a system of evaporative design sprays water directly on the condenser coils to discharge condenser heat.

Unit Twelve Key Terms

Air-Cooled Condenser

Water-Cooled Condenser

Double Tube Condenser

Shell and Coil Condenser

Shell and Tube Condenser

Water Cooling Towers

Natural Draft Cooling Tower

Forced Draft Cooling Tower

Water Circulating Pump

Makeup Water

Bleed Line

Drift

Tower Sump

Tower Fill

Spray Headers

Distribution Pan

Tower Drain

Eliminator

Water Regulating Valve

Pressure Regulating Valve

Temperature Regulating Valve

Electric Regulating Valve

Evaporative Condenser

UNIT TWELVE REVIEW

1. What is the purpose of a condenser?

2. Join the three general classifications of condensers with the letters for their advantages and disadvantages.

   TYPE OF CONDENSER   ADVANTAGE   DISADVANTAGE

   Air-Cooled Condenser

   Water-Cooled Condenser

   Evaporative Condenser

   a. Lower condensing pressure than air-cooled condensers.

   b. Unacceptable noise level from condenser fans.

   c. Self-contained.

   d. Less water loss.

   e. Lower-horsepower compressor than air-cooled condensers.

   f. In cool weather, can function as an air-cooled condenser.

   g. Water disposal may be a problem.

   h. Easy and economical to install.

   i. Coil tubing and fins must be kept clean.

   j. Cost of water may be excessive or restrictive.

3. What are the two types of airflow designs by which condensers can be classified?

4. Several factors affect the ability of a condenser to transfer. Fill in the blanks.

   a. The _____ the surface area, the _____ heat will be dissipated from the condenser.

   b. The _____ the temperature difference between the refrigerant and the condenser cooling medium, the _____ the rate of heat transfer.

   c. Efficient conductors of heat, such as aluminum, steel, and copper will _____ the capacity of heat transfer.

   d. _____ of the condenser coil and all heat transfer surfaces is important. _____ and _____ can affect heat transfer.

   e. As the velocity of the cooling medium over or through the condenser _____ , the amount of heat transferred _____.

5. Name three advantages of an air-cooled condenser.

6. Name three disadvantages of an air-cooled condenser.

7. In the application of air-cooled condensers, how can discharge pressure be controlled in low-ambient conditions?

8. In the application of air-cooled condensers, how is discharge pressure determined?

9. What is the airflow requirement for air-cooled condensers?

10. Describe the two ways in which water may be used as a condensing medium.

11. Describe the counterflow principle in the application of water-cooled condensers.

12. What are the advantages of water-cooled condensers?

13. What are the disadvantages of water-cooled condensers?

14. Match the following types of water-cooled condensers with their description: double tube condenser, shell and coil condenser, shell and tube condenser.

    a. The refrigerant discharge vapor enters the condenser at the top, passes around the water tubes and condenses and exits the condenser at the bottom.

    b. The refrigerant discharge vapor enters the outside tube of the condenser at the top, and flows around the inside water tube.

    c. The refrigerant discharge vapor enters the outside tube of the condenser, passes around the cold water coils within the shell and condenses.

15. What are the advantages of the double tube condenser?

16. What are the advantages of the shell and coil condenser?

17. Several variables affect the rate of heat transfer between the refrigerant and water in a water-cooled condenser. Add the terms *greater* or *smaller* as required.

    a. The _____ the temperature difference, the _____ the rate of heat transfer.

    b. The _____ the condenser surface, the _____ the rate of heat transfer.

    c. The _____ the velocity of water flow, the _____ the rate of heat transfer.

18. What is the average water flow requirement for a wastewater system?

19. What is the average temperature difference of the water through the condenser in a wastewater application?

20. What is the average water flow requirement for a recirculating water tower system?

21. What is the average temperature difference of the water through the condenser in a recirculating water tower system?

22. How is a water-cooled condenser leak-tested?

23. Where are most of the leaks in water-cooled condensers found?

24. Why do leaks occur in the water circuit?

25. Describe the operation of a water tower in the application of water-cooled condensers.

26. How is the natural draft cooling tower different from a forced air cooling tower?

27. Pair the following water cooling tower terms with their descriptions: water circulating pump, makeup water, tower bleed, tower sump, spray headers/distribution pan system.

    a. Maintains a constant level and supply of water.

    b. Used to reduce or control particle buildup.

    c. Delivers the proper gallonage of water per minute for the application.

    d. Allows the circulating water to enter the tower and evenly distributes the water over the tower fill.

    e. Used to replace water lost through evaporation, drift, or bleed.

    f. Installed to reduce the amount of water loss as a result of the forced convection air movement.

28. What is the prescribed bleed rate for a tower system?

29. Describe drift as it applies to water cooling towers.

30. What purpose do eliminators serve in a water cooling tower?

31. What is the purpose of a water regulating valve?

32. What are the three types of water regulating valves in common use today?

33. Pressure regulating valve, temperature regulating valve, and electric regulating valve are three types of valves used in conjunction with water-cooled condensers. What purposes do these valves have?

34. What are the advantages of an evaporative condenser?

35. Write the procedure to be followed when testing for leaks in a water-cooled condenser when no isolation valves are available.

36. Write a paragraph to explain how an evaporative condenser works.

UNIT THIRTEEN

Evaporators

OBJECTIVES

After completing this unit, the student will be able to:

1. Describe the function of an evaporator in a refrigeration system.

2. Differentiate between the various types of evaporators used in refrigeration and air-conditioning systems.

3. Describe proper design, installation, and maintenance procedures for evaporators.

In the evaporator the refrigerant absorbs heat from the conditioned space. As with a condenser, some sealed system problems are misdiagnosed by inexperienced technicians because they lack a firm understanding of the function of an evaporator and how the different designs are intended to perform in a given system.

In this unit, we'll discuss the different types of evaporators found in domestic, commercial, and industrial refrigeration systems and how they're designed to function. When troubleshooting a malfunctioning refrigeration system, always be sure to consider the installation and condition of an evaporator.

EVAPORATOR SYSTEMS

An **evaporator** is a vessel used to vaporize a refrigerant for the purpose of removing heat. The two general classifications of evaporators used in refrigeration and air-conditioning applications are dry expansion evaporators and flooded evaporators, referred to in Figures 13–1A and B.

Dry Expansion Evaporators

In a **dry expansion evaporator**, the refrigerant is in a constant state of vaporization as it passes through the evaporator coils. On some systems, the amount of refrigerant fed into the evaporator is controlled by the load demand on the coil. When the evaporator load is low, less refrigerant is fed into the coil. When the evaporator load is high, more refrigerant is fed into the coil.

In dry expansion evaporator applications, the metering device allows only enough refrigerant to enter the evaporator in order to satisfy the load condition without starving or flooding the coil. Liquid refrigerant enters the metering device as a high-pressure, high-temperature liquid. As it passes through the metering device and is introduced into the evaporator, a certain percentage of the liquid will vaporize. This is referred to as **flash gas** at the metering device. This percentage of vaporization, usually around 15 percent, will reduce the remaining 85 percent of the liquid refrigerant to the existing evaporator temperature and pressure. As the refrigerant flows through the

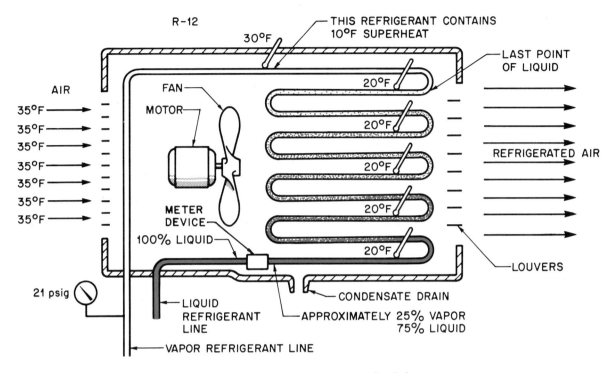

Fig. 13-1A Dry expansion evaporator and flooded evaporator.

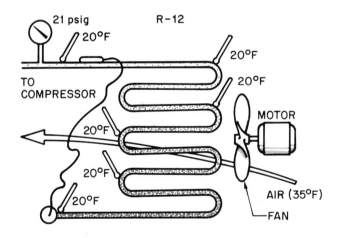

Fig. 13-1B Dry expansion evaporator and flooded evaporator.

evaporator it will continue to absorb heat, vaporize, and leave the evaporator as a complete superheated vapor.

Flooded Evaporators

In a **flooded evaporator**, a float type of arrangement maintains a constant level of liquid

refrigerant. When under a flooded condition, the evaporator contains a mass of boiling liquid refrigerant. Vapor is drawn off the top of the evaporator from the suction created by the compressor. As the level of the liquid refrigerant decreases due to its vaporization, the float arrangement will open to allow more refrigerant to enter the evaporator.

Flooded evaporators have the advantage of providing excellent heat transfer efficiency, allowing more heat transfer per square foot than dry expansion evaporators. As a result, smaller flooded coil can be used for equal capacities of a dry expansion coil. Flooded evaporators are also more flexible. A number of parallel coils can be connected to a common header. One single float-type of metering device can be used to control several evaporators.

The disadvantages of flooded evaporators are that they are bulky in size and require more refrigerant charge. Oil circulation may also become a problem because of low refrigerant velocities in the coil.

EVAPORATOR HEAT TRANSFER CAPACITY

Several factors affect and control the rate and quantity of **evaporator heat transfer**. We'll discuss these factors specifically in the following sections.

Evaporator Surface Area and Condition

If the exposed area of the evaporator is small, the capacity of the evaporator is reduced. For this reason, fins are added to increase the surface area and heat transfer efficiency. Accumulations of ice, frost, or dirt on the evaporator fins or tubes will reduce airflow across the coil and greatly reduce the rate of heat transfer.

Evaporator Construction

Because it is desirable to have a rapid heat transfer, a good conductor of heat is required in evaporator construction. Materials that are not affected by the refrigerants used should be considered and the materials should be strong enough so as not to collapse under a vacuum or rupture when exposed to pressure.

Dew Point of Entering Air

If the evaporator temperature is to be maintained below the dew point of the entering air, the removal of latent and sensible heat must be considered.

Velocity and Rate of Airflow

The velocity and rate of airflow over the evaporator surface has a considerable influence on evaporator capacity. Inadequate air circulation will prevent heat from being carried to the evaporator at a sufficient rate to maintain peak efficiency. When the velocity of the air passing across the coil is low, the air stays in contact with the coil longer and is cooled through a greater range. As the velocity of the air passing across the coil increases, more air is brought into contact with the coil at a given time and the rate of heat transfer increases. In addition, the higher velocity

of air moving across the coil will help break up surface films of static air that build up around the evaporator surfaces and act as a thermal barrier. The key is balance—the correct amount of airflow according to the design of the system.

Velocity of the Refrigerant

If the velocity of the refrigerant through the evaporator becomes too low, oil will accumulate in the tubing and will not be returned to the compressor. Consequently, the heat transfer capacity of the tubing will be greatly reduced. Proper oil circulation can be ensured by maintaining high enough refrigerant velocities to scrub the tubing walls and carry the oil back to the compressor.

Medium Being Cooled

Heat transfer is about five times more effective from a liquid to an evaporator than it is from air to an evaporator. This principle applies to external evaporator surfaces when the outside of the coil is in contact with a liquid or solid medium and flooded coils, which are always filled with liquid refrigerant.

The temperature difference between the evaporating refrigerant and the medium being cooled has a direct effect on evaporator capacity. The colder the refrigerant, with respect to the temperature of the medium being cooled, the greater the capacity of the evaporator.

Evaporator Temperature Difference

The **temperature difference (TD)** is the difference between the temperature of the air entering the evaporator and the saturation temperature of the refrigerant, which corresponds to the pressure at the outlet of the evaporator. The relationship between evaporator capacity and evaporator temperature difference is proportional. The capacity of an evaporator in Btu's/hr varies directly with the evaporator TD. An evaporator with a temperature difference of 1° at a certain capacity will have ten times that capacity if the TD is increased to 10°F, providing that other variables affecting capacity remain the same. Based on this relationship, a relatively small evaporator coil with a high-temperature difference can have the same capacity as a larger coil with a low-temperature difference. What must be considered, however, is the effect the temperature difference has on the stored products and operating efficiency of the system.

Humidity Control

Air contains varying amounts of moisture. The ratio of the actual amount of moisture in a given amount of air at a certain temperature to the amount it can actually hold at that temperature is defined as **relative humidity**.

Warm air can hold more moisture than cold air, and when a quantity of warm air is reduced in temperature, its capacity to hold moisture decreases and its relative liquidity increases. When the temperature of the air is reduced to the point where the relative humidity is 100%, the air is said to be saturated and cannot hold any more moisture. This temperature is referred to as the **dew point**.

When the temperature of the air is reduced below the dew point, the moisture it cannot hold will be released and deposited on a colder surface, such as an evaporator coil in refrigeration applications. If the evaporator temperature is above 32°F but below the dew point, moisture will collect on the coil as water. If the evaporator temperature is below 32°F, moisture will collect and freeze on the evaporator coil.

The rate of moisture removal from the air surrounding an evaporator largely depends upon the temperature difference between the air and the evaporator surface. An evaporator operating with a high temperature difference will tend to produce a condition of low humidity in the refrigerated space. If the product contained in the conditioned space is affected by space humidity, an evaporator TD that will provide optimum humidity conditions for that product should be considered.

For instance, in the storage of perishable items, such as leafy vegetables, meats, and fruits,

conditions of low humidity will result in excessive dehydration. For perishable commodities where a high relative humidity (90%) is required, a TD of 8°F to 12°F is recommended. For slightly lower relative humidities (80%), a TD of 12°F to 16°F is usually adequate. Temperature differences of 5° to 20° are commonly used. The temperature difference should be kept as low as possible to ensure more efficient compressor operation at higher suction pressures.

In applications where humidity is of no concern, the factors governing evaporator selection should focus on cost, system efficiency, and available space for coil installation.

Evaporator Pressure Drop

An excessive pressure drop in the evaporator will result in a reduction of system capacity due to the lower pressure at the evaporator outlet. When there is a reduction in suction pressure, the specific volume of the refrigerant vapor returning to the compressor will increase. The weight of the vapor pumped by the compressor will decrease.

To avoid losses in compressor capacity and efficiency, the evaporator design should allow for only a minimum drop in refrigerant pressure. The length and size of evaporator tubing should also be kept to a minimum. Although a certain amount of pressure drop is required to allow the refrigerant to flow through the evaporator, a high enough velocity must still be maintained to sweep the tubes of oil and vapor bubbles. In addition, the evaporator circuitry must be designed with a minimum pressure drop through the evaporator so that the necessary refrigerant velocities are maintained to provide for proper oil return and a high rate of heat transfer.

The final evaporator design is a compromise between maintaining a minimum pressure drop and maximum refrigerant velocity. Pressure drops of 1 to 2 PSIG are generally acceptable in medium- and high-temperature applications, and pressure drops of ½ to 1 PSIG are common in low-temperature applications.

DESIGNS AND TYPES OF EVAPORATORS

Evaporators are designed in many styles and shapes to fit almost any refrigeration and air-conditioning application. Evaporator designs range from the small, compact units found in domestic refrigerators to the units with mazes of piping found in commercial and industrial equipment. There are three basic requirements for selecting an evaporator for a specific need or requirement:

1. The evaporator must have a sufficient internal volume to accommodate the liquid refrigerant required to meet the cooling demands placed on the system.

2. The evaporator must allow the refrigerant to flow with a minimum pressure drop.

3. The evaporator must be designed for maximum heat transfer.

Because evaporators are manufactured in a variety of styles and shapes, they are classified in a number of different ways:

1. *Type of construction*—bare tube, plate, or finned.

2. *Operating condition*—frosting, defrosting, or nonfrosting.

3. *Method of circulation*—gravity or forced.

4. *Type of refrigerant control*—expansion valve, capillary tube, low-side float, or high-side float.

Bare Tube Evaporators

Bare tube evaporators are used in a wide variety of refrigeration applications. They are made in square, round, oval, or rectangular shapes. Many designs are custom-made for a particular cabinet application and often serve a dual role as evaporator and shelving, such as the type commonly found in freezer cabinets and hardening rooms. While most bare tube evaporator coils are designed as one continuous coil from the expansion valve to the compressor intake valve, some may be designed to form a series of coils fed from a common header. A bare tube evaporator is illustrated in Figure 13–2.

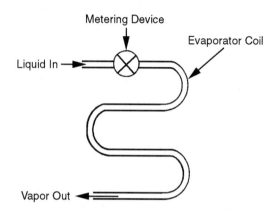

Fig. 13–2   Bare tube evaporator.

The type of material used to construct bare tube evaporators depends upon the type of refrigerant being used and the service application for which it is intended. Steel pipe is used on ammonia refrigerant systems and for large coil applications. Copper is used in manufacturing smaller coils intended for use with refrigerants other than ammonia. Carbon, steel, and stainless steel coils may also be found in some applications.

Plate Evaporators

Plate evaporators are manufactured in various designs. One of the primary advantages of a plate evaporator is that it may be used where space is an important factor. It can form walls, ceilings, or shelving and is particularly adaptable to applications such as small coolers and domestic refrigerators and freezers.

One of the most widely used applications of a plate evaporator is in household refrigerator and freezer applications, as shown in Figure 13–3. The evaporator is constructed of two flat sheets of metal, usually aluminum, embossed and welded together in such a way that a path is provided for the refrigerant to flow between the sheets. This design has the advantages of being easy to clean, cheap to manufacture, and easily formed to fit any shape required.

Another extremely useful application of the plate evaporator is in liquid cooling applications where high load conditions are experienced intermittently. During periods of normal evaporator

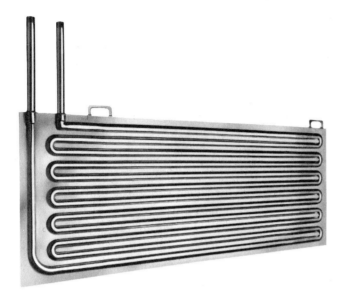

Fig. 13–3   **Plate-type evaporator.**   *Courtesy Tranter Inc., Platecoil-Division.*

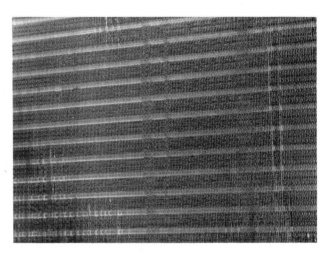

Fig. 13–4   **Finned evaporator.**   *Courtesy Frigidaire Company of Dublin, OH.*

load, an ice bank builds up on the plate surface. During periods of heavy evaporator load, the ice bank serves in a holdover capacity, helping the system carry the additional load. The advantage of this type of operating design is that the condensing unit need be sized only for normal evaporator loads, thus reducing installation and operating costs.

Plate evaporators are usually of the dry expansion design. They may be used as single coils or in banks. They may be used in a series of coils fed by a single metering device or connected in parallel with a header to provide refrigerant distribution.

Finned Evaporators

Finned evaporators are bare tube evaporators upon which fins have been installed to transfer heat from the air to the vaporizing refrigerant contained within. The fins act as a secondary heat-absorbing surface, improving evaporator efficiency by increasing overall evaporator surface area while reducing coil size, weight, and cost. A finned evaporator is illustrated in Figure 13–4.

Finned evaporators are used quite extensively in residential and commercial refrigeration and air-conditioning applications. In order for them to be effective, there must be good thermal contact between the evaporator fins and the tubing surface. This can be accomplished in several ways. One method is to solder the fin directly to the tubing. Another method is to slip the fin over the tubing and expand the tubing so that the fin locks onto the tubing surface. Or, the fin hole can be flared slightly to allow the fin to slip over the tubing. Once in place, the flare can be straightened out, securing the fin to the tubing.

Maximum heat transfer efficiency is related to the spacing of the fins as well as to the height of the fins in proportion to the size of the tubing being used. Small evaporator tubing requires small fins. As the tubing size increases, the size of the fins must also be increased. Fin spacing depends primarily upon the operating temperatures of the evaporator. On coils operating at low temperatures, the accumulation of frost or ice will restrict the circulation of air through the coil. In these applications the spacing of the fins must be wide, usually two to four fins per inch. On air-conditioning applications or applications where the temperature of the coil is high enough that no frost or ice accumulates, there can be as many as fourteen fins per inch.

Frosting Evaporators

Frosting evaporators operate at temperatures below 32°F. Because of this low operating temperature, a continuous frosting of the coil surface is experienced. This accumulation of ice and frost greatly reduces heat transfer efficiency and therefore requires the periodic defrosting of the coil.

The disadvantage of frosting evaporators is that the frost that forms on the coil surfaces and fins comes from the moisture in the air, which leaves the conditioned space dry and results in dehydration of the stored product. For this reason, frosting evaporators are used in low-temperature applications, such as for storing frozen foods.

Defrosting Evaporators

Defrosting evaporators operate at a temperature of 20°F to 22°F during the normal running cycle of the compressor. When the compressor cycles off, the temperature of the evaporator will rise to slightly above 32°F, allowing any frost that may have accumulated to melt before the unit cycles back on.

Defrosting evaporators provide excellent heat transfer efficiency and maintain high relative humidities of usually 90% to 95% in the conditioned space. The disadvantage of defrosting evaporators is that a larger evaporator surface is required, thus sacrificing space for temperature and humidity control. When natural convection is the only means of air circulation across the evaporator surface, a greater amount of fin and tubing area is required. Fin spacing is usually wider to allow for less resistance to airflow across the coil.

When finned evaporators are used in applications where frost or ice may accumulate on the coil surface, some means of manual or automatic defrosting must be employed periodically to defrost the coil.

Nonfrosting Evaporators

Nonfrosting evaporators operate at a temperature above 32°F and therefore do not experience a buildup of frost or ice on the evaporator surface. In some applications a light accumulation of frost may build up on the evaporator surface during the running cycle, but will quickly melt during the off cycle period. The nonfrosting coils generally operate at a temperature of 33°F to 34°F, and the temperature of the refrigerant within the coil is 20°F to 22°F.

The nonfrosting coils draw very little moisture from the conditioned space and maintain a relative humidity of 75% to 80%, which helps keep produce fresh and prevent weight shrinkage. Nonfrosting evaporators have relatively large coil surfaces and are bulkier than the frosting evaporators of equal capacity.

Unit Thirteen Summary

- The purpose of an evaporator is to vaporize refrigerant and remove heat from the conditioned space. Refrigerant changes in state from a liquid to a vapor as it passes through a dry expansion evaporator, while a flooded evaporator maintains a constant liquid level.

- Evaporators must be manufactured of a material that is a good conductor of heat. This factor, along with the amount of airflow across the surface of an evaporator and the degree of pressure drop within the coil, determines the efficiency of an evaporator.

- Frosting evaporators operate below 32°F while nonfrosting evaporators operate above 32°F. Defrosting evaporators operate below 32°F, but will warm up above 32°F and defrost during the off cycle of the refrigeration system.

Unit Thirteen Key Terms

Evaporator

Dry Expansion Evaporator

Flash Gas

Flooded Evaporator

Evaporator Heat Transfer

Temperature Difference

Relative Humidity

Dew Point

Bare Tube Evaporator

Plate Evaporator

Finned Evaporator

Frosting Evaporator

Defrosting Evaporator

Nonfrosting Evaporator

UNIT THIRTEEN REVIEW

1. What is the purpose of an evaporator?

2. The two classifications of evaporators are the dry expansion evaporators and the flooded evaporators. Match the evaporator with the following phrases:

   a. A float type of arrangement maintains a constant level of liquid refrigerant.

   b. Bulky in size and requires more refrigerant charge.

   c. The refrigerant is in a constant state of vaporization as it passes through the evaporator coils.

   d. A certain percentage of the liquid will vaporize as it is introduced into the evaporator.

3. What purpose do fins have in an evaporator?

4. Why is the selection of materials used in evaporator construction important?

5. The velocity of the airflow over an evaporator surface affects the capacity of the evaporator. Which of the following statements are true?

   a. If the velocity of the air is low, the air stays in contact with the coil longer.

   b. If the velocity of the air passing over the coil increases, more air is brought into contact with the coil.

   c. Higher velocity air moving across the coil will help break up surface films of static air.

6. Define flash gas.

7. What two things will happen if the velocity of the refrigerant in the tubing becomes too low?

8. The _____ the refrigerant with respect to the temperature of the medium being cooled, the _____ the capacity of the evaporator.

9. What is the dew point?

10. What are the advantages of a flooded evaporator?

11. What are the disadvantages of a flooded evaporator?

12. Describe three factors that influence evaporator heat transfer capacity.

13. How does temperature difference affect evaporator heat transfer capacity?

14. An evaporator operating with a high temperature difference will tend to produce low humidity. What will be the result for foods like leafy vegetables, meats, and fruits?

15. How does the length and size of the evaporator tubing affect refrigerant pressure?

16. Where the accumulation of frost or ice will restrict the circulation of air through the evaporator coil, how are fins affected?

17. Connect the type of evaporator with its description: frosting evaporators, defrosting evaporators, nonfrosting evaporators.

    a. Operates at temperatures below 32°F.

    b. When the compressor cycles off, the temperature of the evaporator will rise slightly above 32°F.

    c. Operates at temperatures above 32°F.

18. How does an evaporator control humidity in an occupied space?

19. What factors govern the selection of an evaporator?

The temperature difference (TD) is the difference between the temperature of the air entering the evaporator and the saturation temperature of the refrigerant at the outlet of the evaporator. The greater the temperature difference the greater the capacity of the evaporator.

For fun and practice, a ratio can be set up between the temperature difference (TD) and the size of the tubing.

20.

	TD	Tubing Size	Ratio
Evaporator A	2° to	1"	2:1
Evaporator B	1° to	½"	?:?

Find the ratio. (Hint: multiply both sides by 2)

$1 \times 2 = 2$. $1/2 \times 2 = 1$. Ratio 2:1

But which evaporator has the greater capacity (based only on temperature difference and tubing size) and why?

21.

	TD	Tubing Size	Ratio
Evaporator C	1°	⅞"	?:?
Evaporator D	2°	⁷⁄₁₆"	?:?

Find the ratios.

Which evaporator has the greater capacity?

22. Write a paragraph using the following terms to describe an evaporator: minimum pressure, tubing, heat transfer, velocity, compressor, and refrigerant.

UNIT FOURTEEN

Metering Devices

OBJECTIVES

After completing this unit, the student will be able to:

1. Describe the function of a metering device in a refrigeration system.

2. Explain the method of operation of a capillary tube, automatic expansion valve, and thermostatic expansion valve.

3. Describe the proper method for determining superheat and explain how it's used in troubleshooting and evaluating a sealed system.

The metering device is the component in a refrigeration system that controls the flow of refrigerant into the evaporator. Some metering devices are constant feed, such as a capillary tube, regardless of the cooling load. Others, such as a thermostatic expansion valve, adjust according to the load on the cooling system at that moment and allow more or less refrigerant into the evaporator. This variable-feed-type of system results in the refrigeration unit being allowed to do more work when necessary and to operate at less than full capacity when possible.

One simple approach to a metering device is to consider it as a component that allows for a controlled restriction in a refrigeration system. This controlled restriction not only meters the correct amount of refrigerant into the evaporator, it also creates the pressure differential that must exist in order for a refrigeration system to operate.

In this unit, we'll discuss the fundamental principles behind the operation of a metering device, the different types of devices used in various types of refrigeration systems, and the ser-vice procedures that, as a service technician, you have to understand.

The two primary purposes of a metering device, or refrigerant flow control device are:

1. To meter the proper amount of liquid refrigerant into the evaporator at such a rate that all load conditions are satisfied.

2. To maintain a pressure differential between the high and low sides of the system. This allows the liquid refrigerant to vaporize at the desired low temperature in the evaporator and also assures that the high-pressure vapor condenses in the condenser.

The amount of refrigerant flow into the evaporator must be controlled carefully. If too much liquid refrigerant is fed into the evaporator, it will not all vaporize. Some refrigerant may enter the suction line and slug back to the compressor, which will cause damage. If too little refrigerant is fed into the evaporator, the evaporator will starve, which will result in a loss of system efficiency.

Refrigerant flow control devices can be cate-gorized as:

1. Control operation, based on temperature changes.
2. Control operation, based on pressure changes.
3. Control operation, based on volume changes.
4. Control operation, based on a combination of the aforementioned changes.

HAND EXPANSION VALVES

One of the earliest types of refrigerant flow control devices was the **hand expansion valve**, which is illustrated in Figure 14–1. This metering device was most commonly used on systems monitored by technicians in applications where the heat load was almost constant and the compressor operated continuously. Today, hand expansion valves are seldom used except for emergency use as bypass valves around automatic valves that need repair or replacement and as throttling valves to prevent the wire drawing of automatic valves.

A disadvantage of the hand expansion valve is that as the evaporator load changes, the valve must be readjusted manually to prevent the evaporator coil from starving under high load conditions or flooding under low load conditions. In addition, when the compressor cycles off, the valve must be closed manually then opened again when the compressor cycles back on.

CAPILLARY TUBES

The **capillary tube** is the simplest type of metering device used in refrigeration and air-conditioning applications. It consists of a small-diameter tubing installed between the outlet of the condenser and the inlet of the evaporator. A capillary tube is shown in Figure 14–2.

The operating principle of the capillary tube is based upon the gradual pressure reduction of the liquid refrigerant flowing through it. The liquid refrigerant will pass through three phases as it flows from the condenser through the capillary tube to the evaporator. The phases are liquid length, bubble point length, and two-phase length.

Refrigerant leaving the condenser will enter the capillary tube in a liquid state. This is referred to as the **liquid length**. As the refrigerant flows through the capillary tube, a pressure drop will occur due to the friction developed by the small-diameter tubing. When the pressure drops below the point where the refrigerant is saturated, some flash gas will form. This is referred to as the **bubble point**. As the refrigerant continues to flow

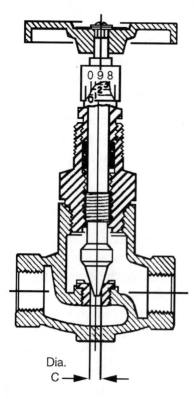

Fig. 14–1   Hand-operated expansion valve. *Courtesy Henry Valve Co.*

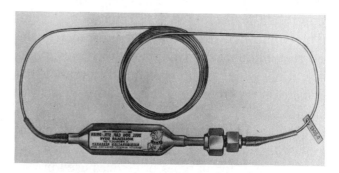

Fig. 14–2   Capillary tube with filter-drier. *Courtesy Refrigeration Research Inc.*

through the capillary tube, the amount of flash gas will increase, creating a greater restriction to fluid flow and further reducing the pressure and the temperature of the liquid refrigerant. The last phase of the capillary tube is referred to as the **two-phase length**. At this point, the liquid refrigerant and vapor are delivered to the evaporator at, or slightly above, evaporator pressure and temperature.

Four basic factors govern the capacity of a capillary tube:

1. Inside diameter of the tubing
2. Length of the tubing
3. Temperature of the tubing
4. Amount and tightness of the tubing coils

At a given inside diameter and length, the resistance of the capillary tube is fixed and the flow of liquid refrigerant is proportionate to the pressure difference between the inlet and the outlet of the tubing. The smaller the inside diameter, or the longer the capillary tube, the greater will be the frictional resistance of the tubing. This will require a greater pressure difference for a given fluid flow rate.

The most efficient operation of the capillary tube will be one combination of suction and discharge pressures. At design conditions, the flow rate of liquid refrigerant through the capillary tube will be exactly equal to the capacity of the compressor when it's operating under the same conditions. At this point, system efficiency will be at design maximum.

If the resistance of the capillary tube is such that the flow rate is greater than or less than the capacity of the compressor at design conditions, a new balance will be established between the capillary tube and the compressor. This new balance will be at a condition other than the original design, however, which will result in a loss of system efficiency. For example, if the resistance of the capillary tube is too great, less liquid refrigerant will flow into the evaporator, resulting in a starved coil condition and a decrease in suction pressure. Liquid refrigerant will start to back up

into the condenser, and the condensing pressure will start to increase. Under these conditions, the flow rate of liquid refrigerant through the capillary tube will increase and compressor capacity will decrease. Eventually a new balance between the flow rate and compressor pumping capacity will be found. It will, however, be at a higher discharge pressure and lower suction pressure than originally designed. As a result, reduced system efficiency will be experienced.

If the capillary tube does not have enough resistance, the flow of liquid refrigerant to the evaporator will be excessive, resulting in the possible floodback of refrigerant to the compressor. The liquid refrigerant will be drained from the condenser, and vapor, along with liquid refrigerant, will enter the evaporator, reducing evaporator efficiency and preventing the compressor from effectively reducing evaporator pressure. This will result in a further loss of system efficiency.

The capillary tube is best suited to applications where the evaporator heat load is relatively constant. It does, however, have the ability to self-compensate over a limited range of pressure differences with a minimal loss of system efficiency. If, for example, the discharge pressure decreases or the suction pressure increases, there will be less of a pressure difference and the flow of liquid refrigerant through the capillary tube will be reduced. Liquid refrigerant will start to back up into the condenser, causing a reduction of effective condenser volume, an increased liquid subcooling, and an increase of discharge pressure. The additional liquid subcooling will cause the bubble point to move toward the evaporator, increasing the liquid length and decreasing the two-phase length. This will reduce the resistance to fluid flow and, combined with a higher discharge pressure, will increase the flow of liquid refrigerant through the capillary tube.

If the discharge pressure increases or the suction pressure decreases, the flow of liquid refrigerant through the capillary tube will increase. This will result in little or no subcooling of the liquid refrigerant entering the capillary tube. The bubble point will now move close to the inlet of the capillary tube, decreasing the liquid length and increasing the two-phase length. Overall restriction to the flow of liquid refrigerant in the capillary tube will increase. Consequently, the flow of liquid refrigerant will decrease.

The advantages of the capillary tube make it one of the more popular metering devices, especially in domestic and fractional horsepower commercial applications. These advantages include:

1. No moving parts
2. No receiver needed
3. Small refrigerant charge
4. Off-cycle equalization of pressures, allowing for the use of low starting torque motors

The largest disadvantage of the capillary tube is that it is subject to plugging. For this reason a strainer should always be installed at the inlet. The system should be clean and completely dehydrated to eliminate any dirt or moisture that may cause capillary tube restrictions.

Because of the off-cycle equalization of pressure, a pressure type of motor control cannot be used with the capillary tube. A thermostatic type of motor control is most commonly used.

The refrigerant charge of a system equipped with a capillary tube is considered to be critical. An overcharge or undercharge by just a few ounces of refrigerant could drastically affect system performance.

Sizing Capillary Tubes

Sizing of the capillary tube for a given application is difficult to calculate accurately. Although it theoretically can be done, most preliminary selections are made from manufacturer's reference charts. Actual testing is then performed on the system to find a tubing size and length that will perform best at expected operating conditions or over the widest range of desired conditions. Once the proper size of tubing has been selected, the same tubing could then be applied to identical equipment, making it a well-adapted metering device to use in production line equipment.

AUTOMATIC EXPANSION VALVES

The **automatic expansion valve (AEV)**, illustrated in Figures 14–3A and B, is a pressure-regulated valve designed to operate in response to changes in evaporator pressure. The valve functions to maintain constant evaporator pressure, regardless of the load placed on the evaporator coil. The basic parts of the valve consist of a nee-

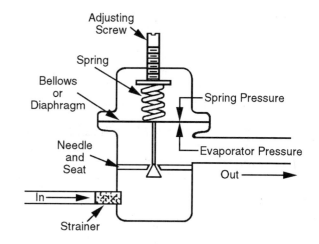

Fig. 14–3A   Automatic expansion valve.

Fig. 14–3B   Diaphragm-type automatic expansion valve. Note refrigerant flow direction.   *Courtesy Parker Hannifin Corporation.*

dle and seat assembly, a bellows or diaphragm, and an adjustable spring. An inlet screen or strainer, usually made of brass or stainless steel, is installed in the liquid inlet of the valve to protect against solid particle contamination in the valve body.

The operation of the automatic expansion valve is based on evaporator pressure, which is applied to the underside of the valve diaphragm, opposed by a spring and atmospheric pressure, which is applied to the opposite side of the diaphragm. The evaporator pressure will act to move the valve in a closing direction, while the spring and atmospheric pressure will act to move the valve in an opening direction. During normal compressor operation, the valve will function to maintain these opposing pressures in a state of equilibrium or balance.

If the evaporator pressure falls below the valve setting, the spring pressure will become greater and force the valve to open, allowing more liquid refrigerant to enter the evaporator. As more liquid refrigerant enters the evaporator, the rate of vaporization will increase, raising the evaporator pressure until a balance with the spring pressure is again reached.

If the evaporator pressure rises above the valve setting, it will override the spring pressure and force the valve to close, restricting the flow of liquid refrigerant into the evaporator. The rate of vaporization will decrease, lowering the evaporator pressure until a balance with the spring pressure is again reached.

When the compressor cycles off, the evaporator pressure will increase, overcome the spring pressure, and keep the valve tightly closed during the off cycle. When the compressor cycles on, the valve will remain in the closed position until the evaporator pressure is reduced below the spring pressure. The valve will then start to open to allow refrigerant to enter the evaporator until an established operating balance between the evaporator and spring pressures is reached.

The automatic expansion valve has limited use in today's refrigeration and air-conditioning applications. Its primary disadvantage is that in order to maintain a constant pressure in the evap-

orator, the rate of liquid vaporization must be kept constant. To do this, liquid refrigerant must be restricted from entering the evaporator when the heat load on the coil is high and the heat transfer capacity of the coil is at a maximum. This will severely limit the capacity and efficiency of a system at a time when they are needed the most.

When the heat load on the evaporator decreases, the heat transfer capacity also decreases and more of the evaporator must be flooded in order to maintain a constant rate of vaporization. This may result in the automatic expansion valve overfeeding the evaporator coil, causing liquid refrigerant to enter the suction line and flood back to the compressor. In properly designed systems, this condition should not be a problem because the thermostatic motor control will cycle the compressor off before this condition can occur.

Automatic expansion valves are used only on dry expansion evaporators. They do not have the ability to properly balance the amount of liquid refrigerant used in flooded evaporators. Because of their constant pressure feature, automatic expansion valves also cannot be used in parallel evaporator applications. For instance, if one valve were set at a higher pressure than another valve, the vapor from the evaporator with the higher pressure would back up into the evaporator with the lower pressure, raising its pressure and closing the automatic expansion valve feeding that coil. This would result in starving the coil with the lower pressure of refrigerant. Moreover, if two valves were adjusted to maintain the same pressure, the pressure would rise and both coils would be starved of refrigerant when the load on either coil was increased.

Another disadvantage of the automatic expansion valve, which is also due to its constant pressure characteristic, is that it cannot be used with a low-pressure control as a temperature control device. The operation of a low-pressure control is dependent upon changes in suction pressure during the running cycle of a unit. Because the automatic expansion valve is designed to maintain a constant evaporator pressure, a thermostatic motor control must be used for temperature control.

The application of an automatic expansion valve is best suited to systems that operate under relatively constant heat loads. Under these conditions the performance characteristic of the valve offers a variety of useful features. In applications such as liquid chillers and water coolers, the valve can be adjusted to maintain an evaporator temperature just above the freezing point of water or the fluid being used, preventing the liquid from freezing or the frost from accumulating on the evaporator coil. In air-conditioning applications, the valve can be adjusted to prevent the evaporator coil from frosting and freezing yet still maintain a low evaporator temperature for comfort cooling and humidity control.

During high load condition the automatic expansion valve will maintain a constant low-side pressure that will not vary as the evaporator load fluctuates. As a result, motor current draw will remain stable, preventing an excessive current draw during high load conditions. Because the evaporator pressure is constant, lower horsepower motors can be utilized, regardless of extreme variations in heat loads. A reserve motor capacity is not required, and the condenser surface area can be reduced.

Valve Adjustment

The operating pressure of an automatic expansion valve can be changed by rotating an adjusting screw, which changes the spring tension and raises or lowers the operating pressure of the valve. When adjusting the valve pressure, turn the adjusting screw one-quarter turn at a time and allow the unit to run for at least 15 minutes between each adjustment. This will allow the system operating pressures and temperatures to stabilize, preventing possible overshooting of the valve adjustment.

Valve Selection

Automatic expansion valves are manufactured in many designs. They are available in solder connections, flare connections, or flange connections. Valve bodies are usually constructed of

aluminum or brass. The capacity of an automatic expansion valve should be matched carefully to the capacity of the condensing unit employed in a given application. An oversized valve will overfeed the evaporator and cause a floodback of liquid refrigerant. An undersized valve will cause a starved coil condition.

The following variables affect the capacity of an automatic expansion valve:

1. Type of refrigerant used
2. Evaporator temperature and pressure
3. Amount of liquid subcooling
4. Condensing temperature
5. Pressure drop across the valve
6. Size of valve orifice

THERMOSTATIC EXPANSION VALVES

The **thermostatic expansion valve (TEV or TXV)** is a frequently used metering device in modern refrigeration and air-conditioning applications. The key to its popularity lies in its high efficiency and ability to keep the evaporator supplied with the correct amount of liquid refrigerant under any load condition. Because of these characteristics, the thermostatic expansion valve is particularly well suited for use on systems that experience wide variations in evaporator loads.

Often referred to as a constant superheat valve, the operation of a thermostatic expansion valve is based upon controlling the flow of liquid refrigerant into the evaporator by maintaining a constant degree of superheated vapor leaving the evaporator. After absorbing heat in the evaporator, the liquid refrigerant will vaporize. As the vapor continues to absorb heat and its temperature is raised above the saturation temperature corresponding to its pressure, it becomes superheated. The degree of superheat will be the temperature increase of the vapor above saturation temperature at the existing pressure. The three pressures that govern the operation of the thermostatic expansion valve are spring pressure, evaporator pressure, and bulb pressure. The principal parts of

Fig. 14–4   Parts of a thermostatic expansion valve. In this valve, thermostatic element is threaded on body of valve. *Courtesy Sporlan Valve Company.*

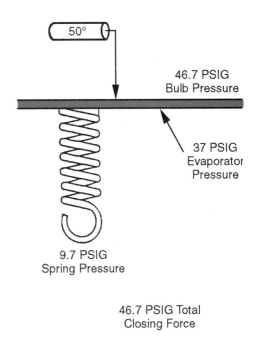

Fig. 14–5   Relationship of TEV operating pressures.

a thermostatic expansion valve are illustrated in Figure 14–4.

The evaporator pressure works under the valve diaphragm in the closing direction of the valve. The spring pressure is also a closing force, and its pressure is transmitted to the underside of the valve diaphragm by means of the pin carrier and push rods. The bulb pressure is transmitted to the top of the valve diaphragm through the capillary tube and is considered an opening force. The relationship of TEV operating pressures is illustrated in Figure 14–5.

In order to sense the variations of superheat in the suction vapor, the feeler bulb is attached to the suction line at the outlet of the evaporator. It is charged partly with liquid and partly with saturated vapor, usually with the same refrigerant that

is being used in the system. Through close thermal contact, the feeler bulb will assume the same temperature as the suction vapor at the point of the bulb contact. The pressure of the refrigerant within the feeler bulb will be in accordance with the saturated pressure/temperature of the refrigerant being used.

In Figure 14–6, Refrigerant-12 is introduced into an evaporator that is maintaining a pressure of 37 PSIG. The saturation temperature of R-12 at 37 PSIG is 40°F. As the refrigerant moves through the evaporator it will absorb heat and vaporize. The amount of liquid refrigerant will diminish. The refrigerant will remain at a temperature of 40°F as long as the liquid remains in the presence of vapor and the pressure remains the same. At point V, all of the liquid refrigerant has evaporated, leaving only a vapor. As the vapor continues to move through the coil it will continue to absorb heat from the surrounding ambient. The evaporator pressure remains at 37 PSIG. At point B, the temperature of the vapor has reached 50°F, 10° above its saturation temperature, and has been superheated 10° (50°F to 40°F).

The degree to which the refrigerant vapor is superheated depends upon the load on the evapo-

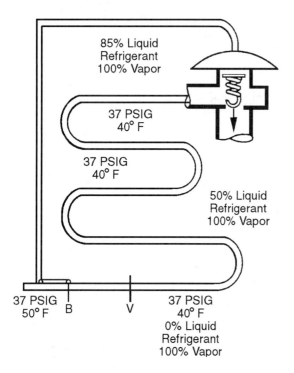

85% Liquid
Refrigerant
100% Vapor

37 PSIG
40° F

37 PSIG
40° F

50% Liquid
Refrigerant
100% Vapor

37 PSIG
50° F   B   V   37 PSIG
40° F
0% Liquid
Refrigerant
100% Vapor

Fig. 14–6   Refrigerant-12 introduced into evaporator.

rator and the amount of liquid refrigerant being fed into the coil. If the evaporator load increases and not enough refrigerant is fed into the coil, the point of vaporization at point V will move farther up the coil. The result will be a starved coil, high superheat, and loss of cooling efficiency. If the evaporator load decreases and too much liquid refrigerant is fed into the coil, the point of vaporization at point V will move farther down the coil. The result will be a low superheat and danger of liquid floodback to the compressor.

The ideal condition is to have the point of complete liquid vaporization at the end of the evaporator coil or at the end of superheat, allowing for the complete use of the evaporator surface for cooling, without the danger of liquid floodback. Although this works in theory, in a practical application this is not possible because it requires several degrees of superheat to operate the thermostatic expansion valve. For this reason, 10°F is an acceptable value for expansion valve settings. The superheat setting of the expansion valve may vary, however, depending upon the application and manufacturer's specifications.

Types of Thermostatic Expansion Valves

Because of the wide range of equipment on which the thermostatic expansion valve can be used, several designs of thermostatic charges have been developed over the years to provide for maximum valve performance on each application. There are basically three types of thermostatic charges in common use today: gas charge, liquid charge, and cross charge.

Gas Charged Valves

The **gas** or **limited charged thermostatic expansion valve** is charged with the same refrigerant that is in the system. The amount of liquid charge is such that, at a predetermined temperature, all of the liquid contained within the bulb will vaporize. Beyond this point the vapor temperature will increase and become superheated, but the pressure within the bulb will remain almost constant.

Figure 14–7 illustrates the relationship of the three controlling pressures of the gas charged valve. During the off-cycle warm-up period, the evaporator temperature and pressure will increase at the same rate as the feeler bulb temperature and pressure. The spring pressure keeps the valve closed tightly. At a predetermined point all of the liquid refrigerant in the feeler bulb will vaporize. Above this point the bulb temperature will continue to increase with virtually no increase in bulb pressure. Evaporator temperature and pressure will continue to increase and, along with the spring pressure, keep the valve closed tightly. The advantage of this characteristic is that the valve will not open during the off-cycle warm-up of the evaporator.

Referring again to Figure 14–7, evaporator temperature and pressure are high when the compressor starts. The valve will remain closed until the evaporator temperature is reduced below the maximum pressure of the feeler bulb. When this occurs the vapor in the bulb will then condense and the valve will start to throttle to the open position. The advantage of this characteristic is that, during the initial pull-down, the valve will remain

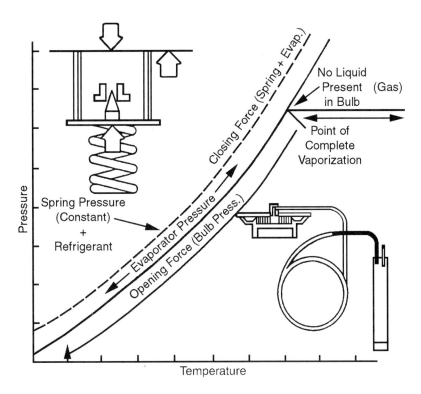

Fig. 14–7   Relationship of the three controlling pressures of a gas charged TEV.

closed, preventing the possibility of liquid refrigerant floodback to the compressor and preventing possible compressor overloading.

The precaution that must be taken with the gas charged valve is to ensure that the valve body and diaphragm are maintained at a warmer temperature than the feeler bulb and capillary tube. This is necessary to prevent migration of liquid refrigerant from the feeler bulb to the valve diaphragm, which would result in a loss of valve control from the feeler bulb. On systems utilizing a pressure drop type of distributor, refrigerant migration is not a problem. The pressure and temperature of the refrigerant at the valve body is higher than the suction vapor and the feeler bulb temperature. A pressure drop refrigerant distributor is illustrated in Figure 14–8.

Liquid Charged Valves

The **liquid charged thermostatic expansion valve** is generally limited to special equipment and large capacity ammonia systems. The feeler bulb is charged with the same refrigerant that is in

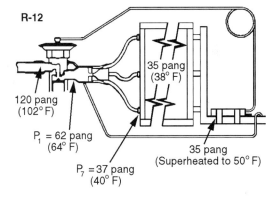

Fig. 14–8   Pressure drop distributor.

the system, but the volume and the amount of the charge is such that a sufficient amount of liquid refrigerant always remains in the bulb under any operating temperature. The primary advantage of this design is that the feeler bulb will always maintain valve control despite a colder valve body or diaphragm.

A disadvantage of the liquid charged valve pertains to the pressure/temperature relationship of the three controlling pressures of the liquid charged valve, as shown in Figure 14–9.

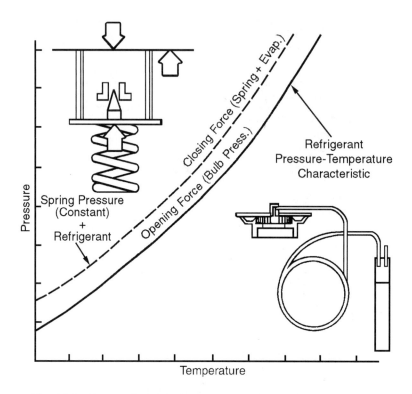

Fig. 14–9   Pressure/temperature relationships of liquid charged TEV.

Evaporator temperature and pressure rise at the same rate as the feeler bulb temperature and pressure. The spring pressure will keep the valve closed. If the feeler bulb is placed in a warm location, the bulb temperature and pressure may become great enough to open the valve. This will cause liquid refrigerant to flood the evaporator, inducing possible floodback on start-up.

When the compressor is started, suction pressure and evaporator pressure will decrease. Because the temperature of the feeler bulb is not immediately decreased, its pressure will remain high. This will cause the valve to open too far, resulting in a low superheat and possible liquid floodback to the compressor. Moreover, the introduction of liquid refrigerant into the evaporator at this time will delay the suction full downtime and may overload the compressor motor.

Cross Charged Valves

The **cross charged thermostatic expansion valve** uses a different fluid in the feeler bulb than the refrigerant that is in the system. The volume

and the amount of the bulb charge is such that liquid will always remain in the feeler bulb under any operating temperature.

Cross charged valves have a flatter pressure/temperature curve than the refrigerant in the system. Because of this, they offer several advantages. Figure 14–10 shows the relationship of the three controlling pressures of the cross charged valve. During the off cycle the evaporator pressure will increase much more rapidly than the feeler bulb pressure. The advantage of this is that the valve will close quickly and stay closed during the entire off cycle. Another advantage is that the valve is able to maintain control at high evaporator temperatures and superheat conditions, preventing liquid floodback on start-up and allowing for faster pull-down of evaporator pressure and temperature. A third advantage is the reduction of hunting or cycling because the cross charged valve is more responsive to changes in suction pressure than to changes in feeler bulb temperature. This is attributed to the flatter temperature/pressure curve of the feeler bulb charge.

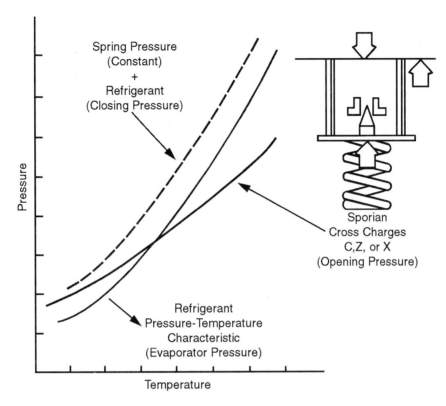

Fig. 14–10   Three controlling pressures of a cross charged valve.

External Equalization

In our previous discussions of the thermostatic expansion valve we didn't mention a pressure drop between the inlet and the outlet of the evaporator coil. Technically, all evaporators experience a pressure drop—especially evaporators utilizing a refrigerant distributor. An evaporator pressure drop in excess of 2 PSIG will cause an increase of superheat and a starved evaporator if an internally equalized valve, which was described earlier, is used on the system.

Figure 14–11 illustrates an evaporator experiencing a pressure drop of 10 PSIG between the inlet and outlet of the evaporator coil. Notice the evaporator inlet pressure is exerting a force of 37 PSIG on the underside of the valve diaphragm combined with a spring pressure of 9.1 PSIG. The total closing force of the valve on the underside of the diaphragm is 46.1 PSIG.

Because of the evaporator pressure drop, the outlet pressure of the coil has decreased to 27 PSIG with a corresponding saturation tempera-

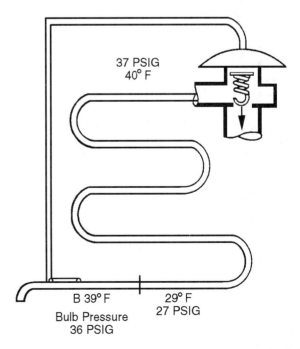

Fig. 14–11   Evaporator experiencing a pressure drop of 10 PSIG.

ture of 27°F. If the expansion valve was adjusted to a 10°F superheat and total vaporization of the refrigerant took place where it normally would at point V, the temperature at the feeler bulb, point B, would be 39°F. This would result in a bulb pressure of 36 PSIG. In Figure 14–12 the closing force of the valve is greater and the valve remains closed.

In Figure 14–13, the point of liquid vaporization has moved farther up the evaporator coil. At a superheat of 21°F, the valve opening and closing forces become equal, as shown in Figure 14–14. However, only a small portion of the evaporator is usable for cooling. The rest of the coil is starved and is used to superheat the suction vapor. The problem is that the valve operation is controlled by the outlet pressure of 37 PSIG and the bulb function is controlled by the outlet pressure of 27 PSIG. Consequently, there is no way the valve can compensate for the pressure drop across the evaporator coil.

In Figure 14–15 an external equalizer line is installed connecting the suction line at the outlet of the evaporator to the underside of the valve diaphragm. Through internal packing around the valve push rods, evaporator inlet pressure is sealed from the underside of the valve diaphragm.

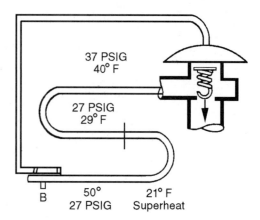

Fig. 14–13   The point of liquid vaporization has moved up farther in the evaporator coil.

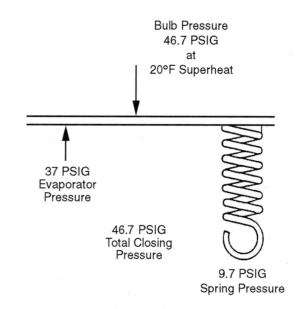

Fig. 14–14   Equalization of valve pressures, but at high superheat.

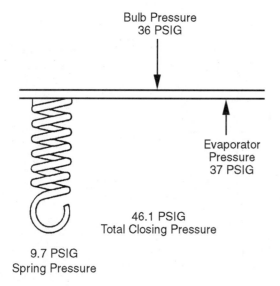

Fig. 14–12   The closing force of the TEV is greater and the valve remains closed.

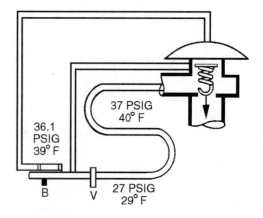

Fig. 14–15   External equalizer line installed.

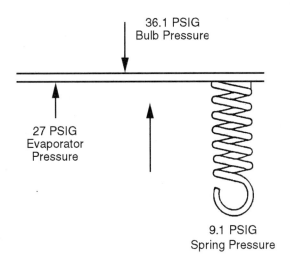

Fig. 14–16 Valve pressures in equilibrium with external equalizer line installed.

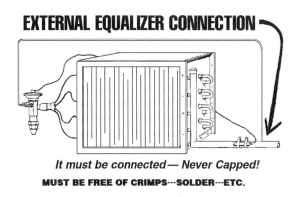

Fig. 14–17 Proper installation of external equalizer line. *Courtesy Sporlan Valve Company.*

The inlet pressure of the thermostatic expansion valve is 37 PSIG, and as the pressure drop across the evaporator occurs, the outlet pressure is reduced to 27 PSIG. The external equalizer line now transmits 27 PSIG to the underside of the valve diaphragm and combines with the spring pressure of 9.1 PSIG for a total closing force of 36.1 PSIG.

At point V, total liquid vaporization takes place, and with a 10°F superheat setting, the temperature at the feeler bulb is 39°F, providing an opening force of 36.1 PSIG. The valve pressures are now in equilibrium and the pressure drop across the evaporator has no effect on valve operation. The equalizer line has provided the valve with the same pressure that the closing force of the valve provides as it controls the opening force. Figure 14–16 illustrates valve pressures in equilibrium.

External Equalizer Connection

In all applications, the external equalizer line of the valve should be connected or improper valve operation and inefficient system performance will result. The line should be connected to the suction line of the evaporator at a point where it can best sense the pressure at the feeler bulb location. The line should be connected down-

stream of the feeler bulb to prevent liquid refrigerant from entering the equalizer line through leaking valve packing and affecting the temperature and operation of the feeler bulb. The equalizer line should also be located on the top of the suction line, as illustrated in Figure 14–17, to prevent refrigerant oil from entering and becoming trapped.

Valve Location and Installation

The location and installation of the thermostatic expansion valve are critical for proper valve operation and system control. The mounting of the valve body is not critical, however, and may be installed in any position. In valve applications utilizing refrigerant distributors, the distributor should be installed directly into the outlet of the valve. In these applications it is recommended to install the valve so that it feeds vertically, either upward or downward, into the distributor. In all installations the valve should be installed as close to the evaporator as possible.

If the valve body temperature becomes lower than the feeler bulb temperature, the refrigerant charge in the bulb may migrate to the top of the valve diaphragm, causing a loss of valve control. This is not usually a problem with the liquid charged valves because they are charged with a sufficient amount of liquid to maintain control over a wide variety of operating temperatures. The exception to this is when the valve body is exposed to extremely low temperatures during the off cycle. Consequently, start-up may be delayed until the bulb and valve diaphragm are sufficiently warmed.

Because of the limited amount of liquid refrigerant contained within gas charged valves, they should be installed so that the valve body is in a warmer location than the feeler bulb.

When installing the solder type of thermostatic expansion valve, it is not usually necessary to disassemble the valve. It is important to ensure, however, that the torch flame does not come into direct contact with the valve body or diaphragm. Wrapping a wet rag around the valve body and diaphragm during any brazing or soldering operations is recommended. If there is a possibility of overheating, take extra precaution and disassemble the valve.

You should follow the equipment or manufacturer's installation instructions when locating and installing the feeler bulb. This will ensure optimum system performance. Many service problems involving improper valve operation are a result of improper bulb location and installation. The following installation guidelines are generally accepted principles of proper suction line piping and bulb installation. In all applications the horizontal suction line on which the feeler bulb is to be installed should be pitched away and slightly downward from the evaporator.

1. The location of the feeler bulb should be as close as possible to the outlet of the evaporator, as shown in Figure 14–18.

2. Installation on a horizontal suction line is generally recommended. The bulb should be installed, however, on a vertical riser if required by the installation.

3. Good thermal contact between the feeler bulb and the suction line is required. The tubing should be cleaned to remove any dirt, corrosion, paint, etc., before installing the bulb. The bulb should be fastened securely with two straps to a straight, round section of suction line.

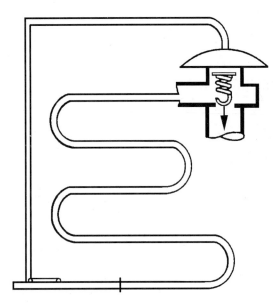

Fig. 14–18   TEV feeler bulb installed as close as possible to the evaporator.

4. The bulb should not be located in any suction line trap or pocket. This will prevent the liquid refrigerant and oil that is leaving the evaporator from affecting the temperature of the feeler bulb and, consequently, valve performance.

5. On suction lines of 5/8" or larger, the temperature of the tubing may vary around the circumference. For this reason the bulb should be installed at a four o'clock or an eight o'clock position, as illustrated in Figure 14–19. On smaller suction lines the bulb may be mounted anywhere around the circumference of the tubing. It should not be installed, however, on the bottom of the tubing. This prevents oil and refrigerant from affecting bulb operation.

6. In applications where the compressor is located below the evaporator and no system pump down is used, a short trap and vertical riser extending to the height of the evaporator should be used to prevent refrigerant from draining to the compressor during the off cycle, as illustrated in Figure 14–20.

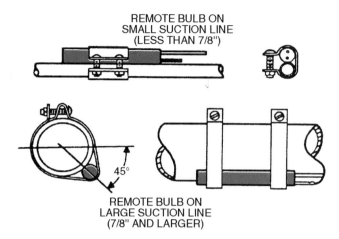

Fig. 14–19   Installation of TEV feeler bulb.   *Courtesy ALCO Controls.*

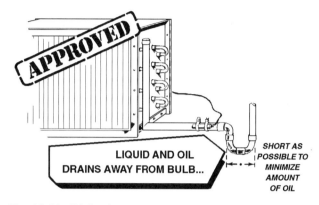

Fig. 14–21   Piping for compressor located above evaporator. *Courtesy Sporlan Valve Company.*

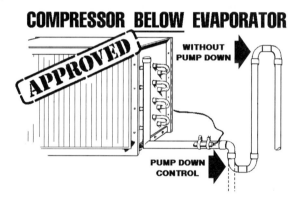

Fig. 14–20   Piping for compressor located below evaporator. *Courtesy Sporlan Valve Company.*

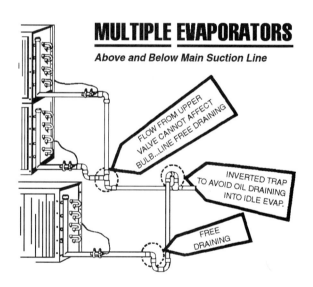

Fig. 14–22   Piping for multiple evaporators.   *Courtesy Sporlan Valve Company.*

7. In applications where the compressor is located above the evaporator, a trap, which is illustrated in Figure 14–21, should be installed to drain any liquid refrigerant and oil from the suction line that may affect bulb operation. The trap will serve to facilitate oil return to the compressor. The traps may be constructed of street ells or may be purchased commercially as a preformed unit. The traps should be as short as possible.

8. On multiple evaporators, illustrated in Figure 14–22, the suction line piping should be arranged so that the flow of refrigerant and oil from one evaporator does not affect the feeler bulb operation of another evaporator.

9. In the application of liquid charged valves, the bulb should be installed in a location where it is the same temperature as the conditioned space. This will ensure that the valve closes tightly and does not open during the off cycle. If this type of installation is not possible, the system should have a solenoid valve installed ahead of the valve in the liquid line and wired to close when the compressor cycles off. An

automatic pump down system may also be considered.

10. Gas charged valves may be installed inside or outside the conditioned space. To prevent migration of the feeler bulb charge to the valve diaphragm, the valve body should not be located in any supply airstream. If there is a possibility that the feeler bulb may be affected by air outside the conditioned space, the feeler bulb should be insulated.

Valve Adjustment

In most service applications, it is not necessary to change the superheat setting of a thermostatic expansion valve. Experience has proven that other factors affect the performance of a refrigeration or air-conditioning system more than valve operation or adjustment. After all other possibilities have been considered and it is decided that an adjustment of the superheat setting of the valve is required, refer to Figure 14–23 and Figure 14–24 and the following procedures:

1. Place a thermometer at the point of bulb contact and record the temperature.

2. With gauges installed, record the suction pressure at the compressor.

3. Add 2 PSIG to the recorded suction pressure to allow for suction line loss. (Note: In some applications a pressure port is available at the outlet of the evaporator to record evaporator pressure. In these cases it is not necessary to add 2 PSIG to the suction pressure reading.)

4. Referring to a temperature/pressure chart, convert the suction pressure to temperature.

5. To obtain the superheat setting of the valve, subtract the temperature of the refrigerant in the evaporator from the temperature recorded at the point of the feeler bulb installation.

When adjusting an expansion valve for superheat settings, turn the valve stem one-quarter turn at a time and allow the system to operate until system pressures and temperatures have stabilized before making another adjustment.

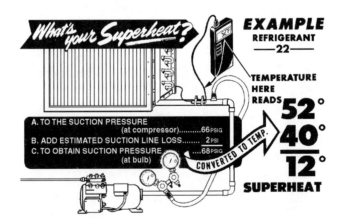

Fig. 14–23   Determining system superheat.   *Courtesy Sporlan Valve Company.*

Thermo® Valve Troubleshooting — Service Hints

Low Suction Pressure... High Superheat	
EXPANSION VALVE LIMITING FLOW	
Probable Cause	**Remedy**
Inlet pressure too low from excessive vertical lift, undersized liquid line or excessively low condensing temperature. Resulting pressure difference across valve is too small.	Increase head pressure. If liquid line is too small, replace with proper size.
Gas in liquid line...due to pressure drop in the line or insufficient refrigerant charge. If there is no sight glass in the liquid line, a characteristic whistling noise will be heard at the expansion valve.	Locate cause of liquid line flash gas and correct it by use of any or all of the following methods: 1. Add charge. 2. Clean strainers, replace filter-driers. 3. Check for proper line size. 4. Increase head pressure or decrease temperature to insure solid refrigerant at the valve inlet.
Valve restricted by pressure drop through evaporator.	Change to an expansion valve having an external equalizer.
External equalizer line plugged, or external equalizer connector capped without providing a new valve cage or body with internal equalizer.	If external equalizer is plugged, repair or replace. Otherwise, replace with valve having correct equalizer.
Moisture, wax, oil, or dirt plugging valve orifice. Ice formation or wax at valve seat may be indicated by a sudden rise in suction pressure after shutdown and system has warmed up.	Wax and oil indicates wrong type of oil is being used. Recover refrigerant and recharge system, using proper oil. Install an ALCO filter-drier to prevent moisture and dirt from plugging valve orifice.
Valve orifice too small.	Replace with proper valve.
Superheat adjustment too high.	See "Superheat Adjustment" in this bulletin.
Power assembly failure or partial loss of charge.	Replace power assembly (if possible) or replace valve.
Gas charged remote bulb of valve has lost control due to the remote bulb tubing or power head being colder than the bulb.	Install small wattage heater on power element. Do not allow capillary tube to come in contact with colder tubing.
Filter screen clogged.	Clean all filter screens.
RESTRICTION IN SYSTEM OTHER THAN EXPANSION VALVE (Usually but not necessarily indicated by frost or lower than normal temperatures at point of restriction.)	
Probable Cause	**Remedy**
Strainers clogged or too small.	Remove and clean strainers. Check manufacturer's catalog to make sure that correct strainer was installed. Add an ALCO filter-drier to the system.
A solenoid valve not operating properly or is undersized.	Refer to solenoid valve's installation sheets for service hints. If valve is undersized, check manufacturer's catalog for proper size and conditions that would cause a malfunction.
King valve at liquid receiver too small or not fully opened. Hand valve stem failure or valve too small or not fully opened. Discharge or suction service valve on compressor restricted or not fully opened.	Repair or replace faulty valve if it cannot be fully opened. Replace any undersized valve with one of correct size.
Plugged lines.	Clean, repair or replace lines.
Liquid line too small.	Install proper size liquid line.
Suction line too small.	Install proper size suction line.

Fig. 14–24A   Probable cause/remedy tables.   *Courtesy ALCO Controls.*

Thermo® Valve Troubleshooting — Service Hints

Fluctuating Suction Pressure

Probable Cause	Remedy
Incorrect superheat adjustment.	See "Superheat Adjustment" in this bulletin.
Trapped suction line.	Install "P" trap to provide free-draining suction line.
Improper remote bulb location or installation.	Clamp remote bulb to free-draining suction line. Clean suction line thoroughly before clamping bulb in place. See figures 2, 3, 4 and 5.
Floodback of liquid refrigerant caused by poorly designed liquid distribution device or uneven evaporator loading. Improperly mounted evaporator.	Replace faulty distributor with a new one. If evaporator loading is uneven, install proper load distribution devices to balance air velocity evenly over evaporator coils. Remount evaporator lines to provide proper angle.
External equalizer lines tapped at a common point although there is more than one expansion valve on the same system.	Each valve must have its own separate equalizer line going directly to an appropriate location on evaporator outlet to insure proper operation response of each individual valve. See figure 4 for an example.
Faulty condensing water regulator, causing change in pressure drop across valve.	Replace condensing water regulator.
Evaporation condenser cycling, causing radical change in pressure drop across expansion valve. Cycling of blowers or brine pumps.	Check spray nozzles, coil surface, control circuits, thermostat overloads, etc. Repair or replace any defective equipment. Clean clogged nozzles, coil surface, etc.
Restricted external equalizer line.	Repair or replace with correct size.

Fluctuating Discharge Pressure

Probable Cause	Remedy
Faulty condensing water regulating valve.	Replace condensing water regulating valve.
Insufficient charge...usually accompanied by corresponding fluctuation in suction pressure.	Add charge to the system.
Cycling of evaporation condenser.	Check spray nozzles, coil surface, control circuits, thermostat overloads, etc. Repair or replace any defective equipment. Clean clogged nozzles, coil surface, etc.
Inadequate and fluctuating supply of cooling water to condenser.	Check water regulating valve. Repair or replace if defective. Check water circuit for restrictions.
Cooling fan for condenser cycling.	Determine cause of fan cycling and correct.
Fluctuating discharge pressure controls on low ambient air-cooled condenser.	Adjust, repair or replace controls.

High Discharge Pressure

Probable Cause	Remedy
Insufficient cooling water due to inadequate supply or faulty water valve.	Start pump and open water valves. Adjust, repair or replace any defective equipment.
Condenser or liquid receiver too small.	Replace with correct size condenser or liquid receiver.
Cooling water above design temperature.	Increase supply of water by adjusting water valve, replacing with a larger valve, etc.
Air or non-condensable gases in condenser.	Recover refrigerant and recharge system.
Overcharge of refrigerant.	Remove excess refrigerant in approved manner.
Condenser dirty.	Clean condenser.
Insufficient cooling air circulation over air-cooled condenser.	Properly locate condenser to freely dispel hot discharge air. Tighten or replace slipping belts or pulleys and be sure blower motor is of proper size.

Fig. 14–24B Probable cause/remedy tables. *Courtesy ALCO Controls.*

Thermo® Valve Troubleshooting — Service Hints

Low Suction Pressure... Low Superheat

Probable Cause	Remedy
Poor distribution in evaporator causing liquid to short-circuit through favored passes and throttling valve before all passes receive sufficient refrigerant.	Clamp power assembly remote bulb to free-draining suction line. Clean suction line thoroughly before clamping bulb in place. See figure 2. Install a refrigerant distributor. Balance evaporator load distribution.
Compressor oversized or running too fast due to improper sizing or poor air distribution or brine flow.	Reduce speed of compressor by installing proper size pulley, provide compressor capacity control.
Uneven or inadequate evaporator loading due to poor air distribution or brine flow.	Balance evaporator load distribution by providing correct air or brine distribution.
Evaporator too small...often indicated by excessive ice formation.	Replace with proper size evaporator.
Excessive accumulation of oil in evaporator.	Alter suction piping to provide proper oil return or install oil separator, if required.

High Suction Pressure... High Superheat

Probable Cause	Remedy
Unbalanced system having an oversized evaporator, an undersized compressor and a high load on the evaporator. Load in excess of design conditions.	Balance system components for load requirements.
Compressor undersized.	Replace with proper size compressor.
Evaporator too large.	Replace with proper size evaporator.
Compressor discharge valves leaking.	Repair or replace valves.

High Suction Pressure... Low Superheat

Probable Cause	Remedy
Valve superheat setting too low.	See "Superheat Adjustment" in this bulletin.
Gas in liquid line with oversized expansion valve.	Replace with proper size expansion valve. Correct cause of flash gas.
Pin and seat of expansion valve wire drawn, eroded or held open by foreign material, resulting in liquid floodback.	Clean or replace damaged parts or replace valve. Install an ALCO filter-drier to remove foreign material from system.
Ruptured diaphragm or bellows in a constant pressure (automatic) expansion valve, resulting in liquid floodback.	Replace valve power assembly.
External equalizer line plugged, or external equalizer connector capped without providing a new valve cage or body with internal equalizer.	If external equalizer is plugged, repair or replace. Otherwise, replace with valve having correct equalizer.
Moisture freezing valve in open position.	Apply hot rags to valve to melt ice. Install ALCO filter-drier to insure a moisture-free system.

Measuring Superheat

1. Determine the suction pressure with an accurate gauge at the evaporator outlet. On close-coupled installations, the suction pressure may be read at the compressor suction connection.
2. From refrigerant pressure-temperature tables, determine saturation temperature at observed suction pressure.
3. Measure temperature of suction gas at Thermo Valve remote bulb location.
4. Subtract saturation temperature (read from tables in step 2) from temperature measured in step 3, the difference is the superheat of the suction gas.

Superheat Adjustment

ALCO Thermo valves are factory set to a specific superheat — however, the superheat should be adjusted for the application. To adjust the valve to other superheat settings:
1. Remove the seal cap from the bottom of valve.
2. Turn the adjustment screw clockwise to increase superheat and counterclockwise to decrease superheat. One turn changes the superheat approximately 3–4°F, regardless of the refrigerant type. As much as 30 minutes may be required for the system to stabilize after adjustment is made.
3. Replace and tighten seal cap.

Warning: There are 10 turns on the adjustment stem. When adjusting superheat setting—when stop is reached, any further turning adjustment will damage valve.

Fig. 14–24C   Probable cause/remedy tables. *Courtesy ALCO Controls.*

Unit Fourteen Summary

- Metering devices control the amount of refrigerant entering the evaporator and maintain a pressure differential between the high and low side of a refrigeration system.

- Hand expansion valves were one of the earliest types of metering devices used and have been used in applications where the load is constant. A capillary tube is considered to be a constant feed metering device and operates on the principle of gradual pressure reduction. The inside diameter, length, and tightness of coils, as well as the ambient temperature, affect the capacity of a capillary tube.

- In a capillary tube, the liquid length is described as the section in which the liquid refrigerant enters. Bubble point is achieved when the liquid refrigerant begins to vaporize in the capillary tube and the two-phase length is the point at which both liquid and vapor exist. Capillary tube systems equalize during an off cycle.

- Automatic expansion valves (AEV) maintain a constant evaporator pressure and are used on equipment such as drinking fountains.

- Thermostatic expansion valves (TEV or TXV) control the amount of refrigerant entering an evaporator based upon the temperature of the refrigerant leaving the evaporator.

- Bulb pressure is the opening force of the TEV, while spring pressure and evaporator pressure make up the closing force of the TEV.

- Gas charged TEVs are charged with a liquid that will vaporize at a given temperature, while liquid charged TEVs will always maintain a liquid charge regardless of temperature.

- Cross charged TEVs are charged with a refrigerant different from the system on which they're installed.

- External equalization compensates for evaporator pressure drop and keeps superheat low. The external equalizer line should be installed on top of the suction line and downstream of the expansion valve sensing bulb. A sensing bulb should be installed as close to the outlet of the evaporator as possible.

- When the compressor is located above the evaporator, the suction line should be trapped to ensure proper oil return to the compressor.

- Superheat in a refrigeration system is the difference between the saturation temperature of the refrigerant in the evaporator and the temperature of the refrigerant vapor at the expansion valve bulb location.

Unit Fourteen Key Terms

Hand Expansion Valve

Capillary Tube

Liquid Length

Bubble Point

Two-Phase Length

Automatic Expansion Valve

Thermostatic Expansion Valve

Gas Charged Thermostatic Expansion Valve

Limited Charged Thermostatic Expansion Valve

Liquid Charged Thermostatic Expansion Valve

Cross Charged Thermostatic Expansion Valve

UNIT FOURTEEN REVIEW

1. What are the purposes of a metering device?

2. Give three reasons why the amount of refrigerant flow into the evaporator must be metered.

3. Three types of metering devices (or refrigerant flow control devices) are the capillary tube, the automatic expansion valve, and the thermostatic expansion valve. Connect each type of metering device with its main principle of operation.

   a. It's control operation is based on temperature changes.

   b. It's control operation is based on pressure changes.

   c. It's control operation is base on volume changes.

4. The _____ the inside diameter, or the _____ the capillary tube, the _____ will be the frictional resistance of the tubing.

5. Describe the three phases of liquid refrigerant flow through a capillary tube.

6. What are the factors that govern the capacity of a capillary tube?

7. What are the advantages of the capillary tube as a metering device?

8. In the automatic expansion valve, what is the relationship of evaporator pressure to spring pressure and atmospheric pressure in the operation of opening and closing the valve?

9. What are the disadvantages of an automatic expansion valve?

10. Why is selection in the size of the automatic expansion valve important?

11. What are the variables that affect the capacity of an automatic expansion valve?

12. How is a thermostatic expansion valve selected for a given application?

13. What are the three pressures that govern the operation of a thermostatic expansion valve?

14. What are the three types of thermostatic expansion valve charges?

15. Match the description with the appropriate type of metering device: capillary tube, automatic expansion valve, thermostatic expansion valve.

    a. Best suited where the evaporator heat load is relatively constant.

    b. Has no moving parts.

    c. Used only on dry expansion evaporators.

    d. High efficiency under a wide variation of evaporator loads.

    e. Bulb pressure is transmitted through a capillary tube.

16. On which side of the metering device, inlet or outlet, is a strainer installed? Explain why.

17. Match the type of thermostatic expansion valve with its description: gas charge, liquid charge, cross charge.

    a. Charged with the same refrigerant that is in the system.

    b. Charged with a different fluid in the feeler bulb than the refrigerant that is in the system.

    c. Charged with the same refrigerant, but a sufficient amount of liquid refrigerant always remains in the bulb.

18. What are the advantages of a gas charged thermostatic expansion valve?

19. In what applications are liquid charged thermostatic expansion valves used?

20. What is the purpose of an external equalizer line?

21. What is the proper location of an external equalizer line?

22. What is the proper location for the installation of the feeler bulb of a thermostatic expansion valve?

23. Describe the procedure to determine the superheat setting of a thermostatic expansion valve.

24. In what direction should the thermostatic expansion valve stem be turned in order to increase the superheat setting?

In a simple mathematical equation, the pressures on the thermostatic expansion valve indicate the position of the valve. In Example A, the valve is in a state of equilibrium because the bulb pressure is equal to the spring pressure plus the evaporator pressure.

EXAMPLE A

(bulb pressure) = (spring pressure) + (evaporator pressure)

45.7 PSIG = 9.7 PSI + 36 PSIG

25. 52 PSIG > 9.7 PSI + 36 PSIG. Is the valve opening or closing?

26. 41.6 PSIG < 9.7 PSI + 36 PSIG. Is the valve opening or closing?

27. Write a description of the technique to be used for installing a solder-type of thermostatic expansion valve to prevent damage to the valve.

28. Write a paragraph describing the operation of the thermostatic expansion valve using the following terms: evaporator pressure, spring pressure, atmospheric pressure, open, close, valve.

29. Explain the procedure to manually change the operating pressure of an automatic expansion valve.

UNIT FIFTEEN

Refrigeration System Accessories

OBJECTIVES

After completing this unit, the student will be able to:

1. Identify the various types of accessory devices used on refrigeration systems.

2. Explain the methods of installation and operation of receivers, oil separators, mufflers, check valves, sight glasses, solenoid valves, filter-driers, strainers, and refrigerant redirection valves.

In all cases the fundamental principles behind the operation of a refrigeration system are unchanged whether the system is small or large, domestic or commercial, high-pressure or low-pressure. In order to create a system that operates at maximum efficiency in a specific situation, accessories must be added to the basic components of a refrigeration system. Safety, convenience, and improved endurance are also considerations for the installation of system accessories.

In this unit, we'll discuss some accessories used in a variety of applications as well as some basic troubleshooting and installation procedures.

LIQUID RECEIVERS

The purpose of a **liquid receiver** is to store the liquid refrigerant until it is needed by the system. The receiver also stores refrigerant while the system is under service or repair.

A liquid receiver should be sized to hold the entire refrigerant charge of a system and still be only 85 percent full. This will allow for the ther-

mal expansion of the stored refrigerant. Receivers are designed in vertical or horizontal mounting positions, as shown in Figures 15–1 and 15–2. The outlet pipe of the receiver must extend below the level of the liquid contained within. This will

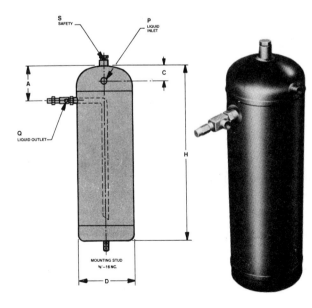

Fig. 15–1 Vertical receiver. *Courtesy Refrigeration Research.*

Fig. 15–2   Horizontal receiver.   *Photo by Bill Johnson.*

act as a liquid seal to prevent vapor from entering the liquid line. Some liquid receivers are equipped with gauge glasses to determine the liquid level in the receiver.

In order to pump down refrigerant and store it in the receiver during service operations, the receiver is equipped with inlet and/or outlet valves, which isolate the receiver from the rest of the system. As a safety precaution, receivers are equipped with pressure relief devices, such as a fusible plug, to relieve excessive refrigerant pressures before they become dangerously high.

SIGHT GLASSES

The purpose of a **sight glass** is to determine the condition of the liquid refrigerant in the liquid line. Under normal operating conditions, the sight glass should show a solid vapor-free column of liquid refrigerant leaving the receiver and entering the metering device. The presence of bubbles in the sight glass indicates the presence of flash gas in the liquid line. While the most common cause of this condition is a shortage of refrigerant in the system, other possible causes, such as low discharge pressure and an undersized liquid line, should not be overlooked. If the sight glass is located after the liquid line, and the filter-drier becomes restricted, bubbles may also appear in the sight glass. The appearance of bubbles in the sight glass on initial start-up is normal in many systems until the operating temperatures and pressures have stabilized. If the bubbles persist, however, you may have to check the system further to determine the cause.

Various types and designs of sight glasses are available commercially to fit almost any instal-

lation. They are available with either flared or soldered fittings, as illustrated in Figure 15–3. On applications with large liquid lines, parallel piping of the sight glass is commonly used. (see Figure 15–4) The best location for the installation of a sight glass is at the inlet of the metering device. Since this may not be practical in some

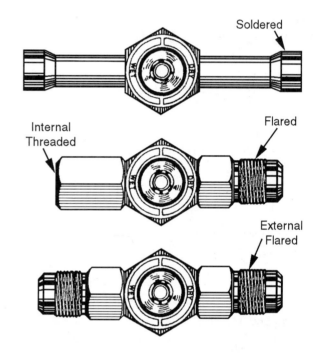

Fig. 15–3   Sight glasses with flared or soldered fittings. *Courtesy Virginia KMP Corp.*

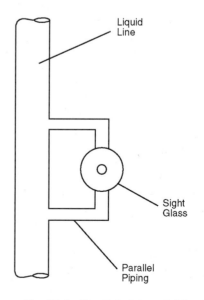

Fig. 15–4   Parallel piping of sight glass.

installation cases, the sight glass can be installed in the liquid line close to the condensing unit.

Many sight glasses are equipped with a moisture indicator, which will indicate the presence of moisture contamination in the sealed system. The presence of moisture will cause a chemical reaction in the moisture indicator creating a color change to indicate that the system is wet. When the sight glass is initially installed in a system, the moisture indicator will indicate that the system is wet, but this will usually change and the indicator will show a dry system after a few hours of operation.

At least eight hours of running time should be allowed before the moisture indicator can give an accurate reading. The moisture indicator should be kept as close as possible to 75°F. Liquid refrigerant at high temperatures will require a higher moisture content to change the color of the indicator. Excessive moisture or contamination due to an acid burnout will ruin a moisture indicator and give a false reading. Excessive oil circulation in the system will cause the moisture indicator to turn brown.

When installing a sight glass, always refer to the manufacturer's instructions for the proper installation procedures. A sight glass should be installed in a vertical position whenever possible. Leak test all soldered and flared connections as well as the sight glass itself.

VIBRATION ELIMINATORS

Vibration eliminators are installed in compressor suction and discharge lines to prevent the transmission of noise and vibration from the compressor to the remainder of the system and/or building structure. On small systems where soft copper tubing is used, compressor vibration may be adequately controlled by forming a coil or loop in the suction line just before it is connected to the compressor, or in the discharge line before the condenser. In applications where rigid tubing is used, a flexible metallic hose vibration eliminator is employed. A flexible hose vibration eliminator is shown in Figure 15–5.

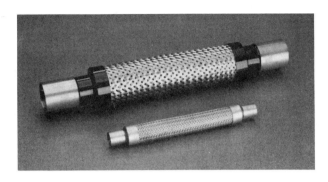

Fig. 15–5   Flexible hose vibration eliminator. *Courtesy Superior Valve Co., Div. Amcast Industrial Corp.*

A vibration eliminator should have the same inside diameter as the tubing in which it is to be installed. Proper installation is important because if stress is placed on the vibration eliminator, it will crack. While a certain amount of lateral movement is acceptable, care must be taken not to install the vibration eliminator in a position where it may become compressed or stretched. This condition may be experienced if the vibration eliminator is installed perpendicular to the compressor crankshaft. As a general rule, the vibration eliminator should be installed parallel to the compressor crankshaft and as close as possible to the compressor. The correct installation of a vibration eliminator is illustrated in Figure 15–6. In low temperature applications, the vibration eliminator should be installed at a slight angle to allow condensation to drain and to prevent moisture from accumulating and freezing, and possibly cracking the vibration eliminator tubing. Recently, vibration eliminators that can be installed to form a bend in the piping system have been developed. Always install a vibration eliminator in accordance with manufacturer's specifications.

OIL SEPARATORS

During normal system operation a percentage of oil leaves the compressor and circulates through the system. In a properly designed system the amount of oil leaving the compressor is equal to the amount of oil returning to the compressor. As a result, a proper oil level is always

NOTE: Mount as close to the
compressor as possible

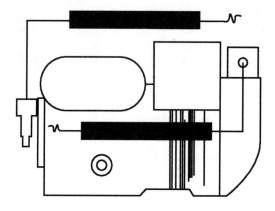

Recommended

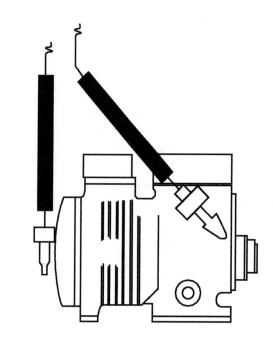

Acceptable

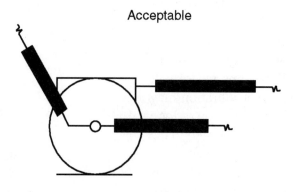

Not Acceptable

Fig. 15–6   Methods of installing vibration eliminator.   *Courtesy Copeland Corp.*

maintained in the compressor. This provides adequate lubrication. In systems where the return of oil is not adequate, an **oil separator** may be installed to ensure that the compressor maintains a proper oil level. Applications where an oil separator may be needed include:

1. Low temperature systems
2. Flooded evaporators
3. Systems utilizing capacity control
4. Systems that have long suction or discharge line risers
5. Systems utilizing a refrigerant that does not mix well with oil

An oil separator is illustrated in Figure 15–7. The oil separator depends upon three basic factors in order to remove oil from the hot discharge vapor leaving the compressor:

1. A decrease in gas velocity
2. A change in gas directional flow
3. A surface to which the oil will cling

Referring to Figure 15–8, oil-laden hot discharge gas enters the inlet port of the oil separator. As the vapor passes through the inlet screen, its velocity is reduced, allowing the screen to collect some of the oil. The oil will then drain to the bottom of the separator. The vapor must now change direction in order to leave the separator. The oil, being heavier than the vapor, cannot change direction as readily and will separate from the vapor and collect in the bottom of the separator. As the vapor leaves the separator the outlet screen will collect more oil, which will also drain to the bottom of the separator. The oil-free discharge vapor will now leave the oil separator through the outlet point to the condenser.

When the oil in the bottom of the oil separator reaches a predetermined level, the ball float will rise, opening the valve port and allowing the oil to return to the compressor through the oil return line.

Fig. 15–7 Oil separator. *Courtesy AC&R Components.*

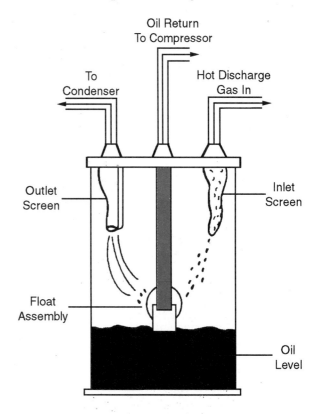

Fig. 15–8 Oil separator operation.

Installation of the Oil Separator

When installing an oil separator, the following factors should be considered:

1. The oil separator should be installed as close as possible to the compressor.

2. It should be mounted in a level position.

3. If the oil separator is installed at a remote location from the compressor, it should be well insulated to prevent the discharge vapor from condensing.

4. The oil return line to the compressor from the oil separator should be connected directly to the compressor crankcase. In sealed unit compressors the oil return line may be installed in the suction line.

5. During normal operation the temperature of the oil return line should intermittently be hot, then be at room temperature. This indicates that the ball float is opening to allow a flow of oil to the compressor and then closing. A continuously hot oil return line indicates that the ball float is stuck in the open position. A cold oil return line indicates that liquid refrigerant is returning to the compressor. In these cases, corrective action should be taken.

MUFFLERS

Mufflers are installed in the discharge line of a system to reduce the resonant noise created from the reciprocating action of the compressor pistons. Mufflers are especially helpful when the discharge line is of considerable length, resulting in a maximum noise level. In this case, the muffler should be installed in the discharge line as close to the compressor as the installation will permit. The muffler serves as an expansion area designed to absorb the pulsations of the compressor discharge gas, thereby minimizing the transmission of hot gas pulsation. Figure 15–9 illustrates a typical hot gas muffler.

In all applications, it is recommended that the discharge muffler be installed in a horizontal position. If it is installed in a vertical position, an oil pickup tube must be installed at the base of the

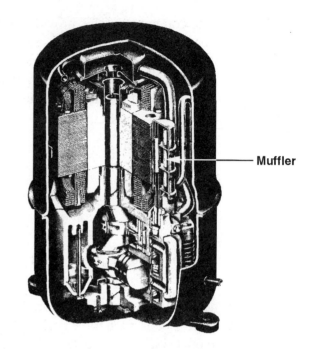

— Muffler

Fig. 15–9 Hot gas muffler. *Courtesy Tecumseh Products Company.*

muffler to reinject any oil, that may have accumulated in the muffler shell, into the system. This is accomplished by the pressure drop through the muffler and the aspirating effect of the tube, which lifts the oil to a point where it will reenter the gas stream of the system.

CHECK VALVES

The purpose of a **check valve**, which is illustrated in Figure 15–10, is to allow refrigerant flow in one direction while restricting refrigerant flow in the opposite direction. The check valve remains open while the refrigerant flows in a normal direction. If the refrigerant attempts to reverse its flow, the valve will close. Spring loading of the valve prevents any noise or chattering created by the pulsation of refrigerant gas from the compressor discharge valves. In all applications it is important to refer to the manufacturer's installation specifications to ensure the proper mounting and directional flow.

SOLENOID VALVES

Solenoid valves are installed in refrigeration and air-conditioning systems to control the flow

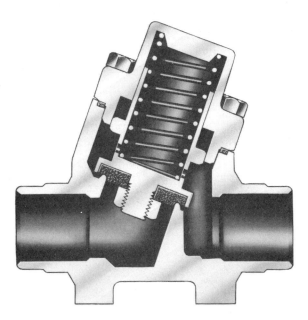

Fig. 15–10   Check valve.   *Courtesy Superior Valve Co., Division of AMCAST Industrial Corporation.*

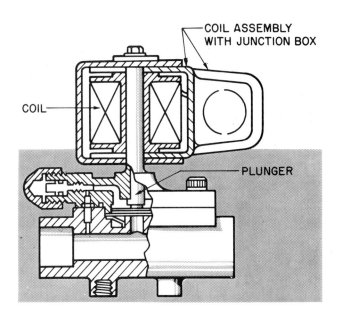

Fig. 15–11B   Cutaway view of solenoid.   *Courtesy Parker Hannifin Corporation.*

Fig. 15–11A   Solenoid valve.   *Courtesy Sporlan Valve Company.*

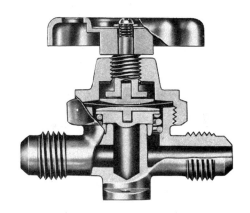

Fig. 15–12   Two-way valve.   *Courtesy Henry Valve Co.*

of refrigerant in a circuit or to isolate components from the rest of the system. The valve is controlled electrically and will open or close in response to an electrical switch, such as a thermostat. A solenoid valve, which is illustrated in Figures 15–11A and B, consists of three major parts: a valve body, a plunger/needle assembly, and a solenoid coil.

The valve is normally closed when there is no current flow through the solenoid coil. The weight of the plunger/needle assembly, sometimes assisted by a spring pressure, will keep it sealed firmly in the valve orifice located in the valve body and will prevent refrigerant flow through the valve. When the solenoid coil is energized, the magnetic field created by the flow of current through the solenoid windings will lift the plunger/needle assembly off of its seat and allow refrigerant to flow through the valve orifice. In some solenoid valve designs, a manual lift stem is provided to manually open the valve in the event of power interruption or in order to perform certain service functions.

Two-Way Valves

A **two-way valve**, which is illustrated in Figure 15–12, has one inlet and one outlet connection and provides for the flow of refrigerant in one line. The valve is normally closed when deenergized and will open when energized. In some applications a two-way valve may be designed so that the valve is open when deenergized and closed when energized.

Three-Way Valves

A **three-way** or **diverting valve**, which is illustrated in Figure 15–13, has one inlet connection that is common to two outlet connections. It is designed to allow for the flow of refrigerant in two directions and lines. These valves are utilized in various types and designs of refrigeration systems where the controlled flow of refrigerant is required in two circuits.

Four-Way Valves

A **four-way valve**, which is illustrated in Figure 15–14, has one inlet connection that is common to three outlet connections. These valves are often referred to as *reversing valves*, and they are commonly used in heat pump systems to redirect the flow of refrigerant through a system. They are also found on commercial refrigeration systems that employ the hot gas defrost method.

Valve Selection and Installation

Solenoid valve selection is based upon the following factors:

1. Capacity, based on system tonnage
2. Type of refrigerant used

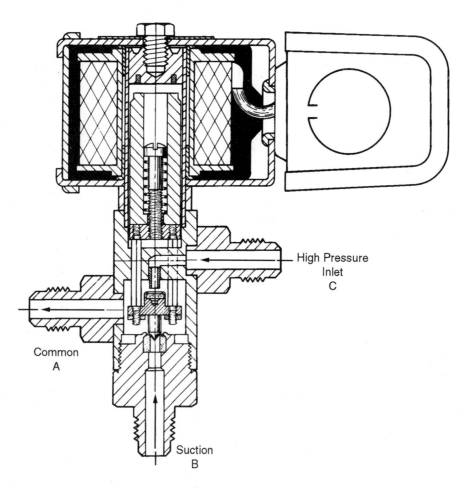

Fig. 15–13 Three-way or diverting solenoid valve. *Courtesy Sporlan Valve Company.*

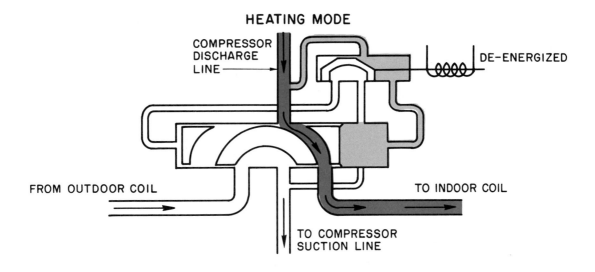

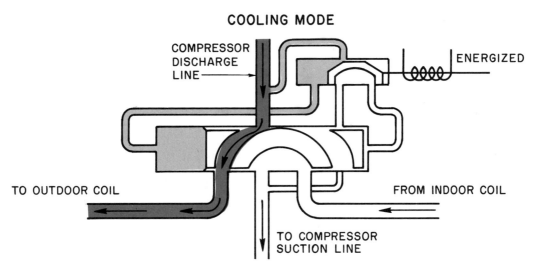

Fig. 15–14 Four-way valve.

3. Operating pressure

4. Electrical characteristics

Check the manufacturer's specifications for the proper mounting position of a solenoid valve. In some applications the plunger may require gravity to close properly. The valve may not seat correctly if it is not mounted vertically to the line in which it is installed. When installing the valve, check for proper direction of flow. An arrow located on the valve's body will indicate the proper direction of refrigerant flow. A proper strainer should be installed ahead of the valve. Carefully disassemble the valve before soldering or brazing it into place. Do not apply heat directly to the valve body or it may warp. Use wet rags to protect the valve body if necessary. Allow it to cool sufficiently before reinstalling the parts and gaskets.

Under normal operating conditions, the proper selection and installation of solenoid valves should provide trouble-free performance.

However, if service is needed, refer to the following problems and possible causes.

Problem: Valve will not open.

Possible causes:

1. Burned out coil
2. Improper wiring
3. Improper voltage
4. Improper valve assembly
5. Corroded or otherwise physically damaged parts
6. Improper valve application
7. Defective components in operating control circuit

Problem: Valve will not close.

Possible causes:

1. Operating control circuit components not opening
2. Physically damaged valve
3. Open manual stem
4. Excessively high pressure in valve outlet

Problem: Refrigerant bleeds through the closed valve.

Possible causes:

1. Foreign matter in valve seat or on plunger/needle assembly
2. Worn or damaged valve seat
3. Improper valve application

Problem: Coil has burned out.

Possible causes:

1. Improper wiring
2. Improper voltage
3. Excessively high ambient temperature
4. Excessively high fluid temperature
5. Binding valve plunger

STRAINERS

Strainers are installed in refrigeration and air-conditioning systems to trap foreign matter such

Fig. 15–15   Strainer for small equipment. *Courtesy Henry Valve Co.*

as scale, metal chips, and dust, and to prevent it from entering the sealed system and causing damage to the compressor or restriction of the system components, such as metering devices, solenoid valves, and pressure control valves. Many system components are already equipped with a factory-installed strainer. If the system component is not equipped with a strainer, an in-line strainer should be installed upstream of the component.

The screening material of a strainer may be constructed of a noncorrosive metal screen, a fiber pad, or a porous ceramic material. The screening ability of a strainer is dependent upon the following factors:

1. Density of the material
2. Depth of the filtering material
3. Mesh size
4. Amount of exposed area
5. Pressure drop across the strainer

Various types and designs of strainers are available to suit any application. In small equipment, the strainer, such as the one illustrated in Figure 15–15, is sealed and cannot be opened for cleaning. When it gets clogged it must be replaced. A larger strainer, such as the one illustrated in Figure 15–16, can be opened for cleaning purposes.

FILTER-DRIERS

A **filter-drier** is installed on the high side of a system at the outlet of the condenser. The purpose of a filter-drier is to filter any particles that could restrict the capillary tube and trap moisture. Some filter-driers, such as the one illustrated in Figure 15–17, are equipped with service ports so that access to the high side of the system may be

Fig. 15–16A and B   Strainers for large equipment. *Courtesy Henry Valve Co.*

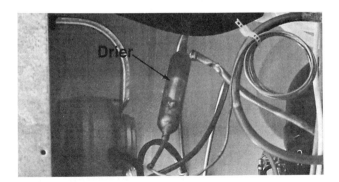

Fig. 15–17   Replacement drier. *Courtesy Maytag Customer Service.*

gained. A filter-drier of this type is commonly found on smaller refrigeration systems and a new one should be installed whenever the sealed system is opened for service. When installing a new

filter-drier, make sure that the direction of flow is proper, as indicated by an arrow inscribed on the filter-drier's body. Do not remove the sealed end caps until the filter-drier is ready for installation. Always cut the old filter-drier out of the system. If the old filter-drier is unsoldered rather than cut out, the applied heat will release trapped moisture back into the system. After the required system repairs have been made, use the following procedures to replace the filter-drier on a fractional horsepower, capillary tube system:

1. Evacuate all refrigerant from the system.

2. If necessary, carefully bend the capillary tube to move the filter-drier in order to gain access and working room.

3. With a fine emery or sand cloth, remove all paint and scale within an area of about three inches from the original joint of the capillary tube.

4. Cut the capillary tube about one inch from the original joint. Use a three-cornered file to score a cut around the tubing, then break it by hand.

5. Cut the filter-drier from the condenser coil and thoroughly clean the end of the tubing.

6. Determine the size of filter-drier required for the application and cut the tubing. Some replacement filter-driers, such as the one illustrated in Figure 15–18, have graduated

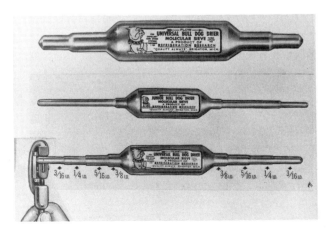

Fig. 15–18   Graduated drier. *Courtesy Refrigeration Research Inc.*

tubing sizes which allow the filter-drier to fit a variety of condenser tubing sizes.

7. Immediately solder the new filter-drier in place. In most cases the condenser tubing will be steel, so silver solder will have to be used. On copper-to-copper applications, a sil-phos brazing alloy can be used.

8. Leak test all joints.

9. Evacuate and charge the system as required.

Suction Line Filters

Suction line filters are installed in the suction line to protect the compressor from foreign particles that could enter the compressor and cause damage. Figure 15–19 illustrates a typical suction line filter.

Suction line filters are installed in new systems as well as existing systems that have experienced extensive service and repair, especially in systems where a hermetic compressor motor burnout has occurred. Suction line filters have the ability to trap and remove solidified particles as well as acid and moisture from the system. The two basic types of suction line filters in common use are a sealed filter and a replaceable core filter. The following factors should be considered when selecting a suction line filter for a particular application:

1. Type of refrigerant used in the system

2. Suction line size

3. Capacity of the system

4. Allowable pressure drop

5. Temperature application

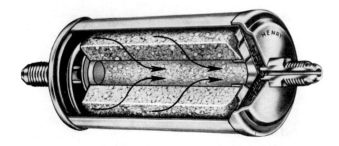

Fig. 15–19   Suction line filter.   *Courtesy Henry Valve Co.*

Always refer to the manufacturer's specifications to ensure that the proper filter is selected for an application.

Many suction line filters are equipped with a pressure port installed at the filter inlet. By taking a pressure reading through this port at the inlet side of the filter and another reading downstream of the filter outlet in the suction line or at the compressor suction service valve port, it can be determined whether the filter is restricted and needs to be replaced. A pressure drop of more than 2 to 3 PSIG across the suction line filter indicates restriction and the need for replacement.

Unit Fifteen Summary

- Liquid receivers are used to store excess refrigerant in a system that uses a thermostatic expansion valve as a metering device and are equipped with a pressure relief device for safety.

- Sight glasses are used to determine the condition of the refrigerant in the liquid line of a refrigeration system. Some sight glasses are equipped with moisture indicators.

- Oil separators allow for the return of oil to a compressor and they should be installed as close to the compressor as possible. Oil separators are often insulated to prevent vapor condensation.

- Check valves allow for one-way flow of liquids. Solenoid valves are electromechanical devices that open or close in response to a control device and allow the valve to control the flow of fluid. Control valves, sometimes referred to as redirection valves, control fluid flow in various directions.

- Suction line filter-driers trap particles, acid, and moisture. Liquid line filters trap particles and moisture.

Unit Fifteen Key Terms

Liquid Receiver

Sight Glass

Vibration Eliminator

Oil Separator

Muffler

Check Valve

Solenoid Valve

Two-Way Valve

Three-Way Valve

Diverting Valve

Four-Way Valve

Strainer

Filter-Drier

Suction Line Filter

UNIT FIFTEEN REVIEW

1. Match the accessory with its function: strainer, liquid receiver, sight glass, suction line filter, vibration eliminator, oil separator, muffler, check valve, solenoid valve, filter-drier.

    a. Installed at outlet of the condenser to filter particles and trap moisture.

    b. Allows refrigerant flow in one direction.

    c. Installed in the discharge line to reduce noise from the compressor pistons.

    d. Installed to ensure that the compressor maintains a proper oil level.

    e. Installed to trap foreign matter.

    f. Installed to control the flow of refrigerant in a circuit.

    g. Used to determine the condition of the liquid refrigerant in the liquid line.

    h. Installed to protect the compressor by removing particles, acid, and moisture.

    i. Stores liquid refrigerant until it is needed by the system.

    j. Installed in the discharge line to prevent the transmission of noise and vibration from the compressor.

2. Why is the refrigerant charge in a liquid receiver only 85 percent full?

3. Three possible causes of bubbles in the sight glass are a shortage of refrigerant, low discharge pressure, and an undersized liquid line. Give another cause.

4. Why would the soft copper tubing of a suction line be shaped into a loop before being connected to the compressor?

5. What does a continuously hot oil return line indicate?

6. What does a continuously cold oil return line indicate?

7. What is normally the position of the two-way solenoid valve when no electrical current flows through the solenoid coil?

8. Because solenoid valves have different capacities, use different types of refrigerant, operate under varying pressures, and have particular electrical characteristics, what should be done to determine the proper mounting position?

9. Diagnose each of four problems (valve will not open, valve will not close, refrigerant bleeds through closed valve, coil has burned out) from the list of symptoms.

    a. Improper wiring, voltage, valve assembly, or valve application.

    b. Improper wiring or voltage; excessively high ambient temperature, or high fluid temperature.

    c. Operating control circuit components not opening or excessively high pressure in the valve outlet.

    d. Worn or damaged valve, foreign matter in the valve seat.

10. What is a service port?

11. What is the recommended storage capacity of a liquid receiver?

12. In the application of liquid receivers, what is the purpose of a fusible plug?

13. What is the best location for the installation of a sight glass?

14. What is the recommended installation procedure for a vibration eliminator?

15. In the application of solenoid valves, what purpose does a manual lift stem serve?

16. What factors affect the screening ability of a strainer?

17. What factors must be considered when selecting a suction line filter?

18. What is the allowable pressure drop across a suction line filter?

19. Write an explanation of how the oil is separated from the vapor in an oil separator.

20. Since the procedure to replace a filter-drier is extensive, write an explanation that only covers the part having to do with the soldering operation.

UNIT SIXTEEN

Sealed Systems

OBJECTIVES

After completing this unit, the student will be able to:

1. Describe the characteristics of a hermetically sealed system.

2. Explain the advantages and disadvantages of a hermetic system.

3. Identify the types of compressors, condensers, evaporators, and metering devices used in a hermetically sealed system.

The intent of this unit is to introduce you to the specifics of hermetically sealed systems such as those commonly found in domestic refrigeration equipment and light commercial equipment in restaurants.

In this unit we'll be identifying some of the specific types of components found in these systems and discussing some of the basics in service procedures.

A **sealed unit** or **hermetic system** is a motor and compressor enclosed in a steel shell or housing that is referred to as a *dome*. The motor drives the compressor by means of direct drive. In comparison to conventional belt-driven units, the sealed unit offers the following advantages:

1. It is smaller and more compact.

2. Its operation is quieter.

3. The direct-drive motor is more efficient.

4. There is less leakage because the dome is sealed and the crankshaft seal has been eliminated.

5. Less maintenance is required.

The sealed unit has the following disadvantages:

1. In the event of internal failure of the compressor or motor, the entire unit must be replaced.

2. It is more difficult to service; some units have no service valves or service points.

3. It is difficult to check the oil level because most units are not equipped with an oil sight glass.

4. A compressor motor burnout could severely contaminate the system.

DOMESTIC SEALED UNIT SYSTEMS

Domestic refrigeration is generally considered to be comprised of household refrigerators and household freezers. Although limited in scope, the servicing of domestic refrigeration systems represents a significant portion of the refrigeration industry. This is due to the large number of units in operation. Domestic systems are limited in size with ratings from 1/20 to 1/3 horsepower.

In 1926, The General Electric Monitor Top was the first hermetically sealed system to appear on the market. From 1926 to 1934, several other manufacturers introduced sealed unit domestic systems. In 1934, Frigidaire introduced the Meter Miser compressor in its domestic systems, and by 1940, sealed unit systems began to take over the market. Today all domestic refrigerators and freezers are of the sealed unit or hermetic design.

PARTS OF A HERMETICALLY SEALED SYSTEM

Compressor

The **compressor** of a sealed system is enclosed within the hermetically sealed shell along with the direct-drive motor. While most of the compressors used in sealed systems are reciprocating compressors, rotary compressors are also popular.

Fully Hermetic Compressor

The **fully hermetic compressor**, illustrated in Figure 16–1A with a cutaway view shown in Figure 16–1B, is characterized by a completely

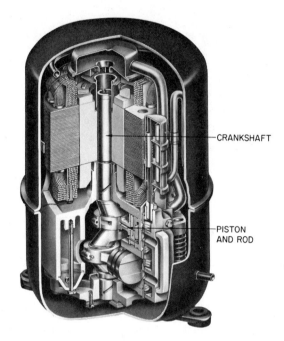

CRANKSHAFT

PISTON AND ROD

Fig. 16–1B   Cutaway view of a fully hermetic compressor. *Courtesy Tecumseh Products Company.*

sealed shell or dome. It has no service valves or service ports through which service operations can be performed. The only means of access is through a process tube usually installed on the low-pressure side of the compressor dome.

Condensers

The two types of condensers used in hermetically sealed systems are **natural convection** or **static condensers**, and **forced convection** or **forced draft condensers**. Natural convection condensers rely upon the natural rising of warm air to remove condenser heat. Force convection condensers utilize a fan to move air across the condenser and dissipate condenser heat. The three designs of condensers used in sealed unit systems are fin and tube condensers, wire and tube condensers, and plate condensers.

Fin and Tube Condenser

The **fin and tube condenser**, illustrated in Figures 16–2A and B, is a forced convection type of condenser. The condenser and fan are usually encased within a shroud to increase and control

Fig. 16–1A   Fully hermetic compressor. *Courtesy Bristol Compressor, Inc.*

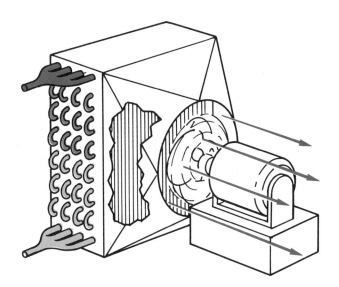

Fig. 16–2A   Fin and tube condenser.

Wire and Tube Condenser

The **wire and tube condenser**, illustrated in Figure 16–3, is usually a natural convection type of condenser. It consists of tubing runs with wire fins soldered or welded across the entire surface of the condenser. The wire and tube condenser is relatively easy to maintain. Ample room, however, must be provided to allow for free air circulation across the condenser surface.

Plate Condenser

The **plate condenser** is also a natural convection type of condenser. Although the plate condenser is easy to manufacture and maintain, it is bulky and takes up a considerable amount of space in order to provide adequate cooling. The plate condenser was popular in the early years of domestic refrigeration and has since been replaced by the fin and tube and the wire and tube condensers.

Metering Device

The capillary tube is the primary **metering device** used in domestic sealed systems to control the flow of refrigerant to the evaporator. Earlier

the airflow across the coil. Ductwork is used on some units to direct the in and out airflow across the coil. This will improve airflow and reduce noise levels. Units equipped with this type of airflow design are installed easily in built-in cabinets or wall recesses where natural convection condensers cannot be used. The ducts and condenser coil should be cleaned regularly to prevent air restrictions.

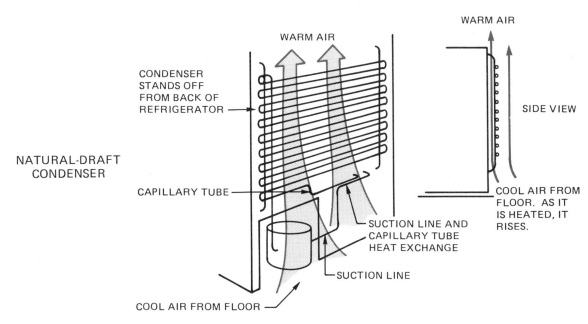

Fig. 16–2B   Finned static natural convection condenser.

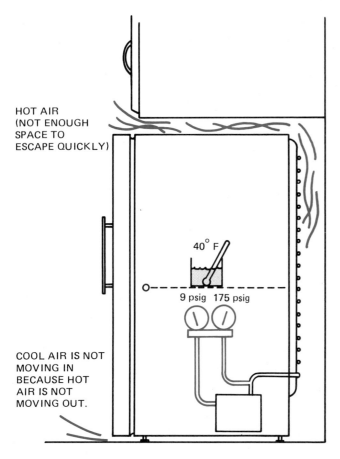

Fig. 16–3   Wire and tube condenser.

sealed systems used the automatic expansion valve. However, the many advantages of the capillary tube make it the ideal metering device for domestic applications today. The capillary tube is usually installed at the outlet of the condenser. Because of the small diameter of the capillary tube, a filter-drier is generally installed in front of the capillary tube at the condenser outlet to prevent restrictions and trap moisture. The length of the tube follows the system suction line through the cabinet to the evaporator. In most applications the capillary tube is soldered to the suction line, creating a heat exchange between the two. This serves to subcool the liquid refrigerant flowing through the capillary tube, thus increasing system efficiency.

Evaporators

Evaporators can be classified as using forced convection or natural convection. Modern domes-

tic evaporator coils are usually constructed of aluminum. Forced convection coils consist of aluminum tubing with aluminum fins. Natural convection coils consist of a bare aluminum tube or an aluminum tube positioned in close thermal contact with an aluminum shell. Another coil design consists of an aluminum shell within which coil sections are stamped or formed into tubing sections. Most domestic evaporators have an accumulator or header installed at the outlet of the coil.

The accumulator acts as a reservoir to prevent liquid refrigerant from entering the suction line and compressor during the running cycle and off cycle. During the normal running cycle of a system equipped with a capillary tube metering device, approximately two-thirds of the system's refrigerant charge is in the evaporator and accumulator.

Numerous types of evaporator designs are used in domestic sealed systems. The design used depends upon the application of the coil and the space allotted.

The simplest and oldest design of evaporator system is the air spill-over system, which is illustrated in Figure 16–4. The evaporator section is located at the top center of the cabinet and acts as a freezer compartment. The refrigerator part of the cabinet is cooled by the spill-over of cold air. The cold air will drop and the warm air will rise, setting up a natural movement of air in the refrigerator cabinet.

The refrigerant spill-over system, which is illustrated in Figure 16–5, utilizes an indepen-

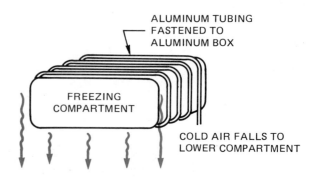

Fig. 16–4   Air spill-over system. *(Continued)*

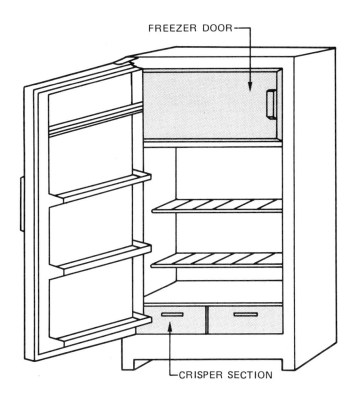

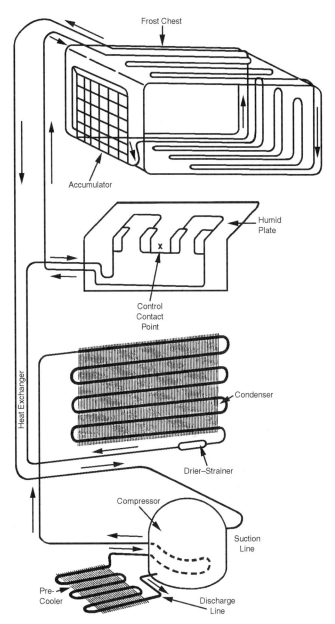

Fig. 16–5 **Refrigerant spill-over into humid plate located in food compartment.**

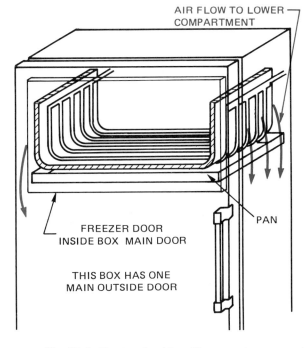

Fig. 16–4 *(Continued)* **Air spill-over system.**

dent freezer compartment as an evaporator. A cold plate that is piped in series with the freezer evaporator is located in the refrigerator compartment. Refrigerant leaving the freezer evapo-

rator will spill-over and feed the cold plate before entering the suction line. The cold plate will cool the refrigerator section by natural convection, usually through a bare tube coil located at the top center of the compartment, or by forced convection, utilizing a fin and duct system to deliver the cold air to the refrigerator compartment.

Another type of evaporator system design, which is illustrated in Figure 16–6, utilizes a

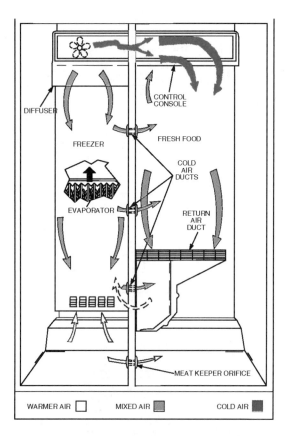

Fig. 16–6 Evaporator system design using a forced convection evaporator coil located in freezer section.

forced convection evaporator coil located in the freezer section. A fan and duct system is used to deliver air to the freezer and refrigerator compartment. Thermostatic dampers are installed in the ductwork to control compartment temperatures.

Unit Sixteen Summary

• A hermetically sealed refrigeration system has some distinct advantages over conventional open drive systems, including less leakage because of the elimination of a crankshaft seal, a more compact design, and a quieter operation.

• Some of the disadvantages of a hermetically sealed system are: the inability to check the oil level; no service valves on some units; and, in the event of a failure of a compressor valve or motor, the entire compressor must be replaced.

• Natural convection condensers and evaporators depend on the natural movement of air across their heat transfer surface to absorb or dissipate heat. Forced convection condensers and evaporators use a fan to move a larger volume of air across their heat transfer surfaces.

• The capillary tube is the most common metering device used in hermetic systems. An accumulator is designed to intercept any liquid refrigerant that may enter the suction line and prevent slugging of the compressor.

Unit Sixteen Key Terms

Sealed Unit System

Hermetic System

Compressor

Fully Hermetic Compressor

Natural Convection Condenser

Static Condenser

Forced Convection Condenser

Forced Draft Condenser

Fin and Tube Condenser

Wire and Tube Condenser

Plate Condenser

Metering Device

Evaporator

UNIT SIXTEEN REVIEW

1. What is a hermetically sealed system?
2. In a comparison of the direct-drive motor to the belt-driven motor, note whether each of the following statements is an advantage or a disadvantage for the direct-drive motor.

a. Efficiency of operation

b. Amount of maintenance

c. Ability to check the oil level

d. Ease of servicing

e. Size of the system

f. Ability to repair the compressor or motor

3. Which system, direct-drive or belt-driven, is more popular in domestic refrigerators and freezers?

4. How is a forced draft condenser different from a natural convection condenser?

5. Name the three designs of the hermetically sealed condensers.

6. Match each of the three hermetically sealed condensers with its description.

a. A natural convection type of condenser, it consists of tubing with soldered or welded wire fins.

b. A natural convection type of condenser, it is bulky and takes up a considerable amount of space.

c. A forced draft type of condenser, the condenser and fan are usually encased within a shroud to increase and control air flow.

7. What are the disadvantages of a sealed system?

8. Describe a fully hermetic compressor.

9. What are the two types of condensers used in sealed systems?

10. What is the primary metering device used in domestic hermetically sealed systems?

11. What are the classifications of evaporators used in sealed systems?

12. What is the purpose of an accumulator?

13. It is time to order some supplies. One supplier has some prices on a boxcar of soft copper refrigeration tubing.

a. The price for ten 100-foot rolls of ½" is $400.

b. The price under 1,000 feet is $2.50 per pound.

c. The price for 50-foot rolls is based on $.43 per foot.
Which is the best price? Which is the worst price? Show your work. (Hint: Weight of ½" copper tubing is .182 pound per foot.)

14. The supplier has air-conditioning and refrigeration, hard temper, type L copper tubing.

a. The price for 500 feet of ½" is $5.55 per pound.

b. The price for 20-foot lengths under 500 feet is $23.65 per length.

Which is the better deal? Show your work. (Hint: Weight is .198 pound per foot.)

15. Write a paragraph to explain the operation of the spill-over evaporator system.

UNIT SEVENTEEN

Refrigerant Recovery

OBJECTIVES

After completing this unit, the student will be able to:

1. Explain the difference between CFC, HCFC, and HFC refrigerants.

2. Explain the proper procedures for refrigerant recovery.

3. Define refrigerant recovery, recycle, and reclaim.

The rules and regulations surrounding the recovery of refrigerants has had the most significant impact of any development in the HVAC/R industry. Environmental Protection Agency (EPA) regulations have dictated that chlorofluorocarbon (CFC) refrigerants that we have used for many years cannot be vented to the atmosphere and must be phased out of production.

Technician certification has also had an effect on those working in the HVAC/R industry. Certification testing is now required of all those who will be involved in the handling of refrigerants.

In this unit, we'll discuss EPA guidelines, proper refrigerant handling procedures, and the requirements for technician certification.

THE EFFECTS OF OZONE DEPLETION

According to certain scientific studies, the effects of ozone depletion cannot be emphasized enough. For each 1 percent of ozone depletion, there can be an increase of 1.5 percent to 2 percent exposure to ultraviolet (UV) radiation. The effects of ultraviolet exposure on humans and the environment are several and could be more as research and study continue.

If the ozone layer is depleted, three types of skin cancer would likely increase—basal cancer, squamous cell cancer, and malignant melanoma. Increased exposure to ultraviolet radiation would increase the number of cataract cases. Research also shows that UV radiation affects the human immune system, weakening the ability to fend off certain diseases.

Other studies show that two-thirds of crops exposed to increased levels of UV proved to be sensitive. Studies of soybeans indicate a 25 percent depletion of ozone decreased yield by over 20 percent.

Other effects of stratospheric ozone depletion would be found in the ozone, a major component of smog, and the weathering and cracking of plastics used in outdoor applications.

CFCs are also considered to be "greenhouse" gases, thus contributing to global warming.

CHLOROFLUOROCARBONS (CFCs)

Ozone is a gas comprised of three atoms of oxygen in each molecule. It is slightly blue in color and has a pungent odor. It is chemically designated as O_3.

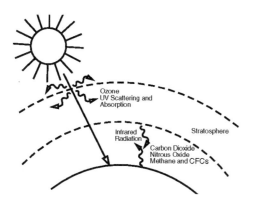

Fig. 17–1   Ozone layer.

The ozone layer, Figure 17–1, is in the stratosphere above the earth at an altitude of 7 to 28 miles. It is formed by ultraviolet light from the sun acting on oxygen molecules. The ozone layer absorbs and scatters the ultraviolet light from the sun preventing harmful amounts of ultraviolet from reaching the earth's surface.

Ozone is also found in the lower atmosphere. It is caused by ultraviolet radiation from the sun acting on air pollutants during hot summer days.

While the ozone layer in the upper stratosphere is protective of the earth's environment, the ozone layer in the lower atmosphere is a pollutant.

Extensive research and study have confirmed that certain chlorine-containing compounds, such as certain refrigerants, can destroy ozone in the stratosphere. Refrigerants that contain chlorine, but no hydrogen, are very stable and do not break up when released into the atmosphere.

Gradually these chemicals work their way into the stratosphere where the chlorine reacts with the ozone, causing it to be changed back into oxygen. Furthermore, the chlorine is not used up during this process; each molecule continues to cause more ozone to change to oxygen.

When the depletion of ozone occurs, more ultraviolet radiation can penetrate to the surface of the earth where it can create health hazards for human beings, damage to crops and aquatic organisms, increases in ground-level ozone, and increased global warming.

CFC Phaseout

These concerns have prompted the United States and other countries to initiate legislation controlling the use of halogenated CFC refrigerants and their phasing out by the turn of and into the next century. The affected compounds are as follows:

CFC-11	R-11	Trichloromonofluoromethane
CFC-12	R-12	Dichlorodifluoromethane
CFC-113	R-113	Trichlorotrifluoroethane
CFC-114	R-114	Dichlorotetrafluoroethane
CFC-115	R-115	Chloropentachloroethane

A reduced production and consumption level schedule is as follows:

By January 1, 1993, a reduction of 20 percent of the total production

By January 1, 1995, a reduction of 50 percent of the total production

By January 1, 1997, a reduction of 85 percent of the total production

By January 1, 2000, an elimination of 100 percent of the total production

Studies have indicated that refrigerants designated as HCFCs, HFC-22, R-22, HFC-502, and R-502 cause much less ozone damage. These refrigerants are not fully halogenated, containing chlorine, fluorine, and hydrogen atoms connected to a carbon atom. These molecules break down in the lower atmosphere before getting into the stratosphere where they can cause ozone damage.

However, a worldwide declaration of intent to phase out HCFCs no later than January 1, 2020, was considered to be essential to the CFC phaseout program. Therefore, an HCFC phaseout schedule was also adopted. It calls for a production freeze and use limitation by the year 2015; a prohibition in new air-conditioning and refrigeration after 2020; and a total phaseout in 2030.

Section 608 of the United States Clean Air Act requires the United States EPA to develop regulations that will limit emissions of ozone-depleting compounds during their use and disposal to the lowest achievable level and maximize recycling.

This act also prohibits the releasing of refrigerants into the atmosphere during the maintenance, service, and disposal of refrigeration and air-conditioning equipment beginning July 1, 1992.

REPLACEMENT REFRIGERANTS

Due to the phaseout schedule of CFCs and HCFCs, new refrigerants to replace these are in the process of being developed and tested. As with all refrigerants, the new refrigerants must have certain characteristics to make them suitable for use. These include chemical stability, low cost, ease of leak detection, toxicity, thermal quality, environmental effects, compatibility with system components, materials, oil, and efficiency.

Applications for Replacement Refrigerants

Equipment	Present	Future
Domestic appliances	R-12	R-134A, R-22
Auto air-conditioning	R-12	R-134A
Centrifugal chillers	R-11 (80%)	R-123
	R-12 (10%)	R-134A
	R-114 (marine)	R-124
Supermarket equipment	R-502	R-125

As an HCFC, R-22 has a small fraction of CFCs that have the ability to affect the stratospheric ozone but will continue to be used in its present applications for many years to come.

The new refrigerants are not "drop-in" refrigerants, in the effect equipment modification will have to take place. R-134A, as an example, having properties similar to those of R-12, does not mix well with petroleum-based lubricants.

The laws governing the release of CFCs and HCFCs into the atmosphere are very strict and can result in heavy penalties if violated. This has led to the development of specific procedures that must be followed to recover, recycle, and reclaim refrigerants.

RECOVERY

The definition of **recovery** is to remove refrigerant in any condition from a system and store it in an external cylinder without necessarily testing or reprocessing it in any way.

RECYCLING

Recycling is defined as a process followed to clean refrigerant for reuse by oil separation and single or multiple passes through devices such as replaceable core filter-driers to reduce moisture, acid, and particle matter. This can be done at the field job site or service shop.

RECLAIM

The definition of **reclaim** is to process refrigerant to new product specification by means that may require distillation. This process requires chemical analysis of the refrigerant to determine that proper specifications are met. Reclaim implies the use of processes or procedures available only at a reprocessing or manufacturing facility.

At the current time "new product specifications" means Air Conditioning Refrigeration Institute (ARI) Standard 700-88, which is the standard for new and reclaimed refrigerant.

ARI 700 Specifications

Physical Properties	ARI 700
Noncondensible	1.5% by volume liquid
Moisture	10 PPM
Boiling point range	0.5°F
Nonvolatile residue	100 PPM
Chloride content	must test for none
Other impurities	less than 200 PPM
Acidity	less than 0.01%

Recovery and Incineration

There are instances when a refrigerant is so badly contaminated, as in the case of a hermetic burnout or when a refrigerant is mixed with other refrigerants to form an azeotropic refrigerant, that effective reclaiming is impossible. The only option is to incinerate the refrigerant at a temperature of approximately 1,200°F.

This process is a very expensive option. CFCs are difficult to destroy because of their stability and the release of fluorine during the incineration process, which must be contained. There are several disadvantages to the incineration process:

1. The refrigerant destroyed must be replaced.

2. The refrigerant is lost.

3. The process is expensive.

Recovery and Reuse

In many cases the refrigerant in a system is not contaminated and can be reused after system repairs have been made. The refrigerant can be removed from the system by means of a recovery or recycling unit, stored in a recovery cylinder, and recharged back into the system.

It is important that the service technician properly handle the transfer process by following specific procedures designed to prevent contamination of the refrigerant and personal injury. The equipment used must be designed for the specific type of refrigerant being handled and the storage cylinder must be clean and designed to contain the refrigerant.

The disadvantage of this process is that because there is no laboratory analysis of the refrigerant, the quality of the refrigerant may be in doubt and this may cause warranty problems. The advantage is that the refrigerant is not lost. In the recovery and reuse process the contractor assumes liability and must ensure compliance with regulations.

On-Site Recovery and Recycle

If there is any doubt as to the quality of the refrigerant in the system, the refrigerant may be processed to remove various contaminants. A variety of equipment is available to perform this task as shown in Figures 17–2A, B, and C. The advantage of on-site recycling is that the refrigerant is not lost and its quality is improved. The disadvantage is that the process takes time and is best suited for systems with small amounts of refrigerant capacities.

Recovery and Reclaim Off-Site

If the contamination of the refrigerant is severe or the system requires the refrigerant to meet a certain specification and certification of its purity, off-site reclamation should be used. Off-site facilities will provide certification that the refrigerant meets ARI standards or Federal Specifications BB-F-1421A. This certification will also protect the equipment warranty.

Fig. 17–2A Refrigerant recovery unit. *Courtesy Robinair Division, SPX Corporation.*

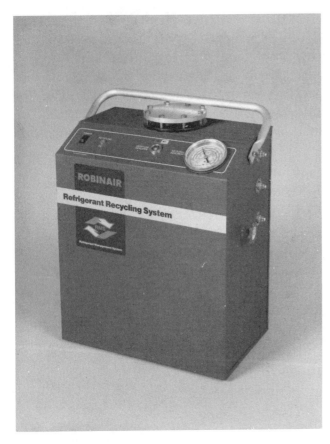

Fig. 17–2B This unit is used to recycle refrigerant.
Courtesy Robinair Division, SPX Corporation.

Although the process can be costly, the advantages are as follows:

1. The off-site facility assumes liability.
2. Refrigerant quality is guaranteed.
3. There is compliance with regulations.
4. Costly refrigerant replacement is avoided.
5. Refrigerant is available for reuse.

REFILLABLE REFRIGERANT CYLINDERS

Refillable refrigerant cylinders, shown in Figure 17–3, are now available for transportation and use in the air-conditioning and refrigeration industry. These cylinders are used for the same refrigerants as are the disposable cylinders commonly used. All refillable cylinders are regulated by the United States Department of Transportation (DOT) in their design, testing, and fabrication. If a cylinder is not DOT approved, it should not be used. Design concerns of cylinders are based on ambient temperature exposure, enclosure of the cylinder, proximity to a heat source, and any other factor that may cause excessive over-pressurization due to loss of vapor expansion space within the cylinder.

The two types of current cylinder specifications are the 4BA and 4BW specifications. The

Fig. 17–2C This modular unit reduces a heavy unit to two manageable pieces. *Courtesy Robinair Division, SPX Corporation.*

Fig. 17–3 Approved Department of Transportation (DOT) Cylinder.

4BA cylinders are generally sized for capacities of 50 pounds or less. The common sizes include 15 lb, 30 lb, 37 lb, and 50 lb. The 15 and 30 lb cylinders have a design pressure of 340 PSI. The 37 and 50 lb cylinders have a design pressure of 302 PSI. Another 50 lb cylinder is available with a design pressure of 400 PSI.

The 4BW cylinders are available in 100 lb, 250 lb, and 1,000 lb capacities. The 100 lb cylinder has a design pressure of 303 PSI while the 250 and 500 lb cylinders have a design pressure of 400 PSI. The 1,000 lb cylinder has a design pressure of 260 PSI. These storage cylinders are painted gray; the shoulder area of the tank and 12 inches down the side is painted yellow.

The refillable cylinders with a 50-pound capacity have a "Y" configuration service valve that provides liquid or vapor flow from a single port, as in Figure 17–4. The valve handles are color coded red for liquid and blue for vapor. They are also labeled. These valves also include a spring-actuated safety relief valve for over-pressure protection.

The larger 100 lb 4BW cylinder is equipped with a combination service valve that has a dual handle and exit port. Integrated in the valve is a spring-type relief valve set to discharge at 515 PSI. The 250 and 500 lb cylinders have individual liquid and vapor valves. The cylinders are also

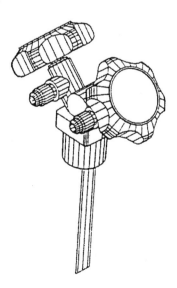

Fig. 17–4   Service port.

equipped with a ¼" NPT rupture disk designed to rupture at 600 PSI. The 1,000 lb cylinder is equipped with a spring-type relief valve and the same type of dual handle exit port valve that is used on the 100 lb cylinder except without the integral relief valve.

Each cylinder has a warning decal to caution against both physical contact with the refrigerant and storing a refrigerant that has a vapor pressure in excess of 318 PSI at 130°F. These conditions are based on Refrigerant 502, which has the greatest vapor pressure.

All cylinders should be tested every five years to guarantee that the interior of the cylinder is free of moisture and the exterior of the cylinder is free of corrosion and physical damage. The valves and safety relief valves should be tested for damage, deterioration, and proper operation.

Recovery Cylinder Safety

1. Keep outlet valve on valve outlet and valve hood securely screwed onto the neck of the returnable cylinder at all times except when performing service procedures.

2. Keep cylinders secured in an upright position.

3. Never drop a cylinder or strike it with a hammer or other tool.

4. Never apply heat or flame to a cylinder

5. Do not lift a cylinder by the valve cover or valve.

6. Never remove a valve from the cylinder or attempt to repair it.

7. Never refill disposable cylinders.

8. Do not alter any permanent cylinder markings because it is illegal.

9. Be careful not to cut or dent the cylinder or valve.

10. Protect cylinders from moisture or other corrosive forms.

11. Always open and close cylinder valves slowly.

12. Do not use a cylinder that is in a rusted or otherwise deteriorated condition.

13. Use only recovery cylinders designed to recover refrigerants.

14. Do not use a recovery cylinder if the date on the cylinder is more than five years past the test date stamped on the cylinder.

15. Wear liquid seal-tite type safety glasses when handling refrigerants and cylinders.

16. Never fill a cylinder that is not labeled for that material.

17. Check to ensure all valves are closed tightly and leak tested after filling a recovery cylinder.

18. Plainly mark the type of refrigerant contained within the cylinder.

19. Label all refrigerant cylinders.

20. Exercise caution when moving and transporting cylinders.

REFRIGERANT RECOVERY PROCEDURES

There are many different ways to recover refrigerant from a system. Most procedures will effectively remove the liquid refrigerant but a certain amount of vapor will still remain in the system. For this reason, manufactured recovery equipment or recovery recycle machines are designed specifically to remove all of the refrigerant from the system.

There are a wide variety of systems available to suit any given application. The four basic designs of equipment in common use today are:

1. Removal of refrigerant in liquid form only.

2. Removal of refrigerant in vapor form only.

3. Removal of refrigerant in vapor and liquid form without oil separation.

4. Removal of refrigerant in liquid and vapor form with oil separation.

There are numerous types of recovery units on the market. Although some recovery units require evacuation before each use, most require evacuation when a different refrigerant is used. Many also require the filter cores in the recovery unit be changed each time the refrigerant is changed. Storage cylinders must be deeply evacuated each time a different refrigerant is recovered. The purpose of this is to prevent two different refrigerants from mixing and forming an azeotropic mixture that cannot be separated during a recycling process and must be burned. For this reason it is advisable to dedicate cylinders to a refrigerant, making sure they are properly tagged.

Setup procedures will vary from one recovery unit to another. It is important to follow the manufacturer's directions carefully. CAUTION: DO NOT USE A RECOVERY UNIT OR ATTEMPT TO RECOVER REFRIGERANT UNTIL YOU RECEIVE THE PROPER TRAINING ON THE EQUIPMENT AND RECOVERY PROCEDURES.

Basic Refrigerant Recovery

Most of the refrigerant can be removed from the system by simply connecting a service hose to a liquid service port on the system; the center hose of the service manifold connects to the service cylinder.

Make sure the service cylinder has been evacuated to at least 1,000 microns to prevent contamination from air, water, or other refrigerants. This is accomplished by merely installing a vacuum pump and micron gauge, opening the service cylinder, and starting the vacuum pump and vacuum according to required specifications. When the process is complete, close the tank and disconnect the vacuum pump.

To expedite the transfer process and gain efficiency, the service cylinder should be placed in a container of iced water. This will lower the pressure of the refrigerant as it enters the cylinder and allow more to be quickly removed from the system.

Referring to Figure 17–5A, which illustrates a basic recovery procedure, proceed as follows:

1. Install the pressure gauge service hose to a LIQUID service port.

2. Install the center service hose of the gauge manifold to the service cylinder.

3. Keep the compound gauge closed.

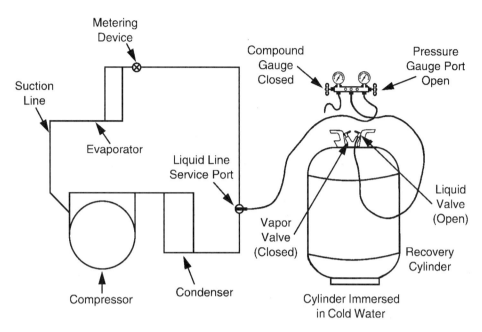

Fig. 17–5A   Basic recovery procedure.

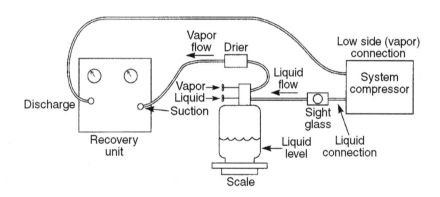

Fig. 17–5B   Liquid refrigerant recovery method.

4. Purge the gauge hoses of air. (NOTE: A slight amount of refrigerant will escape during this procedure, however U.S. federal regulations state that minimal releases associated with good-faith attempts to recapture, recycle, or safely dispose of any substance shall not be subject to the prohibition.)

5. Open the service cylinder.

6. Open the pressure gauge on the gauge manifold.

Two methods of refrigerant recovery, liquid and vapor recovery, are shown in Figures 17–5B and 17–5C.

Liquid refrigerant will now enter the service cylinder from the system. When the pressure in the system and the service cylinder become equal, no more refrigerant will flow. Allow the system to sit for a few more minutes before closing off the cylinder and manifold gauge valves and disconnecting. The remainder of the refrigerant in the system will have to be removed by means of a recovery unit.

Recovery Equipment

When determining what type of recovery-recycle unit is best suited, consider the type of

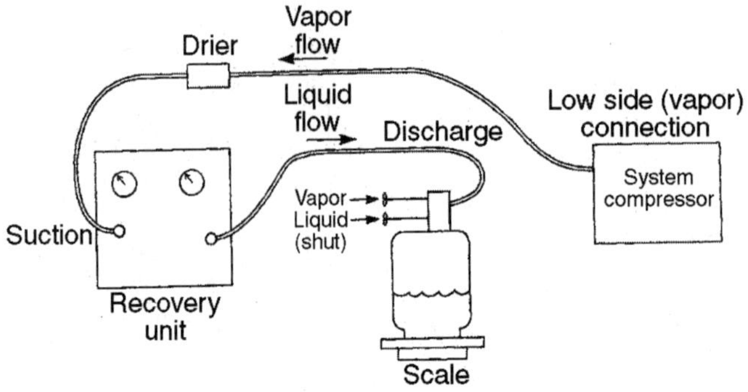

Fig. 17–5C   Vapor refrigerant recovery method.

equipment being serviced, the amount of refrigerant contained within the system, the size and weight of the recovery-recycle unit, the amount of time it will take on a job to remove the refrigerant from the system, and numerous other factors. Units are available to remove refrigerant from a system in the vapor form, the liquid form, or both. Single-pass units process the refrigerant through filter-driers or a distillation process once and then into a storage cylinder. Multiple-pass units recirculate the refrigerant several times through filter-driers before transferring it into a storage cylinder.

Vapor recovery systems are designed to remove the refrigerant from the system as a vapor, recondensing it, and transferring it into a storage cylinder. Because the vapor removal process takes a longer period of time to evacuate a system, it is used on small applications equipment, such as appliances, auto air-conditioning, and small residential and commercial equipment.

Liquid recovery is much faster and may be preferred. Many units are available to recover liquid refrigerant initially, then complete the recovery process by removing remaining vapor.

Many units are equipped with automatic shut-off devices to shut the unit down when the service cylinder becomes 80 percent full or in the event excessive pressure is experienced in the cylinder or unit.

Recovery-Recycling Hints

1. Use only DOT-approved containers.
2. Mark all containers.
3. Change filters according to manufacturer's specifications.
4. Do not mix refrigerants.
5. Use short hoses with large inside diameters to gain efficiency.
6. Keep all hoses, valves, and ports clean and evacuated.
7. Drain oil as required after use.
8. Keep the storage container cooled to reduce heat and pressure buildup.
9. Avoid contact with the refrigerant and oil.
10. Follow manufacturer's instructions.
11. Never modify a recovery-recycle unit.

REFRIGERANT RECLAIM PROCEDURES

There is an increasing number of recycling facilities becoming available to dispose of waste refrigerants or reclaim the refrigerants to specifications for future reuse. Many industry wholesale houses are serving as a medium to the technician to facilitate this process, providing storage cylinders and arranging transportation of the refrigerant to the recycling facility.

The refrigerant is collected and stored in an approved DOT cylinder and is labeled with the type of refrigerant contained within. Gross weight is stenciled on the cylinder, which is the total weight of the cylinder including content and cylinder weight. A used refrigerant label is applied to the shoulder of the cylinder and appropriate DOT tags are applied to the cylinder safety hood. A bill of lading is made out and the cylinder is prepared for transportation after it is leak tested and properly secured.

Recovery cylinders are painted yellow on the shoulder area and 12 inches down the side. The remainder of the tank is painted in accordance to the color coding used on new refrigerant cylinders, for example, white for R-12 and green for R-22. All cylinders have gold-colored safety hoods.

A recovery cylinder that is damaged or more than five years past the test date stamped on the shoulder is nonserviceable and should be returned to the manufacturer.

TECHNICIAN CERTIFICATION

All HVAC/R technicians must be certified through an EPA-sanctioned testing procedure. There are four categories of certification:

1. Type 1   Small appliances
2. Type 2   High-pressure and very high-pressure appliances
3. Type 3   Low-pressure appliances
4. Type 4   Universal (Type 1, 2, and 3)

If you work with equipment that is manufactured, charged, and hermetically sealed with five pounds or less of refrigerant, you must be certified as a Type 1 technician. This classification covers technicians who work with refrigerators, freezers, room air conditioners, dehumidifiers, under-the-counter ice makers, vending machines, and drinking fountains.

For those technicians who work with high-pressure appliances (equipment that uses a refrigerant with a boiling point between $-58°F$ and $50°F$ at atmospheric pressure), a Type 2 certification is necessary. This would include comfort cooling equipment that uses R-22 as the refrigerant and other equipment that uses R-12, R-114, R-500, and R-502 refrigerants.

A Type 3 certification is required for those technicians who service low-pressure equipment that uses a refrigerant with a boiling point above $50°F$ at atmospheric pressure. This would include low-pressure chillers and other equipment that use R-11, R-123, and R-113 refrigerants.

Table 17–1 shows the chemical name, formula, and boiling points for various replacement refrigerants.

A universal certification that covers all the equipment and refrigerants described in the Type 1, 2, and 3 certifications is available for the technician who wishes to be certified in all areas.

Several associations, schools, and agencies have developed and offer EPA-approved certification testing. The most common procedure for certification is to attend a one-day training seminar, then take the test the same day.

REFRIGERANT	CHEMICAL NAME	FORMULA	BOILING PT., °F
HCFC 123	Dichlorotrifluoroethane	$CHCl_2CF_3$	82.2
HCFC 124	Chlorotetrafluoroethane	$CHClFCF_3$	12.2
HFC 125	Pentafluoroethane	CHF_2CF_3	−55.3
HFC 134a	Tetrafluoroethane	CH_2FCF_3	−15.7

Table 17–1   Reference data for replacement refrigerants.

Vapor Pressures of the Replacement Refrigerants			
Temp (°F)	HCFC 123	HCFC 124	HFC 125	HFC 134a
−100	29.9*	29.2*	24.4*	27.8*
−90	29.8*	28.8*	21.7*	26.9*
−80	29.7*	28.2*	18.1*	25.6*
−70	29.6*	27.4*	13.3*	23.8*
−60	29.5*	26.3*	7.1*	21.5*
−50	29.2*	24.8*	0.3	18.5*
−40	28.9*	22.8*	4.9	14.7*
−30	28.5*	20.2*	10.6	9.8*
−20	27.8*	16.9*	17.4	3.8*
−10	27.0*	12.7*	25.6	1.8
0	26.0*	7.6*	35.1	6.3
10	24.7*	1.4*	46.3	11.6
20	23.0*	3.0	59.2	18.0
30	20.8*	7.5	74.1	25.6
40	18.2*	12.7	91.2	34.5
50	15.0*	18.8	110.6	44.9
60	11.2*	25.9	132.8	56.9
70	6.6*	34.1	157.8	70.7
80	1.1*	43.5	186.0	86.4
90	2.6	54.1	217.5	104.2
100	6.3	66.2	252.7	124.3
110	10.5	79.7	291.6	146.8
120	15.4	94.9	334.3	171.9
130	21.0	111.7	380.3	199.8
140	27.3	130.4	430.2	230.5
150	34.5	151.0	482.1	264.4
160	42.5	173.6		301.5
170	51.5	198.4		342.0
180	61.4	225.6		385.9
190	72.5	255.1		433.6
200	84.7	287.3		485.0
210	98.1	322.1		540.3
220	112.8	359.9		
230	128.9	400.6		
240	146.3	444.5		
250	165.3	491.8		
260	185.8			
270	207.9			
280	231.8			
290	257.5			
300	285.0			

Vapor pressures are PSIG, except (*), which are inches of mercury vacuum.

Table 17–2   Vapor pressures of the replacement refrigerants.

Unit Seventeen Summary

- Due to environmental concerns regarding the depletion of the ozone layer and the resulting effects, legislation regarding the use, production, and phasing out of CFC and HCFC refrigerants has been created. The three procedures established in regard to EPA guidelines are recovery, recycle, and reclaim.

- HCFC refrigerants that are designed as replacement refrigerants are not "drop-in" replacements. New systems and modifications to existing systems are required to use the new chemicals.

- On July 1, 1992, the venting of refrigerants became illegal and those violating the law became subject to fines. Contaminated refrigerants that cannot be recycled must be incinerated at a temperature above 1,200°F.

- Refillable cylinders are specifically designed to meet Department of Transportation (DOT) standards. Recovery methods include removing the refrigerant as a vapor or liquid, with or without oil separation.

- Single-pass recovery systems recover the refrigerant from the system, pass it through filters, and process it into the recovery cylinder. Multi-pass systems recover the refrigerant and cycle it through filters for a length of time before processing it into the recovery cylinder.

- Proper maintenance of recovery systems is critical to ensure efficiency and long service life. Filters must be changed regularly.

- Service technicians must be properly trained and accomplish a certification test approved by the EPA that covers the safe handling of refrigerants and proper use of cylinders.

Unit Seventeen Key Terms

Ozone	Recycle
Recovery	Reclaim

UNIT SEVENTEEN REVIEW

1. What is the ozone layer?

2. What is good ozone?

3. What is a CFC?

4. What is an HCFC?

5. How does a CFC act upon the ozone layer?

6. Why is an HCFC less harmful to the ozone layer?

7. Name three types of CFCs.

8. Which is the most popular HCFC used today?

9. Match CFC, HCFC, and HFC refrigerants with the following information:

   a. Contains chlorine compounds that destroy ozone.

   b. Total phaseout by the year 2000.

   c. The refrigerant used in domestic appliances and auto air-conditioning.

   d. R-11 and R-12 refrigerants.

   e. Total phaseout by the year 2030.

   f. R-134a does not mix well with petroleum-based lubricants.

10. What is the role of the United States Environmental Protection Agency (EPA)?

11. Name three of the several characteristics that the development of any new refrigerant must have.

12. What types of safety equipment should be worn when handling refrigerants?

13. What is the difference between stratospheric ozone and surface ozone?

14. Name three effects the depletion of the ozone layer has on human health.

15. Match recover, recycle, and reclaim with their descriptions.

   a. To remove a refrigerant in any condition from a system and store without testing or reprocessing.

   b. To clean a refrigerant for reuse, removing moisture, acid, and particle matter.

   c. To process a refrigerant by means that may require distillation.

16. Why must a service cylinder be evacuated prior to use?

17. What is the minimum vacuum required to evacuate a service cylinder?

18. What is the maximum amount a refrigerant service cylinder should be filled?

19. Why shouldn't refrigerants be mixed?

20. What is an azeotropic refrigerant?

21. At what temperature does waste refrigerant have to be incinerated?

22. When is destruction of a refrigerant by incineration carried out?

23. What is the advantage of liquid refrigerant recovery as opposed to vapor recovery?

24. What is the determining factor in whether or not a refrigerant is processed on site or recovered and reclaimed off site?

25. Identify the system by its description.

    a. Designed to remove the refrigerant as a vapor.

    b. Recondensing the refrigerant before transferring into a storage cylinder.

26. What color is the refillable refrigerant cylinder?

27. What color is the storage cylinder being returned for recycling?

28. Describe a single-pass recovery system.

29. Describe a multiple-pass recovery system.

30. Why is it beneficial to use short service hoses with a large inside diameter?

31. Why is it important to frequently change the filters on a recovery-recycling unit?

32. Why is it desirable to place the recovery cylinder in cold water?

33. Name five safety procedures to be followed when handling refrigerant cylinders.

34. Match the category of certification (Type 1, 2, 3, or 4) with its description. A technician must be certified to work on the appliances.

    a. High-pressure appliances using a refrigerant with a boiling point between −58°F and 50°F at atmospheric pressure.

    b. Low-pressure appliances using a refrigerant with a boiling point above 50°F.

    c. Appliances manufactured, charged, and hermetically sealed with five pounds or less refrigerant.

    d. Covers all appliances and refrigerants.

35. What is the purpose of two cylinder valves on the recovery cylinder?

36. What safety devices for over-pressure incidents are provided on refrigerant recovery systems?

37. Why should all refrigerant cylinders be properly tagged as to their contents?

38. A technician has an opportunity to bid on the central air-conditioning for 10 homes in a housing development. Two different suppliers have the same compressor, condenser, and evaporator units, but offer different proposals. What is the total cost in each package? Which package is the better deal?

    a. One supplier said the cost would be 85 percent of the retail price of $3,000 for each unit.

    b. One supplier said the cost would be nine units at $2,845 each with one free unit.

39. Write a description of what has to be done to a cylinder holding a refrigerant before transporting.

40. Write a paragraph using the following terms in the sentence: CFC, EPA, recovery, reclaim, and HCFC.

UNIT EIGHTEEN

System Service Procedures

OBJECTIVES

After completing this unit, the student will be able to:

1. Describe the proper procedure for gaining access to a sealed system.

2. Identify the types of access valves used by technicians when servicing sealed systems.

3. Describe the proper procedure for installing gauges and diagnosing sealed system problems.

4. List the necessary steps in evacuating a system, replacing a compressor, adding oil to a sealed system, and repairing aluminum tubing.

As a service technician, your responsibility is to properly diagnose and perform repairs in a safe and effective manner on HVAC/R equipment. That means being able to differentiate between problems that exist within the electrical system and those that exist within the sealed system.

In this unit we'll outline service procedures for both areas and discuss some of the step-by-step procedures to follow when servicing the electrical and sealed systems.

ACCESS TO THE SEALED SYSTEM

On sealed systems that are not equipped with standard service valves, an alternate means of system access must be used. Some units are equipped with hermetic service ports that require special adapter kits. If the unit has no service valves or ports, piercing valves must be installed in order to gain system access. When process tubes are provided for system entry on full hermetic compressors, process tube adapters must be used. A variety of access valves and service adapters are available commercially to fit almost any application.

Piercing Valves

Piercing valves or line taps are used on full hermetic systems where no service valves or ports are available through which to gain system entry. They are installed on the system suction and discharge lines or the process lines located on the compressor dome. There are two predominant designs of piercing valves in common use: the bolted-on valve and the brazed-on valve. The bolted-on valve is available for specific tubing sizes or it can fit various tubing sizes through the use of shims supplied with the valve. Figure 18–1 illustrates a variety of bolted-on piercing valves.

Fig. 18–1 Assorted bolt-on piercing valves. *Courtesy Robinair Division, SPX Corporation.*

Fig. 18–2   Braze-on piercing valve.   *Courtesy J/B Industries.*

The brazed-on valve, which is illustrated in Figure 18–2, is designed for a specific tubing size only.

Installing the Bolted-On Piercing Valve

Always refer to the manufacturer's instructions for the proper installation procedure of a specific valve. The following procedures can serve as a guide for general installations:

1. Before installing the piercing valve make sure the location has ample working room for service lines and valve adapters.

2. Make sure the tubing is straight, round, and clean.

3. Check that the valve piercing needle is recessed in the valve body.

4. Place a small amount of refrigerant oil on the tubing and mount the valve.

5. Tighten the valve clamping screws evenly until the valve is secure on the tubing.

Installing the Brazed-On Piercing Valve

The brazed-on piercing valve requires more care in installation. Precautions must be taken to ensure that adjacent areas are protected from heat. Make sure that there are no soft solder connections near the area of the installation. Safety glasses should be worn during this procedure. Proceed as follows:

1. Make sure that the tubing is straight and round.

2. Clean the tubing and the piercing valve mating surface.

3. Remove the piercing valve needle and gasket.

4. Mount the valve on the tubing and secure it with a C-clamp to prevent it from moving during the brazing operation.

5. Heat the tubing and the valve uniformly until the brazing alloy starts to flow.

6. Allow the valve and tubing to cool before installing the piercing needle and gasket.

Schraeder Valves

Schraeder valves are similar to the type of valve cores used on automobile tires. They are available in clamp-on and brazed-on designs. When using a Schraeder valve, gauge manifold service lines must be equipped with a Schraeder valve depressing pin. Figure 18–3 illustrates a typical Schraeder valve.

Figure 18–4 illustrates a Schraeder valve core removing tool, which is designed to remove the Schraeder valve depressing pin and allow full flow access to the system during servicing procedures.

Process Tube Adapters

The manufacturer installs a process line on the hermetic compressor shell in order to evacuate and charge the system at the factory. As the service technician, this provides you a convenient location for installing a piercing valve or **process tube adapter** in order to service the sealed sys-

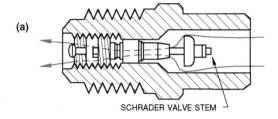

THIS SCHRADER VALVE HAS BEEN DEPRESSED, BUT IT STILL CAUSES A PRESSURE DROP. IT IS A RESTRICTION TO GAS FLOW.

(a)

SCHRADER VALVE STEM

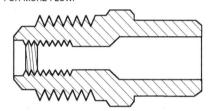

(b) THE SCHRADER VALVE HAS BEEN REMOVED FOR MORE FLOW.

Fig. 18–3 Typical Schraeder valve. *Courtesy Robinair Division, SPX Corporation.*

tem. In most cases the process tube allows access to the low side of the system. In some applications where two process tubes are provided, access to the high side of the system can also be gained.

The advantage of the process tube adapter is that the system can be completely sealed after service operations have been completed. A process tube adapter kit, which is illustrated in Figures 18–5A and B, can be used on various tubing sizes, ranging from ³⁄₁₆″ to ³⁄₈″ OD tubing. In some applications where the existing process tube is too short to use, it can be extended by brazing an extension tube on the existing stub. To install a process tube adapter, follow these procedures:

1. Clean the process line tubing thoroughly to remove all paint, oil, and dirt.

2. Carefully cut the tip of the process line to evacuate refrigerant from the system. Wear safety glasses when performing this task and make sure that the work area is well ventilated.

3. Attach the process tube adapter to the process tube.

Fig. 18–4 Schraeder valve core removal tool, (a) core installed, (b) core removed.

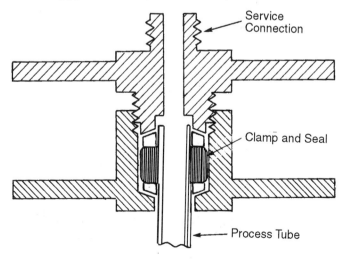

Service Connection

Clamp and Seal

Process Tube

Fig. 18–5A Process tube adapter cutaway. *Courtesy Robinair Division, SPX Corporation.*

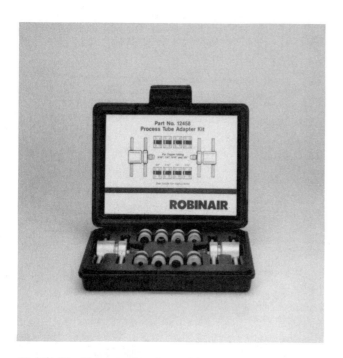

Fig. 18–5B   Process tube adapter kit.   *Courtesy Robinair Division, SPX Corporation.*

4. When service operations are completed, use a pinch-off tool to seal the process tube, as shown in Figure 18–6.

5. Remove the process tube adapter and braze the opening closed.

6. Remove the process tube adapter and restore the stem to normal operation.

Hermetic Service Valve Adapters

On many partially sealed systems the only means of system access is through a hermetic service port, which requires a special **hermetic**

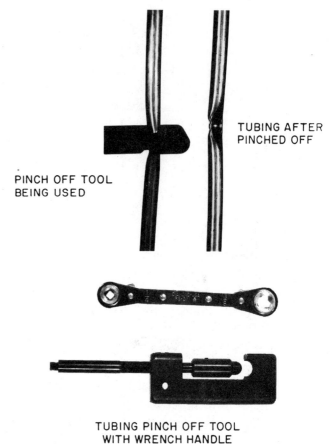

Fig. 18–6   Pinch-off tool.   *Photo by Bill Johnson.*

service valve adapter, such as the ones illustrated in Figures 18–7A and B. The hermetic valve adapter kit contains a variety of adapter connections designed to fit almost any application. Refer to the manufacturer's instructions for the proper installation procedures of the specific adapter being used.

Fig. 18–7A   Valve adapters that can be used on various makes of semihermetic or hermetic refrigeration units. *(continued)*   *1, 2 & 3: Photos by Bill Johnson. 4: Courtesy J/B Industries.*

(2)

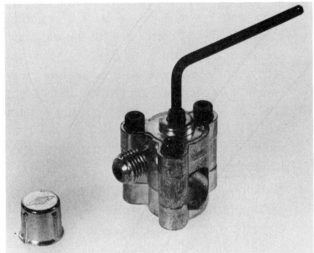

(3)

(4)

Fig. 18–7A *(continued)* **Valve adapters that can be used on various makes of semihermetic or hermetic refrigeration units.** *1, 2 & 3: Photos by Bill Johnson. 4: Courtesy J/B Industries.*

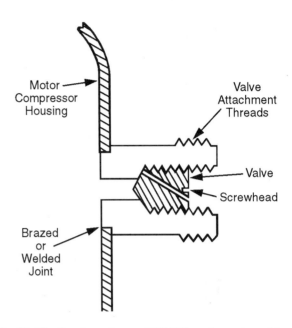

Fig. 18–7B Service valves on HVAC/R equipment must be soldered in place to prevent the escape of refrigerant.

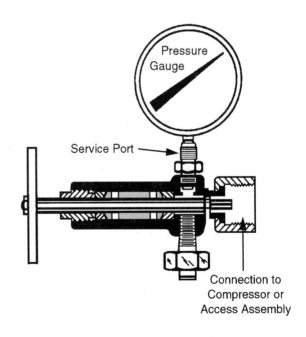

Fig. 18–7C On this type of HVAC/R service valve attachment, one port is used for a pressure gauge and the other is used to accomplish service procedures.

System Diagnosis and Service

In order to effectively diagnose problems in a sealed system you must develop and follow a logical, step-by-step troubleshooting procedure. In all applications, the troubleshooting procedure should start with a thorough understanding of the system components and their purpose and theory of operation. The second step in the process is to determine what is or is not happening in the system. The third step is to establish a logical process of elimination until the source of the problem can be determined. The final step in the trouble-shooting process is to determine the cause of the initial failure so that repairs or adjustments can be made to prevent repeated failures.

This diagnostic procedure is highly subjective and based on the service technician's knowledge, experience, and preference. The following list can serve as a guide in diagnosing system problems by system pressures:

1. If the discharge pressure is near normal, if the suction pressure is lower than normal (possibly in a vacuum), or if the current or wattage draw is lower than normal, check the system tubing for possible kinks or restrictions—especially the capillary tube at exposed areas outside of the cabinet. If system pressures do not equalize within the prescribed time, usually five minutes, add refrigerant to the system. If suction pressure continues to remain low or in a vacuum accompanied by an increase in discharge pressure and unit current draw and no increase in evaporator refrigeration effect, look for a buildup of frost at the point of restriction.

2. If the discharge pressure is lower than normal, if the suction pressure is lower than normal (possibly in a vacuum), or if the current or wattage draw is lower than normal, there may be a leak in the system. System pressures will become progressively lower as more refrigerant escapes from the system. Add refrigerant. If the refrigeration effect on the evaporator improves and if suction pressure starts to increase, then the system was low on refrigerant.

3. If the discharge pressure is higher than normal, if the suction pressure is normal or higher than normal, or if the current or wattage draw is higher than normal, there may be air in the system. To confirm this possibility, allow the system to stand idle and system pressures to equalize. Then take a standing pressure check of the system's high- and low-side pressures. The standing pressures should correspond to the saturated pressure-temperature relationship for the refrigerant being used. If the standing pressures are higher than prescribed, the system contains air. If this is the case, the system will have to be evacuated, vacuumed, and recharged. Replacing the system filter-drier is also recommended.

4. If the discharge pressure is higher than normal, if the suction pressure is higher than normal, or if the current or wattage draw is higher than normal, check for an overcharge of refrigerant. The extent of pressure increase will depend on the degree of the overcharge. During the running cycle, the suction line will probably have a frost pattern extending back to the compressor. Evacuating some of the refrigerant from the system should result in improved operating conditions. The system should be evacuated and recharged with the correct amount of refrigerant.

5. If the discharge pressure is lower than normal and the suction pressure is higher than normal, or if the current or wattage draw is lower than normal, check for an inefficient compressor. This is also indicated by a compressor that runs continuously and a thin pattern on the evaporator coil. To verify the condition, crimp the suction line ahead of the compressor and observe suction pressure. If the compressor does not pull a vacuum of at least 25 inches Hg, the compressor is inefficient and needs to be replaced.

When diagnosing system problems by observing system pressures, refer to the manufacturer's service manual for the specific operating wattage, system pressures, and temperatures for the system being serviced. During the diagnostic procedure, check for the proper operation of evaporator and condenser fans and observe evaporator and condenser coil conditions for the proper airflow, thermostat feeler bulb locations, and so forth.

Troubleshooting Guide

The information that follows gives problems and possible causes and solutions for various service situations related to domestic and light commercial refrigeration systems.

Problem: The compressor will not run.

Possible causes and solutions:

1. Disconnected service cord—Reconnect service cord.
2. Tripped circuit breaker—Reset the circuit breaker and determine the cause of the trip.
3. Low-voltage condition—Check the voltage, determine the source and cause of the problem, and correct as required.
4. Defective compressor motor—Check the compressor motor and repair or replace as required.
5. Defective relay or overload device—Check the relay or overload device and replace as required.
6. Defective defrost clock-timer—Check the timer and replace as required.
7. Defrost clock-timer stuck in defrosting cycle—Manually place the timer into the refrigeration cycle. If the unit cycles on, check the timer for proper operation and replace as required.
8. Defective capacitor—Remove the capacitor and replace it with a good one. Make sure to replace the capacitor with one of equal value.
9. Defective wiring harness—Check the wiring harness and repair as required.
10. Defective thermostat—Check the thermostat and repair or replace as required.
11. Improper thermostat setting—Check the thermostat and make the required adjustments.

Problem: The freezer compartment is too warm.

Possible causes and solutions:

1. High temperature control setting—Adjust the temperature control as required.
2. Defective temperature control—Check and repair or replace the temperature control as required.
3. Poor condenser air circulation—Check the condition of the condenser coil for proper air circulation. Check the condenser fan motor for proper operation and rotation direction.
4. Poor evaporator airflow—Check the condition of the evaporator for proper air circulation. Check the evaporator fan motor for proper operation and rotation direction.
5. Poor door gasket—Check the condition of the door gasket and repair or replace as required. Check to ensure that the fixture cabinet is leveled properly and adjust if necessary.
6. Defective compressor—Perform an efficiency test and repair as required.
7. Defective defrost heater—Check and repair as required.
8. Defective defrost limit switch—Check the limit switch and replace as required.
9. Defective defrost clock-timer—Check the timer operation and repair or replace as required.
10. High evaporator heat load—Allow sufficient time for the system to reduce cabinet heat load.

Problem: The food compartment is too cold.

Possible causes and solutions:

1. Temperature control setting too cold—Check and adjust the temperature control as required.

2. Defective temperature control—Check the temperature control and repair as required.

3. Improperly installed temperature control feeler bulb—Check the location and installation of the feeler bulb and repair as required.

4. Short in wiring harness—Check the system wiring and repair as required.

Problem: The food compartment is too warm.

Possible causes and solutions:

1. High temperature setting—Check and adjust the temperature control as required.

2. Defective temperature control—Check the temperature control and repair as required.

3. Defective door gasket—Check the door gasket and repair as required.

4. Defective compressor—Perform a compressor efficiency test and repair as required.

5. Unit stuck in defrosting cycle—Manually place the system in the defrosting cycle. Check unit operation and repair as required.

6. High compartment heat load—Allow sufficient time for the system to reduce the heat load.

7. Improper air circulation—Check the entire system to ensure that the airflow throughout the cabinet is unrestricted, that the evaporator coils are free of ice buildup, and that all fans are operating properly.

Problem: The cabinet light will not operate.

Possible causes and solutions:

1. Burned-out bulb—Check and replace the bulb as required.

2. Defective light switch—Check and repair the light switch as required.

3. Defective lamp socket—Check and repair the lamp socket as required.

4. Defective wiring—Check and repair the wiring as required.

Problem: The cold plate drips water on the cabinet and food.

Possible causes and solutions:

1. Product containers touching the cold plate—Relocate the products.

2. Greasy cold plate—Clean the cold plate.

Problem: There is excessive moisture in the refrigerator cabinet.

Possible causes and solutions:

1. Worn or improperly-fitted door gasket—Check the door gasket and repair or replace as required.

2. Cabinet not level—Check the level of the cabinet and adjust as required.

3. Clogged drain tube—Check the drain tube and repair as required.

Problem: Moisture is dripping on the floor.

Possible causes and solutions:

1. Overcharged system—Check for frost on the suction line. Check for the correct refrigerant charge and repair as required.

2. Overflowing or misplaced defrost drain pan—Check drain pan or position as required.

3. Defective cabinet heaters—Check all heaters and replace as required.

Problem: Operation is noisy.

Possible causes and solutions:

1. Cabinet not level—Check the cabinet position and adjust as required.

2. Loose tubing—Check all tubing and repair as required.

3. Defective compressor—Check the compressor operation and repair as required.

4. Condenser fan striking shroud—Check the operation of the condenser fan and repair as required.

Problem: Water freezes in the drain trough.

Possible causes and solutions:

1. Defective drain heaters—Check all drain heaters and replace as required.

2. Defective defrost clock-timer—Check the timer and repair as required.

3. Defective wiring—Check all wiring harnesses and repair as required.

4. Restricted drain tube—Check the drain tube and repair as required.

Problem: The door does not seal properly.

Possible causes and solutions:

1. Cabinet not properly leveled—Level cabinet as required.

2. Worn door gasket—Check the door gasket and replace as required.

3. Cabinet doors not properly aligned—Check the cabinet doors and adjust as required.

4. Improperly adjusted door hinge—Check the door hinge and adjust as required.

Problem: The evaporator fan will not operate.

Possible causes and solutions:

1. Unit in defrosting cycle—Manually place the system into the refrigeration cycle and check the fan operation.

2. Defective fan switch—Check the door switch and repair as required.

3. Defective fan thermostat switch—Check the operation of the switch and replace as required.

4. Defective fan motor—Check the operation of the fan motor and replace as required.

5. Defective wiring—Check the wiring and repair as required.

Problem: The freezer compartment will not defrost.

Possible causes and solutions:

1. Defective defrost clock-timer—Check the operation of the timer and replace as required.

2. Defective defrost heater—Check the operation of the heater and replace as required.

3. Defective defrost limit switch—Check the limit switch and replace as required.

4. Defective wiring—Check the system wiring and repair as required.

BACKSEAT AND CRACKING A SERVICE VALVE

Installing Gauges

Installing the manifold gauges is the first step taken when entry into a sealed system is required. It is important to remove all air from the gauge manifold and service hoses to prevent system contamination. The two methods used to purge air from the gauge manifold and service lines are the external service cylinder and system high-side pressure.

To install gauges using an external service cylinder, refer to Figure 18–8 and the following procedures:

1. Make sure that the compressor suction and discharge service valves are in the backseated position.

2. Make sure that the compound and pressure gauge manifold valves are in the closed position.

3. Connect the compound gauge service hose to the compressor suction service valve port.

4. Connect the pressure gauge service hose to the compressor discharge service valve port.

5. Connect the center service hose of the gauge manifold to the refrigerant service cylinder.

6. Open the refrigerant cylinder service valve and the compound gauge manifold valve. Refrigerant will now flow from the service cylinder, through the center hose of the gauge manifold, and to the compressor suction service valve port. Loosen the hose connection at the compressor suction service valve port and allow the refrigerant to purge for a second. Retighten the hose connection and close the compound gauge manifold valve.

7. Open the pressure gauge manifold valve. Refrigerant will now flow across the bar manifold to the compressor discharge service valve point. Loosen the hose connection at the compressor discharge service valve port and allow the refrigerant to purge for a second. Retighten the hose connection and close the pressure gauge manifold valve.

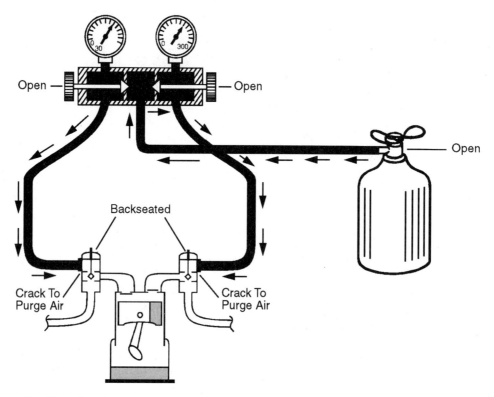

Fig. 18–8   Purging air from service hoses and gauge manifold using external cylinder.

8. Close the refrigerant service cylinder valve. The gauge manifold and service hoses are now purged of air.

9. Crack the compressor suction service valve and discharge service off of the backseated position to read system standing pressures.

System pressure is used to install gauges on a system when there is no convenient service cylinder to purge the gauge manifold and service lines of air. Because the low side of the system may be in a vacuum, the high side of the system is used in order to prevent air from entering the system. To install gauges using system high-side pressure, refer to Figure 18–9 and the following procedures:

1. Make sure that the compressor suction and discharge service valves are in the backseated position.

2. Make sure that the compound and pressure gauge manifold valves are in the closed position.

3. Connect the pressure gauge service hose to the compressor discharge service valve port.

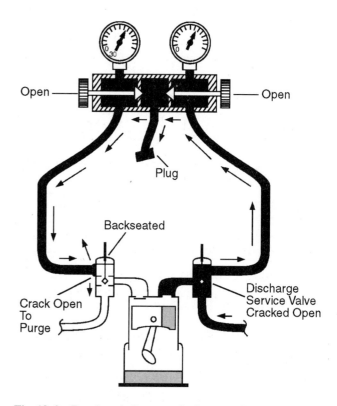

Fig. 18–9   Purging air from service hoses and gauge manifold using system pressure.

4. Connect the compound gauge service hose to the compressor suction service valve port.

5. Plug the center hose of the gauge manifold.

6. Crack the compressor discharge service valve off of the backseated position. The pressure gauge should indicate a pressure reading.

7. Open the pressure gauge manifold valve. Refrigerant will flow from the compressor discharge service valve port, across the gauge manifold bar, and to the center hose connection of the gauge manifold. Loosen the plug on the center service hose and allow the refrigerant to purge for a second. Retighten the plug.

8. Open the compound gauge manifold valve. Refrigerant will flow across the gauge manifold bar to the service hose connection at the compressor suction service valve port. Loosen the hose connection at the compressor suction service valve port and allow the refrigerant to purge for a second. Retighten the hose connection.

9. Close the compound and pressure gauge manifold valves. The gauge manifold and service hoses are now purged of air.

10. Crack the compressor suction service valve off of the backseated position to read standing pressure on the low side of the system.

Many systems utilize a Schraeder valve as a means to gain access into a sealed system. The Schraeder valve is similar in design to the type of valve used on an automobile tire. When servicing a system equipped with a Schraeder valve, the gauge manifold service hoses must have a Schraeder valve depressing pin; if they do not, an adapter must be used. To install gauges on systems equipped with a Schraeder valve, refer to Figures 18–10A and B and the following procedures:

1. Make sure that the compound and pressure gauge manifold valves are in the closed position. This is important. If the valves are in the open position, the center service hose may "whip" and cause injury when the gauge hoses are connected to the system.

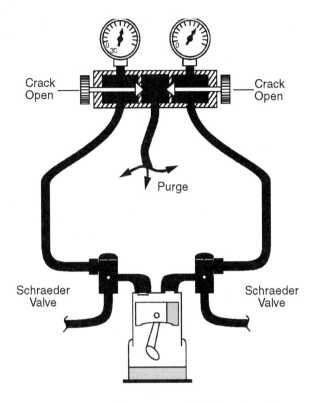

Fig. 18–10A   Installing gauges with Schraeder valves.

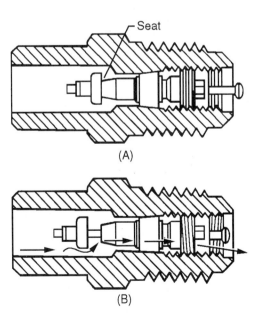

Fig. 18–10B   How the Schraeder valve works. It is similar to the valve on a car tire except that it has threads to accept the service technician's gauge line threads. *Courtesy J/B Industries.*

2. Connect the pressure gauge service hose to the Schraeder valve installed on the high side of the system. System high-side standing pressure should be indicated on the pressure gauge.

3. Connect the compound gauge service hose to the Schraeder valve installed on the low side of the system. System low-side standing pressure should be indicated on the compound gauge.

4. To purge the gauge manifold and service hoses of air:

   a. Open the pressure gauge manifold valve and allow the refrigerant to purge through the manifold bar center service hose for a second. Close the pressure gauge manifold valve.

   b. Open the compound gauge manifold valve and allow the refrigerant to purge through the manifold bar center service hose for a few seconds. Close the compound purge manifold valve.

The gauge manifold bar and service hoses are now purged of air and the system can be serviced as required.

VAPOR CHARGING A SYSTEM

Charging a system with vapor refrigerant is done through the system's low side, using the compressor to draw the refrigerant into the system from the refrigerant service cylinder. The primary advantage of **vapor charging** is that it ensures that a clean refrigerant is charged into the system. During the charging process the vaporization of the liquid refrigerant in the service cylinder distills the refrigerant, thus preventing contamination from entering the system. Liquid refrigerant should never be introduced into the low side of the system since damage to the compressor could result.

During the vapor charging process, the pressure in the refrigerant service cylinder will decrease and reduce the vaporization process of the refrigerant. Alternating the service cylinders

or placing the cylinder in a container of warm water will help maintain cylinder pressure. The service container should never be exposed to a direct flame or other heat source since excessive pressures could result and rupture the cylinder.

To vapor charge a system, refer to Figure 18–11 and the following procedures:

1. Make sure that the compound and pressure gauge manifold valves are in the closed position.

2. Make sure that the compressor suction and discharge service valves are in the backseated and cracked positions.

3. Connect the center service hose of the gauge manifold to the refrigerant service cylinder.

4. Start the compressor and slowly open the compound gauge manifold valve allowing gaseous refrigerant to enter the low side of the system. (Note: During the charging procedure it is important to maintain a charging pressure that is equal to the pressure under which the compressor normally operates. If the charging pressure is too high, the compressor may cycle on overload. If the charging pressure is too low, the compressor may pump oil.)

5. Continue the charging process until the system is charged with the correct amount of refrigerant.

LIQUID CHARGING A SYSTEM

Charging the system with liquid refrigerant is much faster than vapor charging. In order to introduce the liquid refrigerant into the system, a pressure differential must be created between the refrigerant service cylinder and the system. On new installations or systems that have been evacuated of refrigerant for service, the pressure differential is created by pulling the system into a deep vacuum. A **liquid charging** process is then implemented through a high-side service port.

When the exact amount of refrigerant that a system requires is known, the refrigerant charge should be weighed to prevent overcharging the

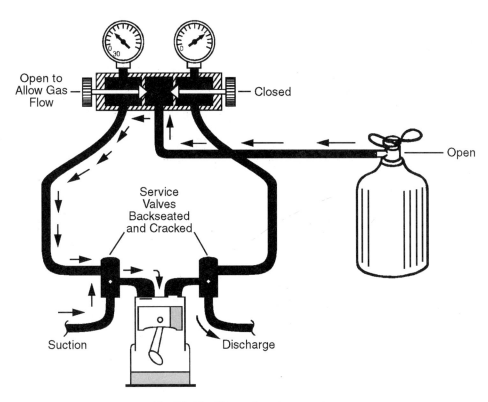

Fig. 18–11   Vapor charging a system.

system. This can be done easily by placing the service cylinder on a scale during the charging process and monitoring the weight. To liquid charge a system, refer to Figure 18–12 and the following procedures:

1. Make sure that the compound and pressure gauge manifold valves are in the closed position.

2. Make sure that the compressor suction service valve is in the backseated and cracked positions.

3. Make sure that the compressor discharge service valve is placed in the midway position.

4. Connect the center service hose of the gauge manifold to the refrigerant service cylinder. Loosen the center hose connection at the manifold bar and open the refrigerant service cylinder, allowing refrigerant to purge air from the center service hose at the manifold bar.

5. Retighten the hose connection at the manifold bar.

6. Invert the refrigerant service cylinder and open the pressure gauge manifold valve. Liquid refrigerant will now flow into the system's high side.

7. When the desired amount of refrigerant has been charged into the system, close the pressure gauge manifold valve and the service cylinder. Backseat and crack the discharge service valve to place the system back into normal operation. (Note: During the charging procedure, system pressure and service cylinder pressure may equalize before the desired amount of refrigerant has been introduced into the system. If the system requires more refrigerant, the charging process should be completed by vapor charging through the system's low side.)

In many large commercial applications a liquid-charging valve is installed in the system's liquid line or at the receiver. When a refrigerant service cylinder is connected to the charging valve, it will act as the system receiver and liquid refrigerant will be

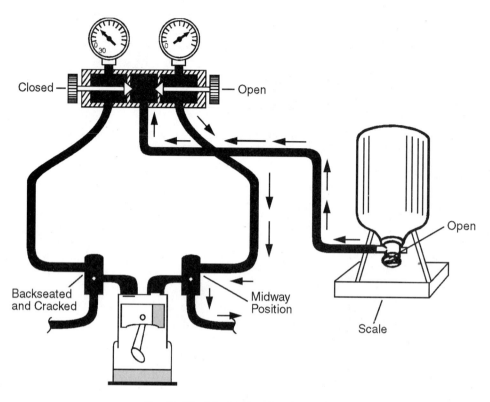

Fig. 18–12   Liquid charging the system.

drawn into the system's high side through the pressure differential created by the system compressor. This process is recommended for charging a large capacity system. Refer to Figure 18–13 and the following procedures:

1. Make sure that the compound and pressure gauge manifold valves are in the closed position.

2. Make sure that the compressor suction and discharge service valves are in the backseated and cracked positions.

3. Connect the refrigerant service cylinder to the system's high-side service valve port. Open the cylinder and loosen the hose connection at the service port. Allow refrigerant to purge the hose of air for a few seconds, then retighten the hose connection. If the system has been evacuated of refrigerant and placed in a vacuum, proceed as follows:

   a. Close the receiver inlet valve. If no receiver inlet valve is provided, frontseat the compressor discharge service valve.

This will prevent liquid refrigerant from entering the compressor.

   b. Invert the refrigerant service cylinder or make sure that the service valve of the cylinder is in such a position that liquid refrigerant can be drawn from the cylinder.

   c. Open the refrigerant service cylinder and the high-side charging valve. Liquid refrigerant will now enter the system receiver or high side.

4. Continue the charging procedure until the correct amount of refrigerant has been introduced into the system or until the pressure between the service cylinder and the system becomes equalized and no more refrigerant will flow into the system.

5. Close the refrigerant service cylinder valve.

6. If applicable, open the receiver inlet valve.

7. If the compressor discharge service valve was frontseated, make sure that it is placed in the backseated and cracked position at this time.

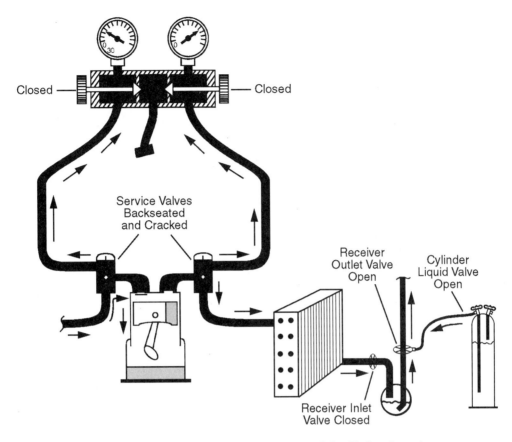

Fig. 18–13   Liquid charging the system through liquid charging valve.

8. Start the compressor and observe system operating temperatures and pressures. If the system requires additional refrigerant, proceed as follows:

   a. Stop the compressor.

   b. Close the receiver outlet valve.

   c. Open the refrigerant service cylinder.

   d. Start the compressor.

The charging cylinder will now supply the system with liquid refrigerant as the system receiver normally would.

During the charging procedure, system operating pressures should be observed carefully. All increases in discharge pressure indicate that the system receiver is full and cannot hold any more refrigerant. At this point the charging process should be stopped. In order to determine if the proper amount of refrigerant has been charged into a system, close the refrigerant service cylin-

der and open the receiver outlet valve. Observe system operating pressures and conditions. Continue the charging process until the system is charged completely.

PUMPING DOWN A SYSTEM

The purpose of **pump down** is to remove all of the refrigerant from a system and store it in the system's receiver so maintenance or service can be performed. To pump down a system with gauges installed, refer to Figure 18–14 and the following procedures:

1. Make sure that the compressor suction and discharge service valves are in the backseated and cracked positions.

2. Make sure that the compound and pressure gauge manifold valves are in the closed position.

3. Close the receiver outlet valve.

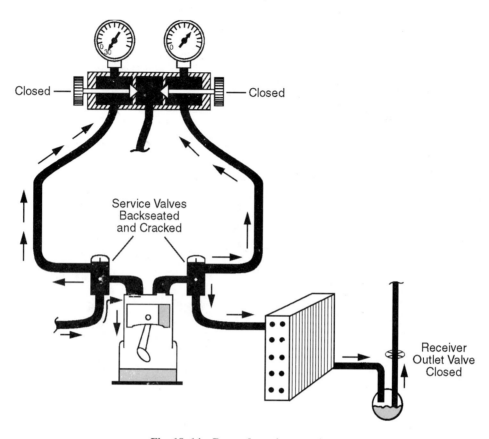

Fig. 18–14   Pump down into receiver.

4. Start the compressor and monitor system pressures. The compound gauge should indicate a decreasing suction pressure. The pressure gauge should indicate a slight rise or a stable discharge pressure.

5. Allow the unit to run until the system's low-side pressure is reduced to 2 PSIG. The unit should then be shut off and the receiver inlet valve closed. If the unit is not equipped with a receiver inlet valve, frontseat the compressor discharge service valve.

An increase in discharge pressure during the pump down procedure indicates that the system receiver is full and cannot hold any more refrigerant. The system should be shut down and the remaining refrigerant evacuated into a recovery system and external service cylinder to prevent contamination of the ozone layer.

VACUUMING A SYSTEM

Any time a sealed system is opened for service and exposed to atmospheric pressure, noncondensible gases and moisture enter the system and must be removed before the system is placed back into operation. The most effective method of removing moisture and noncondensible gases is to place the system under a deep **vacuum** by using a vacuum pump. A vacuum pump capable of reducing system pressure to at least 500 microns should be used. Because the compound gauge is incapable of measuring a vacuum this deep, an electronic vacuum gauge or manometer should be used to measure the degree of vacuum. The length of time a vacuum pump is used to obtain the desired vacuum depends on the efficiency of the vacuum pump, the degree of system contamination, and the size of the system. On smaller systems where a small displacement vac-

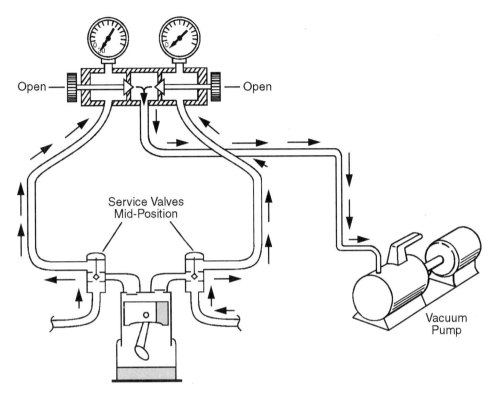

Fig. 18–15   Vacuuming system.

uum pump is used, standard manifold gauge service lines of at least ¼" ID should be used. On large systems or when a high displacement vacuum pump is used, copper tubing or special service lines of at least ½" ID should be used to prevent line restrictions, which would affect the ability of the vacuum pump to draw a deep vacuum.

To vacuum a system with gauges installed, refer to Figure 18–15 and the following procedures:

1. Connect the center service hose of the gauge manifold to the inlet of the vacuum pump. (Note: Some vacuum pump manufacturers recommend starting the vacuum pump before connecting the center service hose to prevent hard starting of the vacuum pump.)

2. Open the compound and pressure gauge manifold valves.

3. Place the compressor suction and discharge service valves in the midway position.

4. Start the vacuum pump.

5. Allow the vacuum pump to run until the desired vacuum is reached.

6. Close the compound and pressure gauge manifold valves.

7. Stop the vacuum pump.

8. Disconnect the center service hose of the gauge manifold from the vacuum pump and connect it to a refrigerant service cylinder. Purge the service line of air as described in the gauge installation procedures. Service the system as required.

EVACUATING A SYSTEM

Evacuating a system removes all the refrigerant from the system and reduces it to atmospheric pressure. It is illegal to purge large quantities of refrigerant into the atmosphere. Whenever possible, the refrigerant from a system should be pumped down into a system receiver or recovery system, or evacuated into an external service cylinder.

Evacuation Into an External Cylinder

When evacuating refrigerant into an external service cylinder, caution should be taken to prevent excessive cylinder pressures. The fusible plugs in service cylinders melt at a temperature of approximately 265°F. Temperatures that reach this level while refrigerant is being discharged into the cylinder can cause injury. When using this procedure, keep several clean, dry service cylinders available to rotate during the evacuation process. It is important not to overfill the service cylinder. If possible, place the service cylinder in a container of cold water. This will aid in the condensing process of the refrigerant as it is pumped into the service cylinder.

To evacuate system refrigerant into an external cylinder with gauges installed, refer to Figure 18–16 and the following procedures:

1. Make sure that the compound and pressure gauge manifold valves are in the closed position.

2. Connect the center service hose of the gauge manifold to the refrigerant service cylinder.

3. Make sure that the compressor suction service valve is in the backseated and cracked position.

4. Frontseat the compressor discharge service valve and back it off one full turn.

5. Open the pressure gauge manifold service valve and crack the center service hose connection at the refrigerant service cylinder for a second to purge the line of air. Retighten the hose connection at the service cylinder.

6. Open the refrigerant service cylinder and start the compressor. A portion of the system discharge vapor will now enter the service cylinder and condense. Continue this process until the service cylinder is filled, observing system discharge pressure. When the dis-

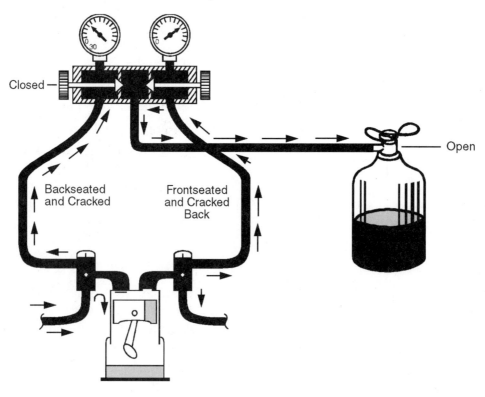

Fig. 18–16   Evacuation into external cylinder.

charge pressure starts to rise, the refrigerant cylinder cannot take any more refrigerant so the refrigerant charge must then be transferred to an external service cylinder.

When the system's low-side pressure reaches the limit of the recovery system, the system is evacuated. System access can be gained.

Triple Evacuation

Triple evacuation is a highly recommended means of dehydrating a refrigeration or air-conditioning system. The process is based on the theory that introducing a dry refrigerant into a vacuumed system will serve to blot the system of moisture. Triple evacuation, combined with the installation of filter-driers, is an effective method of system dehydration under normal service conditions. Refer to the following procedures:

1. Vacuum the system to about 1,500 microns.

2. Break the system vacuum by introducing dry refrigerant into the system, bringing the system to a positive pressure of about 2 PSIG. Remember that the refrigerant should not be purged into the open atmosphere but into a certified recovery cylinder.

3. Vacuum the system again to about 1,500 microns.

4. Break the vacuum again to 2 PSIG.

5. Now vacuum the system to 500 microns or as low as the vacuum pump will reduce the system pressure.

6. Charge the system as required.

The longer the dry refrigerant is allowed to blot the system between vacuums, the more effective the procedure will be.

TESTING COMPRESSOR EFFICIENCY

An inefficient compressor does not operate at normal suction and discharge pressures. Under a full-load condition, the symptoms of an inefficient compressor are higher than normal suction pressures and/or lower than normal discharge

pressures. The compressor efficiency test is designed to test the internal discharge and suction valves of a compressor to determine whether they are sealing properly. To test compressor efficiency with gauges installed, refer to Figure 18–17 and the following procedures:

1. Make sure that the compound and pressure gauge manifold valves are in the closed position.

2. Make sure that the compressor discharge service valve is in the backseated and closed position and the compressor suction service valve is in the frontseated position.

3. Start the compressor and allow the suction pressure to decrease to 25 inches Hg or lower.

4. When the suction pressure registers 25 inches Hg or lower, stop the compressor. Observe the following:

   a. If the pressure on the compound gauge does not rise above 0 PSIG within one

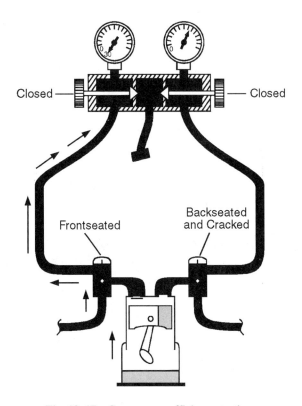

Fig. 18–17   Compressor efficiency testing.

minute, the suction and discharge valves of the compressor are good.

b. If the pressure on the compound gauge rises rapidly to 0 PSIG or higher, the compressor discharge valves are leaking.

c. If the compressor will not decrease the suction pressure to 25 inches Hg after about two minutes of operation, the compressor discharge or suction service valves may be defective.

d. If the suction pressure holds at the lowest vacuum reached after two minutes of operation, the discharge valves are good but the suction valves are defective.

e. If the suction pressure rises rapidly, the discharge valves are defective.

5. After testing is completed, place the compressor suction and discharge service valves into the backseated and cracked positions to place the system back into a normal operating mode.

TESTING AND CHARGING THE OIL LEVEL

It is important to maintain the correct amount of oil in the compressor crankcase at all times. An undercharge of oil will result in higher than normal compressor operating temperatures (especially critical in hermetic compressors), noisy operation, and increased friction between the internal parts of the compressor, which will cause premature wear and component failure. An overcharge of oil will cause the compressor to pump excessive amounts of oil into the system. This will reduce the efficiency of the compressor and cause possible valve damage. Excessive circulation of oil in the system will reduce overall system efficiency, especially in the evaporator where the oil may stratify and reduce heat transfer ability. Oil trapped in the suction line may result in oil slugging back to the compressor causing intake valve damage.

Several methods are used to test compressor oil level. Some compressor crankcases are equipped with an oil sight glass. Under normal conditions, after the system has been in operation long enough to equalize operating pressures and temperatures, the oil level should be at about one-half to three-fourths of the sight glass. Another method used to test compressor oil level is the dipstick method, in which a port is provided in the compressor crankcase. The oil level is checked with a dipstick that is placed through the port. This method requires the compressor crankcase to be evacuated of refrigerant and opened to the atmosphere.

There are several methods of adding oil to a compressor. One method is to pour oil into the compressor through the oil plug of the crankcase. Another method is to place the compressor crankcase into a vacuum and allow the oil to be drawn in by the pressure differential created. A manual oil pump, such as the one illustrated in Figure 18–18, can also be used to inject oil into the compressor crankcase.

To check compressor oil level by the dipstick method with gauges installed, refer to Figure 18–19 and the following procedures:

1. Make sure that the compound and pressure gauge manifold valves are in the closed position.

2. Make sure that the compressor discharge service valve is backseated and cracked.

3. Frontseat the suction service valve.

4. Start the compressor and allow the unit to run until the suction pressure is reduced to 2 PSIG.

5. Stop the compressor and frontseat the discharge service valve to isolate the compressor from the system.

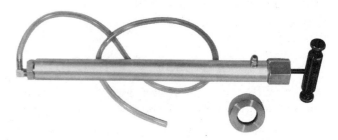

Fig. 18–18   Compressor oil charging pump. *Courtesy Robinair Division, SPX Corporation.*

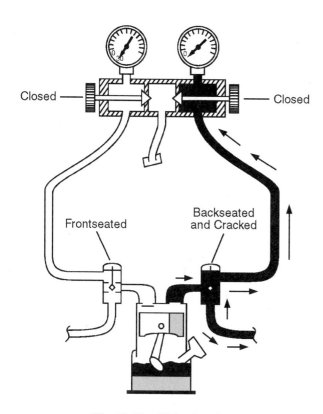

Fig. 18–19   Oil level testing.

6. Slowly open the compressor oil fill plug, allowing residual crankcase pressure to vent. Remove the oil fill plug after all pressure has dissipated.

7. Insert a dipstick into the fill hole and measure the oil level.

If the manufacturer's specifications for the proper oil level are not known, the general rule is that the oil level should be about one-half the distance from the crankshaft center to the bottom of the crankcase. Before the system is placed back into operation, the compressor crankcase must be purged of any air that may have entered it during the service procedure. To place the system back into operation follow these procedures:

1. Reinstall the oil fill plug but do not tighten it.

2. Slowly crack the suction service valve off the frontseated position. This will allow refrigerant to enter the compressor crankcase.

3. Allow the refrigerant to purge through the oil fill plug for about 15 seconds to ensure that all air has been removed from the crankcase. Tighten the oil fill plug and return the compressor service valves to the backseated and cracked positions.

4. Place the system back into normal operation.

LEAK TESTING

It is essential that the sealed system be absolutely leak proof. Any loss of refrigerant will reduce system efficiency and allow moisture and noncondensible gases to enter the system. **Leak testing** is done by first pressurizing the system with refrigerant then checking for leaks. The system should have at least 35 PSIG of pressure. In some applications the system is pressurized with refrigerant. An inert gas, such as dry nitrogen, is then introduced to increase system pressure to approximately 150 PSIG for testing purposes. A typical setup of leak testing is shown in Figure 18–20.

Never use oxygen to pressure test a system or an explosion may occur. Always use a pressure regulator and gauges to control and monitor the pressures being introduced into the system. When checking for leaks, check all mechanical joints

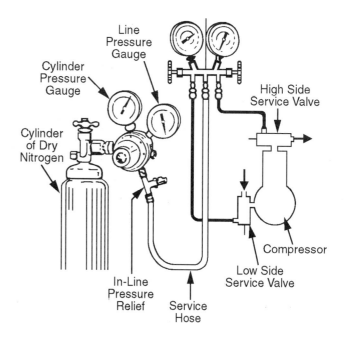

Fig. 18–20   A typical setup for leak testing.

thoroughly. Look for oil leakage around the joints. These are also obvious areas for refrigerant leaks.

Halide Leak Detector

A halide leak detector is one of the most popular leak detectors. Properly used, it can detect a refrigerant leak as small as one pound per year. A halide leak detector, which is illustrated in Figure 18–21, uses propane or acetylene gas to heat a copper reactor plate located in the burner head. A sniffer tube draws air into the burner head chamber where the air and gas are mixed. A control valve controls the flow of gas to the burner.

When the burner is ignited, the flame heats the copper reactor plate. When the sniffer tube picks up a halogen refrigerant leak, such as R-12, R-22, or R-502, a color change will take place in the flame. This is due to the refrigerant being exposed to the flame and hot copper reactor plate. A small leak will produce a blue-green flame. A large leak will produce a blue-violet flame. When the burner is first ignited, allow the reactor plate to heat until it is a cherry red, then adjust the flame to about one-half inch above the reactor plate.

Take care when using a halide leak detector around combustible materials. Make sure the work area is well ventilated and avoid breathing the fumes. Burning Freon, for example, produces a toxic phosgene gas.

Soap Bubbles

The use of soap bubbles is one of the oldest methods of leak detection and can best determine the exact location of a leak. There are numerous brands of commercially available soap bubbles on the market; however, any solution of soap and water that will produce suds when mixed is suitable. Apply the solution over all joints, connections, and areas where leaks are suspected. The appearance of bubbles will indicate the precise location of the leak. Leak detection using a soap solution is illustrated in Figure 18–22.

Electronic Leak Detector

The electronic leak detector, illustrated in Figure 18–23, is the most sensitive leak detector in use. It is capable of detecting a leak as small as

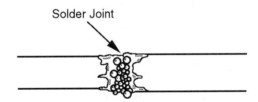

Fig. 18–22   Bubble leak test. Foam is placed on the solder joint. Bubbles indicate refrigerant leakage.

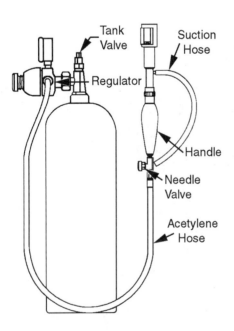

Fig. 18–21   Halide leak detector.   *Courtesy Prest-O-Lite®, Product of The Esab Group.*

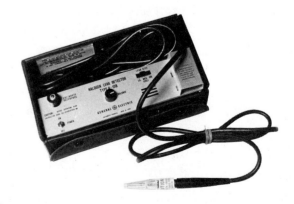

Fig. 18–23   Electronic leak detector.   *Courtesy Yokogawa Corp.*

one-half ounce per year. When halogenated refrigerant is drawn into a sensing probe, the electronic leak detector will emit a sound or a visual signal such as a flashing light. Electronic leak detectors are very sensitive and require more maintenance than other types of detectors. They are available in AC- or DC-powered units.

REMOVING MANIFOLD GAUGES

After the system has been serviced the manifold gauges must be removed before the system can be returned to normal operation. To remove the manifold gauges, refer to Figure 18–24 and the following procedures:

1. Stop the unit.

2. Backseat the discharge service valve.

3. Open the compound and pressure gauge manifold valves. This will allow the equalization of pressure between the manifold gauges and the system's low side. This step is done to prevent the purging of high-

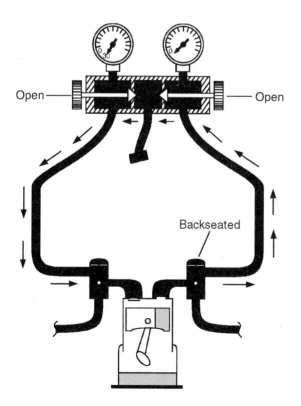

Fig. 18–24   Removing gauges.

pressure discharge gas when the service line is removed from the high side of the system.

4. After pressures have equalized, backseat the suction service valve.

5. Remove the service lines from the compressor service valve ports. When performing this step it is important to bleed pressure slowly from the service lines. When the manifold gauges read 0 PSIG it is safe to remove the service hoses. Safety glasses should be worn during this step.

6. Replace all service valve stem covers and service port caps.

FOLLOWING SAFETY PROCEDURES

Working on a refrigeration or air-conditioning system can present many hazards that you, as a service technician, need to identify and avoid in order to prevent injury or equipment damage. Although the refrigerants commonly used in most systems are considered nontoxic, they could displace the oxygen in the work area and create a health hazard when in a gaseous form. For this reason it is important that good ventilation is provided in the work area. In normal operating conditions, system pressures may be high but not necessarily dangerous. However, if a malfunction occurs or if you inadvertently position a valve improperly, high pressures that could cause injury might be experienced. Thus, gloves, safety glasses, and protective clothing should be worn any time the potential for injury exists. Heat should never be applied to a refrigerant service cylinder to increase its pressure. Gauges should always be installed on the system to observe operating system pressures as well as to perform leak tests when pressurizing a system. When performing soldering or brazing operations it is important that you wear proper protective clothing and safety glasses. It is also important to protect the surrounding work area from the flame and heat generated during these operations.

You, the service technician, should follow these guidelines when performing any service or repair procedures on refrigeration and air-conditioning equipment:

1. Wear safety glasses when working with refrigerants.

2. Wear gloves and protective clothing when potential hazards exist.

3. Always use gauges and pressure regulators when pressurizing a system.

4. Never use oxygen to pressurize a refrigeration system.

5. Work in a well-ventilated area.

6. Properly secure refrigerant cylinders during service operations to prevent an accident.

7. Keep a fire extinguisher handy during brazing or soldering operations.

8. Never face a cylinder gauge when opening a cylinder valve.

9. Open all cylinder valves slowly.

10. Do not warm a refrigerant cylinder with a torch.

11. Do not refill disposable refrigerant cylinders.

12. Check the position of all system valves before operating equipment.

13. Do not jumper system safety controls.

14. Purge refrigerant into an open and well-ventilated area.

15. Do not use a sharp object to remove ice from an evaporator coil.

16. Check the type of refrigerant contained in a system before adding refrigerant.

17. Keep protective caps on all cylinders when not in use.

18. Do not overpressurize a system when checking for refrigerant leaks.

19. Do not store refrigerant cylinders where they may be exposed to high temperatures.

20. Store refrigerant only in identified and properly designed storage containers.

SERVICE PROCEDURES

Moisture in the Sealed System

Moisture in the sealed system is directly or indirectly the cause of many service problems, perhaps more than all other causes combined. Moisture in the system will cause a freeze-up at the metering device and loss of the refrigeration effect. The moisture will react with the refrigerant and oil to create acids, sludge, and corrosion, which will result in premature compressor motor failure. The most common way that moisture enters a sealed system is through improper service procedures—not purging the gauge manifold and service lines of air, using contaminated oil, and generating low-side leaks where the system pressure is below atmospheric pressure and air is drawn into the system.

It is easier for moisture to enter a sealed system than it is for the technician to remove it. After finding the source of the leak, a good vacuum, combined with triple evacuation and replacement of the filter-drier, will effectively resolve most moisture problems. When vacuuming, apply heat to the various components of the system, such as the evaporator and condenser. This will facilitate moisture removal. A heat lamp can also be very effective in accomplishing this. Caution should be taken to prevent damage to adjacent plastic components of the fixture due to overheating.

Adding Oil to the Sealed System

The general rule to follow when dealing with oil in sealed systems is that if you don't know how much oil has been lost from the system, don't add any. A disadvantage of the hermetic compressor is that the oil level cannot be checked. When changing or adding oil, be careful not to add too much. If the compressor is not equipped with an oil sight glass, the only accurate method of adding oil is to remove the compressor from the system, drain the compressor oil, and add the correct amount by referring to the manufacturer's specifications. In rare situations where oil is to be added to the system while the compressor is installed, place the system in a vacuum and draw the oil in through the system's low side. This procedure cannot be used unless you know how much oil has been lost.

Servicing Hermetic Motor Burnouts

A hermetic motor burnout usually results from overheating of the motor windings. The wire insulation is burned and the motor windings are either shorted to each other, grounded, or burned open. The compressor oil will also be burned and is indicated by a strong acrid odor when the system is opened. A hermetic motor burnout can be attributed to one or more of the following conditions:

1. Internal motor defects
2. Defective starting relay
3. Defective overload protector
4. Defective start or running capacitors
5. Incorrect external wiring
6. Improper system voltage
7. Contamination in the system
8. High discharge pressures
9. Low refrigerant charge

The procedure for replacing a burned-out compressor requires additional care and precautions. Depending upon the severity of the burnout, the sealed system may be contaminated with a mild to heavy concentration of acid and sludge. Most of this contamination will be in the old compressor oil, system filter-drier, and condenser coil. Before installing the new compressor, flush the system to remove these contaminants, or the new compressor will also burn out. It is important, if this procedure is used, that the refrigerant be recovered in an authorized storage container.

When servicing a burned-out compressor, use caution to prevent skin and eye contact with the refrigerant and oil. The acids produced during the burnout can cause severe burns. Wear safety glasses and gloves when servicing a contaminated system and handling its components. In addition, protect the adjacent work area from possible acid and sludge damage. The following guidelines can be used when replacing a burned-out hermetic compressor:

1. Work in a well-ventilated area.
2. In the initial evacuation of refrigerant from the sealed system, carefully contain and control the escaping refrigerant in an authorized service cylinder to prevent personal injury damage to the work area and prevent purging refrigerant into the atmosphere.
3. Cut out and carefully remove the old compressor and filter-drier.
4. Reverse flush the condenser coil with liquid R-12 or pressurized R-11. Use quick-connect couplings to expedite this process. If quick-connect couplings are not available, install a flare nut on the filter-drier end of the condenser. To this connection attach the service hose connected to the refrigerant cylinder. During this service process, it is extremely important that no refrigerant is allowed to be vented into the atmosphere.
5. Connect the opposite end of the condenser to a service hose that has been wrapped in a clean cloth and placed in a container.
6. Open the refrigerant cylinder and carefully flush the coil for a short period of time. Inspect the cloth for debris and discoloration. Repeat the procedure until the cloth shows clean.
7. Reverse flush the oil cooler lines in the same manner. Some procedures recommend back flushing the evaporator in the same manner; however, this requires removing the capillary tube, which is not practical in most modern refrigerator and freezer applications. The flushing procedure just described and the

installation of a new filter-drier will usually prove to be satisfactory. After the replacement compressor and filter-drier have been installed, leak test, triple evacuate, and charge the system as required.

Servicing a Capillary Tube Restriction

A capillary tube restriction may be caused by the following conditions:

1. Moisture in the system

2. A kink in the capillary tube

3. Foreign particles trapped in the capillary tube

Each of these conditions will cause similar results in the system: the evaporator will have little or no frost accumulation; the compressor will run continuously and eventually cycle on the overload protector; and cabinet temperatures will be higher than normal.

If moisture is present in the system, it will usually freeze at the outlet of the capillary tube. This will restrict the flow of refrigerant into the evaporator and reduce, if not eliminate, the refrigeration effect of the system. The compressor will continue to run until it cycles on the overload protector due to the lack of returning suction vapor required to cool the compressor motor. When the compressor cycles on overload, the frozen moisture in the capillary tube will melt. When the compressor cycles back on, the refrigerant will again circulate until another freeze-up occurs. This condition can be confirmed by applying heat to the evaporator at the capillary tube inlet when a freeze-up occurs. If a gurgling sound (a result of the refrigerant surging through the evaporator) is heard as heat is applied, the system is probably moisture bound. This can be confirmed further by installing gauges on the system.

When a freeze-up occurs, the system's low-side pressure will indicate a vacuum, and refrigeration effect in the evaporator will not exist. When the frozen moisture melts, the system's low-side pressure will increase and the refrigeration effect in the evaporator will return to normal until another moisture freeze-up occurs. Evacuating

and vacuuming the system and installing a new filter-drier should resolve the problem.

A kinked capillary tube will totally restrict the flow of refrigerant to the evaporator, resulting in a lack of refrigeration effect, extended compressor running time, and high cabinet temperatures. If system gauges are installed on the system, the low-side pressure will be in a vacuum. In these cases, the entire length of the capillary tube should be checked for possible damage and the capillary tube should be replaced if necessary. In most applications, capillary tube damage will be outside of the fixture cabinet where the liquid line enters the filter-drier and the capillary tube.

If foreign particles are lodged in the capillary tube the system will exhibit the same symptoms as a kinked capillary tube. After checks have been made to eliminate the possibility of moisture in the system or a kink in the capillary tube, it may be assumed that the capillary tube is restricted by foreign particles. In most cases the capillary tube restriction will be only a few inches upstream of the filter-drier. To unrestrict the capillary tube, cut it one or two inches upstream of the filter-drier. Replace the filter-drier and, after vacuuming the system, place it back into normal operation.

Leak Testing the Sealed System

If a sealed system is low on refrigerant there may be a leak. A system should also be checked for refrigerant leaks after major service has been performed. Any leak must be located and repaired. After the leak has been located it must be decided whether the system can be repaired or whether the defective component should be replaced. An electronic leak detector, halide torch, and soap bubble solution are effective devices in locating refrigerant leaks. However, the urethane foam insulation used in many modern refrigerators and freezers contains an expandable material that is made up of the same ingredients used in refrigerants. Consequently, the halide torch and the electronic leak detector will sense the insulation and react as though a refrigerant leak has been detected. In applications where urethane foam is used, leak testing is best accomplished through the use of soap bubbles.

When leak testing a sealed system, a pressure of at least 40 PSIG is recommended. A few ounces of refrigerant added to a system at room temperature should be sufficient. Make a visual inspection of the entire system. Check all joints, lines, and fittings for traces of oil, which will usually indicate the source of the leak. If traces of oil are found, verify the leak by using soap bubbles, a halide torch, or an electronic leak detector. When using a halide torch or electronic leak detector, check all tubing connections, and return bends and tubing surfaces carefully. Move the leak detector slowly, allowing time for the leak detector to react. Once a leak has been confirmed, use soap bubbles to determine its exact location.

Servicing an Inefficient Compressor

An inefficient compressor does not produce adequate capacity during normal operation. The results are an evaporator covered with a thin film of frost and a compressor that operates continuously. Moreover, evaporator temperatures will not descend to the cut out temperature of the thermostat even with the compressor operating continuously. To test for an inefficient compressor, place a hand on the evaporator or accumulator for a few seconds and then examine the surface. If the frost has melted, install gauges on the system and check operating pressures. If the suction pressure is higher than normal and if the discharge pressure is lower than normal, the compressor is inefficient and needs to be replaced. This condition is normally accompanied by a lower than normal amperage or wattage reading. To confirm that the compressor is inefficient, install a pinch-off tool on the suction line of the compressor. Start the compressor and observe the system's low-side pressures. If the compressor cannot reduce the low-side pressure to 25 Hg it is defective and needs to be replaced.

Before installing a replacement compressor, check to ensure that it was not damaged during shipment. Installation instructions are usually supplied by the manufacturer. Read them carefully for detailed information and installation procedures of the particular unit being serviced. Most replacement compressors are charged with the correct amount of oil. If the compressor is shipped dry, the correct amount of oil will have to be added during installation. All replacement compressors contain a holding charge of dry nitrogen or refrigerant to prevent contamination. If a replacement compressor shows no indication of internal pressure when the lines are opened, it should be returned to the manufacturer. Check the type and size of fittings that will be required to install the new compressor. Sometimes the fittings are supplied with the new compressor, sometimes they are not. Many replacement compressors will include a new relay and overload to be used with the installation.

Use the following guidelines for removing and replacing an inefficient compressor:

1. Disconnect all power to the unit.

2. Remove all electrical connections from the old compressor.

3. Evacuate the system of all refrigerant. (Note: This step can be accomplished by using existing or field-installed service valves. Some manufacturers' procedures require that evacuation be accomplished by cutting the refrigerant lines. If this method is used, cut the suction line first. When the refrigerant starts to escape, stop the cut until all pressure has dissipated from the system. Hold a cloth over the cut to catch any oil that might escape. Once all the system pressure has been released, finish the cut. Use the same method to cut the discharge line. When cutting the refrigerant lines, make the cut as close as possible to the compressor so that the replacement compressor can be installed without adding any more tubing. During this procedure always wear safety glasses, work in a well-ventilated area, and protect the surrounding area from oil spillage.

4. Lift out the old compressor and place the new one in position. On some installations

the original mounting springs or grommets must be transferred to the new installation.

5. Connect the tubing and solder all joints.

6. Install the new filter-drier.

7. Connect all the necessary electrical connections.

8. Leak test, vacuum, and charge the system as required.

9. Test run the system to ensure that it operates according to the manufacturer's specifications.

10. After the installation is complete, thoroughly clean the work area. Clean and paint the tubing connections to restore the system to its original condition.

Charging the Sealed System

Most domestic sealed systems use the capillary tube as a metering device; therefore, it is important that the system is charged with the exact amount of refrigerant, to within one-half ounce of accuracy, as specified by the manufacturer. An overcharge or an undercharge of refrigerant could seriously affect the performance of the system and result in component damage.

When the exact amount of refrigerant charge is unknown, one of the most common methods used to charge the capillary tube system is the frost-back method. In this method, a small amount of refrigerant is added to the system through the low side while the compressor is in operation. The frosting of the evaporator indicates the correct refrigerant charge. When the evaporator is completely frosted and the cabinet is at the desired temperature, the system has been charged with the correct amount of refrigerant. If the evaporator is not completely frosted, add more refrigerant. When the frost line extends along the suction line outside the fixture cabinet, purge some refrigerant from the system until the proper frost line (just a few inches along the suction line leaving the fixture cabinet) is obtained. Compressor running amperage should also be in accordance with the compressor motor specifications.

When the exact amount of refrigerant charge is known, the recommended and most accurate method of charging the system is by using a charging column. A charging column, illustrated in Figure 18–25, is designed to meter the correct amount of refrigerant into the system by weight. Taking into consideration the prevailing temperature and humidity, the charging column will accurately introduce the correct amount of refrigerant into the system to within one-fourth ounce of the required refrigerant charge. Charging columns are available in various refrigerant capacities ranging from several ounces to several pounds. Most manufacturers offer charging columns equipped with heating elements to overcome the pressure equalization between the system and the charging column. Charging columns with heating elements reduces the time required to charge a system and

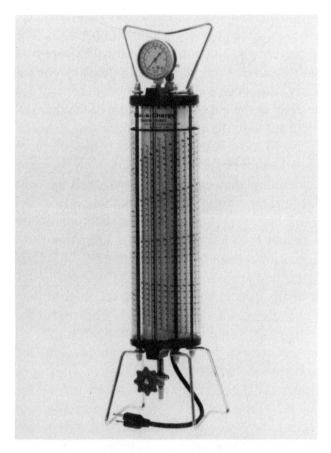

Fig. 18–25 Charging column. *Courtesy Robinair Division, SPX Corporation.*

eliminates the need to warm the charging cylinder in order to overcome system pressures.

To prepare a charging column for the charging process, check the unit's specifications or service manual for the proper type and amount of refrigerant charge required and follow these procedures:

1. Referring to Figure 18–26, connect the service hose from the refrigerant service cylinder to the service valve located at the bottom of the charging column.

2. Before tightening the connection, open the refrigerant service cylinder and purge the service hose of air.

3. Invert the refrigerant service cylinder and open the refrigerant service cylinder valve and the bottom service valve of the charging column to allow liquid refrigerant to enter the charging column, as shown in Figure 18–27.

4. As the charging column begins to fill with liquid, open the top service valve of the charging column to purge refrigerant vapor from the charging column and hasten the filling operation.

5. When liquid refrigerant is visible in the charging column's sight glass, dial the plastic shroud to the point where the pressure reading on the shroud corresponds to the pressure indicated on the charging column's gauge for the refrigerant being used.

6. Fill the charging column with more refrigerant than the system requires.

7. During the filling process, redial the plastic shroud to compensate for changes in pressure.

8. When the charging column has been filled with the desired amount of refrigerant, close the refrigerant service cylinder valve and the charging column valve.

9. Remove the service hose from the refrigerant service cylinder. When disconnecting the service hose from the refrigerant cylinder, wear safety glasses to prevent injury from the liquid refrigerant.

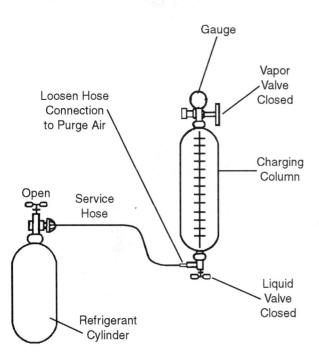

Fig. 18–26 Connecting refrigerant cylinder to charging column.

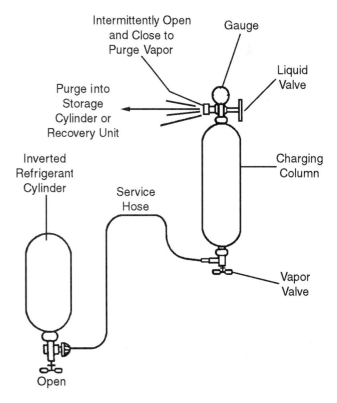

Fig. 18–27 Filling charging column purging into storage cylinder or recovery unit.

The charging column is now ready for use. If the unit is equipped with a heating element, plug it in and allow the pressure in the column to build. Redial the plastic shroud to compensate for pressure changes.

The charging column can be used to vapor charge or liquid charge a system. Vapor is drawn from the top service valve of the charging column and liquid is drawn from the bottom service valve. After all system repairs have been made and the system is under a proper vacuum, with gauges installed, follow these procedures to charge the system:

1. Connect the center service hose of the gauge manifold to the service valve located at the bottom of the charging column.

2. Open the charging column service valve and purge the service hose of air at the gauge manifold.

3. Open the pressure gauge manifold valve and allow the liquid refrigerant to enter the system's high side. Control the flow of refrigerant into the system by the valve on the charging column, not by the gauge manifold. If the pressure reading on the charging column changes, redial the scale on the plastic shroud accordingly. (Note: The compressor is not in operation at this time.)

4. When the desired amount of refrigerant has been introduced into the system, close the charging column service valve.

The charging process is now complete. During the charging process, the glass of the liquid gauge in the charging column may become full of bubbles as the liquid refrigerant vaporizes in the column. This condition may prevent an accurate reading of the amount of refrigerant being charged into the system. The bubbles may be cleared by inverting the charging column for a few seconds. During the charging process, system pressure and charging column pressure may equalize and prevent the transfer of refrigerant from the charging column to the system. If the charging column is equipped with a heating element, this is not a problem. The heater will create

a pressure difference and allow refrigerant to flow into the system. If the charging column is not equipped with a heater, the refrigerant remaining will have to be charged into the system through the low side with the compressor running.

Another extremely accurate method used to charge a sealed system is with an automatic charging meter. This instrument will dispense a preset amount of refrigerant into the system by means of a microcomputer, which monitors the amount of refrigerant charged into the system and stops the flow automatically when the programmed amount of refrigerant has been charged into the system. An automatic charging meter is illustrated in Figure 18–28.

Incorrect Refrigerant Charge

A system that is undercharged will produce various conditions, depending on the degree of undercharge. During normal system operation the evaporator will have a full covering of frost. Any degree of undercharge will result in a reduced frost pattern on the evaporator and the accumulator. As the leakage of refrigerant progresses, the frost pattern on the evaporator will diminish. The compressor will run continuously and cabinet temperatures will be higher than normal. Eventually the compressor will cycle on overload due to the lack of suction vapor required to cool the compressor motor. System pressures will indicate lower than normal suction and discharge pressures and compressor motor running amperage or wattage will be lower than normal.

On overcharged systems the suction line will be frosted back to the compressor. Suction and discharge pressures will be higher than normal as well as compressor motor running amperage and wattage.

If it has been determined that the system is undercharged, the entire system must be leak tested and the necessary repair procedures must be performed. If it has been determined that the system is overcharged, the system should be evacuated of all refrigerant and recharged with the correct amount of refrigerant.

Fig. 18–28   Automatic charging meter.   *Courtesy TIF Instruments, Inc.*

Repairing Aluminum Tubing

Many times a refrigerant leak will be found in an aluminum evaporator coil or tubing. While most of these leaks can be repaired by soldering, the process is difficult. Epoxy repair is a quick and effective alternate method. (see Figure 18–29) When properly applied the epoxy will last indefinitely. Several types of epoxy repair kits are available for aluminum tubing. Refer to the manufacturer's instructions for the specific repair kit needed. Use the following procedures as a general guide:

1. Pressurize the system and locate the area of the leak.

2. With the system still under pressure, thoroughly clean the area around the leak with a

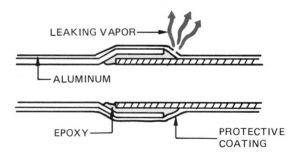

Fig. 18–29   Epoxy repair of aluminum evaporator.

sand cloth and the acetone solution that is normally supplied with the epoxy repair kit.

3. Thoroughly evacuate the system of all refrigerant, reducing the system pressure to 0 PSIG. Keep the system vented to prevent any pressure buildup due to any residual refrigerant remaining in the system.

4. Using a clean surface, mix equal amounts of epoxy hardener and resin until the color of the mixture is uniform.

5. Immediately apply the epoxy mixture over the area of the leak. Feather it out about one-half inch around the area of the leak. Allow it to cure and set. To cure the epoxy patch when a hot patch epoxy repair kit is used, apply a low flame to the repair area in accordance with the manufacturer's instructions. Be careful not to overheat or burn the epoxy. To expedite the curing process of the cold epoxy patch, use a heat lamp on the repaired area. Make sure that the adjacent plastic areas of the cabinet are protected from the heat. Total curing time of the cold epoxy patch with the application of heat is usually one to two hours.

6. Once the repair has been completed, pressurize and leak test the system.

Checking for Air in the System

Air introduced into the sealed system contains gases that are noncondensible under the normal operating temperatures and pressures of a refrigeration system. When air is introduced into a sealed system, it will become trapped in the upper tubes of the condenser. This will reduce the capacity of the condenser and cause the system to operate at higher than normal discharge pressures, accompanied by higher than normal compressor running amperage, longer compressor running times, and a loss of refrigerant effect.

To check for air in the system, follow these procedures:

1. Turn off the unit and allow the suction and discharge pressures to equalize and stabilize.

2. When the system standing pressures have stabilized, take a standing pressure check. The pressure readings should correspond to the saturated pressure and temperature relationship as indicated on a temperature/pressure chart. If the standing pressure is higher than normal for the temperature and the

refrigerant recorded, the system contains air and needs to be evacuated, vacuumed, and recharged. Replacing the filter-drier is also recommended.

Unit Eighteen Summary

- Hermetically sealed systems require the installation of a piercing valve in order to gain access to the system. Process tube adapters are used on systems that are equipped with process tubes. Some hermetically sealed systems are equipped with special service ports and an adapter is used to access the unit.

- Troubleshooting a sealed system is a logical process of elimination and system diagnosis is effectively accomplished by observing system operating procedures.

- Evacuation of a sealed system removes air and other noncondensible gases.

- A hermetic motor burnout occurs when the motor windings are shorted either to each other or to ground. Burnouts may cause acid to form in a sealed system and technicians must be careful to prevent burns when working with a system that has experienced a burnout.

- An inefficient compressor has failed suction and discharge valves that won't allow the compressor to pull a good suction pressure.

- An accurate method of charging a refrigeration system is to weigh the refrigerant and charge in the liquid state. Liquid refrigerant may be charged into the high-side access of a system when the compressor is off.

- Epoxy can be used to repair an aluminum evaporator puncture.

- Air in a sealed system will cause higher than normal discharge pressure, long running time, and the loss of cooling capacity.

- When performing any service procedures on a refrigeration system, it's important to ensure that no refrigerant is vented into the atmosphere. When installing gauges on a system it's important that air is purged from the hoses and gauge manifold.

- During a pump-down procedure a rise in discharge pressure would indicate that the system receiver is full and that the remainder of the refrigerant will have to be evacuated into an external cylinder.

- The proper oil level of a compressor should be at one-half on the oil sight glass.

- Leak testing of a refrigeration system can be done by using soap bubbles, a halide leak detector, or an electronic leak detector.

Unit Eighteen Key Terms

Piercing Valve

Schraeder Valve

Process Tube Adapter

Hermetic Service Valve Adapter

Vapor Charging

Liquid Charging

Pump Down

Vacuum

Evacuating

Triple Evacuation

Leak Testing

UNIT EIGHTEEN REVIEW

1. What is the function of a piercing valve?

2. What types of piercing valves are used on sealed systems?

3. Describe the procedure for installing a bolted-on piercing valve.

4. Describe the procedure for installing a brazed-on piercing valve.

5. Describe a Schraeder valve.

6. What is the purpose of a Schraeder valve core removing tool?

7. What purpose does the process tube of a system serve?

8. What is one advantage the process tube adapter has over the piercing valve?

9. Diagnose the possible cause (leak, low on refrigerant, air in the system, overcharge of refrigerant, inefficient compressor) in the following situations:

    a. Discharge pressure is higher than normal, suction pressure is higher than normal, or the electrical current draw is higher than normal. During the running cycle, the suction line will probably have a frost pattern extending back to the compressor.

    b. Discharge pressure is lower than normal, and the suction pressure is higher than normal, or the electrical current draw is lower than normal. The compressor is running continuously.

    c. Discharge pressure is lower than normal, the suction pressure is lower than normal, or the electrical current draw is lower than normal. System pressure continues to drop.

    d. Discharge pressure is higher than normal, the suction pressure is normal or higher than normal, or the electrical current draw is higher than normal. A measurement of the saturated pressure-temperature relationship is higher than prescribed by the refrigerant being used.

   e. Discharge pressure is lower than normal, the suction pressure is lower than normal, or the electrical current draw is lower than normal. Adding refrigerant improves the suction pressure.

10. Identify the problem by possible causes.

   a. Defective thermostat

      Improper thermostat setting

      Low voltage condition

   b. Poor evaporator airflow

      High temperature control setting

      Defective door gasket

      Defective defrost heater

11. Fill in the blanks with the correct problem and solution for the given cause.

	Problem	Solution
a.	_____. Cabinet not properly leveled.	_____.
b.	_____. Overcharged with refrigerant.	_____.
c.	_____. Defective defrost heater.	_____.
d.	_____. Clogged or restricted drain tube.	_____.

12. Installing the manifold gauges, what is the compound gauge service hose of the compound gauge manifold valve connected to? And what is the pressure gauge service hose of the pressure gauge manifold valve connected to?

13. When is a plug placed in the center hose of the gauge manifold?

14. What are three methods for installing the manifold gauges?

15. What is the advantage of vapor charging a system over liquid charging a system?

16. Why is liquid charging a system done through a high-sided service port?

17. Match the leak-detection method (pressurizing the system, halide leak detector, soap, electronic leak detector) with its description.

   a. Uses propane or acetylene gas to detect a color change in the flame.

   b. An inert gas such as dry nitrogen is used to raise system pressure to approximately 150 PSIG.

   c. The detector will emit a sound or a visual signal.

   d. Bubbles will locate the leak.

18. What is the procedure to be followed if it is not known how much oil has been lost from a system?

19. Describe the procedure for installing a process tube adapter.

20. What is a hermetic service valve adapter?

21. What is the purpose of applying heat to a system while it is being vacuumed?

22. How is oil added to a sealed system?

23. Describe a hermetic motor burnout.

24. What are the causes of a hermetic motor burnout?

25. What safety procedures should be followed when servicing a burned-out compressor?

26. Along with replacing a burned-out compressor, what else must be replaced rather than flushing the system to remove contaminants that could burn out a new compressor?

27. What are the three causes of capillary tube restriction?

28. Describe the effect of moisture in a capillary tube system.

29. Describe the system pressures of a system with a restricted capillary tube.

30. What is the minimum pressure recommended for the leak testing of a sealed system?

31. Describe the condition of an inefficient compressor.

32. Describe the procedure used to check for an inefficient compressor.

33. Describe three safety procedures to be followed when replacing a compressor.

34. Describe the frost-back method of charging a sealed system.

35. What is the most accurate method used to charge a sealed system?

36. What conditions indicate an undercharge of refrigerant in a sealed system?

37. What conditions indicate an overcharge of refrigerant in a sealed system?

38. What methods can be used to repair a leak in aluminum tubing?

39. Describe the procedure for the epoxy repair of aluminum tubing.

40. What happens when air is introduced into a sealed system?

41. Describe the service procedure used to check for air in a sealed system.

42. What is the purpose of purging the manifold gauges and service hoses?

43. Describe the two methods of purging the gauge manifold and service hoses.

44. Why should a service cylinder never be exposed to a direct flame or heat source?

45. If the discharge pressure starts to increase during the pump-down procedure, what is the problem?

46. How is a vacuum measured?

47. Define triple evacuation.

48. What safety precaution should be taken when evacuating refrigerant into an external service cylinder?

49. What is the purpose of introducing dry refrigerant into a system during the triple evacuation process?

50. Define an inefficient compressor.

51. What is the result of an overcharge of oil in a compressor?

52. Define the methods used to determine the proper oil level in a compressor.

53. Describe the methods used to add oil to a compressor.

54. What is the minimum test pressure for leak testing a system?

55. What is the advantage of the electronic leak detector?

56. What is the most frequent cause of service problems in the sealed system?

57. One micron is one one-millionth of a meter, or:

   1 millimeter = .001 meter = .039 inch

   Hg = column of mercury.

   Answer the following questions.

   a. 5 mm Hg = ? microns

   b. .5 mm Hg = ? microns

   c. .01 mm Hg = ? microns

58. Write an explanation of the difference between pumping down a system and evacuating a system.

59. Write a paragraph using four of the twenty safety guidelines on the service and repair of refrigeration and air-conditioning equipment, explaining why each safety recommendation is important.

UNIT NINETEEN

Thermostats and System Defrosting

OBJECTIVES

After completing this unit, the student will be able to:

1. Explain the method of operation of a thermostat in controlling a refrigeration system.

2. Differentiate between the different types of thermostats used in refrigeration equipment.

3. Identify the components in a defrosting system.

4. Explain the method of operation of a hot gas defrost system.

5. Explain the sequence of operation in an electric defrost system.

Cycling a refrigeration system off and on and keeping a low-temperature system coil frost-free are basic principles in maintaining the effective control and operation of equipment. In this unit, we'll discuss the method of operation of control thermostats as well as the accomplishment of the defrosting process you'll have to deal with when working with some systems. Step-by-step diagnostic procedures used in troubleshooting these electrical components are also discussed.

THERMOSTATS AND SYSTEM DEFROSTING

Thermostatic Motor Controls

In order to provide sufficient refrigeration during periods of high operating temperatures and high load conditions, the refrigeration system should be designed with enough capacity to meet extreme operating conditions. A means of capacity control is needed, as well, when the system is operating under reduced load conditions. The most common method of controlling system capacity in refrigeration and air-conditioning systems is to start and stop the compressor as needed. This can be accomplished by an automatic temperature control device known as a **thermostat**.

A thermostat is basically a single-pole, single-throw, temperature operated switch that will close on a rise in temperature and open on a decrease in temperature. This action will start and stop the compressor, thus maintaining the desired temperature.

Temperature Sensing Elements

There are basically two types of temperature sensing elements in common use: the bulb or fluid thermostat and the bimetallic thermostat.

Bulb Thermostat

Figure 19–1 illustrates the schematic of a simplified **bulb thermostat**. The control bulb con-

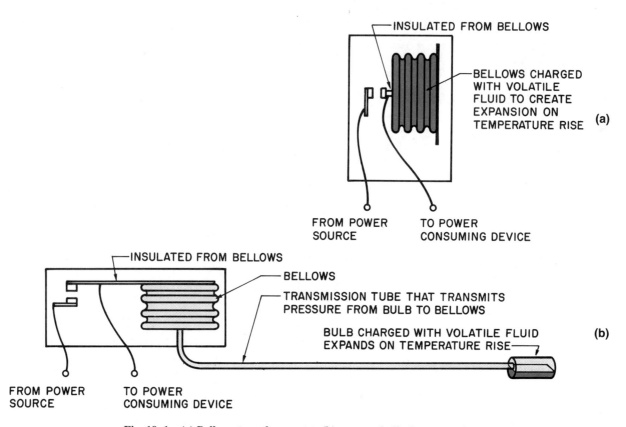

Fig. 19–1   (a) Bellows type thermostat, (b) remote bulb thermostat.

tains a gas, a liquid, or a saturated mixture of both and is connected to a bellows or diaphragm assembly by means of a capillary tube. The bulb may be clamped in close thermal contact with the evaporator coil to sense evaporator temperature or it may be placed in an airstream to sense the circulating air temperature. The bellows or diaphragm movement is opposed by an adjustable spring pressure.

As the temperature of the control bulb increases, the pressure of the confined fluid will increase, forcing the diaphragm or bellows assembly to expand. Through a system of mechanical linkage the diaphragm or bellows assembly will close a set of electrical contacts to complete the electrical circuit. Decreasing the temperature of the control bulb will produce the opposite effect.

A screw adjustment is used to change the spring pressure. An increase in spring pressure will require a greater pressure in the diaphragm or bellows assembly to close the control contacts,

resulting in a higher cabinet temperature. A decrease in spring pressure will result in a lower cabinet temperature. This temperature adjustment is referred to as the *range adjustment* and is built into the control.

The control bulb, the connecting capillary tube, and the diaphragm or bellows assembly are referred to as the *control power element*. The capillary tube length will vary with the application. The fluid used to charge the power element must be able to provide a positive pressure at low temperatures without an excess of pressure at high temperatures.

Bimetallic Thermostat

The **bimetallic thermostat** has a sensing element comprised of two dissimilar metals, usually Invar and brass or Invar and steel, bonded together. The Invar has a very low coefficient of expansion, whereas the brass and steel have a

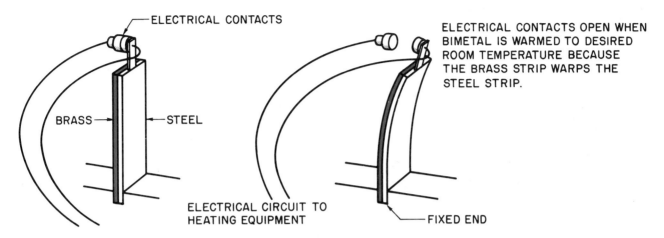

Fig. 19–2   Action of bimetallic strip.

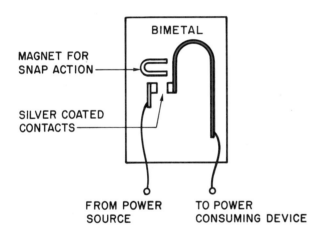

Fig. 19–3   Bimetal type thermometer.

very high coefficient of expansion. Thus, with a change in temperature, the Invar will experience a lesser change in length per degree of temperature change than the brass or steel. Increasing the temperature of the bimetallic strip will cause the bimetal to warp in the direction of the inactive metal, the Invar in this case.

Decreasing the temperature of the bimetallic strip will cause the bimetal to move in the direction of the active metal, or the steel or brass. This movement of the bimetal with changes in temperature is used to open or close the electrical contacts positioned on the bimetallic strip. Figure 19–2 illustrates the operation of a bimetallic strip. Figure 19–3 illustrates the operation of a bimetal in a thermostat application.

Control Contacts

The slow opening and closing of the electrical **control contacts** will cause arcing between the contact points and result in burnt or welded contacts. Consequently, the control contacts are designed to open and close rapidly. This can be accomplished in several ways.

One design of control contacts uses a magnetic snap action operation. As the pressure in the bellows increases and the movable contact arm moves toward the stationary contact, the strength of the magnetic field between the armature and the magnet increases. When the movable contact arm comes within a predetermined distance of the stationary contact, the magnetic field becomes strong enough to overcome the opposing spring tension and the armature is pulled rapidly into the magnet, closing the contact points with a snapping action. As the pressure in the bellows decreases, the spring tension will open the contacts. The force of the spring is opposed by the force of the magnetic attraction between the armature and the magnet. Therefore, when a considerable amount of force has developed in the spring, the contacts will snap open quickly to avoid arcing.

Another design of control contacts uses various types of springs and toggles to produce the snap action movement. As the bellows expand, a control lever will move the spring. When the spring has passed the midpoint of its travel, the contact on the toggle will snap to its opposite position in a motion quick enough to prevent the control contacts from arcing.

Control Differential

The point at which the thermostatic contacts close to start the compressor is referred to as the *cut-in point*. The point at which the thermostatic contacts open to stop the compressor is referred to as the *cut-out point*. The difference between the cut-in and the cut-out points of the control contacts is referred to as the **control differential**. For example, if a control is set to cut in at 40°F and cut out at 20°F, the differential is 20°F. Figure 19–4 illustrates different types of differential control adjustments.

Because a thermostat cannot stop and start a compressor at the same temperature, a difference in temperature, however small, must exist for the control to function. If the differential is too small the system will have a tendency to short cycle. This will reduce the life of the compressor and other system components as well as create other unsatisfactory operating conditions. If the differential is too large, the on and off cycles will be too long, resulting in wide temperature fluctuations.

Thermostats used in modern domestic applications are factory set to provide for optimum performance over a wide range of operating conditions. Manufacturers of thermostats do not recommend recalibrating the thermostatic controls in the field. If the control does not operate in accordance with specifications, it should be replaced.

Range Adjustment

The **range adjustment** of a thermostat is associated with the cut-in and cut-out settings of the control contacts. The range adjustment of a thermostat refers to the minimum and maximum temperature settings of the control contacts. For example, in a control that is set to cut in at 40°F and cut out at 20°F, the differential is 20°F and the range is 20°F to 40°F. Figure 19–5 illustrates typical range adjustments in a control. The range adjustment of a thermostat is an adjustable force that acts directly on the bellows or diaphragm assembly. Depending upon the design of the control contacts, the range adjustment will operate in one of the following ways:

1. Changing the range adjustment will change both the cut-in and cut-out points of the con-

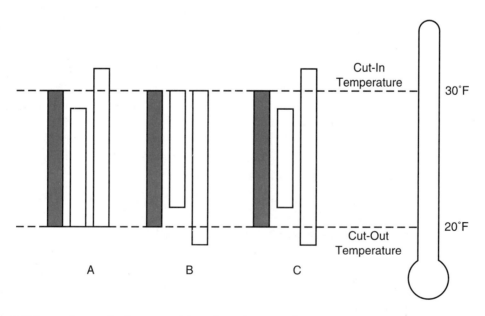

Fig. 19–4   Types of differential control adjustments (a) cut-in setting type, (b) cut-out setting type, (c) cut-in and cut-out type. In each illustration, normal setting is shown in black.

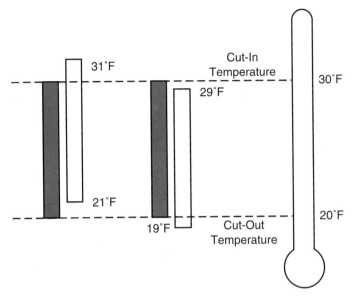

Fig. 19–5   Typical range adjustments.

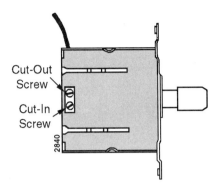

Altitude Adjustment Chart

Counterclockwise Turns			
Altitude In Feet	Cut-In Screw	Cut-Out Screw	
2000	7/60	7/60	
3000	13/60	13/60	
4000	19/60	19/60	
5000	25/60	25/60	
6000	31/60	31/60	
7000	37/60	37/60	
8000	43/60	43/60	
9000	49/60	49/60	
10000	55/60	55/60	1/60 of a Turn Equals 6° of Rotation

Fig. 19–6   The temperature control has two adjustment screws; both must be turned to compensate for altitude. The chart gives the exact turns for each 1,000 ft. The screw rotations are in graduations of sixtieths. These screw turns are critical, so follow the circular scale as a guide.

trol contacts. The differential will remain the same.

2. Changing the range adjustment will change only the cut-in point of the control contacts. The cut-out point will remain the same, but the differential will change.

3. Changing the range adjustment will change only the cut-out point of the control contacts. The cut-in point will remain the same, but the differential will change.

Altitude Adjustment

At altitudes above sea level, the atmospheric pressure decreases and the pressure exerted on a thermostat bellows or diaphragm assembly may be reduced enough to affect the control operation. A refrigerator adjusted for operating at sea level would operate at too low a temperature if moved to 6,000 feet above sea level. As a result, many thermostatic controls are equipped with altitude adjustment screws to compensate for variances in altitude. The adjustment of a control to compensate for altitude variations is critical. Because it is not a common service problem, always refer to the manufacturer's service manual for the particular calibration instructions for the specific con-

trols. Most service manuals have an altitude adjustment chart that indicates the exact calibration procedures for each 1,000 feet of elevation. Figure 19–6 illustrates a typical altitude adjustment chart.

Diagnosing Thermostatic Problems

The thermostatic motor control regulates the compressor running time in order to maintain the proper temperature range in the conditioned cabinet. If the conditioned area is too cold or too warm, an adjustment to the control knob may easily remedy the problem. If further service is required, it will be necessary to check the control operation and/or the operating temperature of the control. A defective thermostat will usually cause the compressor to operate continuously or not at all.

The problem of the compressor not running is usually a result of the thermostatic contacts being burnt open or the power element losing its charge and not allowing the switching contacts to close or open, depending on the design of the control.

If the system is equipped with a condenser fan motor, place the thermostatic control at a low-temperature position where a call for cooling is required. If the condenser fan operates, the thermostat is allowing current to flow to the compressor circuit. If the compressor does not start, refer to the compressor electrical test procedures.

If the system is not equipped with a condenser fan or if the fan motor is not wired through the thermostatic motor control, follow these procedures:

1. Disconnect all power to the system.

2. Remove the thermostat from its mounting position.

3. Remove all electrical connections from the control.

4. Take a continuity reading across the control terminals with an ohmmeter. If no continuity exists, the thermostat has failed and needs to be replaced. If continuity is indicated, check the cabinet wiring harness for possible problems.

5. To confirm that the thermostat has failed, cross the two wires that were initially removed from the control and apply power to the unit. If the compressor starts and continues to run, the thermostat is defective and needs to be replaced. If the compressor does not start, the problem is elsewhere in the electrical system.

If the refrigerated compartments are experiencing temperatures lower than normal and if the compressor is running continuously, the thermostat may be defective. This problem is usually a result of the thermostatic contacts being welded or stuck in the closed position. In applications where the thermostatic sensing bulb is clamped directly to the evaporator surface, make sure that the connection is tight and that a good thermal contact exists between the thermostatic sensing bulb and the evaporator surface. To check the operation of the thermostat, follow these procedures:

1. Connect the temperature sensing probe of a remote reading thermometer securely to the evaporator surface where the sensing bulb of the thermometer is located.

2. Refer to the manufacturer's performance specifications for the system being serviced.

3. Start the unit and allow it to operate for the specified time period. Record the temperature reading of the evaporator at the point of thermostatic sensing bulb contact. If the temperature of the evaporator has been reduced to factory specifications and the thermostat has not cycled the compressor off, the thermostat needs to be replaced. If the temperature of the evaporator cannot be reduced to factory specifications, further testing of the refrigeration system is required.

If factory specifications are not available, the operation of a thermostat can be checked easily. With normal cabinet temperatures prevailing, follow these procedures:

1. Slowly rotate the temperature control knob to the higher temperature setting. The compressor should cycle off.

2. If the compressor does not cycle off proceed as follows:

   a. Disconnect all power to the unit.

   b. Remove the thermostat from its mounting.

   c. Remove one of the leads from the control terminals.

   d. Apply power to the unit. If the compressor does not start, the thermostat is defective and needs to be replaced. If the compressor starts and runs, a defect in the cabinet wiring is causing the problem.

Replacing a Thermostat

When replacing a thermostat, it is important that an exact replacement of the original is used. Replacement thermostats in the form of the manufacturer's original equipment or general replacements are readily available. When selecting a replacement control, refer to the model number and, if applicable, the bill of material number for the unit being serviced. The removal and replacement procedure for specific brands and models will vary. Refer to the manufacturer's service manual for the specific step-by-step replacement procedures. In general, the following procedures can be used:

1. Disconnect all power from the system.
2. Remove the control knob and other hardware to gain access to the control.
3. Remove the control from its mounting.
4. If applicable, remove the clamps securing the thermostatic sensing bulb to the evaporator.
5. Remove the thermostat.

On some refrigerators it may be necessary to remove the cabinet breaker strips before removing the thermostatic capillary tube and sensing bulb. On units where the thermostatic capillary tube is routed between the inner liner and the outer cabinet shell, tie a heavy cord or wire to the end of the capillary tube before pulling it out of the cabinet. Carefully fasten the new thermostatic capillary tube to the cord or wire and slowly pull it through the cavity. Reverse the procedure to install the new control.

DEFROSTING SYSTEMS

Excessive buildup of frost or ice on the evaporator will greatly reduce heat transfer and system efficiency. Periodically defrosting the evaporator coil is necessary to maintain a maximum heat transfer surface. An automatic defrosting system is determined by four principal ways:

1. Daily by clock-timing—where an electric clock-timer places the system into a defrosting cycle at a predetermined time interval. This is a common method employed in some refrigerators and freezers and many types of low-temperature units you'll find in restaurants.
2. By accumulated compressor running time— where an electric clock-timer is wired in parallel with the compressor motor circuit. The timer motor runs whenever the compressor motor runs and activates a defrosting cycle after a certain amount of compressor running time, usually every 6 to 12 hours, depending on the type of equipment and the manufacturer's design.
3. By accumulated open-door time—where an electric clock-timer is wired in parallel with the door switch. Every time the door is opened, the clock-timer is energized. After a predetermined accumulation of open-door time, the system will cycle into defrosting.
4. By accumulated door-opening count— where the system is cycled into defrosting based upon the number of times the door is opened. When the predetermined door-opening count is reached, the system will cycle into defrosting. The predetermined door-opening count is usually 60 times.

Methods of Defrosting

The evaporator coil can be defrosted using either hot gas defrosting or electric defrosting.

Hot Gas Defrosting

As shown in Figure 19–7A, **hot gas defrosting** depends on the discharge gas from the compressor being bypassed from the condenser and directly routed to the evaporator coil. When the clock-timer initiates the defrosting cycle, the hot gas solenoid valve will open and the evaporator fans will cycle off. In some applications, the condenser fan motor will cycle off in order to maintain the discharge gas temperature. Drain heaters will be energized. The compressor discharge gas will bypass the condenser and flow directly into the evaporator coil to defrost the coil. The flow of the discharge gas into the evaporator is facilitated by the following:

1. An easier path through which the gas can flow—The hot gas bypass line offers less restrictions than the condenser coil. The discharge gas will take the path of least resistance, which is directly into the evaporator coil.

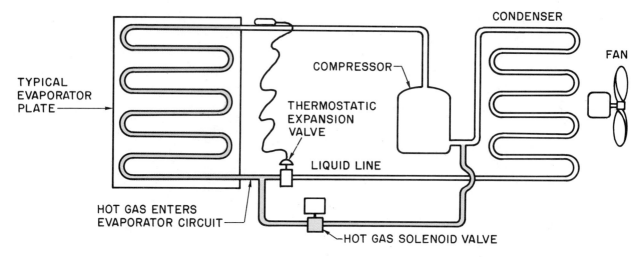

Fig. 19–7A   Hot gas defrost.

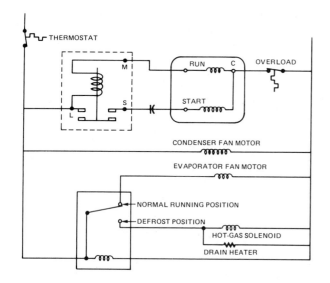

Fig. 19–7B   Wiring for hot gas defrost.

2. A pressure-temperature differential—In comparison to the condenser, the evaporator coil has a lower temperature and pressure. This differential will facilitate the flow of hot gas to the evaporator.

As the hot gas flows through the evaporator, latent and sensible heat will defrost the coil. If the hot gas condenses into a liquid after passing through the evaporator, a suction accumulator located at the outlet of the evaporator will collect and vaporize the liquid, preventing floodback to the compressor. After the defrosting cycle is ter-

minated by the clock-timer, the solenoid valve will close and the system will be placed back into normal operation. On some systems, the energizing of the evaporator fans and drain heaters will be delayed to prevent the circulation of warm air and to allow the condensate water to drain. Figure 19–7B shows a hot gas defrost electrical circuit.

Electric Defrosting

Electric defrosting is the most popular method used to defrost the evaporator coil in modern refrigerators and freezers. The clock-timer (also known as the defrost timer) controls the complete operation of the normal refrigeration and defrosting cycles and is considered to be the heart of the system. When the clock-timer initiates the defrosting cycle, the compressor and, on most models, the condenser fan, will also be cycled off. The evaporator fan will also be shut down and the main defrost heater and drain heaters will cycle on.

During the defrosting cycle the high wattage resistance heaters used to defrost the evaporator coil may cause the coil to overheat. To prevent this, a bimetallic switch (often referred to as a defrost termination thermostat) is wired in series with the main defrost heater and is placed in close thermal contact with the evaporator coil to sense the temperature of the coil. When the evaporator

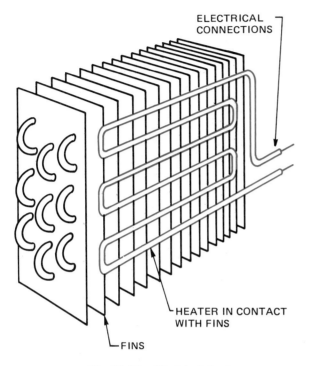

Fig. 19–8A   Electric defrost.

coil temperature reaches a predetermined maximum, the bimetallic switch will open and remove the main defrost heater from the circuit. The defrost timer continues to run even though the heater is deenergized. During a defrost cycle of 25 minutes the heater may be energized for only 10 minutes. The remaining 15 minutes of the defrost cycle time allows the water to drip from the coil and run down the drain system before the refrigeration system starts again.

Some systems will delay energizing the evaporator fans and drain heaters to prevent the circulation of warm air and to allow condensate water to drain. Electric defrosting is illustrated in Figures 19–8A and B.

Components of Defrosting Systems

The operation of defrosting systems relies on defrost clock-timers, defrost limit switches, defrost heaters, and drain heaters.

Defrost Clock-Timers

The heart of the automatic defrosting system is the **defrost clock-timer**. The purpose of the defrost clock-timer is to initiate the defrosting cycle and after a predetermined length of time, place the system back into the normal refrigeration cycle. The operation of the defrost clock-timer and its related components will vary from one manufacturer to another but its basic operation for both domestic and commercial applications is illustrated in Figures 19–9A and B.

In many cases a defrost limit switch, wired in series with the main defrost heater, will deenergize the heater to prevent the evaporator coil from overheating. All other defrost electrical components will continue to operate as long as the system is in the defrosting cycle. This will allow the areas affected to remain warm, permitting the defrost water to drain without freezing.

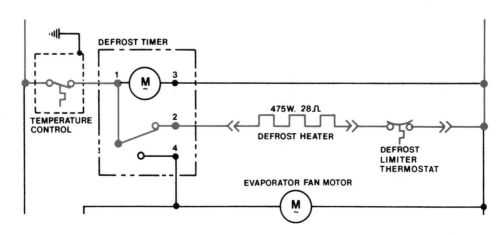

Fig. 19–8B   Wiring for electric defrost. *Courtesy Frigidaire Company.*

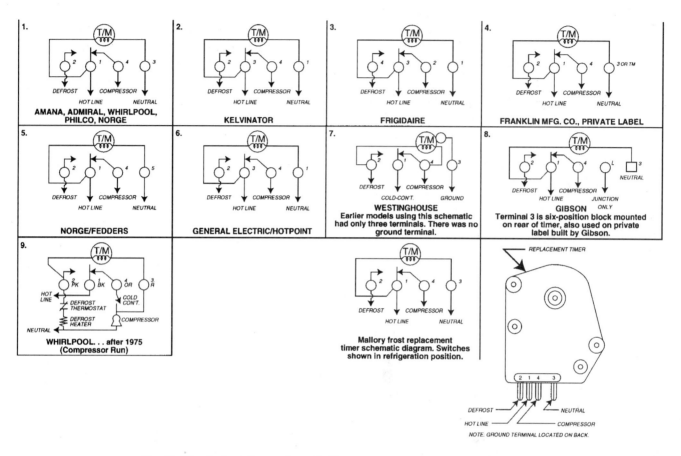

Fig. 19–9A Defrost timer schematic diagrams. *Courtesy Paragon Electric Co., Inc.*

At the end of the defrosting cycle the clock-timer will switch off the defrost circuit and energize the compressor and evaporator fan motor circuits. On some models, when the system switches back into the normal refrigeration cycle, the energizing of the evaporator fan may be delayed until the evaporator coil temperature is reduced sufficiently to prevent the circulation of warm air. Also, on some models, the drain heaters may remain energized for a short period of time after the normal defrost cycle is established to allow the defrost water to drain completely.

If the defrost clock-timer is suspected to be the cause of a problem, check to see if the timer motor is operating. Most clock-timers have a small window located on the back of the timer motor housing through which motor operation, indicated by a rotating wheel, can be observed. If the gear is not rotating and power is applied to the timer motor, the timer needs to be replaced.

To check the defrost clock-timer switching circuits, disconnect all power to the unit and remove all wires from the timer harness. If the wiring connections are comprised of individual wires rather than a quick disconnect harness, make sure that each connection is labeled in order to identify from which terminal of the timer it was removed. This will ensure easy reinstallation. A control shaft is provided on the timer body to manually set the timer and cycle it through the refrigeration and defrosting cycles. When making adjustments, some control shafts are designed to rotate only in a counterclockwise direction, while others are designed to rotate only in a clockwise direction.

The first click will place the clock-timer in the defrosting position. The second click will place the clock-timer back into the normal refrigeration cycle. If the control shaft is rotated too rapidly, the second click may not be heard or the defrost-

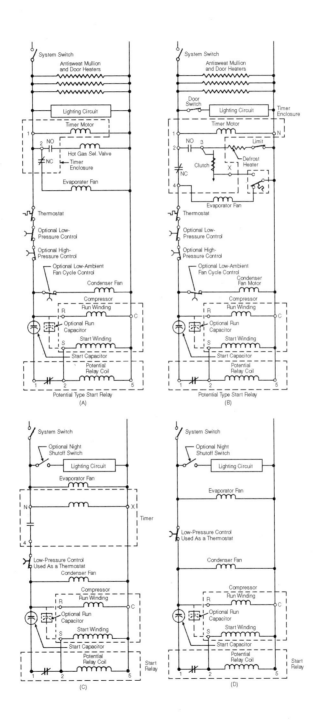

Fig. 19–9B   Defrost timer and operating sequence.

ing cycle may be passed over. Use an ohmmeter to take a continuity reading and determine if the contact points inside the timer are making and breaking. If no continuity is indicated between the designated terminals, the timer needs to be replaced. Replacement timers are readily available from the original manufacturer, or they can be found in the form of general replacement kits. Always refer to the model number of the unit being serviced for the proper replacement component.

Defrost Limit Switch

The purpose of the **defrost limit switch** is to deenergize the main defrost heater after ice and frost have melted from the evaporator coil. The defrost clock-timer initiates and terminates the defrosting cycle; the defrost limit switch controls the duration of time the main defrost heater remains energized.

The defrost limit switch, illustrated in Figure 19–10, consists of a bimetallic switch located on or near the evaporator coil. It is wired in series with the main defrost heater. After all the ice and frost have melted from the evaporator coil, the temperature of the evaporator will start to increase. When the temperature limit of the switch is reached, the bimetallic switch will warp in such a direction as to open the switch contacts, removing the main defrost heater from the circuit before the evaporator coil can overheat. The temperature at which a bimetal opens will vary. After the temperature of the evaporator has cooled sufficiently, usually to around 15°F, the bimetallic switch will warp in the opposite direction, closing the switch contacts.

A defective defrost limit switch will prevent the main defrost heater from energizing. This can result in an ice-restricted evaporator coil, causing higher than normal operating temperatures because air cannot circulate properly through a clogged coil. An ohmmeter can be used to test the defrost limit switch. Make sure that the defrost limit switch is chilled below its cut-in point and follow these procedures:

1. Disconnect all power from the system.

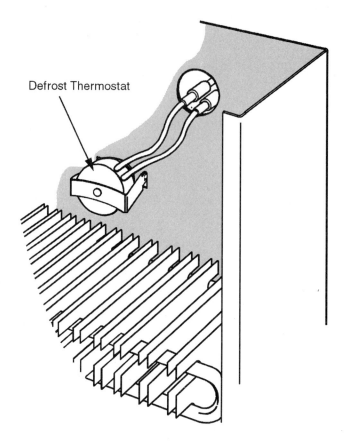

Fig. 19–10   Defrost thermostat or limit switch.

2. Remove the defrost limit switch leads from the circuit.

3. Take a continuity test across the switch leads. If continuity is indicated, the switch is good. If no continuity is indicated, the switch is defective and needs to be replaced.

Always replace a defective defrost limit switch with an exact replacement part. Make sure that the switch is mounted firmly and in good thermal contact with the mounting surface. Splice the wire leads with waterproof terminal connectors. The defrost limit switch should never be bypassed; this could result in damage to the unit.

Defrost Heaters

Frost-free automatic defrosting systems utilize one or more high wattage **defrost heaters** to melt the ice or frost that accumulates on the evaporator during the normal running cycle of the compressor. Some models use a radiant heater where the heating element is encased in a glass tube; other models use a tube resistance heater located in the recesses between the fins of the evaporator coil.

A defective defrost heater will result in the accumulation of ice and frost on the surface of the evaporator coil, restricting airflow and resulting in increased compressor running times and higher than normal cabinet temperatures. When replacing a defrost heater, an exact duplicate of the original must be used. If one heater in a series of several fails, it is recommended that all of the heaters in the series be replaced. Each model will utilize a different procedure in removing and replacing the defrost heater. Refer to the manufacturer's instructions whenever possible. In general, the following guidelines can be used in checking a defrost heater:

1. Disconnect all power from the system.

2. Carefully remove all evaporator covers in order to gain access to the defrost heater.

3. Remove all ice and frost from the evaporator in order to service the system and prevent residual icing of the coil when the system is placed back into normal operation.

4. Once exposed, make a visual inspection of the heater. Look for broken glass, opaque glass, a broken or burnt heater element, or corroded terminal end caps on the glass-encased heater.

5. Disconnect the terminal leads to the heating element.

6. Using an ohmmeter, take resistance readings across the heating element leads. The meter should indicate a resistance reading as specified by the manufacturer. If it does not, the defrost heater needs to be replaced. If no resistance reading is indicated, the heater is open and needs to be replaced.

The removal and replacement of a defrost heater must be done carefully. On some applications the entire evaporator coil must be lifted out of the cabinet. Handle the sharp coil and fin edges with care. In addition, it is important not to twist, kink, or break the tubing.

Handle the glass-encased heater by the end caps only. Do not touch the glass tube itself. Salt deposits from your hand could cause premature failure of the heater. If the glass-encased heater is touched, clean it with a paper towel, not a cloth.

The tube resistance heater can be replaced easily by carefully pressing it back into the fin recesses. Handle the sharp coil fins with care. Reconnect all electrical leads using waterproof connectors and arrange the heating element leads so that they are not in direct contact with the heating element before replacing the trim and covers.

Drain Heaters

Drain heaters are used to prevent the drain troughs and drain tubes from freezing. This will allow the melted ice and frost to drain completely from the evaporator during the defrosting cycle. The drain heater is energized when the defrost clock-timer initiates the defrosting cycle. The drain heater remains energized throughout the defrosting cycle. Many models utilize the main defrost heater in directing heat to the drain area; in these cases, a separate drain heater may not be required.

A failed drain heater will usually result in an ice-clogged drain. This will eventually cause an ice buildup that will restrict airflow across the evaporator coil and result in higher than normal cabinet temperatures. The drain heater can be checked easily by taking a resistance reading across the heater leads. The meter should indicate a resistance reading in accordance to the manufacturer's specifications. If it does not, or if no resistance is indicated, the heater needs to be replaced.

Unit Nineteen Summary

- In order for a refrigeration system to operate within the design conditions intended, some method of capacity control is necessary. A thermostat is used to start and stop a refrigeration system.

- A thermostat is basically a single-pole, single-throw temperature-operated switch that will close on a rise in temperature and open on a drop in temperature. Two types of thermostats are the bulb or fluid thermostat and the bimetal thermostat.

- Within a thermostat, the control contacts open and close to make and break a circuit to the compressor and, if applicable, the evaporator and condenser fans. The point at which the contacts close is the cut-in point and the point at which the contacts open is the cut-out point. The difference between the cut-in and cut-out points is referred to as the control differential.

- The range adjustment of a thermostat refers to the maximum and minimum temperature settings of the control contacts.

- In low temperature refrigeration systems, some method of defrosting the evaporator must be employed because a buildup of frost will reduce heat transfer and system efficiency. Two methods of defrosting a coil are the hot gas defrost system and the electric heater system.

- In a hot gas defrost system the refrigerant is redirected to warm the evaporator and accomplish the defrosting process.

- In an electric defrost system, a timer is used to cycle the equipment to the defrost mode and a bimetal thermostat is used to break the circuit to the electric heating element when a pre-determined temperature has been achieved. Drain heaters are also used to prevent an ice-clogged drain.

Unit Nineteen Key Terms

Thermostat

Bulb Thermostat

Bimetallic Thermostat

Control Contacts

Control Differential

Range Adjustment

Hot Gas Defrosting

Electric Defrosting

Defrost Clock-Timer

Defrost Limit Switch

Defrost Heater

Drain Heater

UNIT NINETEEN REVIEW

1. What is the purpose of a thermostat?
2. What are the two types of temperature sensing elements in common use?
3. Describe the operation of the bulb thermostat.
4. Describe the operation of the bimetallic thermostat.
5. Why are electrical control contacts designed to open and close rapidly?
6. What is the purpose of snap action in thermostatic control contacts?
7. What two methods are used to obtain a snap action movement of the control contacts?
8. Define control differential.
9. Define what range adjustment is in a thermostat.
10. If the altitude above sea level increases, what happens to atmospheric pressure? What adjustment must then be made to the altitude adjustment screw?
11. How may the compressor be responding if the thermostatic contacts are burnt open, or the power element is losing its charge and not allowing the switching contacts to close?

12. If the system is equipped with a condenser fan motor, the thermostatic control is placed at a low temperature that calls for cooling, and the condenser fan motor does not start, what is the problem?

13. Describe the procedure for testing a thermostat.

14. What is the purpose of defrosting an evaporator?

15. What are the four methods by which an automatic defrosting cycle is determined?

16. What methods are used to defrost an evaporator coil?

17. What are the two major factors that facilitate the flow of hot discharge gas into the evaporator?

18. Why is the condenser fan cycled off during the hot gas defrosting cycle?

19. What purpose does a suction accumulator serve in a hot gas defrosting system?

20. Describe the sequence of operation of an electric defrosting system.

21. What purpose does the bimetallic switch serve in an electric defrosting system?

22. What is the purpose of delaying the operation of the evaporator fans upon termination of the defrosting cycle?

23. What purpose does the defrost clock-timer serve in an automatic defrosting system?

24. What is the difference between a defrosting system by accumulated open door time and accumulated door opening count?

25. Why is the condenser bypassed in the hot gas method of defrosting so that hot gas passes from the compressor to the evaporator?

26. Match the following terms with their descriptions: defrost clock-timer, defrost limit switch, defrost heater, drain heater.

    a. Melts ice or frost that accumulates on the evaporator.

    b. Melts ice and frost to drain from the evaporator.

    c. Deenergizes the main defrost heater after ice and frost have melted from the evaporator coil.

    d. Initiates the defrosting cycle.

27. Describe the procedure used to check the defrost clock-timer.

28. Describe the method used to test the defrost limit switch.

29. Describe the procedure used to test the defrost heater.

30. Why shouldn't the glass-encased heater be handled with bare hands?

31. The control contacts were set to cut in at 40°F and cut out at 20°F.

    a. What is the temperature differential in Celsius?

    b. What is 40°F in Celsius?

32. Write a paragraph using the following terms: cut-in point, cut-out point, differential, and control contacts.

SECTION FOUR

Domestic Refrigeration Systems

Refrigeration and electrical systems, as they apply to refrigerators, freezers, and room air conditioners, are one aspect of a service technician's responsibility. In this section we'll review some of the service procedures related to domestic refrigeration systems.

While subtle differences exist from manufacturer to manufacturer and even from model year to model year in the domestic refrigeration/appliance field, there are some basic processes that apply to this segment of the service industry. Electrical and refrigeration principles apply to this type of equipment in the same way that they apply to HVAC and commercial refrigeration systems.

If you plan a career in this area, further study and concentrating your work experience in this area will be required in order to become proficient as a technician servicing domestic systems.

UNIT TWENTY

Refrigerators and Freezers

OBJECTIVES

After completing this unit, the student will be able to:

1. Describe the method of construction of a domestic refrigerator.

2. Explain the methods used to test light switches, elements, and fan motors.

As a service technician you may be expected to service a refrigerator or freezer. In this unit, we'll discuss some of the basic service procedures related to domestic refrigeration equipment and also discuss step-by-step service procedures for switches, evaporator fan motors, and condenser fan motors.

REFRIGERATOR CABINETS

The construction, design, and features of refrigerator cabinets vary, depending on the manufacturer and model of unit. Generally the cabinets used in domestic refrigerators and freezers are constructed of a steel sheet outer cabinet with a separate inner liner of painted steel porcelain on steel or plastic. The outer cabinet and inner liner are separated by an insulating material, such as fiberglass or urethane foam. Plastic breaker strips, which are installed between the inner liner and the outer cabinet, act as an insulator to prevent the conduction of heat from the outer cabinet to the inner liner. The door construction and cabinet hardware are also designed to minimize the leakage of heat into the conditioned area.

To prevent moisture from entering the insulated space between the inner liner and the outer cabinet, moisture barriers, such as plastic, are placed over the outside of the insulation. In addition, the holes where the refrigerant and electrical lines were run through the cabinet are sealed with permagum putty and a sealant is used in the cabinet and seams.

The cabinet shelving is constructed of aluminum, stainless steel, or tinned steel. It may be of an open grill design to allow for the circulation of air through the cabinet. The doors of modern refrigerators and freezers are sealed by a rubber magnetic gasket designed to form an airtight seal between the door and the cabinet when the doors are closed, as illustrated in Figure 20–1. In most applications the cabinet must be leveled properly to allow the doors to close when released, sealing the cabinet by virtue of their weight and the magnetic gasket.

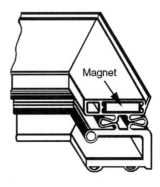

Fig. 20–1   Cross section of magnetic door gasket.

The removal, replacement, and service procedures for refrigerator and freezer cabinet accessories and devices vary with each manufacturer and model of unit. Refer to the unit's service manual for specific step-by-step service and repair procedures.

ELECTRICAL COMPONENTS

Lights and Switches

Light accessories are installed in the food compartments as well as in the freezer compartments of many refrigerators. A **switch** located in the breaker panel on the hinge side of the door controls the light operation, illustrated in Figure 20–2. If a cabinet light fails to operate when the door is opened, first check the bulb and replace it if necessary. To check the light switch, follow these procedures:

1. Disconnect all power to the unit.

2. Remove the switch from the cabinet.

3. Disconnect the leads to the switch.

4. Using an ohmmeter, check for continuity with the switch in the released position. If no continuity is indicated the switch has failed and needs to be replaced. If continuity is indicated, check the bulb socket and unit wiring.

Heating Elements

A variety of cabinet **heaters** are installed in modern refrigerator and freezer cabinets to prevent moisture condensation on certain areas of the fixture. The location of these heaters varies with each brand and model. Heaters may be located around the door frame, the divider panel between the food and freezer compartments, or the sides and top of the cabinet, as illustrated in Figures 20–3, 20–4, and 20–5. The failure of these heaters is usually indicated by a collection of moisture condensation on the surface of the cabinet in the area where the heater is located. To check the cabinet heaters, follow these procedures:

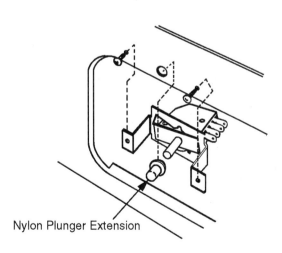

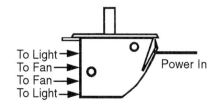

Nylon Plunger Extension

To Light
To Fan
To Fan
To Light
Power In

Fig. 20–2 Exploded view of door switch and switch terminals.

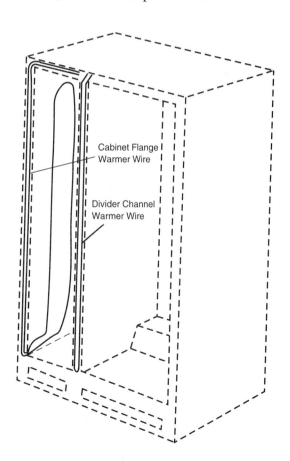

Cabinet Flange
Warmer Wire

Divider Channel
Warmer Wire

Fig. 20–3 Cabinet heaters.

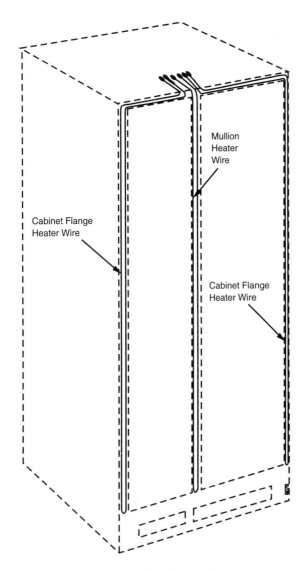

Fig. 20–4   Cabinet heater wiring.

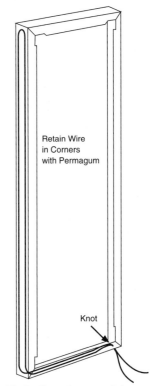

Fig. 20–5   Door heater wiring.

1. Disconnect all power to the unit.

2. Disconnect the electrical leads to the heater.

3. Using an ohmmeter, check resistance across the heater. If the resistance value of the heater is in accordance with the manufacturer's specifications, the heater is good and the cabinet wiring should be replaced. If the resistance value of the heater is not in accordance with the manufacturer's specifications, the heater needs to be replaced.

Butter Conditioners

Some older refrigerators have a separate compartment to store butter. Usually located in the refrigerator door, this compartment is isolated from the food compartment by a small door. In some applications a resistance heater is used to maintain the butter compartment's temperature above the food compartment's temperature so that the stored butter can be spread easily when used. To check the resistance heater, follow these procedures:

1. Disconnect all power to the unit.

2. Disconnect the electrical leads to the butter conditioner.

3. Using an ohmmeter, take a resistance reading across the butter conditioner leads. If the resistance value is in accordance with the manufacturer's specifications, the resistance heater is good and the cabinet wiring should be checked. If the resistance value is not in accordance with the manufacturer's specifications, the heating element needs to be replaced.

Evaporator Fan Motors

Circulating fans are installed in forced-convection evaporator systems to circulate air through the food and freezer compartments. If the fan fails to operate or if it runs erratically, the reduced air circulation will cause unsatisfactory temperature conditions in the food and freezer compartments. To check the **evaporator fan motor**, follow these procedures:

1. Disconnect all power to the unit.

2. Remove the cabinet covers and trim as required to gain access to the evaporator fan motor.

3. Disconnect the electrical leads to the fan motor.

4. Using a test cord, apply power to the fan motor. If the fan motor operates, check the door switch and any thermostatic devices that may control the fan operation. If the fan motor fails to operate, it needs to be replaced.

Exact replacement fan motors are available from the original manufacturer. General replacement motors are also available from various sources. When replacing an evaporator fan motor, make sure that the fan blade is installed properly. Many replacement motors allow the service technician to change the direction of motor rotation, depending on the application. In these cases make sure that the motor rotation is in accordance with the manufacturer's specifications.

Condenser Fan Motors

Most domestic units utilize a low wattage shaded pole motor to circulate air across the condenser, as in Figure 20–6. Proper airflow across the condenser is important for efficient system operation. A reduction in airflow will result in higher than normal compartment temperatures. When higher than normal system discharge pressures, operating temperatures, and compressor motor current draw are experienced, the compressor will cycle off on the overload protector.

The **condenser fan motor** is usually wired in parallel with the system compressor. If the compressor will run but the condenser fan motor will not, check the motor as follows:

AIR COOLED CONDENSER AND FAN

Fig. 20–6 Condenser fan motor. *Photo by Bill Johnson.*

1. Disconnect all power to the unit.

2. Remove the motor from its mounting.

3. Apply a test lead directly to the fan motor. If the fan motor operates, check the wiring to the motor. If the fan motor does not operate, it needs to be replaced.

Some older condenser fan motors are equipped with an oil port for lubrication purposes. However, most are considered to be permanently lubricated. In fan motors without an oil port, the motor is permanently sealed and no lubrication is required. When replacing a condenser fan motor, make sure that the wattage and the direction of motor rotation matches the original motor.

Drain Trough Heaters

Drain trough heaters are installed to prevent the defrost water from freezing before it leaves the refrigerated cabinet, as in Figure 20–7. These heaters are usually controlled by the defrost clock-timer and are energized only during the defrosting cycle. Many units are equipped with an auxiliary heater that is installed beside the original heating element. The auxiliary heater serves as a replacement should the original heater become inoperative. To test the drain trough heater, follow these procedures:

1. Disconnect all power to the unit.

2. Disconnect the electrical leads to the heater element.

3. Using an ohmmeter, take a resistance reading across the heater lead terminals. If the heater shows no sign of resistance or if the resistance reading is not in accordance with the manufacturer's specifications, the heater needs to be replaced.

Unit Twenty Summary

• Construction, design, and cabinet features vary from manufacturer to manufacturer and from model year to model year. The inner liner of a refrigerator can be steel or plastic and the insulation between the inner and outer liner may be foam or fiberglass.

• Light and fan switches are installed in the breaker panel on the hinge side of the cabinet door. Butter conditioners may be found on some older refrigerators. They are resistance elements located in a section of the door and are designed to soften the butter.

• Evaporator fan motors are used to circulate air through a frost-free refrigerator and condenser fan motors may be found on some units. The condenser fan motor has two purposes: to assist in dissipating the heat from the refrigerant as it passes through the condenser and to help keep the compressor cool.

• Drain trough heaters are used in some units to prevent an ice buildup in the trough under the evaporator.

Unit Twenty Key Terms

Light Accessories

Switch

Heater

Evaporator Fan Motor

Condenser Fan Motor

Drain Trough Heater

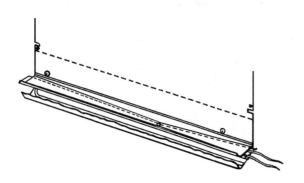

Fig. 20–7   Drain trough heater.

UNIT TWENTY REVIEW

1. How does the technician begin the process to service and repair a refrigerator, its accessories, and devices?

2. The first step, if the cabinet light fails, is to check the bulb. What will ultimately indicate the need to replace the light switch?

3. What types of insulation are used in refrigerator cabinets?

4. What is the purpose of heating elements?

5. What will cause a heating element to fail?

6. Describe the method used to check cabinet heaters.

7. What is the purpose of a butter conditioner?

8. What is the next step to check the resistance heater of the butter compartment if the resistance valve for the heater meets the manufacturer's specifications?

9. Give the general purpose for using a test cord to power an evaporator fan motor.

10. Describe the testing procedure for an evaporator fan motor.

11. What will happen if a condenser fan motor fails?

12. Describe the testing procedure for a condenser fan motor.

13. When replacing a condenser fan motor, give two things that must be checked.

14. What is the purpose of a drain trough heater?

15. What should the technician do if the resistance reading across the lead terminals of the drain trough heater do not meet the manufacturer's specifications?

16. Describe the procedure for checking a drain trough heater.

17. Ohm's Law: $E = I \times R$

    What do the symbols E, I, and R represent?

18. The heater has 9.6 ohms of resistance, using 12 amperes of current. What is the voltage?

19. The heater of the refrigerator's defroster is a resistance wire. If the heater uses 110 volts and draws 5 amperes, what is the resistance wire?

20. (P)ower $= I \times E$

    Watts $= I \times E$

    A central air conditioning unit is rated at 4,600 watts. The unit uses 230 volts. How many amperes are flowing through the electrical cable?

21. In a series circuit the total resistance is equal to the sum of each device's resistance.

    $$R_T = R_1 + R_2 + R_3 = ...$$

    In a parallel circuit the total equivalent resistance is equal to each value of resistance added together and then divided into 1.

    $$R_T = \frac{1}{1/R_1 + 1/R_2 + 1/R_3}$$

Two resistors are in a parallel circuit. The resistors have resistances of 20 ohms and 25 ohms. What is the equivalent resistance of the circuit?

22. Three panel heaters in a refrigerator are wired in parallel, drawing 1,000 ohms, 500 ohms, 1,000 ohms. What is the equivalent resistance of the circuit?

23. Write a paragraph describing the refrigerator cabinet.

UNIT TWENTY-ONE

Room Air Conditioners

OBJECTIVES

After completing this unit, the student will be able to:

1. Describe the method of operation of a room air conditioner.

2. Identify the electrical components used in a room air conditioner.

3. Explain the steps to follow when installing a window or through-the-wall air conditioner.

This unit is intended to give you an overview of the operation of a room air conditioner. Sometimes referred to as window units or "window shakers" by some technicians, room air conditioners are found in older homes that are not equipped with a central comfort cooling system.

A room air conditioner is capable of circulating, cleaning, dehumidifying, and cooling the air in an occupied space. In some applications it can also provide ventilation and heating. It is designed primarily to cool the area of one room only and does not require any duct work. Figures 21–1A and B illustrate typical room air conditioners.

A room air conditioner operates by drawing warm, moist air from the conditioned space into the unit, passing it through a filter that removes dust, lint, and other impurities, and then passing it through the evaporator coil where the heat and moisture in the air are released to the evaporator coil. The cool air is then recirculated back into the conditioned space. Figure 21–2 illustrates a typical room air conditioner airflow.

FEATURES OF ROOM AIR CONDITIONERS

Fresh Air Ventilation

Many room air conditioners have provisions to bring in a small percentage of fresh outside air to mix with the recirculating room air. The percentage of outside air brought in is usually around 20 percent of the total amount of the circulating air volume. The fresh air is brought in through a cable-controlled damper which opens a passage-

Fig. 21–1A   Room air conditioner. *Courtesy Frigidaire Company.*

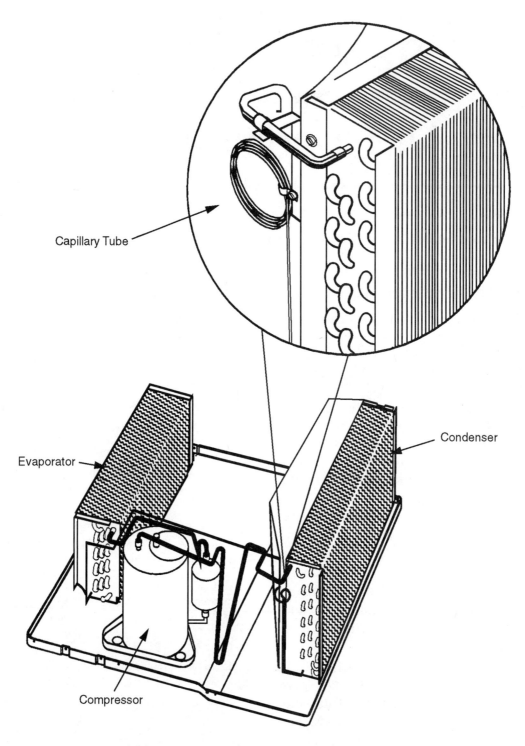

Capillary Tube

Condenser

Evaporator

Compressor

Fig. 21–1B   The components of a room air conditioner.   *Courtesy Fedders North America.*

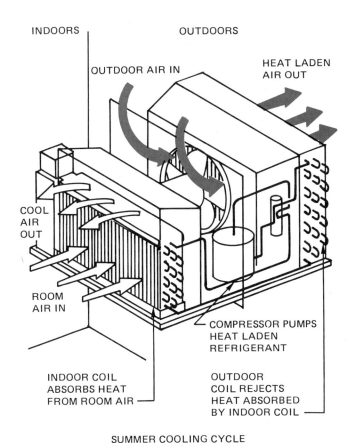

Fig. 21–2   Room air conditioner air flow.

way, allowing fresh air and the room air to mix before being cooled and returned to the conditioned space. This feature prevents the room air from becoming fouled by cooking, smoking, and other odors. Figure 21–3 represents a typical room air conditioner fresh air and exhaust system.

Introducing warm outside air into the conditioned space imposes an additional heat load on the air conditioner. The outside air must not only be cooled, but the moisture contained in it must be condensed and removed. Depending upon the prevailing temperature and the humidity, the capacity of the unit may not have the ability to reduce the conditioned room air to the desired temperature. For this reason, the fresh air vent should be kept closed during normal operation and opened only far enough to permit the removal of odors.

Exhaust

Many room air conditioners are equipped with an **exhaust** feature that removes air from the conditioned space. Approximately 25 percent of the circulating air volume will be exhausted to the outdoors, quickly removing any foul air from the room. During normal operation the exhaust damper should be kept in the closed position. The air exhausted from the room will be replaced by heat-and-moisture-laden outside air which will place an additional heat load on the unit and reduce the cooling capacity.

Capacity

The cooling and heating **capacity** of the room air conditioner is measured in Btu's per hour. System capacities range from approximately 4,000 Btu's per hour to 30,000 Btu's per hour. Unit application charts are available from the manufacturer to determine the proper size unit required for a specific application.

Room air conditioners are available in a variety of types and designs that will fit almost any application. They may be mounted in windows or installed in the wall of the area to be conditioned. The main advantage of window installation is that the unit can be removed easily when the cooling season is over. With the variety of installation kits available, a unit can be easily adapted to fit various window sizes and types. Modern room air conditioners are light in weight, compact, and attractive. Figure 21–4 illustrates a typical window-mount room air conditioner. Figure 21–5 shows a through-the-wall installation.

Voltage Requirements

The **voltage required** for room air conditioners ranges from 120 volts to 240 volts. Most of the units rated at 120 volts are designed to plug into a standard 120-volt receptacle. The units rated at 208/240 volts will require a special receptacle. Room air conditioners should be wired into an

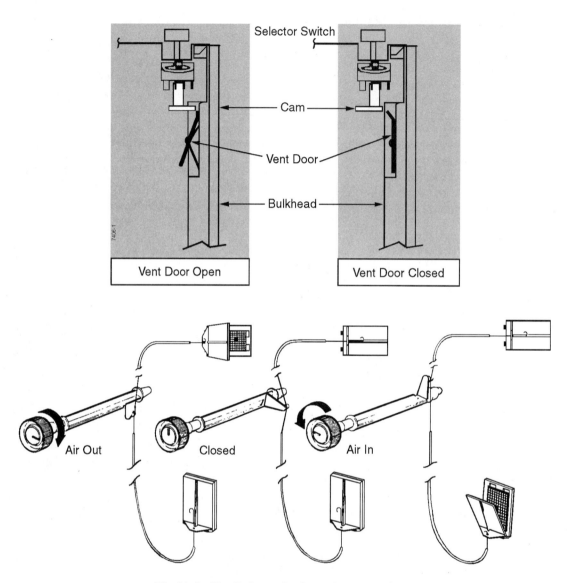

Fig. 21–3   Ventilation and exhaust door operation.

independent voltage source in order to prevent nuisance tripping of the circuit breakers caused by overloading the circuit. The supply of voltage must be within plus or minus 10 percent of the equipment requirements. Local codes should be followed in all installations. Figure 21–6 illustrates line cords and plugs for the various voltages used on room air conditioners.

Energy Efficiency Ratio

The **energy efficiency ratio (EER)** or the seasonal energy efficiency ratio (SEER) is the measurement of the amount of energy consumed by the air conditioner as opposed to the amount of cooling capacity provided. The ratio is calculated by dividing the cooling capacity of a system in Btu's per hour by the total power input in watts at any given set of rating conditions. The total power input includes compressor and fan motor wattage. The EER is expressed in Btu's per hour per watt, or Btu/watt. The higher the EER rating the more energy efficient the unit is considered to be. For example, the total cooling capacity of a unit is 36,000 Btu/hour at 95°F outdoor temperature, 67°F indoor wet bulb temperature, and at 80°F indoor dry bulb temperature with 50 percent rela-

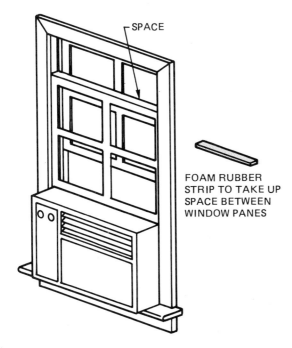

Fig. 21-4   Window unit showing necessary parts to mount unit safely in window and seal openings.

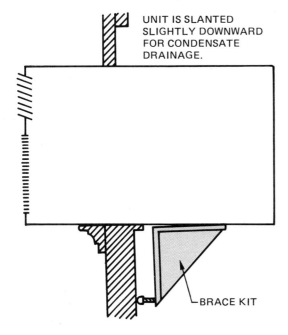

Fig. 21-5   Typical through-the-wall unit air-conditioning installation.

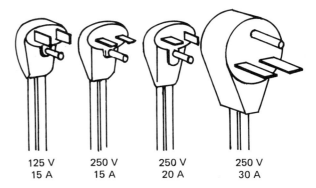

Fig. 21-6   Line cords and plugs for various voltages used on room air conditioners.

tive humidity. The total power input is 5.1 kilowatts or 5,100 watts. The energy efficiency ratio can be calculated as follows:

$$EER = \frac{\text{cooling capacity in Btu/hr}}{\text{power input in watts}}$$

COMPONENTS OF A ROOM AIR CONDITIONER

The components of a room air conditioner are similar to those of any other sealed system, and the service procedures are also basically the same. The refrigerant used in a room air conditioner is usually R-22. The metering device is usually a capillary tube equipped with a factory-installed strainer at the inlet. Depending on the size and design of the evaporator coil, a single capillary tube may be used, or several capillary tubes manifolded off of the liquid line feeding parallel refrigerant circuits into the evaporator may be used. This allows for the use of a smaller evaporator coil, reduces refrigerant pressure drop across the evaporator, and allows for a uniform coil temperature.

Compressor

A room air conditioner usually uses a reciprocating **compressor**. However, rotary compressors are also popular with some manufacturers. The unit is usually of the full hermetic design with process tubes installed to gain sealed system access. The compressor has a permanent split capacitor motor that utilizes a running capacitor. In some applications, a hard start kit in the form of a relay and starting capacitor is used when a high starting torque is required. The motor in a reciprocating unit is cooled by allowing the cool suction vapor returning from the evaporator to pass around the motor windings before being compressed and discharged.

Evaporator and Condenser Coils

The **evaporator** and **condenser coils** in a room air conditioner are usually of the fin and tube design. However, some manufacturers utilize a tube and spiral-spine fin arrangement. The tubing coils are usually manufactured of copper and the fins are usually made of steel or aluminum. In some applications the use of aluminum evaporator coils may be found.

Evaporator and condenser coils must be kept clean. Periodic maintenance and clean evaporator filters will help maintain system efficiency. If the evaporator or condenser fins become damaged, a fin comb can be used to straighten the bent fins. In using a fin comb, select the proper tool for the coil's fins per inch application and carefully rake the coil to straighten the damaged fin areas. Care must be taken during this process to prevent further damage to the fins as well as personal injury from the sharp fin edges.

Fan Motor and Air Movement Selections

Most room air conditioners utilize a single fan motor to provide air circulation across the evaporator and condenser coils. The motor is usually of the permanent split capacitor design and may be single-speed or multi-speed. In some applications a shaded pole motor may be used. In most applications, a forward curved squirrel cage blower is used to move air across the evaporator coil. Axial or propeller fans are usually used to move air across the condenser coil.

Various terms are used by manufacturers to designate air movement selections. Terms such as "quiet cool" and "low cool" are used to designate the low-speed operating mode of the fan. Terms such as "high cool" and "super cool" refer to the high-speed operating mode of the fan. Whichever mode is selected, the only thing that changes is the operating speed of the fan motor; neither compressor operation, speed, nor capacity is affected. When replacing a failed room air conditioner fan motor, the replacement motor should be an exact duplicate of the original, otherwise system performance will be seriously affected.

Evaporator Coil Air Filters

All room air conditioners are equipped with an **evaporator coil air filter** and should not be operated without one. Most units utilize a permanent filter installed in the return air grill of the unit. The filter should be cleaned and inspected regularly. Most permanent filters can be washed in a solution of liquid detergent and water, then rinsed in clean water and dried. Dirty filters will greatly reduce the cooling capacity of the unit and will result in the creation of pungent odors in areas of high humidity. Fiberglass disposable filters are also commercially available. Figure 21–7 represents a typical room air conditioner filter arrangement.

Push-Button Switches

Push-button or **rotary switches** are used to control the operation of a room air conditioner. The off button deenergizes all fan and cooling operations. Other positions can provide either cooling or air circulation at various fan speeds. Figure 21–8 illustrates the typical wiring of a two-speed, push-button-controlled window unit. Figure 21–9 illustrates the typical wiring for various modes of a window air conditioner.

Thermostats

Most room air conditioners utilize a **thermostat** to cycle the compressor off after the desired temperature in the conditioned space is reached. The thermostat feeler bulb is usually installed in front of the evaporator coil or where it can best

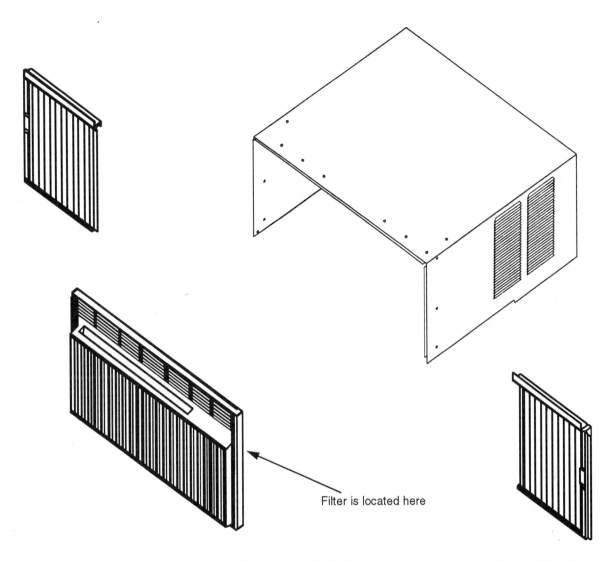

Fig. 21–7   In a room air conditioner the filter is located behind the front cover.   *Courtesy Fedders North America.*

Filter is located here

sense the temperature of the conditioned room air returning to the evaporator coil. To prevent the compressor from short cycling, the feeler bulb of the thermostat should not come in close thermal contact with the cold evaporator surface.

In some applications the thermostat is in a constant cool position. In this position the thermostat switch contacts are kept manually closed, and the compressor will continue to operate regardless of the return air temperature. In low ambient conditions, this may result in an icing condition of the evaporator coil. Customer education in the proper settings of the thermostat will usually alleviate this problem.

Installation of a Room Air Conditioner

Window air conditioners are mounted in several ways. Different types of windows require different installation procedures and kits. In most applications the unit is simply placed into a window cavity and the expanding side panels are opened to seal the installation from the outdoors. Usually installation kits include everything required to make a safe, watertight, and satisfactory installation. Follow the manufacturer's instructions when installing a window air conditioner. The following list, however, can serve as a general guide:

1. Make sure that the unit is leveled properly to ensure adequate condensate drainage.

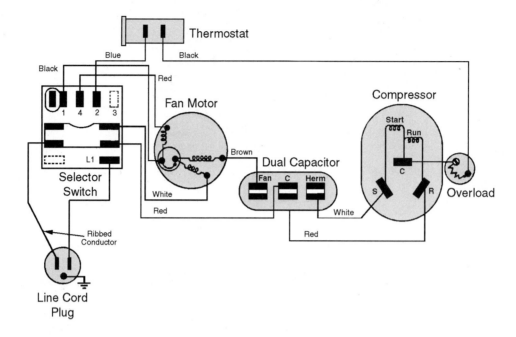

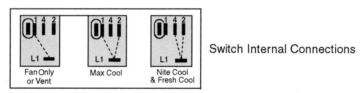

Fig. 21–8   Typical wiring of a two-speed, push-button-controlled window unit.

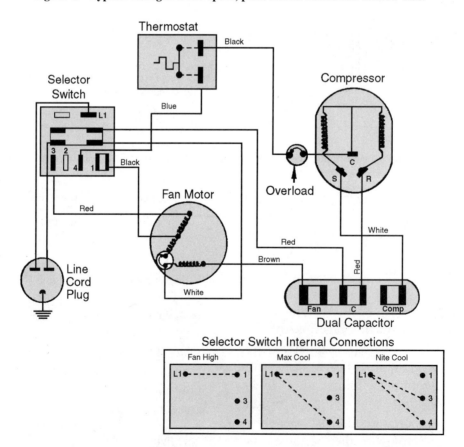

Fig. 21–9   Typical wiring for various modes of window air-conditioning.

2. Make sure that all openings are sealed properly to prevent the entry of warm air, insects, or dirt into the conditioned space.

3. Secure the unit properly to prevent noise from the unit's vibration.

4. Secure the installation properly.

5. Make sure that all local codes and ordinances have been followed.

Through-the-wall installations require cutting an opening in the existing structure of the area to be conditioned. Each installation will vary with the construction of the structure and local codes and ordinances. Following the manufacturer's installation instructions will help ensure a trouble-free installation. The following list, however, can serve as a general guide:

1. Locate the unit near the center of an outside wall approximately 30 inches up from the floor at a point where there is no plumbing or electrical wiring. If the unit is intended to cool more than one room, it should be located so the airstream will flow into the entire area to be conditioned.

2. Check the outside wall for interfering objects, such as telephone wires, electrical lines, or drain pipes.

3. Mark off the planned opening on the inside wall. If the wall is constructed of brick or cinder block, plan the opening so that the seams are cut. It is easier to cut into the seams of a construction material than the actual block or brick.

4. Drill pilot holes in each corner of the marked-off opening. These holes should extend to the outside wall of the installation.

5. Cut the inside wall and remove all wall studs as required. When wall studs are cut, the wall structure must be secured by framing the wall opening as required.

6. Cut through the outside wall. If the wall is of a brick or block construction, use masking tape to mark off the opening and a brick chisel to open the wall.

7. Install the chassis of the air conditioning unit as instructed in the manufacturer's specifications.

8. Install the air conditioning unit into the chassis sleeve.

9. Make sure that all openings are sealed properly to prevent the entry of outside air, water, or dirt.

10. Make sure that the unit is leveled properly to ensure adequate condensate drainage.

Troubleshooting Room Air Conditioners

If a room air conditioner does not appear to be performing satisfactorily, a systematic check of all components should be performed to determine if the cause of the problem is in the electrical parts or the sealed system. Assuming all electrical checks have been performed and are functioning correctly, use the following guidelines to check the operation of the unit.

1. Make sure that the air conditioner has the proper cooling capacity for the room being cooled.

2. Make sure that all doors and windows are closed.

3. If the unit has a vent or exhaust, make sure that it is closed.

4. Check the selector switch to ensure that it is in the cool position and that the thermostat is set to the maximum cooling position.

5. Make sure that the airflow across the evaporator and condenser coils is proper and unrestricted.

6. Check the air filter to ensure that it is clean.

7. Check the location of the installation. If the condenser coil is exposed to direct sunlight for a long period of time, it should be shaded or relocated. The first choice for unit location is in a northern exposure; the second choice is in an eastern exposure. Excessive exposure to the direct rays of the sun will increase the operating discharge pressure of

the system, resulting in longer running times and reduced cooling efficiency.

8. Start the unit and allow it to run for a few minutes until the system operating temperatures and pressures stabilize. Using a thermometer, take a temperature reading of the inlet air returning to the air conditioner and the outlet air leaving the unit. The temperature difference should be 15°F to 20°F. This will vary with the prevailing ambient temperature and humidity conditions. A high temperature and humidity will result in a lower temperature difference between the unit's supply air and return air. Make sure that there is a full airflow across the evaporator coil. A restricted airflow will result in a lower discharge air temperature, and the reduced volume of airflow will affect the overall cooling capacity.

A system that is operating normally will have a good temperature difference across the evaporator coil and normal compressor current draw. Furthermore, the evaporator will be uniformly cool with a full condensate or sweat pattern on each tubing run and return bend.

The following are problems and possible causes that a technician may encounter when servicing room air conditioners.

Problem: Compressor and fan motor will not start.

Possible causes:

1. No voltage to the unit receptacle
2. Disconnected or defective line cord
3. Defective selector switch
4. Incorrect internal wiring
5. Defective capacitor
6. Improper line voltage

Problem: Compressor and fan operates but does not cool.

Possible causes:

1. Dirty filter or restricted airflow
2. Dirty evaporator or condenser coils
3. Thermostat set too high

4. Fan blades slipping on shaft
5. Iced evaporator coil
6. Undercharged or overcharged system

Problem: Compressor and fan operates but provides insufficient cooling.

Possible causes:

1. Dirty filter or restricted airflow
2. Low voltage condition to the unit
3. Unit too small for the conditioned space
4. Condenser exposed to a high heat source

Problem: Fan operates, but compressor will not start.

Possible causes:

1. Low voltage at power supply
2. Defective thermostat or wiring
3. Defective selector switch
4. Incorrect system wiring
5. Defective capacitors
6. Defective compressor

Problem: Air circulation is insufficient.

Possible causes:

1. Dirty filter or restricted airflow
2. Loose fan blades on shaft
3. Iced evaporator coil
4. Fan motor cycling on overload
5. Low voltage condition at power supply

Problem: Compressor operates, but fan motor will not operate.

Possible causes:

1. Defective fan motor
2. Defective selector switch
3. Defective internal wiring
4. Defective fan motor capacitor

Problem: Evaporator is icing.

Possible causes:

1. Defective thermostat
2. Thermostat set too low

3. Dirty filter or restricted airflow

4. Unit too large for the conditioned space

5. Defective evaporator fan motor

Problem: Unit trips the circuit breaker or blows a fuse.

Possible causes:

1. Defective compressor

2. Defective capacitors

3. Low voltage condition

(Note: If an attempt is made to start the unit too soon after it has cycled off, the compressor may stall and trip the circuit breaker or blow a fuse. This condition is a result of the compressor attempting to start before the system pressures have equalized. When the compressor cycles off, the low-side and high-side pressures will equalize through the capillary tube. When this condition occurs the system is said to be "unloaded." Depending upon the system pressures, equalization will take from two to five minutes. Many units have a warning tag that instructs the operator to wait at least three minutes before attempting to start a unit once it has been turned off.)

Problem: Water drips into the conditioned room.

Possible causes:

1. Improperly pitched installation

2. Clogged drain tube

3. Improperly sealed installation

Unit Twenty-One Summary

- The room air conditioner is a unitary component designed to be installed in a window or through an opening in the wall. They are often equipped with fresh air ventilation and exhaust features to allow for outside air to be brought into the conditioned space and also to allow the exhaust of odors to the outside.

- The cooling capacity of a room air conditioner is rated in Btu's per hour and the EER of a unit is determined by the energy consumed.

- It's common to find a room air conditioner compressor to be a PSC (permanent split capacitor) type. The evaporator and condenser coils are usually of a fin and tube design.

- Room air conditioners employ a single fan motor to provide air circulation across the evaporator and condenser coils.

- It's important that the filter in a room air conditioner be kept clean. A dirty filter will affect the efficiency of the unit and result in customer complaints of inefficient cooling.

Unit Twenty-One Key Terms

Fresh Air Ventilation

Exhaust

Capacity

Voltage Required

Energy Efficiency Ratio (EER)

Compressor

Evaporator Coil

Condenser Coil

Evaporator Coil Air Filters

Push-Button Switches

Rotary Switches

Thermostat

UNIT TWENTY-ONE REVIEW

1. While a room air conditioner is primarily designed to cool, what are its other capabilities?

2. Define air-conditioning.

3. Match the following terms having to do with room air-conditioning with their description: thermostat, cable-controlled damper, exhaust, push-button switches, evaporator coil air filters, process tube, fin comb, fan motor.

   a. Used to straighten bent fins.

   b. Installed to gain sealed system access.

   c. Allows fresh air and room air to mix before being cooled.

   d. Removes air from the conditioned space.

   e. Cleaning can improve cooling capacity.

   f. Provides air circulation across the evaporator and condenser coils.

   g. Can provide either cooling or air circulation at various fan speeds.

   h. Cycles the compressor off after desired temperature is reached.

4. How does fresh air ventilation affect the capacity of a room air conditioner?

5. What is the main advantage of a window-installed room air conditioner?

6. What can cause an icing condition of the evaporator?

7. What will usually eliminate the icing problem?

8. Why is installation in direct sunlight a bad idea?

9. A _____ temperature and a _____ humidity will result in a _____ temperature difference between the unit's supply air and return air.

10. What is the energy efficiency ratio of an air-conditioning system?

11. How is the energy efficiency ratio of a room air conditioner determined?

12. What type of metering device is commonly used in a room air conditioner?

13. What is the purpose of parallel refrigerant circuits in the evaporator coil of a room air conditioner?

14. What design of compressor motor is commonly used in the application of a room air conditioner?

15. How is the compressor motor cooled in a room air conditioner?

16. Match the following problems with their possible causes: (a) compressor and fan motor will not start, (b) evaporator is icing, (c) fan operates but compressor will not start, (d) compressor and fan operate but do not cool.

   1. Dirty filter

      Thermostat too high

      Dirty evaporator or condenser coil

   2. No voltage to unit receptacle

      Defective capacitor

      Defective selector switch

3. Defective capacitor

   Defective selector switch

   Defective compressor

4. Dirty filter

   Defective thermostat

   Thermostat set too low

Btu = the amount of heat necessary to raise 1 pound of water 1°F.

1 watt (power) = 3.413 Btu

1 kilowatt = 1,000 watts = 3,413 Btu

17. The capacity of room air conditioners ranges from approximately 4,000 Btu's per hour to 30,000 Btu's per hour. What is the range in kilowatts?

18. If the cost is $.05 per kilowatt hour, what will it cost to operate a 30,000 Btu unit running four hours per day for 30 days?

Energy Efficiency Ratio measures the energy efficiency of air conditioners. The higher the EER rating, the more energy efficient the air conditioner.

$$EER = \frac{cooling\ capacity\ in\ Btu's\ per\ hour}{power\ input\ in\ watts}$$

19. Unit A has a capacity of 24,000 Btu's an hour with an input of 2,700 watts. Unit B has a capacity of 24,000 Btu's an hour with an input of 3,000 watts. What are the EERs of units A and B? Which is more efficient?

20. Unit C has a capacity of 30,000 Btu's an hour with an input of 3 kilowatts. Unit D has a capacity of 30,000 Btu's an hour with an input of 3.2 kilowatts. What are the EERs of units C and D? Which is more efficient?

21. Write an explanation covering some of the points for installation of a through-the-wall room air conditioner using the following terms: 30 inches, outside wall, brick, wall studs, chassis, and level.

SECTION FIVE

Residential Heating and Cooling Systems

A significant segment of the refrigeration industry is based in the design, installation and operation of residential heating and cooling systems. In this section, we'll discuss the fundamentals of comfort cooling systems, gas and electric furnaces, heat pumps, and the control systems used by residential and light commercial HVAC systems.

UNIT TWENTY-TWO

Comfort Cooling Systems

OBJECTIVES

After completing this unit, the student will be able to:

1. Explain the operation of a residential comfort cooling system.

2. Explain the difference between a split system and a package unit.

3. Outline proper start-up procedures for a comfort cooling system.

4. Describe troubleshooting procedures related to residential air-conditioning systems.

In order to be able to take a practical approach to servicing a residential comfort cooling system, the service technician must understand the relationship between the electrical, refrigeration, and airflow systems that make up the equipment. In this unit, we'll offer an overview of the different types of systems used in residential comfort cooling applications and outline step-by-step procedures for evaluating and troubleshooting these systems.

TYPES OF RESIDENTIAL AIR-CONDITIONING SYSTEMS

A properly designed and installed residential air-conditioning system is necessary for human comfort. If the system is oversized or undersized for the application, the proper temperature and humidity conditions will not be maintained, resulting in uncomfortable environmental conditions. An improperly designed and/or installed air delivery system will result in disturbing noise levels and improper air movement. The best designed and installed air-conditioning system is one that an individual never notices because he or she is comfortable. The two types of systems in residential air-conditioning applications are packaged systems and split systems.

Packaged Systems

The **packaged air-conditioning system** illustrated in Figures 22–1 and 22–2 is commonly found in residential and commercial applications. The basic design of the unit makes it adaptable to many installation applications. The unit is self-

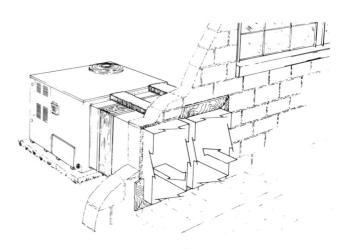

Fig. 22–1 Typical packaged system installation. *Courtesy Inter-City Products (U.S.A.)*

Fig. 22–2   Self-contained packaged air conditioner.   *Courtesy Lennox Industries, Inc., Worldwide manufacturer of comfort equipment.*

contained with all of its parts in a single housing. Supply and return air duct connections are provided on the front of the unit. Warm return air from the conditioned space is usually drawn through the evaporator coil, cooled, and then discharged back into the conditioned space. Condenser air is usually drawn through the sides of the unit and then discharged through the condenser coil by means of a propeller-type fan. This arrangement also allows for cooling of the compressor as outside air passes over the compressor before being discharged through the condenser coil.

A control box located within the system contains all of the electrical components—for example, relays, capacitors, contactors—all of which are prewired at the factory. The installer simply needs to wire line voltage to the system and install a remote thermostat.

Cooling capacities of packaged systems usually range from 2.5 tons to 50 tons. Refrigerant-22 is the most commonly used refrigerant in packaged systems. Lower capacity units use a capillary tube as a metering device, while higher capacity units use a thermostatic expansion valve.

The simplicity of the packaged system requires a minimum amount of field-installed plumbing, wiring, and ductwork. The unit can be installed in an attic with ductwork delivering the cool air to conditioned space below. It can also be installed on a concrete slab and connected to an existing warm air heating duct or in a basement or crawl space. Another application of the packaged system is in mobile home air-conditioning systems. Many manufacturers have complete installation kits available for quick and easy installation. Furthermore, the system can be easily disconnected and reinstalled in a new location.

Split Systems

A **split system** is one in which the evaporator or air handling unit is installed separately from the condensing unit. The condensing unit may be installed on the ground adjacent to the conditioned space or on the roof of the structure. The evaporator may be installed in the plenum or supply duct of an existing heating system, in an attic or crawl space, or on the roof, depending upon installation requirements. The condensing unit and evaporator are connected to each other by means of field-installed piping or precharged tub-

ing kits. Figure 22–3 illustrates a typical installation of a split system.

COMPONENTS OF A RESIDENTIAL AIR CONDITIONER

Condensing Unit

The **condensing unit** consists of the compressor, condenser coil, condenser fan and motor, and related electrical controls. Figure 22–4 illustrates a typical condensing unit of a residential air conditioner.

The case construction of the condensing unit varies with each manufacturer. Most units are made of steel, rustproofed, and painted with a weatherproof enamel. The use of plastics has become increasingly popular in the construction of condensing unit cases. In most cases, the condensing unit case is divided into compartments— one for the compressor, one for the condenser coil and fan, and one for the electrical controls of the system. Access panels to the compartments are provided for ease of maintenance.

Compressor

A fully hermetic **compressor** is usually used in residential applications. In larger commercial

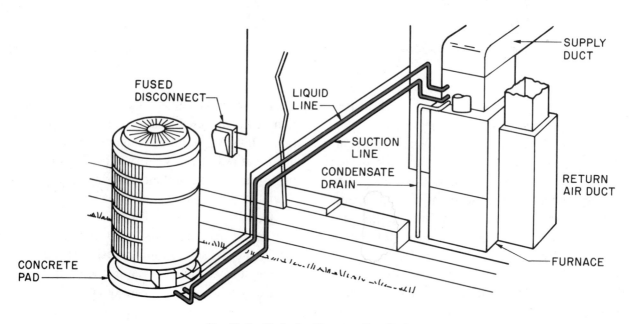

Fig. 22–3   Typical split system installation.

Fig. 22–4   Condensing unit of a residential air conditioner.   *Courtesy Lennox Industries, Inc., Worldwide manufacturer of comfort equipment.*

applications, a semihermetic compressor is used. Most systems utilize R-22 as the refrigerant. The compressor motor windings are cooled by the returning suction vapor, which passes over and around the motor windings. Compressor lubrication is provided internally by the splash method. Some larger capacity units use an oil pump to ensure positive lubrication.

The permanent split capacitor motor is the most common compressor motor used in residen-

tial and light commercial applications. In some applications where a high starting torque is required, a hard start kit consisting of a starting capacitor and relay is used.

Motor overload protection is usually in the form of a bimetallic internal line break overload protector that will interrupt power at the common terminal of the motor when the motor experiences excessive current or heat. Some systems utilize a pilot duty internal overload protector that will interrupt power in the low voltage control circuit of the system, stopping motor operation.

Condenser Coil

The **condenser coil** is usually manufactured of copper or steel tubing with attached aluminum fins, which increase the area of heat transfer. Generally, the design of the coil allows the ambient air to be drawn through the condenser coil, then discharged vertically or horizontally. The amount of maintenance required on condenser coils is minimal; simply keep the coil and fin surfaces clean to prevent excessive system discharge pressures.

Condenser Fan and Motor

The **condenser fan** is usually of an axial or propeller design. It may be installed in a vertical or horizontal position depending upon condenser design. The **condenser fan motor** is usually a permanent split capacitor motor. In some applications the motor may be a two-speed unit. A thermostatic control will allow the motor to operate at a low speed when the ambient temperature is low. As the ambient temperature increases and more condenser air is required, the motor will operate at a higher speed. Most motors used in residential applications are permanently lubricated and require no maintenance other than keeping the fan blades clean.

Evaporator Coils

The **evaporator coils** used in residential air-conditioning applications are available in a variety of designs that will fit almost any installation. Generally, the coils are constructed of copper or aluminum tubing with attached aluminum fins, which increase the area of heat transfer. Depending upon the design of the system, air may be drawn through or blown through the coil. Condensate water is collected in a tray at the base of the coil and drained by gravity.

"A" Coil

An **"A" coil** is designed to be used in up-flow or down-flow furnace installations. The shape of the coil, as shown in Figure 22–5, permits the furnace plenum to house a greater coil area. Refrigerant flow is metered into the coil by capillary tubes manifolded off of the liquid line and feeding both sides of the coil. Two, four, or more capillary tubes may be utilized, depending upon the size of the coil. Coil outlets are attached to a suction header that directs the flow of refrigerant vapor into the suction line.

Slant Coil

The **slant** or **sloped coil**, which is illustrated in Figure 22–6, is generally used in small capacity installations where a large coil surface is not required. Basically, the slant coil is half of an "A" coil and is used in up-flow or down-flow air movement applications. In some installations where the plenum height of the furnace is restricted, the slant coil may be used in a horizontal duct system.

Fig. 22–5   "A" coil evaporator.   *Courtesy Carrier Corp.*

Fig. 22–6   Slant coil.   *Courtesy Carrier Corp.*

Horizontal Coil

A **horizontal** or **vertical coil** is used in horizontal flow, warm air furnace applications. These units are also equipped with their own housing and, in conjunction with a blower module, form an air handling system designed for separate air delivery systems. A horizontal coil is illustrated in Figure 22–7.

Metering Devices

Most residential and light commercial air-conditioning systems utilize the capillary tube as a **metering device**. The capillary tube is maintenance-free and economical, and allows system pressures to equalize when the compressor cycles off, allowing for the use of a low starting torque motor.

In order to keep the physical size of the evaporator coil to a minimum and still maintain the maximum area of heat transfer, many evaporators are designed with parallel refrigerant circuits. In these applications, several capillary tubes are manifolded off of the liquid line to provide an equal distribution of refrigerant to all of the coil circuits. Parallel circuiting also reduces pressure drop across the evaporator coil. Figures 22–8A and B illustrate a typical application utilizing a refrigerant distributor and a thermostatic expansion valve as a metering device.

In applications where the heat load on the evaporator coil is relatively constant, the capillary tube serves as the ideal metering device. However, in applications where the heat load on

Fig. 22–7 Horizontal-type evaporator coil. *Courtesy Carrier Corp.*

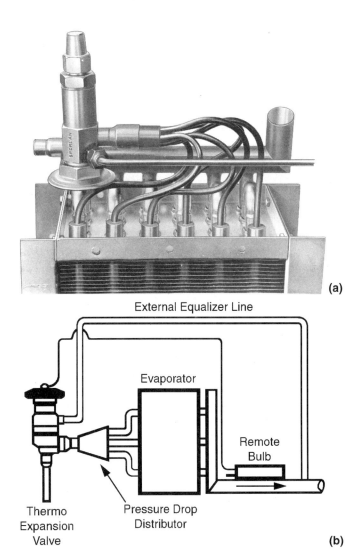

Fig. 22–8 Refrigerant distributor. *a-Courtesy Sporlan Valve Company. b-Courtesy Alco Controls.*

the evaporator fluctuates, the capillary tube cannot compensate and adjust refrigerant flow accordingly. In these applications, a thermostatic expansion valve is the best-suited metering device.

A gas-charged thermostatic expansion valve, usually of the pressure-limiting type, is used. Because of the pressure drops experienced across the evaporator and refrigerant distributors, the valve is equipped with an external equalizer. When parallel evaporator coil circuits are used, a refrigerant distributor, similar to the type used in the capillary tube application, is used to provide an equal distribution of refrigerant to all of the coil circuits.

When the compressor cycles off, the expansion valve will immediately close and prevent refrigerant flow into the evaporator. System high-side and low-side pressures will not immediately equalize. If the compressor attempts to cycle back on before the system pressures equalize, the motor may cycle on overload or trip a circuit breaker. This condition is a result of the low starting torque characteristic of the permanent split capacitor motor. Expansion valve manufacturers who are aware of this situation recommend the installation of an expansion valve equipped with internal bleed ports that allow system pressures to equalize gradually. If the problem is still experienced, however, the installation of a hard start kit is recommended.

EVAPORATOR AIRFLOW

The comfort level of a conditioned space can be greatly improved when the volume of the circulating air is balanced properly. If the air delivery system is designed properly, the system should not need to be balanced. Unfortunately, this is not the case in many installations. The two ways in which airflow across the evaporator can be changed are by using a multi-speed evaporator fan motor or by manually adjusting the pitch of the blower motor pulley.

In air-conditioning applications the average capacities for evaporator airflow are based on 350 to 400 cubic feet of air per minute (cfm) across the evaporator coil per ton of refrigeration capacity. For example, an evaporator coil with a cooling capacity of 36,000 Btu's per hour, or 3 tons, will require an evaporator airflow of about 1,200 cfm.

The actual airflow requirements depend on the humidity and the daily temperature range under which the system operates. The daily temperature range is the difference in number of degrees between the maximum and the minimum temperatures experienced during the cooling season. When the difference between the daily maximum and minimum temperatures is less than 15°F, the range is considered to be low. The temperature range is considered to be medium when the temperature difference is 15°F to 25°F and high when over 25°F. For average humidity conditions of 50 percent to 60 percent, it is usually recommended to circulate 350 cfm per ton of system capacity in the medium temperature ranges and 425 cfm in the high temperature ranges. In areas that experience high humidity, it is recommended to reduce airflow to approximately 300 cfm to allow for more dehumidification of the circulating air.

When there is proper airflow across the evaporator coil, the following temperature differences between the supply and return air should be experienced:

1. Low daily temperature range 15°F to 20°F
2. Medium daily temperature range 20°F to 22°F
3. High daily temperature range 22°F to 25°F

Before taking temperature readings, make sure that all system dampers are opened wide and that the system air filter is clean. If the system has been balanced, do not change the position of the dampers. Allow the system to operate for at least twenty-four hours in order to stabilize system operating temperatures and pressures. Temperature readings should be taken in the airstream in front of and behind the evaporator coil.

On systems equipped with multi-speed motors, select the motor speed that will deliver the best airflow for the application. On systems equipped with an adjustable sheave, increase the diameter of the sheave to increase the airflow and decrease the diameter of the sheave to decrease the airflow until the desired airflow is achieved.

Once the proper evaporator airflow has been established, the air delivery system can be balanced. Adjust the room thermostat to a normal comfort cooling setting, usually 72°F to 75°F. Take and record the temperature of each room with a dry bulb thermometer. Take the temperature readings three to four feet from the floor and mark the original position of each damper before any adjustments are made. This will serve as a reference point during the balancing operation.

Begin with the coldest rooms by partially closing the dampers. Adjust the dampers in the warmer rooms to an open position. Check the temperature in each room about every half hour, record the readings and make the necessary adjustments. For example, open the dampers in rooms that are too warm and close the dampers in rooms that are too cool. Make small adjustments until the system is balanced.

When the airflow is properly balanced, check the relative humidity of the conditioned area. The average person is most comfortable when the room temperature is 72°F to 80°F and the relative humidity is 50 percent to 60 percent. If the humidity is above or below this amount by 10 percent, the evaporator blower speed should be adjusted. The humidity can be lowered by reducing the blower speed or raised by increasing the blower speed. As mentioned before, if the system is sized properly for the application and if the air delivery system is designed properly, it should not be necessary to balance the air delivery system and the humidity in the conditioned space.

INSTALLATION OF A RESIDENTIAL AIR CONDITIONER

The installation of any air-conditioning system will vary according to the type of structure in which the system is installed, the local codes that must be followed, and the type of equipment that is installed. This section will serve only as a guide for general installation procedures.

Before installing a system, a thorough inspection of the installation site should be conducted to confirm where the condensing unit should be placed, how the tubing, the electrical, and the drain lines should be run, and so forth. Review the manufacturer's specifications and installation guide to ensure that the installation will conform to the recommended procedures and local codes. Check the equipment and make sure that all components of the system are properly matched for the application.

Do not remove any of the protective caps from the tubing couplers until the tubing is ready for final installation. Unpack the equipment at the job site to reduce the possibility of damage from excessive handling. Once uncrated, check all the components for possible damage.

Installation of the Condensing Unit

The condensing unit should be installed outdoors as close as possible to the evaporator and electrical power source, keeping in mind the desires of the customer, installation requirements, and local codes. Consider the following factors:

1. The unit should be located in an area that is not subject to direct sunlight.

2. There should be enough space between the unit and any adjacent walls or overhangs to ensure proper condenser airflow and avoid restrictions or short circuiting.

3. The unit should be mounted on a sturdy base at least four to six inches above the ground. To prevent the possible transmission of vibration noise, the base should not be in contact with the building foundation.

4. The unit should be positioned to facilitate access to service panels and the easy installation of refrigerant and electrical lines.

Figure 22–9 illustrates typical condensing unit installations. The base may be constructed of

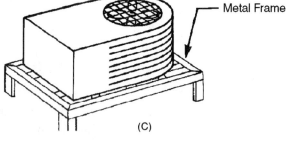

Fig. 22–9   Various types of unit pads.

a precast concrete slab, timbers, or concrete block. Plastic bases that are specially designed for condensing unit installations are also available. If concrete blocks or timbers are used, make sure that the base area is well tamped and that a gravel fill is used to prevent settling.

Electrical Connections

All electrical work must comply with local codes and the manufacturer's specifications. Consider the following factors:

1. Low-voltage control wiring should not be run in the same conduit as high-voltage wiring.

2. Wire and fuse sizes must be as specified by the equipment manufacturer.

3. An outdoor service disconnect should be installed near the condensing unit as specified by local code.

4. All outdoor connections and enclosures must be weatherproof.

Installation of the Evaporator Coil

Depending on the installation, the evaporator coil may be installed in the plenum or supply duct of an existing warm air heating system where the furnace fan and existing ductwork will be used to deliver the air to the conditioned space, or it may be installed in conjunction with an air handling unit connected to field-installed ductwork.

If the furnace and the evaporator coil are made by the same manufacturer, a furnace plenum installation should be relatively simple, with little or no modification of the existing ductwork. The primary concern in this type of installation is to ensure that the evaporator coil is located directly over the air opening of the furnace so that all of the supply air passes over the evaporator coil surface.

For most installations, consider the following factors:

1. It may be necessary to cut open or remove the furnace plenum in order to fit the evaporator coil on the plenum base.

2. If the coil base is smaller than the furnace opening, it will be necessary to fabricate sheet metal baffles so that the circulating air does not bypass the evaporator coil.

3. If the coil base hangs over the furnace plenum opening, an extended plenum will have to be manufactured to fit the application.

4. Support flanges may be required to ensure that the coil is secured on the top of the furnace plenum. The flanges may be provided in the installation kit, or they may have to be fabricated in the field.

5. The manufacturer's installation instructions should be followed. Most coils should be

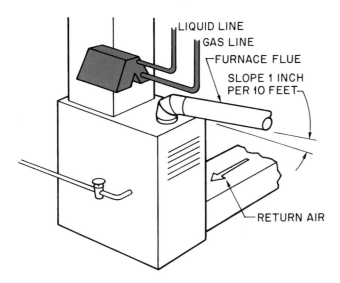

Fig. 22–10   "A" coil evaporator installed in furnace plenum.

mounted levelly in the plenum. The manufacturer usually makes provisions for properly pitching the condensate pan for condensate water drainage.

6. The evaporator should be positioned to allow for easy servicing and the installation of refrigerant and condensate lines.

7. When replacing the plenum, a removable access panel may be considered for future service accessibility.

Figure 22–10 illustrates a typical evaporator coil installation in the plenum of a warm air heating system.

Fan Coil Air Handlers

Fan coil air handlers are usually installed in an attic or other area above the conditioned space; they use field-installed ductwork to deliver the air to the conditioned space below. The air handling unit usually consists of the following parts:

1. A horizontal flow evaporator coil

2. A metering device

3. A squirrel cage blower

4. Related electrical components, such as a blower motor and fan relay center

5. Condensate pan and drain lines

The fan coil air handler may be suspended by rods or chains from the structure's rafters, or it may be installed on the attic floor or ceiling joists. The primary advantage of the suspended installation is the elimination of vibration noise and the drainage by gravity of the condensate water. When the air handler is installed on the attic floor or ceiling joists, a vibration pad should be used to eliminate noise transmission. In all cases, a secondary drain pan should be installed to prevent the possibility of water damage if the primary drain pan becomes restricted. Figure 22–11 illustrates typical fan coil air handler installations.

Connecting the System

Copper liquid and suction lines are used to connect the condensing unit to the evaporator. The tubing connections can be field-installed by

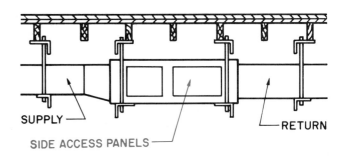

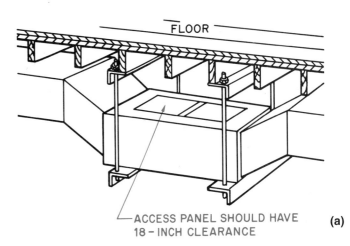

Fig. 22–11   Air handler installations (a) crawl space installation. *(continued)*

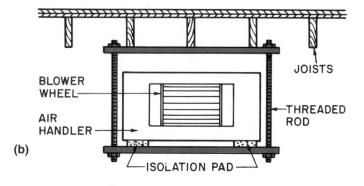

(b)

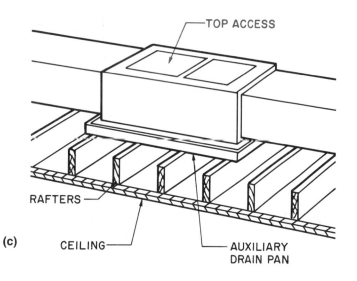

(c)

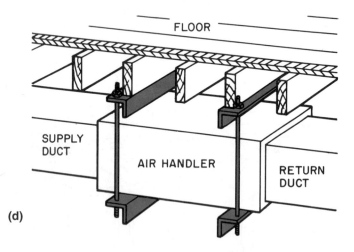

(d)

Fig. 22–11 *(continued)* Air handler installations (b) attic trapeze installation, (c) attic floor installation, (d) crawl space installation.

brazing or flared connections. In most residential applications, precharged tubing kits are used to make the installation. These kits connect a precharged liquid and suction line to a precharged

condensing unit and evaporator coil, eliminating the need to evacuate or charge the system after the installation is complete. This type of installation also eliminates the possibility of system contamination from air, moisture, or brazing operations.

Precharged refrigerant lines are available in various lengths and sizes as specified by the manufacturer for a specific installation. The lines are made of a soft annealed copper that can be easily bent and formed to fit various installation configurations. To facilitate installation, some precharged tubing kits are equipped with mechanical spring benders at the tubing connection ends.

Most precharged tubing kits are equipped with a female fitting on the precharged line and a stationary male fitting on the evaporator and condensing unit connections. Both fittings have a metal diaphragm that seals the refrigerant inside the tubing or component. The male fitting also has a blade device within it. As the male and female fittings are joined, the blade in the male connection will pierce the diaphragm of the female connection, opening refrigerant to the lines and system component (see Figure 22–12).

When installing systems equipped with precharged tubing, the final tubing connections should not be made until all components of the system are in their permanent positions. Once the line connections are made, they cannot be disconnected without losing the refrigerant charge. Some precharged connections utilize an O-ring to help seal the connection. To prevent damage to the seal, the O-ring should be lubricated with refrigerant oil before the final connection is made. Furthermore, suction lines should be insulated to prevent condensation damage to surrounding areas and excessive superheating of suction vapor. On installations where short tubing runs are involved, the excess tubing should be coiled and secured in a horizontal position to prevent oil from becoming trapped in the suction line.

In applications where field-charged tubing is not used, the system may be installed with soft or hard drawn copper tubing. It is recommended that solder joints be used when making the connections to minimize the possibility of refrigerant

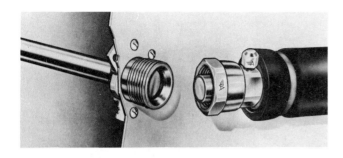

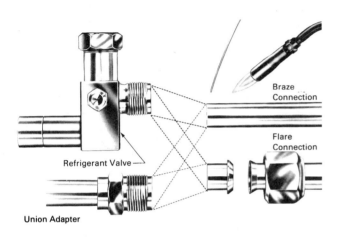

Fig. 22–12 Precharged tubing connections. *Courtesy Aeroquip Corporation.*

leaks, which are more likely to be experienced with mechanical flare connections. A tubing bender may be used with soft drawn copper tubing to form the required sweeps and bends; elbows, couplings, and other types of fittings may be used with hard drawn copper tubing to form the sweeps and bends. After the installation is completed, the system should be properly vacuumed to remove air and other noncondensible gases as well as moisture. Installing a liquid line filterdrier and a suction line filter is also recommended. Clamps or straps should be used to secure the tubing to prevent vibration and movement. All openings where the tubing enters the structure should be sealed.

Installation of the Condensate Removal Lines

Evaporator coils used in furnace plenums and fan coil applications are equipped with drain pans and drain connections to collect and drain mois-

ture that has condensed on the evaporator during the cooling process. The proper installation of condensate lines is important in preventing damage to the equipment and the surrounding structure. The installation of the condensate lines should conform to the manufacturer's specifications and local codes. The lines should be pitched to allow for the positive gravity drainage of the water. All condensate lines should be run to an open drain or sump. When this is not possible, a condensate pump should be installed to collect the water and direct it to an open drain.

Most evaporator coils are equipped with a drain connection to which a copper, brass, or plastic fitting may be adapted. Some units require the connection to be soldered. A plastic pipe, such as PVC, is commonly used because it is less expensive and easy to fabricate. A "P" trap should be installed at the drain outlet to ensure positive water drainage. This is especially important on installations where the condensate drain is located on the negative pressure side of the coil. The "P" trap may be fabricated by using pipe and fittings, or it may be purchased as a preformed unit ready for installation.

It is generally recommended to install a secondary drain pan, especially when the evaporator is installed above a conditioned space. If a secondary drain pan is not provided by the manufacturer, one can be easily fabricated by a sheet metal shop. In some installations, a float switch is installed in the condensate pan and wired in series with the compressor contactor holding coil circuit. If the condensate water in the drain pan rises to an overflow condition, the float switch will open, interrupt power to the compressor contactor holding coil, and stop compressor operation. Figure 22–13 illustrates typical condensate drain installations.

PRESEASON STARTING AND CHECKING PROCEDURES

The preseason starting and checking of an airconditioning system will help ensure a trouble-free season of system operation as well as prolong the equipment's life. Using a checklist will ensure

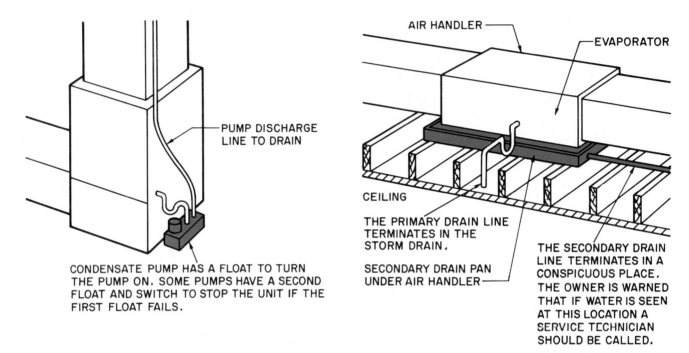

Fig. 22–13   Installation of condensate removal lines.

that all parts of the system are checked. A visual inspection of the entire system is usually a good place to start. This will give the service technician an opportunity to become familiar with the system and the location of its components. The sections that follow can serve as reference points in checking an air-conditioning system.

Checking the Condensing Unit

A condensing unit should have enough clearance from adjacent walls and overhangs to ensure a free airflow across the condenser coil. A good rule of thumb is to allow one foot from the wall for each six inches of fan blade radius. The condensing unit should sit on a firm and level base that is a few inches above the ground and that is larger than the condensing unit. To avoid the transmission of vibration noise to the building interior, the base should not be in physical contact with the building.

Checking the Tubing

The exterior wall or soffits where the suction line and liquid line are routed should be large enough to accommodate the tubing without kinking or chafing the lines. All tubing access holes should be sealed. If there is an excessive amount of refrigerant tubing, as may be experienced on installations utilizing precharged tubing, make sure that the excess tubing is coiled and secured in a horizontal position to prevent oil from becoming trapped in the suction line. All tubing should be neatly run and properly secured to prevent damage and vibration.

Check the Air Handler

Depending on the type of installation and the design of the air handling system, the following should be checked: (1) that the air handler is properly secured and that provisions have been made to prevent vibration noise; (2) that the condensate lines are clean, properly pitched, and trapped; (3) that the condensate line is routed to an approved drain; and (4) that the condensate pump is operative if required for the application. The procedure for checking an air handling system follows.

1. Make sure that all power to the condensing unit has been turned off. Check the supply

voltage at the unit disconnect box. The voltage should conform to the unit's nameplate rating. Voltage variations of more than 10 percent of a system's requirements should be corrected or damage to the system's electrical components may result.

2. Check that all fuses and circuit breakers are the correct size. Also check for proper wire size and types to ensure that they meet unit specifications and conform to local codes. Install a weatherproof disconnect near the condensing unit where it is immediately accessible to anyone servicing the system.

3. Remove all panels from the condensing unit and check the following:

   a. The condition of the condenser coil to ensure that it is free of dirt, scale, or anything that may affect condenser airflow and heat transfer.

   b. The condition of the fan, making sure that the blades are clean and spin freely.

   c. The lubrication of the condenser fan motor, although most newer motors are permanently lubricated. However, if lubrication is necessary, use a 20-weight oil or one prescribed by the manufacturer to lubricate the motor. Do not over lubricate.

   d. The condition of all electrical connections, making sure that they are tight and free of corrosion. Remove the compressor motor terminal cover and inspect the motor terminal connections for loose or corroded terminals.

   e. The compressor mounting springs and/or bushings, if applicable, to make sure that they are secure.

   f. The overall condition of the condensing unit. Remove any debris that may have accumulated and inspect water drainage holes to ensure that they are clean.

   g. After checking, replace all panels.

4. Remove all panels from the air handler, if applicable, and check the following:

   a. The condition of the evaporator coil to ensure that it is clean.

   b. The air filters. Replace or clean them as required.

   c. All electrical connections to ensure that they are free of corrosion and are secure.

   d. The blower motor. Lubrication of the motor is necessary.

   e. The blower blades to ensure that they are clean and spin freely. If the blades are belt driven, check the condition and tension of the belt.

   f. The air delivery duct system to ensure that there are no air leaks. After checking, replace all panels.

5. Check the room thermostat to ensure that it is installed properly and that it is free of dust and dirt.

Starting the System

With all inside and outside checks completed, the system is ready to be started and operationally checked. To start the system, follow these steps:

1. Turn the thermostat selector switch to the cool position and the fan selector switch to the on position. Turn the temperature setting until it is below the prevailing temperature of the room.

2. Apply power to the unit. The compressor and condenser fan should start.

On some systems, there may be a time delay before the compressor starts to allow the condenser fans to start and come up to speed. Also, some systems may have a time delay system that will start the compressor a few minutes after the thermostat calls for cooling.

After the compressor has started, listen for any unusual noises coming from the compressor, fan motor, or contactor. Some systems may be equipped with a crankcase heater. If the ambient temperature is below 70°F, the crankcase heater should be energized at least six to eight hours before starting the compressor. Allow the system to operate for at least 15 minutes to stabilize operating pressures and temperatures. Then perform the following steps:

1. Place the thermostat in the auto position. The condensing unit and the indoor fan motor should continue to operate. Slowly raise the temperature setting of the thermostat. The indoor fan motor and the condensing unit should shut off. Wait at least three minutes before attempting to restart the condensing unit.

2. Place the fan selector switch in the on position. The indoor fan motor should cycle on. Slowly turn the thermostat setting until it is below the prevailing temperature of the room. The compressor should cycle on.

This procedure will verify if the thermostat control center is functioning properly. The rest of the system can now be checked as follows:

1. Using a voltmeter, check the voltage at the compressor motor terminals during initial start and during normal running operation. Voltage variations should not exceed 10 percent of the unit nameplate rating.

2. Using an ammeter, check the current draw of the compressor and compare it with the unit nameplate specifications. Note that the total current draw reading will be the combination of the compressor current draw and condenser fan motors if both are wired off of the same contactor. The current draw of each device should be checked individually and compared to unit specifications.

With the manifold gauges properly installed, system operating pressures can now be checked. Suction and discharge pressures will vary, depending on the prevailing ambient conditions and the evaporator heat load. Refer to the manufacturer's charging charts for the proper system operating pressures under different operating conditions.

Air temperatures across the evaporator coil can also be checked. The average air-conditioning system is designed to deliver about 350 to 400 cfm per ton of system cooling capacity across the evaporator coil. This will result in a temperature difference of 15°F to 20°F between the entering and supply air of the evaporator.

Check the condensate drain after 15 to 20 minutes of operation to ensure that it is draining properly.

It could take several hours of operation before the system is able to reduce the occupied space to the desired temperature. The last step in the start-up procedure is to walk the customer around the installation and explain proper operating and preventive maintenance procedures. Figure 22–14 illustrates a checklist that the technician may use in starting and checking an air-conditioning system.

TROUBLESHOOTING THE RESIDENTIAL AIR-CONDITIONING SYSTEM

In order to effectively diagnose problems in a residential air-conditioning system, the technician should develop and follow a logical step-by-step troubleshooting procedure. The following list contains problems along with possible causes and solutions that a technician may encounter when servicing a residential air conditioner.

Problem: The furnace fan, condenser fan, and compressor will not operate.

Possible causes and solutions:

1. Thermostat set too high—Reset thermostat.

2. Blown fuse or tripped circuit breaker—Replace fuse and reset breaker; determine and correct cause of failure.

3. Defective control transformer—Check transformer and replace as required.

4. Defective thermostat—Check thermostat and replace as required.

Problem: The compressor will not start or hum, and the condenser fan is running.

Possible causes and solutions:

1. Compressor off on overload—Check to verify that the compressor is off on overload; repair or replace as required.

CHECKLIST FOR STARTING AND CHECKING AN AIR-CONDITIONING SYSTEM

Prestarting Checks

1.	Is the condensing unit level? _____
2.	Is there free airflow across the condenser? _____
3.	Are wiring connections tight? _____
4.	Are condensate lines clear and properly installed? _____
5.	Is condensate pump operable? _____
6.	Are filters installed and clean? _____
7.	Is thermostat properly installed? _____
8.	Is tubing neatly run, secured and insulated? _____
9.	Is condensing unit properly fused? _____
10.	Are all motors lubricated? _____
11.	Are all belts serviceable? _____

Start–Up Procedure and Checklist

1.	Are all covers in place to prevent air leaks? _____
2.	Is evaporator sealed to prevent air loss around coil? _____
3.	Do all fans run in the proper direction? _____
4.	Is blower speed adjusted for proper CFM? _____
5.	Is amperage draw of blower motor proper? _____
6.	Is suction pressure adequate? _____
7.	Is discharge pressure adequate? _____
8.	Is outside air temperature entering condenser adequate? _____
9.	Record voltage at contactor when unit is:
	a. Off _____
	b. Starting _____
	c. Running _____
10.	Are there air leaks in ductwork? _____
11.	Is there vibration in tubing and condensing unit? _____
12.	Is air handling system properly balanced? _____
13.	Has operation of thermostat been checked? _____
14.	Has homeowner been instructed about proper system operation? _____

Fig. 22–14   Checklist for starting and checking an air-conditioning system.

2. Broken electrical connection between contactor and compressor—Check and replace or repair wiring as required.

3. Defective compressor wiring—Check the compressor wiring and repair or replace as required.

Problem: The compressor has a low hum, will not start, and cycles on overload. The run winding is energized, but the start winding is not.

Possible causes and solutions:

1. Defective start capacitor—Check and replace the start capacitor as required.

2. Defective starting relay—Check and replace the starting relay as required.

3. Open start winding—Check the starting winding and replace the compressor if defective.

Problem: The compressor hums but will not start. There is a high current draw across both motor windings, and the compressor cycles on overload.

Possible causes and solutions:

1. Excessive discharge pressure—Allow the system pressures to equalize before attempting to start the compressor. Check for refrig-

erant overcharge, dirty condenser, and proper condenser fan operation.

2. Seized compressor—Verify and replace the compressor as required.

3. Defective start or run capacitor—Check and replace the capacitors as required.

4. Low line voltage—Check the line voltage.

Problem: The system runs continuously. There is little or no cooling.

Possible causes and solutions:

1. Low refrigerant charge—Check the system pressures, locate and repair any leaks, then recharge the system.

2. Defective compressor—Check and replace the compressor as required.

3. Undersized unit—Check the system capacity and load requirement of the conditioned space.

Problem: The evaporator freezes up.

Possible causes and solutions:

1. Dirty filters—Check and replace the filters.

2. Evaporator fan running too slow—Check and increase the fan speed as required.

3. Evaporator fan motor belt slipping—Check and tighten the belt.

4. Dirty evaporator coil—Check and clean the coil as required.

Problem: The suction line is frosted.

Possible causes and solutions:

1. Improperly adjusted expansion valve— Adjust the expansion valve to 10°F superheat.

2. Overcharge in capillary tube system—Check the pressures and purge refrigerant as required.

Problem: The liquid line is sweating or frosted.

Possible causes and solutions:

1. Restricted liquid line drier or strainer— Check and replace the drier or strainer.

2. Liquid line service valve partially open— Check the service valve position and open if necessary.

Problem: The system operates normally, but the circuit breaker or fuse trips occasionally.

Possible causes and solutions:

1. Low voltage condition—Check the voltage condition during the running cycle.

2. Undersized fuse or breaker—Check for proper size of fuse or breaker and replace as required.

3. Loose electrical connections to compressor—Check the electrical connections and repair as required.

4. Burnt, pitted, or corroded contactor contacts—Check the contactor contacts and repair or replace as required.

Problem: The system cycles off on low-pressure control.

Possible causes and solutions:

1. Low refrigerant charge—Check the refrigerant charge and the system for a refrigerant leak. If a leak is found, repair and recharge.

2. Insufficient airflow across evaporator— Check the filters, fan speed, and evaporator coil condition; repair or replace as required.

Problem: The system cycles on high-pressure control.

Possible causes and solutions:

1. Dirty condenser—Check and clean the condenser as required.

2. Restricted condenser airflow—Check for the proper condenser airflow and for bent or damaged coil fins.

3. Inoperative or defective condenser fan motor—Check and repair or replace the condenser fan motor.

4. Condenser airflow short-circuited—Check for a full airflow across the condenser.

5. Refrigerant overcharge—Check the operating pressures and purge refrigerant as required.

6. Noncondensible gases in system—Check the standing pressures and evacuate and recharge the system as required.

Problem: The compressor cycles on internal overload protection. (The internal overload protector senses the motor winding temperature and will open when the winding temperature increases.)

Possible causes and solutions:

1. Low refrigerant charge, low voltage conditions, high discharge pressures, or defective electrical accessories. The accessories include relays, capacitors, and electrical connections. In order to isolate the problem, observe the system operating pressures, take voltage and amperage readings, and evaluate the overall system operating conditions.

Problem: The compressor starts, but will not come up to speed, and cycles on overload.

Possible causes and solutions:

1. Defective capacitors—Check and replace the capacitors as required.
2. Low line voltage—Check the line voltage.
3. Defective starting relay—Check and replace the relay as required.
4. High discharge pressure—Check the discharge pressure and repair as required.
5. Defective compressor—Check the compressor and replace as required.

Problem: The evaporator fan runs, but the compressor and condenser fans will not run.

Possible causes and solutions:

1. System off on safety control—Check the safety control and determine the cause of control opening.
2. Defective contactor or contactor coil—Check the contactor and replace as required.
3. Defective condensing unit transformer—If the system has a separate power supply for the condensing unit control circuit, check the transformer and replace as required.

Problem: The fuse or circuit breaker trips when the system is energized.

Possible causes and solutions:

1. Short circuited compressor windings—Check the compressor windings and replace as required.
2. Internal electrical short circuit in the system wiring or in a component, such as the capacitors—Check the system electrical wiring and components and repair or replace as required.

PSYCHROMETRICS

To effectively service comfort cooling equipment, the service technician has to have at least a basic understanding of **psychrometrics**. "Psychro" means properties of air and "metrics" means measure of. Air conditioning is defined not only as a process that cools the air, but also that circulates and controls the moisture content of the air. To be comfortable in a room served by a comfort cooling system, the temperature and the level of moisture in the air (known as relative humidity) must be within certain parameters.

Two tools that help evaluate the performance of a comfort cooling system are the psychrometric chart and the sling psychrometer. The sling psychrometer, shown in Figure 22–15, contains two thermometers. One is used to measure what is known as the dry bulb temperature and the other is used to measure wet bulb temperature. The fundamental difference between the two thermometers when they're in use is that one uses a damp wick in order to read a wet bulb temperature.

The sling psychrometer is simple to use. By grasping the handle and swinging the assembly around, as shown in Figure 22–16, it reads both temperatures in a given area. Swing the sling psychrometer at a rotation of about two to three times per second in order for it to record properly. When you take wet bulb and dry bulb readings, you'll find that the dry bulb reading will be higher than the wet bulb reading.

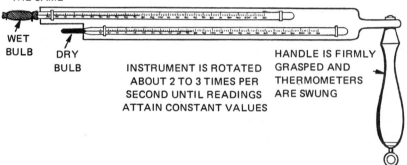

THE TWO THERMOMETERS SHOULD BE MATCHED
BY THE MANUFACTURER; THAT IS, WHEN THE
WICK IS REMOVED AND BOTH THERMOMETERS
ARE EXPOSED TO DRY-BULB CONDITIONS THE
READINGS OBTAINED SHOULD BE EXACTLY
THE SAME

WET
BULB

DRY
BULB

INSTRUMENT IS ROTATED
ABOUT 2 TO 3 TIMES PER
SECOND UNTIL READINGS
ATTAIN CONSTANT VALUES

HANDLE IS FIRMLY
GRASPED AND
THERMOMETERS
ARE SWUNG

Fig. 22–15   Sling psychrometer.

Fig. 22–16   Technician spinning a wet bulb and dry bulb
thermometer, known as a sling psychrometer.

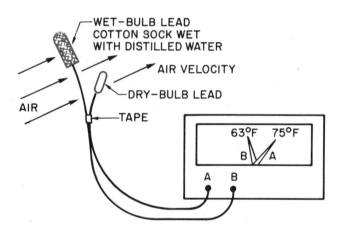

WET–BULB LEAD
COTTON SOCK WET
WITH DISTILLED WATER

AIR VELOCITY

DRY–BULB LEAD

AIR

TAPE

63°F   75°F

B   A

A   B

Fig. 22–17   The speed of the condenser fan is controlled by a
device that senses the changes in condenser pressure.

In the event that you don't have a sling psychrometer and wish to determine the wet bulb and
dry bulb temperatures in an area, you can modify
a standard electronic temperature-reading device
that has two probes. Simply wrap one of the probe
leads with a cotton sock, tape the two probes
together in such a way that they can be near each
other but still read independently, then sling the
probes in the same fashion as you would a sling
psychrometer (see Figure 22–17).

The fundamental purpose behind the exercise
of determining the dry bulb and wet bulb

temperatures of a given area is to determine the
relative humidity in that area. In some cases, sling
psychrometers come equipped with slide scales
that allow you to determine the relative humidity
by lining up the wet bulb and dry bulb readings
rendered by the psychrometer. Another way to
determine the relative humidity in a conditioned
space is to use a psychrometric chart.

The psychrometric chart, shown in Figure
22–18, was developed to aid in the evaluation of
the performance of an air-conditioning system
and is also used as a tool in load estimating.

The various lines and data on the chart are
used to track and record the following:

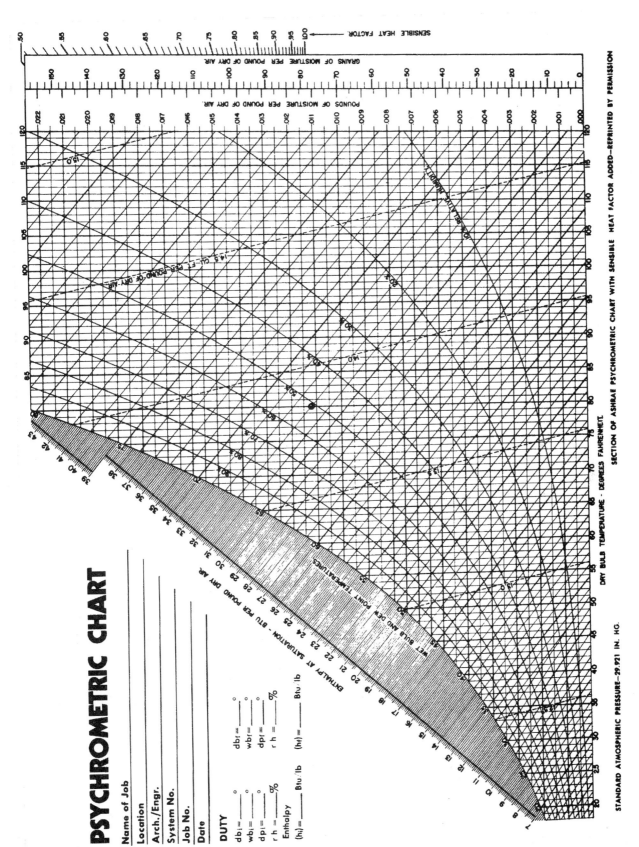

Fig. 22–18 Psychrometric chart. *Courtesy Inter-City Products Corporation (U.S.A.)*

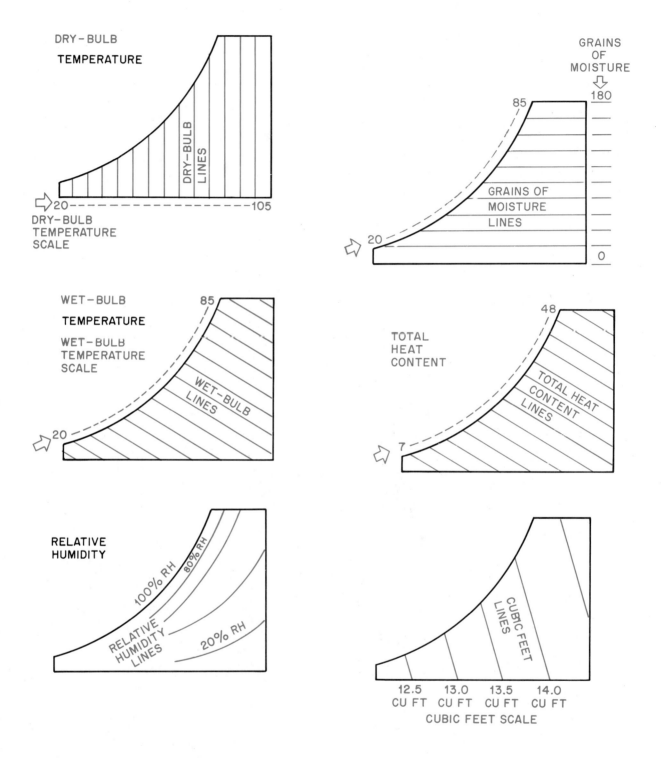

Fig. 22–19   Skeleton chart showing the relative humidity lines.

1. Dry bulb temperature

2. Wet bulb temperature

3. Moisture content of air, expressed in grains of moisture per pound of air

4. The total heat content of air in Btu/lb

5. The relative humidity of an air sample

6. The specific volume of air at different conditions

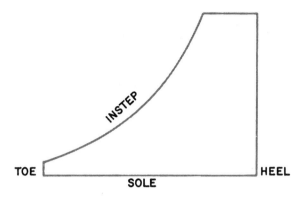

Fig. 22–20 Picturing the psychrometric chart in the shape of a shoe makes it easier to understand.

Figure 22–19 shows a skeleton chart for each of the processes listed on the previous page.

Another method that will help make the psychrometric chart less intimidating is to picture it in the shape of a shoe, as it's shown in Figure 22–20. Recalling, for example, that the dry bulb lines of a psychrometric chart run from the sole of the shoe to the instep of the shoe makes the chart easier to understand and use.

The process of using the psychrometric chart to determine the relative humidity level in a given area can be accomplished through a series of simple steps. For our example, we'll assume that the

temperature readings rendered by the sling psychrometer are:

Dry bulb.....78°F.

Wet bulb.....65°F.

Refer to the series of drawings in Figure 22–21 as you go through the steps listed.

STEP ONE: Read along the bottom of the chart (the sole) until you find 78°.

STEP TWO: Use a pencil to draw a line from the bottom of the chart to the instep.

STEP THREE: Read along the instep of the chart to locate the recorded wet bulb temperature (65°).

STEP FOUR: Use a pencil to draw a diagonal line from the wet bulb reading toward the heel of the shoe. Make sure that you cross over the dry bulb line drawn in Step Two.

STEP FIVE: Locate the relative humidity line at which the drawn dry bulb and wet bulb lines intersect.

STEP SIX: Read along the relative humidity line until you can identify the percentage reading. The relative humidity line tells you the actual amount of moisture in the air com-

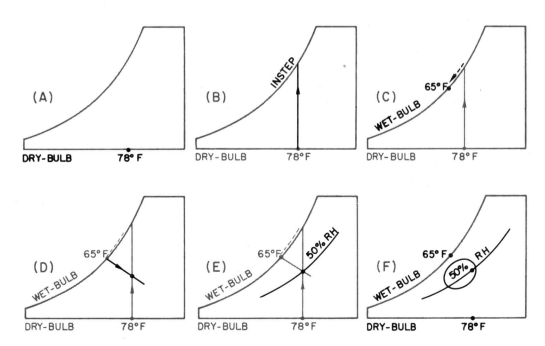

Fig. 22–21 Using the psychrometric chart to determine the relative humidity level.

pared to the total amount of moisture the air can hold (100 percent).

In this case, the relative humidity is determined to be 50 percent when the dry bulb temperature in a given area is 78° and the wet bulb temperature is 65°.

To understand what this information means to you as a technician, consider that the factors we determined are close to the conditions desired in an area served by a comfort cooling system: 78° and 50 percent relative humidity. Airflow problems can prevent the unit from performing properly and achieving these desired conditions.

A dirty filter or clogged evaporator coil could, for example, cause a condition of higher-than-desired humidity, leaving the area uncomfortable. If a unit was moving too much air too fast, the humidity level would be too low—a condition that would also cause discomfort.

The fundamental thing to keep in mind about psychrometrics is that, in addition to the refrigeration and electrical system functioning properly, the airflow system in a comfort cooling system must also function according to design in order for the system to perform as intended.

Unit Twenty-Two Summary

- Two types of systems commonly found in residential comfort cooling applications are the package unit and the split system. The split system has an evaporator positioned in the air handler inside the house and the condensing unit is located outside, either on a pad next to the house or on the roof. A package unit is self-contained with all components contained in a single housing, and ductwork connects it to the conditioned space.

- Air-conditioning systems typically allow for an airflow rate of 350 to 400 cfm per ton of

refrigeration. The average comfort range for air-conditioning is 72°F to 80°F, with a humidity level of 50 percent to 60 percent.

- Condensing units should be installed as close as possible to the evaporator and the electrical supply. A secondary drain system should be installed to prevent water damage in the event the primary drain becomes clogged.

- Precharged refrigerant lines are available for split system air-conditioning systems.

- A preseason and initial start-up check of an air-conditioning system is necessary to ensure trouble-free, efficient operation of the unit. Diagnostic procedures require a systematic elimination of the possibilities based on problem cause and effect.

Unit Twenty-Two Key Terms

Packaged Air-Conditioning System

Split System

Condensing Unit

Compressor

Condenser Coil

Condenser Fan

Condenser Fan Motor

Evaporator Coils

"A" Coil

Slant Coil

Sloped Coil

Horizontal Coil

Vertical Coil

Metering Device

Fan Coil Air Handler

Psychrometrics

UNIT TWENTY-TWO REVIEW

1. What is the difference between the packaged system and the split system types of residential air-conditioning applications?

2. What is the advantage of the packaged system?

3. Describe the installation applications of the split system.

4. List the parts that make up the condensing unit.

5. What is the function of the two-speed condenser fan motor?

6. What are the advantages of utilizing the capillary tube as a metering device?

7. What type of compressor motor is commonly used in residential applications?

8. How is the determination made to use the capillary tube or the thermostatic expansion valve?

9. What type of thermostatic expansion valve is commonly used in air-conditioning applications?

10. What are the two ways that airflow across the evaporator can be changed?

11. Airflow across the evaporator coil depends upon what two requirements?

12. What is the average airflow for an air-conditioning system?

13. What factors affect the airflow for a given application?

14. Describe daily temperature range.

15. What are the temperature differences of airflow across the evaporator for low, medium, and high temperature ranges?

16. To balance the air delivery system, the dampers in the coldest rooms are partially _____, and the dampers in the warmer rooms are adjusted to the _____ position.

17. How is the humidity affected by the evaporator blower speed?

18. What is the average comfort level of humidity in a conditioned space?

19. What factors should be considered when installing a condensing unit?

20. What factors should be considered when making electrical connections in the installation of an air-conditioning system?

21. What factors should be considered in the installation of an evaporator coil in the furnace plenum?

22. What is the primary advantage of suspending a fan coil air handler?

23. What is the purpose of installing a secondary drain pan in a fan coil air handler installation?

24. Match the following terms with their descriptions: crankcase heater, control box, "A" coil, slant coil, fan coil air handlers, float switch.

    a. Use field-installed ductwork to deliver the air to the conditioned space.

    b. Installed in the condensate pan for the purpose of interrupting power to the compressor contacts.

c. Used in small-capacity applications where a large coil surface is not required.

d. Contains all of the electrical components.

e. Designed to be used in up-flow or down-flow furnace installations.

f. Should be energized six or eight hours before starting the compressor.

25. What is the advantage of precharged tubing kits?

26. Why are soldered joints recommended where field-charged tubing is not used?

27. Why are suction lines on an air-conditioning system insulated?

28. What is the purpose of securing excess tubing in a horizontal position?

29. What precautions should be made to prevent the tubing from vibrating or moving?

30. What is the purpose of installing a "P" trap in the condensate line?

31. Match the (p)roblem and its possible (c)ause with the solution.

a. (p) System runs continuously. (c) Low refrigerant charge.

b. (p) Suction line is frosted. (c) Overcharge in capillary tube system.

c. (p) Furnace fan, condenser fan, and compressor will not operate. (c) Thermostat set too high.

d. (p) System cycles off on low-pressure control. (c) Insufficient airflow across evaporator.

e. (p) System operates normally but the circuit breaker trips occasionally. (c) Burnt contactor contacts.

f. (p) System cycles on high-pressure control. (c) Fan motor problem.

g. (p) Circuit breaker trips when system is energized. (c) Short circuited compressor windings.

Solutions:

1. Reset thermostat.

2. Check system pressure, locate and repair, then recharge the system.

3. Check and repair or replace the motor.

4. Check compressor windings and replace as required.

5. Check pressures and purge.

6. Replace as required.

7. Check the filters, fan speed, and evaporator coil condition.

32. fpm = velocity of air

A velometer measures the velocity of airflow.

area = the cross sectional area measured in square feet (a rectangular duct in this case)

cfm = the air measured in cubic feet per minute

cfm = fpm × area

Normal cooling requires 400 cfm per ton.

a. If an evaporator coil has a cooling capacity of 3 tons, how many outlets (ducts) requiring 120 cfm could this system handle?

b. If the size of the duct is 15 inches by 20 inches and the average velocity is 800 fpm, what is the cfm of this system? (Hint: Convert inches to feet.)

Psychrometrics is an examination of air and its properties.

one cubic foot of air weighs 0.075 lb/ft^3

33. How many cubic feet of air is required for a weight of one pound?

34. What is the weight of air in a room 15 feet wide by 15 feet long by 10 feet high?

UNIT TWENTY-THREE

Gas Furnaces

OBJECTIVES

After completing this unit, the student will be able to:

1. Identify the components in a forced air gas furnace.

2. Explain the method and sequence of operation of a gas furnace.

3. Differentiate between a standing pilot and electric ignition system.

When working as an HVAC/R technician, you'll be servicing many different types of heating systems. One of the most common types of heating equipment is the forced air gas furnace, which is found in residential and light commercial applications. In this unit we'll identify the components found in gas furnaces and outline the method of operation of the burner, valve, heat exchanger, and electrical and venting systems.

THE GAS FURNACE AIR HANDLING SYSTEM

The **air handling system** of a gas furnace uses a fan assembly to pull the air from the conditioned space into the furnace cabinet, then circulates it back into the living area. A direct drive air handler is the most common one found on the standard gas furnace, although you may also see some belt-driven fan assemblies. For the most part, newer units will use the direct drive fan assembly, which consists of a motor and squirrel cage fan assembly mounted in a housing. Figure 23–1 shows a typical direct drive blower assembly. Direct drive motors will almost always be a PSC-type motor.

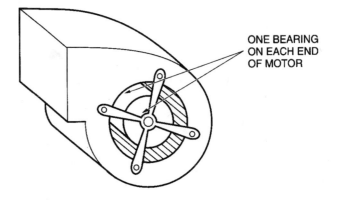

ONE BEARING ON EACH END OF MOTOR

Fig. 23–1 Direct drive blower assembly has only two bearings that will wear.

Belt-driven fan assemblies, such as the one shown in Figure 23–2, are also commonly found in gas furnaces. For the most part, you can expect to find this type of air handler in an older furnace or in some newer units in the 100,000 Btu range.

One main difference between the direct drive air handler and the belt-driven type of blower is that one has more bearings that may need attention during routine preventive maintenance. In the case of the direct drive blower, only the motor bearings need to be considered. With a belt-driven unit, the motor bearings as well as the blower bearings may

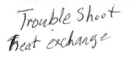
Trouble Shoot
Heat exchange

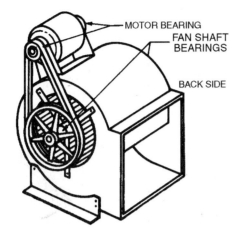

Fig. 23–2 Belt-drive fan assembly has four bearings, two pulleys, and a belt that will wear out.

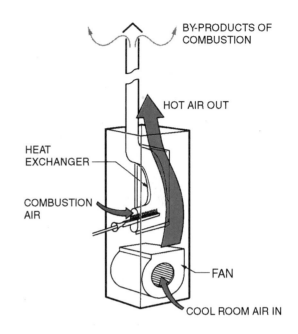

Fig. 23–3 Heat exchange system.

need periodic lubrication. In either case you may find that the manufacturer has provided a motor with permanently lubricated bearings.

HEAT EXCHANGERS

Regardless of the design, the blower assembly is assigned the task of circulating air through the furnace cabinet. A fundamental fact to keep in mind about airflow through the cabinet of a gas furnace is that a **heat exchanger** is an integral part of the system. In other words, the air circulating through the cabinet doesn't come in direct contact with the flame from the burners. A heat exchange system such as the one shown in Figure 23–3 warms the air as it passes across the outer surface of the heat exchanger.

FLAME SYSTEM COMPONENTS

Gas Burners

Gas burners are positioned inside the heat exchanger of a gas furnace. There are several different types of burners used by various manufacturers in their different models of furnaces. Figure 23–4 shows some examples of burners.

Cast iron burners may still be found in some furnaces, usually in an older model unit. Single-port burners, referred to as *upshot* and *inshot burners*, will usually be found in furnaces in

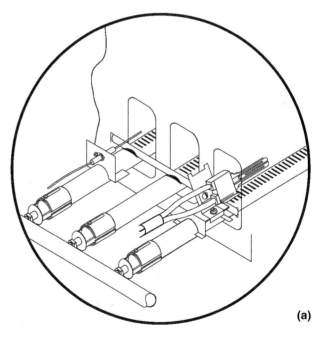

Fig. 23–4 Various types of burners (a) ribbon burner.
Courtesy Inter-City Products Corporation (U.S.A.) (continued)

smaller homes. Stamped steel and ribbon burners are popularly used in a wide range of furnaces.

The number of burners found in a furnace depends on the heating capacity of the unit.

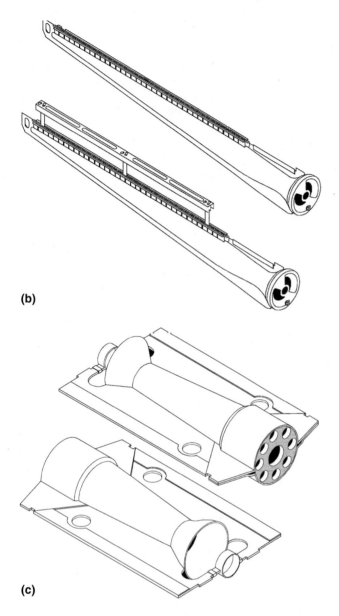

(b)

(c)

Fig. 23–4 *(continued)*   Various types of burners (b) stamped steel slotted burner, (c) inshot burner.   *Courtesy Inter-City Products Corporation (U.S.A.)*

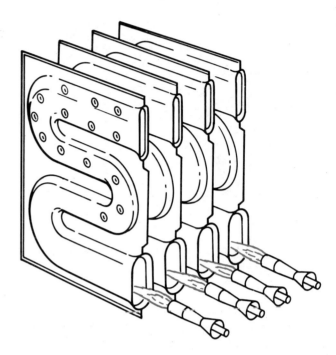

Fig. 23–5   Modern heat exchanger with four sections.

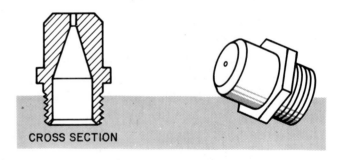

CROSS SECTION

Fig. 23–6   Spud showing the orifice through which the gas enters the burner.

Smaller units having a heating capacity of 40,000 to 50,000 Btu's will usually have two burners while a furnace in a larger home requiring a heating capacity of 80,000 to 100,000 Btu's will have four burners. In any case, a section of heat exchanger is provided for each burner. (see Figure 23–5)

Spuds

A **spud** is the orifice that determines the volume of gas flow to a burner. Threaded on one end

and screwed into the main manifold of the gas furnace, it's a brass fitting with a small opening. Figure 23–6 shows a gas furnace spud.

Factors that you as a technician must be concerned about when considering a gas furnace spud are the size of the orifice, which is designed to precisely match the capacity of the burner, and the cleanliness of the orifice after the furnace has been in service for an appreciable length of time.

When performing routine maintenance of a gas furnace in preparation for seasonal start-up, make sure the burner orifices are clean and clear. As far as orifice size, most modern furnaces will handle natural or LP gas without the need for changing spuds. In these cases it would only be necessary to adjust the air shutter on the burners to ensure complete combustion and proper burner operation. In rare cases, you may find an older furnace or a unit designed for a mobile home that requires a change of orifice to switch the operation from natural to LP gas.

The Gas Valve

Described in its simplest form, a **gas valve** is a device that, when energized by an electrical circuit, opens and allows gas to flow to the burners. Modern furnaces use valves that appear much more complicated than this simplified explanation because in addition to the task of allowing gas flow to the main burners, gas valves found in today's equipment also control the pilot assembly and are an integral part of the operation of an electric ignition system. Another feature of gas valves found in modern equipment is that of a pressure regulator.

Called *combination gas valves,* they contain separate coils for the pilot and main burner flames and wiring connections for the electric ignition systems. Older units with a standing pilot system will not be as complicated, having only the pilot and main burner gas connections and two wire connections for the main valve electrical operation.

Two popular types of valves are the solenoid valve (shown in Figure 23–7) and the heat motor valve (shown in Figure 23–8). Both types of valves are energized by the 24-volt circuit of the furnace.

A fundamental fact to keep in mind about the two different kinds of valves is the speed at which either will open when energized. A solenoid operated valve will open immediately and the gas burners will ignite very quickly. A heat motor type of valve takes a bit longer to open and there will be a slight delay in the establishment of a main burner flame.

Pilot Assemblies

In gas furnaces two distinct types of **pilot assemblies** are found. Older units will have a standing pilot system in which the pilot flame burns constantly, and newer furnaces will employ an electric ignition system to ignite the pilot when the thermostat calls for heat. In either case, the size of the pilot flame is critical to the operation of the furnace. In some cases, a pilot assembly will need to be cleaned periodically to ensure proper flame level. It may also be necessary to

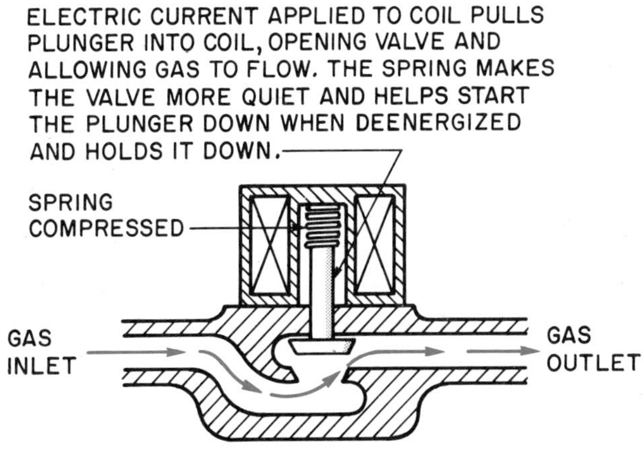

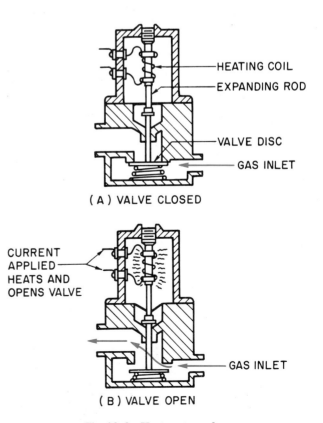

ELECTRIC CURRENT APPLIED TO COIL PULLS PLUNGER INTO COIL, OPENING VALVE AND ALLOWING GAS TO FLOW. THE SPRING MAKES THE VALVE MORE QUIET AND HELPS START THE PLUNGER DOWN WHEN DEENERGIZED AND HOLDS IT DOWN.
SPRING COMPRESSED
GAS INLET
GAS OUTLET

Fig. 23–7 Solenoid valve.

HEATING COIL
EXPANDING ROD
VALVE DISC
GAS INLET
(A) VALVE CLOSED
CURRENT APPLIED HEATS AND OPENS VALVE
GAS INLET
(B) VALVE OPEN

Fig. 23–8 Heat motor valve.

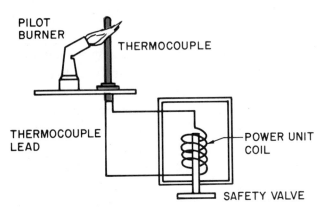

Fig. 23–9   **The thermocouple generates electrical current when heated by the pilot flame.**

Fig. 23–10   **Spark-to-pilot ignition system.**   *Courtesy Robertshaw Controls Company.*

adjust and redirect a pilot flame to ensure proper operation of a gas furnace.

In the case of a standing pilot assembly such as the one shown in Figure 23–9, the pilot flame contacts a device known as a *thermocouple.*

The heat-sensing section of a thermocouple consists of two dissimilar metals that, when heated, generate a small amount of current, which is measured in millivolts. This small amount of electrical energy is enough to keep the pilot coil energized, thus allowing gas flow to the main burners when the 24-volt coil of the gas valve is energized. In the event the pilot flame goes out, the thermocouple acts as a safety device, closing the valve so the gas will not flow to the burners.

In regard to electric ignition systems for gas furnaces, the most popular type of system is one that uses a spark ignition device. (Manufacturers are careful to refer to these systems as "electronic" ignition systems when they want to impress upon their customers that it is state-of-the-art equipment.)

With a spark ignition system, the pilot is lit with every call for heat from the thermostat. At the end of the run cycle, the pilot flame is extinguished until the temperature in the conditioned space again drops below the thermostat set point and the call for heat is again established. Figure 23–10 shows a popular spark ignition device.

You'll note that the tip of the electrode is positioned at a point that will allow a spark to jump between the electrode itself and the mounting bracket. With a call for heat, the spark is initiated and, at the same time, gas flow begins from the pilot tube. In a short time (usually four to eight clicks of the spark system), the gas flow is ignited by the spark and the pilot flame is established.

This spark is created by positioning the electrode, for all practical purposes, to arc to ground. Figure 23–11 shows a wiring diagram for a spark ignition furnace. Note the spark electrode and how its position is shown as leading to ground.

Once the pilot flame is established in a spark ignition system, the arcing ceases through a process known as *flame rectification.* Flame rectification is described as a process through which a flame conducts electric current when voltage is applied across two electrodes immersed in the flame. In other words, the flame is considered to be a conductor of electricity.

The reason this is referred to as flame rectification is because the AC current is rectified and changed to DC current by the heat of the pilot burner. So once the flame is established, instead of AC current creating an arc from the electrode to the pilot assembly bracket (which is fundamentally "ground" or "frame"), current flow is direct and, although the circuit remains, no arc occurs.

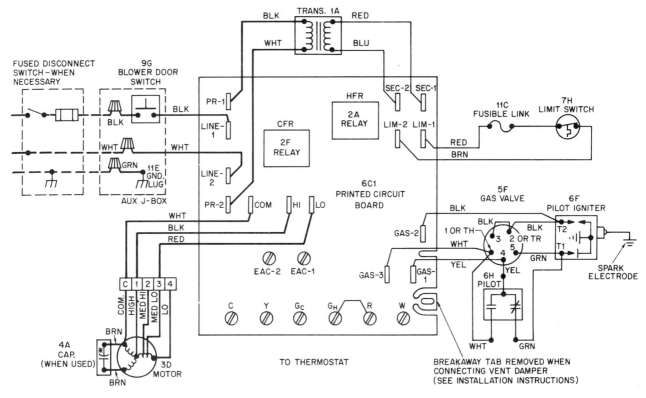

Fig. 23–11   Wiring diagram for a spark ignition furnace.   *Courtesy Carrier Corp.*

Hot Surface Ignition Systems

Another type of electric ignition system you may find in gas furnaces is the **hot surface ignitor**. Instead of a spark device, this type of system uses an ignitor (sometimes referred to as a glow bar), which, when energized, glows bright red and ignites the flow of gas. The ignitor itself is made up of a fragile material known as Carborundum. It's resistive to current flow and, like any resistive material, it gets hot when energized. Several different configurations of Carborundum ignitors are shown in Figure 23–12.

One main difference between a spark ignition system and a hot surface ignition system is that the flame rectification process takes place through a flame sensor. The glow bar itself doesn't create the DC circuit, a separate component must be added to the system to drop the circuit to the hot surface ignitor. Figure 23–13 is an example of a wiring diagram that illustrates a hot surface ignition system.

Fan Switches and Limit Switches

Within a gas furnace electrical system, a **fan and limit switch** is used to control the operation of the blower assembly and to act as a safety system in the event of blower motor failure. A fan switch can be a time delay type relay or it can be a device activated by the temperature rise of the furnace after the burner has been on for a short period of time. Figure 23–14 shows a typical temperature-activated fan switch.

The purpose of the limit switch is to break the circuit to the gas valve in the event the blower assembly doesn't start. A failure of the blower assembly would result in overheating because the burner would continue to operate but there would be insufficient airflow over the heat exchanger of the furnace.

In many cases it's common for manufacturers to use a fan/limit switch combination component to perform the two necessary functions of cycling the blower assembly and providing the safety circuit. Refer to Figure 23–15.

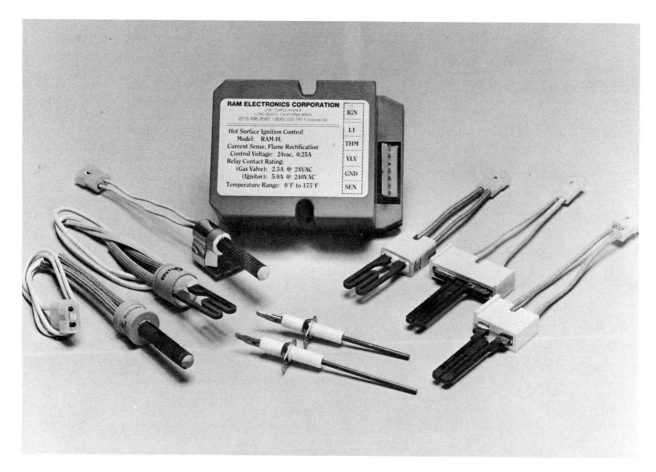

Fig. 23–12 Various Carborundum ignitors. *Courtesy RAM Electronics Corporation.*

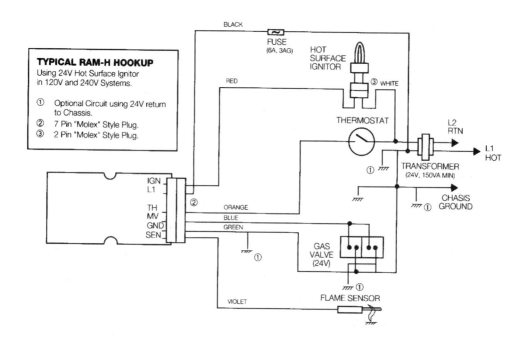

Fig. 23–13 Wiring diagram illustrating a hot surface ignition system. *Courtesy RAM Electronics Corporation.*

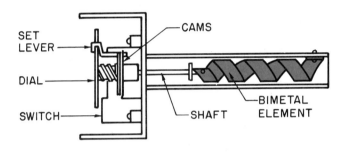

Fig. 23–14   Temperature on/temperature off fan switch.
Photo by Bill Johnson.

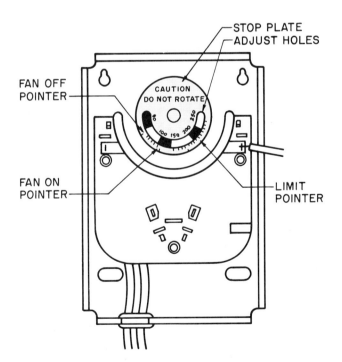

Fig. 23–15   Fan on-off and limit switch.

For an electrical overview of the fan/limit switch circuitry in a typical gas furnace, refer to Figure 23–16. These two diagrams show a simplified illustration of a fan switch cycling system and a limit switch protection system.

GAS FURNACE SEQUENCE OF OPERATION

In a typical residence, the sequence of operation of a gas furnace is as follows:

1. The temperature of the conditioned space falls below the thermostat set point and the thermostat calls for heat. In the case of an electric ignition system, the spark or hot surface ignition system lights the pilot, then after a delay, the main burner is ignited. If you're working with a standing pilot system, the main burner is ignited when the thermostat calls for heat.

2. The main burner heats up the furnace heat exchanger.

3. The fan switch closes after the main burner has been on for a short period of time and the blower assembly begins to operate.

4. The main burner continues to burn until the thermostat is satisfied. When the thermostat set point is reached, the circuit to the gas valve is broken and the main burner shuts down.

5. Because the heat exchanger is still hot, the blower assembly continues to operate until either the temperature-sensing fan switch breaks the circuit to the fan motor or the time delay assembly drops out and shuts the motor down.

The Heat Anticipator

When servicing gas furnaces, it's important that you have an understanding of the heat anticipation system. A **heat anticipator** is a component within the heating thermostat and it's job is to fool the thermostat into thinking that the set point has been reached, breaking the circuit to the main burner just a couple of degrees prematurely.

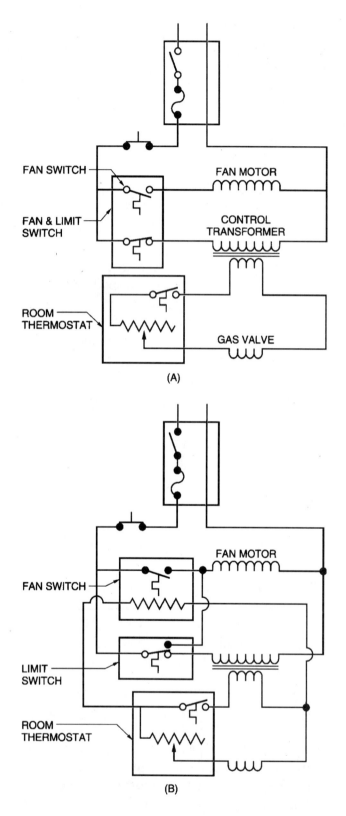

FAN SWITCH

FAN MOTOR

FAN & LIMIT
SWITCH

CONTROL
TRANSFORMER

ROOM
THERMOSTAT

GAS VALVE

(A)

FAN MOTOR

FAN SWITCH

LIMIT
SWITCH

ROOM
THERMOSTAT

(B)

Fig. 23–16   Switch circuitry in a gas furnace: (a) fan (b) limit.

The reason this system exists is because a furnace contains a good deal of residual heat even after the main burner has been shut down. This heat is enough to bring the actual temperature of the conditioned space up to the thermostat set point.

In some cases, you'll find that the heat anticipator within a thermostat is fixed and therefore can't be adjusted. This would be a bottom-of-the-line thermostat found in an installation that was accomplished as inexpensively as possible. (Read it—cheap.) In most situations, however, an adjustable heat anticipator is used and the bottom line on this type of system is that the anticipator must be adjusted to match the current draw of the gas valve or the furnace may short cycle, causing a customer to complain that the unit isn't functioning properly.

Figure 23–17 shows one type of adjustable heat anticipator found in thermostats and Figure 23–18 illustrates the method of determining the current draw of the gas valve and deciding where the heat anticipator should be set for a particular furnace.

Furnace Venting

The **vent system** of a gas furnace is extremely important. When a gas furnace operates, the by-products of combustion (carbon monoxide, for example) are poisonous and must be vented to the outside atmosphere. To accomplish this, the heat exchanger sections that house the burner assemblies open into a vent box, which is connected to a flue. The flue carries the by-products of combustion outside, as shown in Figure 23–19.

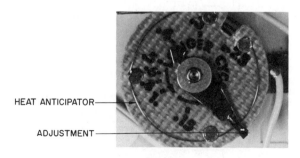

HEAT ANTICIPATOR

ADJUSTMENT

Fig. 23–17   Adjustable heat anticipator.   *Photo by Bill Johnson.*

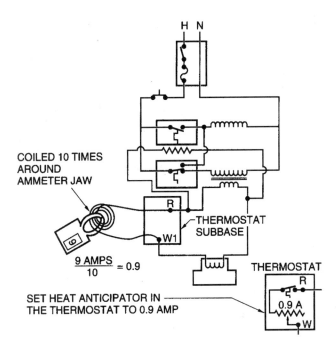

Fig. 23–18   Setting the heat anticipator.

High Efficiency Furnaces

In a standard furnace, the venting system that is necessary for safety also cuts down on the actual efficiency of the furnace. In other words, a good deal of the heat the customer pays for winds up going out the vent system. In most standard furnaces, an efficiency rating of about 65 percent could be assigned, meaning that 35 percent of the energy used is "wasted." High efficiency furnaces can be rated as high as 98 percent efficient.

One type of high efficiency furnace is known as the condensing furnace. A condensing furnace uses what is referred to as a *secondary heat exchanger*, which is a stainless steel component within the furnace that allows for additional heat dissipation. Figure 23–20 shows the fundamental idea behind the secondary heat exchanger. The hot flue gas is routed into the coil instead of being vented straight into the atmosphere.

To understand how this system provides extra heat, think for a moment about one of the basic laws of thermodynamics—the theory of heat. This basic law of physics says that when a substance condenses, changes from a vapor to a liquid, the substance will reject (or, dissipate, if you

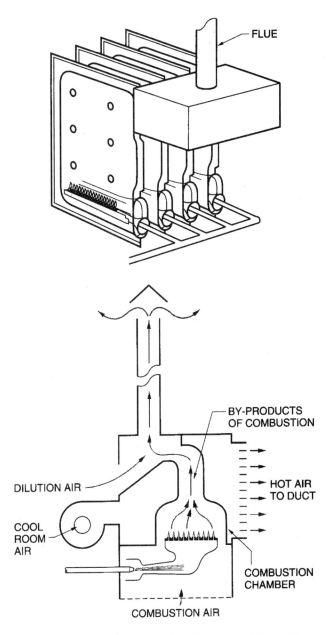

Fig. 23–19   By-products of combustion are carried outside by the flue.

prefer to use that term) heat into the surrounding atmosphere as long as the surrounding atmosphere is cooler than the substance. (Remember, heat always moves from warmer to cooler.)

What this boils down to is that flue gases that are vented to the atmosphere in a standard furnace give up their heat when routed through a coil designed as a secondary heat exchanger. Since the temperature of the flue gases is much higher than

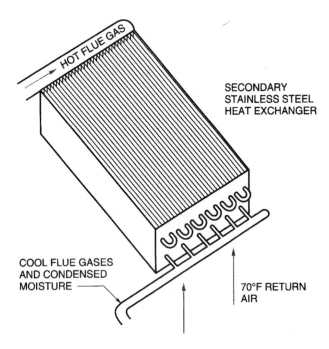

Fig. 23–20 A secondary heat exchanger resembles a hot water coil.

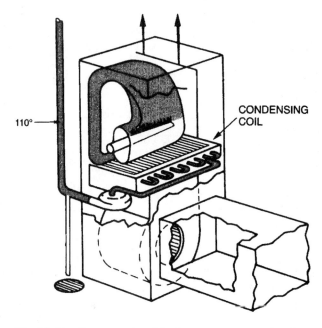

Fig. 23–21 Power vent and drain system on a condensing furnace.

the temperature in the conditioned space, this heat exchange can occur.

What this system creates is the need for some components not found on a standard gas furnace; one is a power vent system, the other is a drain system.

Since the flue gases are cooled in the process of condensing, they can no longer be expected to flow naturally up the flue system because of their high temperatures. This means that some type of blower is necessary to vent the now cooled vapors that remain to the outside. Because the flue gases are cooled, plastic pipe is often used for the venting system on a condensing furnace.

A drain system is necessary because the condensing furnace has created liquids. In conjunction with the drain system, some type of neutralization of the liquids must be accomplished before the substance is routed down the drain. A neutralization cartridge is necessary because the substance contains the acidic by-products of combustion (now in liquid form) and this situation must be dealt with before the disposition of the substance is accomplished. Figure 23–21 shows a power vent and drain system on a condensing furnace.

Unit Twenty-Three Summary

- The air handling system of a gas furnace uses a fan assembly to pull the air from the conditioned space into the furnace cabinet, then circulates it back into the living area. Two types of air handler assemblies are the belt-driven and the direct drive type.

- On a belt-driven assembly the blower bearings as well as the motor bearings must be lubricated. On a direct drive system only the motor bearings must be considered for lubrication. Some motors are permanently lubricated and require no periodic maintenance.

- A heat exchanger is the heart of a gas furnace. The burners are mounted inside the heat exchanger and the air from the conditioned space is circulated past the exterior surface of the heat exchanger.

- Several different types of burners may be found in a residential gas furnace and the number of burners used depends on the capacity of the furnace. A furnace that has a capacity of 80,000 to 100,000 Btu's will typically have four burners.

- A spud is a brass fitting that contains an orifice that determines the volume of gas flow to a burner. A gas valve is an electrically operated device that allows gas flow to both the pilot assembly and the main burner. In newer furnaces the pressure regulator is built into the gas valve. On electric ignition systems a combination gas valve is used.

- Older gas furnaces use a standing pilot system while newer units use an electronic ignition system. In some cases, a pilot assembly needs to be cleaned to ensure proper flame level.

- A thermocouple is used to maintain a pilot flame. It consists of two dissimilar metals that, when heated, generate a small amount of current.

- Two types of electronic ignition systems are the spark ignition and hot surface ignition systems. Both use a method known as flame rectification in order to terminate the ignition once the pilot flame has been established.

- In a gas furnace, a fan switch is used to control the operation of the blower assembly on a delay and the limit switch is a safety device.

- A heat anticipator is a component within a thermostat that shuts the burner down just prior to the actual set point of the thermostat.

- Some thermostats have fixed anticipators while others can be adjusted to match the current draw of the gas valve. A heat anticipator that isn't properly adjusted can cause short cycling of the burners in a furnace.

- In a standard gas furnace, the vent system carries the poisonous by-products of combustion to the outside. High efficiency furnaces have a secondary heat exchanger that allows the flue gases to condense and additional heat gain is accomplished. A condensing gas furnace must use a power vent system and a drain system.

Unit Twenty-Three Key Terms

Air Handling System

Heat Exchanger

Gas Burner

Spud

Gas Valve

Pilot Assembly

Hot Surface Ignitor

Fan and Limit Switch

Heat Anticipator

Vent System

UNIT TWENTY-THREE REVIEW

1. Match each of the following terms with its function or description: gas valve, flame rectification, Carborundum ignitor, thermocouple, heat exchange system, heat anticipator, spud, fan assembly.

    a. Pulls air from the conditioned space into the furnace cabinet; the air is then circulated back into the living area.

    b. The orifice that determines the volume of gas flow to a burner.

    c. Warms the air as it passes across the outer surface of the heat exchanger.

    d. Consists of two dissimilar metals that, when heated, generates current to keep the pilot coil energized.

    e. Opens and allows gas to flow to the burners.

   f. When energized, glows bright red.

   g. Process of changing (current) AC to DC by the heat of the pilot burner.

   h. Has the job to fool the thermostat in order to break the circuit to the main burner.

2. What are two applications for forced air gas furnaces?

3. What is one main difference between the direct drive air handler and belt-driven blower?

4. _____ _____ _____ are referred to as upshot or inshot burners.

5. What determines the number of burners found in a furnace?

6. Name two factors in considering a gas furnace spud.

7. When would the spud require changing?

8. Which of the following is not a function of a gas valve?

   a. Allows gas to flow to the burners

   b. Controls the pilot assembly

   c. Integral part in the operation of an electric ignition system

   d. A pressure regulator

9. What is a fundamental difference between a solenoid valve and a heat motor valve?

10. The electric ignition system of a newer furnace will ignite the _____ when the _____ calls for heat.

11. Why is a limit switch necessary?

12. What is the difference (a percentage number) in efficiency between a standard furnace and a high efficiency furnace?

13. Millivolt = 1 thousandth of a volt

$$= 1/1000$$

$$= .001$$

$$= 10^{-3} \text{ (scientific notation)}$$

   a. Give the fraction, the decimal, and the scientific notation for 7 millivolts.

   b. Give the fraction, the decimal, and the scientific notation for 100 millivolts.

14. Write a paragraph to describe how a spark ignition system works.

UNIT TWENTY-FOUR

Electric Furnaces

OBJECTIVES

After completing this unit, the student will be able to:

1. Identify the components in an electric furnace.

2. Describe the sequence of operation of an electric furnace.

3. Explain the purpose of a sequencer in an electric furnace.

Electric furnaces, while they are not as plentiful as gas furnaces in residences, are an item that HVAC technicians are required to service from time to time. Relatively simple in their construction and method of operation, the components are simply described and easily explained. In this unit, we'll discuss the basic electrical circuits in an electric furnace and how a unit operates to provide heat in a residence.

ELECTRIC FURNACE DESIGN

The most popular configuration of an electric furnace is the up-flow design. In this type of unit, the air handler is located in the same position as the blower motor in a gas furnace, located in the bottom section of the cabinet. The main difference between the method of operation of a gas furnace and an electric furnace is that the airflow is directly across the source of heat. Unlike a gas furnace, an electric furnace has no heat exchanger.

The heating elements are a ribbon-type element made of a high-resistance wire that glows red hot when power is applied, as you can see in Figure 24–1.

An electric furnace is traditionally more expensive to operate than a gas furnace, even

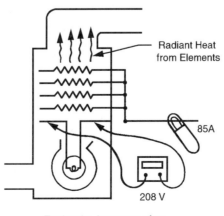

Radiant Heat
from Elements

85A

208 V

Testing the Amperage of an
Electric Furnace

Fig. 24–1   Electric furnace with ribbon-type heating elements.

though an electric furnace is described as being 100 percent efficient. In other words, a gas furnace wastes some of its heat due to the necessity of a venting system to get rid of the by-products of combustion, while an electric furnace has no waste.

Despite that fact, the cost of operation of an electric furnace is higher than that of a gas furnace due to the difference in costs from the supplying utility. Electric furnaces are commonly found in homes where natural gas is not available and the customer prefers not to use propane fuel for heat.

Some residential subdivisions also used electric furnaces because in some areas the gas utility declared a moratorium on construction of homes with gas service. During a situation such as this, the contractor's next choice was to equip the homes with an electric furnace.

A residential electric furnace, such as the one shown in Figure 24–2, is wired to a 240-volt circuit because the higher voltage is necessary to operate the heating elements. It's common to find the blower motor to be a 230-volt component also. Like a gas furnace, an electric furnace contains a step-down transformer that provides the low-voltage control circuit.

Depending on the capacity of an electric furnace, it can have only three heating elements if it's a smaller unit, or it can be equipped with up to six elements. An important fact for you to keep in mind about an electric furnace is that regardless of the number of heating elements in a given unit, some system of energizing the elements in a sequence rather than all at once must be employed. If all of the elements in a furnace were energized at the same time on a call for heat from the thermostat, a severe voltage drop problem would occur.

One common method of operation of an electric furnace is to start the fan motor immediately when the thermostat calls for heat. For this reason, a customer whose past experience has been with a gas furnace may complain about the operation of his or her electric furnace because the initial airflow from the unit at the beginning of a cycle is not as warm as a gas furnace. You'll recall that a gas furnace air handler doesn't operate until the heat exchanger has warmed sufficiently and a fan switch or time delay relay energizes the fan motor.

ELECTRIC FURNACE COMPONENTS
Heating Elements

Heating elements in an electric furnace are made up of a material commonly referred to as *nichrome*, meaning that they are a combination of nickel and chromium resistance wire. They are sometimes referred to as *ribbon-type elements* due to their appearance. The elements themselves are fastened within insulators and fixed to a metal plate, often referred to as a *housing*. The metal plate is fastened to an opening in the furnace that allows the element to be suspended in the airflow through the cabinet. (see Figure 24–3)

Fig. 24–2   Central forced-air electric furnace with multiple sequencers.   *Courtesy Rheem Air Conditioning Division.*

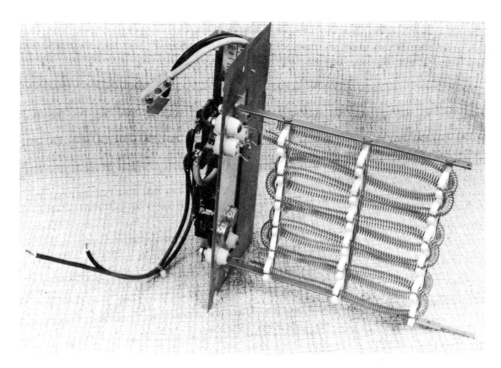

Fig. 24–3   Heating elements.

Fuse Link

A **fuse link** is a one-time protective device that opens in the event the furnace overheats for any reason. If the fan motor should fail, for example, or if the customer doesn't keep the filter clean and the airflow across the heating elements is restricted, the fuse link can open.

A fuse link such as the one shown in Figure 24–4 is mounted in a ceramic base that may also be fastened to the same metal plate that supports the heating element. The fuse link itself is positioned in the airflow and the wiring connections are made in such a way that you can troubleshoot and test without removing the component. When the fuse link is mounted in the same plate with the heating element, the wiring connections will be in close proximity to the element wiring connections.

Reset-Type Safety Controls

The one-time fuse link isn't the only type of safety control found on an electric furnace. Some manufacturers will use a device that reacts to the overheat situation and then resets when the fur-

Fig. 24–4   Fuse link mounted in a ceramic base. *Photo by Bill Johnson.*

nace cools down. Two types of reset-type safety controls are the **manual reset** and the **automatic reset**.

A reset-type safety control is a bimetal device that is in the normally closed (NC) position when it's not being affected by heat. When the heat rises to the level above the set-point temperature of the control, the bimetal reacts and breaks the contact points inside the switch. Figure 24–5 shows both an automatic and a manual reset safety control.

Sequencers

A **sequencer** is a device used in an electric furnace to cycle the heating elements on through

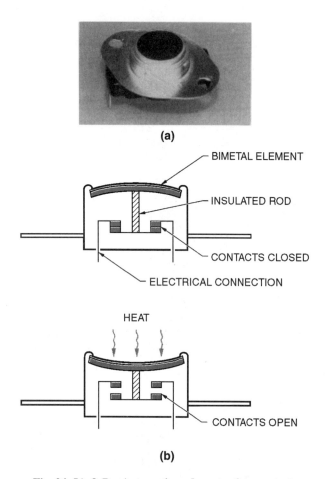

(a)

- BIMETAL ELEMENT
- INSULATED ROD
- CONTACTS CLOSED
- ELECTRICAL CONNECTION

HEAT

- CONTACTS OPEN

(b)

Fig. 24–5A & B   Automatic and reset safety controls.

- MANUAL RESET BUTTON. PUSH TO RESET.

Fig. 24–5C   Manual reset safety control.

a step process in order to prevent a heavy electrical load at the time the thermostat calls for heat. There are several different configurations of sequencers used by the various manufacturers and some of them are shown in Figure 24–6.

Some sequencers have only one set of switching contacts while others may have as many as four sets. Regardless, they all operate in the same manner. Twenty-four volts is applied to the heater

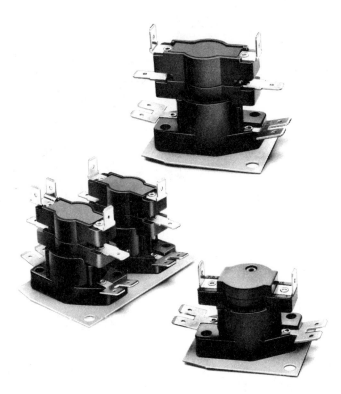

Fig. 24–6   Sequencers.   *Courtesy Therm-O-Disc, Inc., Subsidiary of Emerson Electric Co.*

portion of the sequencer (often referred to as H1 and H2), and that voltage causes the contacts in the switching portion of the sequencer to close. It's common to find the first set of switch contacts designated as M1 and M2 (for main switch), while other sets of switching contacts may be identified as A1 and A2 (auxiliary switch).

One thing to keep in mind about sequencers is that the switching contacts of one unit may be used to control the heater portion of another.

Fan Relays and Controls

Not all furnaces are designed in a manner that allows the air handler to start immediately when the thermostat calls for heat. Many units will be designed with some kind of **time delay system** that allows the fan to come on in a time frame near the start-up of a heating element. The electric furnace time delay system differs from the bimetal type of fan switch found in a gas furnace in that the electric furnace time delay fan control doesn't react to temperature, but rather to time.

In a manner similar to a sequencer, 24 volts is applied to the element portion of the time delay relay. After a time period of approximately 30 seconds, the switching contacts of the relay close and carry the high voltage to the fan motor windings.

Electric Furnace Wiring Circuit

A simplified wiring diagram shown in Figure 24–7 illustrates the components discussed and one method of wiring the circuits.

Note the five heating elements as they are shown boxed on the high-voltage portion of the diagram. The fuse links to the left of the elements and the open-on-temperature-rise safety controls located to the right of the heating elements are simply wired in series with the resistance heaters.

The heater section of the timed fan control is located on the low-voltage section of the diagram. This load, as well as the heater section of the sequencer identified as SEQ #1, is energized simultaneously when the thermostat calls for heat.

The main switch contacts of sequencer #1 (M1 and M2) carry the circuit to heating element #1. The auxiliary contacts of SEQ #1 (A1 and A2)

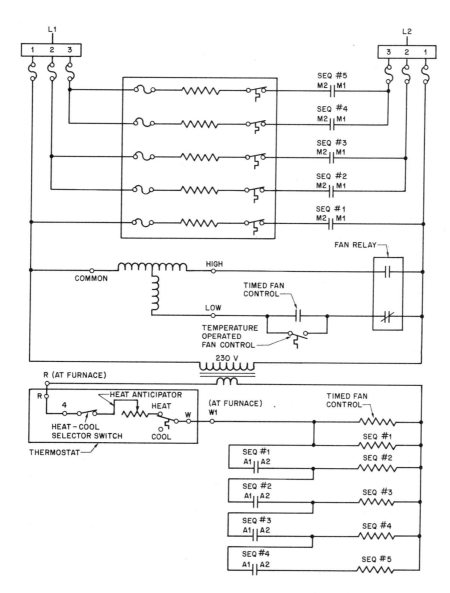

Fig. 24–7   One method of wiring the sequencer control circuit where all amperage passes through the thermostat.

carry the low voltage to the heater portion of sequencer #2. Energizing the heater section on SEQ #2 causes the M1 and M2 contacts on that switching assembly to close and complete a circuit to heating element #2. The auxiliary contacts on SEQ #2 (A1 and A2) then close, completing a low-voltage circuit to the heating section of SEQ #3 and the furnace heating element start-up sequence continues. Eventually, all of the furnace heating elements will be energized through this method of circuitry.

When the thermostat is satisfied, the heating elements will cycle off in an irregular pattern.

Heating Thermostats

Electric furnace **thermostats**, like a thermostat used to control a gas furnace, have heating anticipators that must be adjusted to match the amperage draw of the load in the low-voltage circuit. If all of the sequencers in an electric furnace are wired in parallel, the load on the thermostat will be high and the heating anticipator must be adjusted accordingly.

In some cases, the sequencer circuitry is wired in such a way that only one sequencer's heating portion is wired through the thermostat. This would allow a smaller load on the thermostat

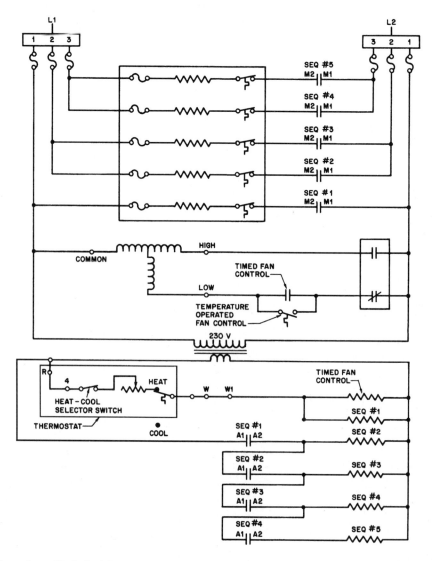

Fig. 24–8 Alternate method of wiring sequencers: power from only one sequencer passes through the thermostat.

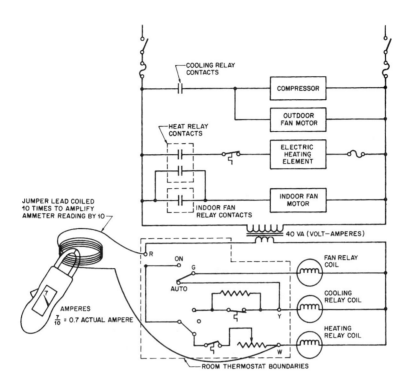

Fig. 24–9   Measuring the amperage in the heat anticipator circuit.

and the heating anticipator would be adjusted to a lower amperage draw due to the difference in wiring design. Figure 24–8 shows an alternate method of wiring sequencers and Figure 24–9 illustrates the method of determining the amperage draw in the low-voltage circuit in an electric heating system.

Unit Twenty-Four Summary

- An electric furnace design differs from that of a gas furnace in that there is no heat exchanger. The conditioned air passes directly across the heat source. An electric furnace is considered to be 100 percent efficient because it doesn't have a vent system and no heat is lost because there is no consideration for combustion.

- Electric furnaces operate on a 240-volt circuit because the higher voltage is necessary to operate the heating elements. It's common to find the blower motor in an electric furnace operating on 240 volts also.

- Some electric furnaces may be designed to start the blower motor when the thermostat calls for heat. With this type of system, cold air would be circulated at the beginning of the cycle. Many electric furnaces are designed to allow a time delay operation of the fan motor that is in accordance with the start-up of the first heating element.

- Heating elements in an electric furnace are a ribbon-type element made up of nickel and chromium (nichrome) wire. This type of material has a high resistance to current flow and glows bright red when energized.

- Safety devices in an electric furnace may be a one-time device such as a fuse link, or they may be a reset type of device. Both automatic and manual reset safety controls are used.

- Sequencers are used in electric furnaces to cycle the heating elements on in a step system. This prevents a heavy load on the electrical circuit when the thermostat calls for heat.

• Heating thermostats used for electric furnaces have heat anticipators that must be adjusted to match the current draw of the low voltage circuit. Some wiring circuits differ in the number of sequencers that are wired through the thermostat.

Manual Reset

Automatic Reset

Sequencer

Time Delay System

Thermostat

Unit Twenty-Four Key Terms

Heating Element

Fuse Link

UNIT TWENTY-FOUR REVIEW

1. What is the main difference between an electric furnace and a gas furnace?

2. Why is an electric furnace described as being 100 percent efficient?

3. Why are the heating elements of an electric furnace energized in a sequence?

4. What is an explanation for the fact that the initial airflow from an electric furnace may not be as warm as from a gas furnace?

5. What is the relationship between the following sets of terms?

    a. heating elements/nichrome

    b. sequencer/thermostat

6. What is a fuse link?

7. A reset-type safety control is normally in the _____ position when it is not being affected by _____ .

8. How does the electric furnace time delay system differ from the bimetal type of fan switch found in a gas furnace?

9. Formula to measure the volume of air across a furnace:

$$\text{cubic feet per minute} = \frac{Q_5}{1.08 \times TD}$$

1.08 = a constant

TD = temperature difference into/out of furnace

Q_5 = heat in Btu's per hour

Watts = amperes $\times$ volts

Btu's per hour = watts $\times$ 3.413 (constant)

Solve for the airflow. (Hint: Three steps—solve for (1) watts, (2) Btu's, and (3) cubic feet per minute.)

a. 230 volts with 82 amperes draw on the furnace, with a measured temperature difference of 52°.

b. 230 volts with 85 amperes draw on the furnace, with a measured temperature difference of 50°.

10. Write a paragraph to explain the type of technician you would like to have servicing your furnace.

UNIT TWENTY-FIVE

Heat Pumps

OBJECTIVES

After completing this unit, the student will be able to:

1. Describe the method of operation of a heat pump.

2. Differentiate between a ground-to-air, water-to-air, and air-to-air heat pump system.

3. Explain the methods of initiating the defrost cycle on a heat pump.

4. Describe the operation of a reversing valve.

The heat pump, due to improvements in design and efficiency, has become a popular method of heating and cooling residences and businesses. While there are specific differences between manufacturers in the way they operate and control a heat pump, there are some fundamental principles that apply to all types of units, regardless of the manufacturer.

In this unit, we'll discuss these fundamental principles and outline some of the factors that have to be considered when evaluating the performance of a heat pump and troubleshooting a customer's complaint.

A **heat pump** is an air-conditioning or refrigeration system that can remove heat from a conditioned space to cool it or add heat to a conditioned space to heat it. Through the use of automatic controls, a heat pump system can be used for environmental control year-round. A heat pump system is often referred to as a *reverse cycle refrigeration system*. However, the cycle is not actually reversed. It is the function of the condenser and evaporator coils that is reversed. Thus, to avoid any confusion in the discussion of heat pumps, we'll refer to the evaporator coil as the *indoor coil* and the condenser coil as the *outdoor coil*.

TYPES OF HEAT PUMPS

Heat pumps are classified according to the medium used to transfer heat to the outdoor coil during the heating cycle. The medium can be air-to-air, ground-to-air, or water-to-air.

Ground-to-Air Heat Pumps

In **ground-to-air heat pump** applications, an outdoor coil with a sufficient heat transfer area is buried in the ground at a depth of four to six feet. The average temperatures experienced at these depths will range from 40°F to 60°F and provide an excellent heat transfer medium. Figure 25–1A illustrates one typical ground-to-air heat pump application and Figure 25–1B illustrates an alternate ground-to-air pump.

Water-to-Air Heat Pumps

In **water-to-air heat pump** applications the outdoor coil is immersed in the bottom of a lake or reservoir. The temperature of the water at the bottom of the lake will be about 39°F or higher, thus providing an excellent heat transfer medium to dissipate heat from the outdoor coil during the

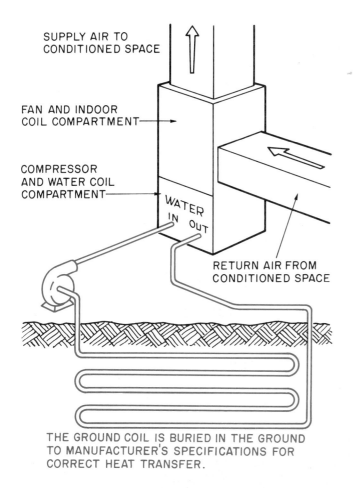

SUPPLY AIR TO
CONDITIONED SPACE

FAN AND INDOOR
COIL COMPARTMENT

COMPRESSOR
AND WATER COIL
COMPARTMENT

WATER
IN   OUT

RETURN AIR FROM
CONDITIONED SPACE

THE GROUND COIL IS BURIED IN THE GROUND
TO MANUFACTURER'S SPECIFICATIONS FOR
CORRECT HEAT TRANSFER.

Fig. 25–1A   One type of ground-to-air heat pump.

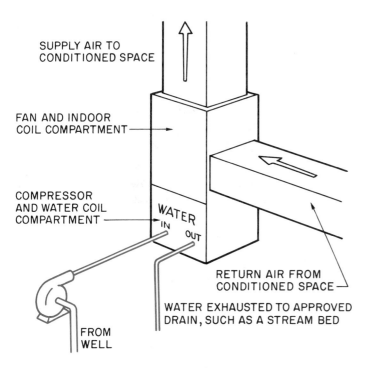

SUPPLY AIR TO
CONDITIONED SPACE

FAN AND INDOOR
COIL COMPARTMENT

COMPRESSOR
AND WATER COIL
COMPARTMENT

WATER
IN
OUT

RETURN AIR FROM
CONDITIONED SPACE

WATER EXHAUSTED TO APPROVED
DRAIN, SUCH AS A STREAM BED

FROM
WELL

Fig. 25–1B   Alternate ground-to-air heat pump.

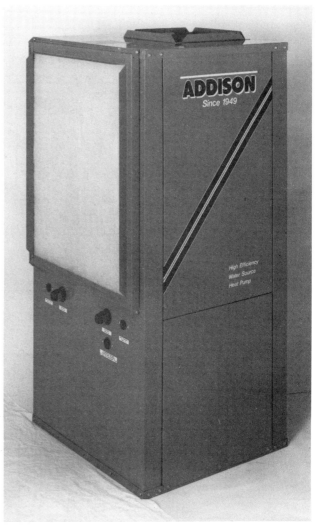

Fig. 25–2   Water-to-air heat pump.   *Courtesy Addison Products Company.*

cooling cycle or absorb heat during the heating cycle. Figure 25–2 shows a water-to-air heat pump.

Air-to-Air Heat Pumps

The **air-to-air heat pump** is the most popular and economical system to run and the least expensive system to install. It uses a standard heat transfer coil along with a blower located within the conditioned space, as in Figures 25–3 and 25–4. During the heating season the outside coil may ice up in certain weather. To prevent this, air-to-air heat pumps reverse the heating cycle to

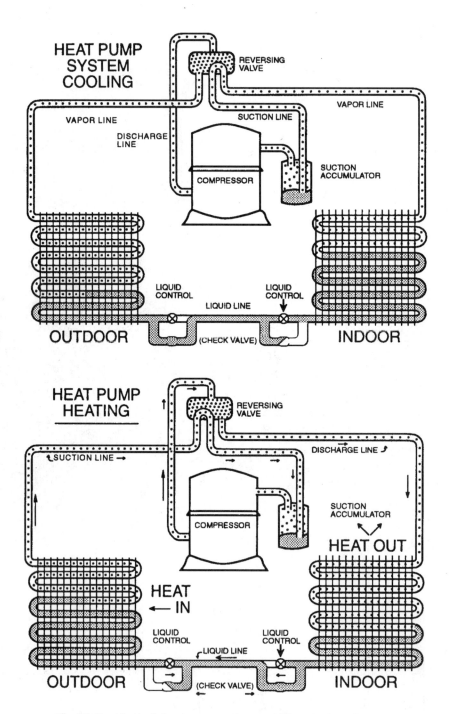

HEAT PUMP
SYSTEM
COOLING — REVERSING VALVE

VAPOR LINE

VAPOR LINE — SUCTION LINE

DISCHARGE LINE

COMPRESSOR

SUCTION ACCUMULATOR

LIQUID CONTROL — LIQUID CONTROL

LIQUID LINE

OUTDOOR — (CHECK VALVE) — INDOOR

HEAT PUMP
HEATING — REVERSING VALVE

SUCTION LINE — DISCHARGE LINE

COMPRESSOR

SUCTION ACCUMULATOR

HEAT OUT

HEAT IN

LIQUID CONTROL — LIQUID CONTROL

LIQUID LINE

OUTDOOR — (CHECK VALVE) — INDOOR

Fig. 25–3   Air-to-air heat pump.   *Courtesy Addison Products Company.*

defrost the outdoor coil. Supplemental heat elements temper the air during the defrost mode.

HEAT PUMP OPERATION

In the conventional cooling cycle of a refrigeration or air-conditioning system, you learned that if the evaporator (or indoor) coil is kept at a lower temperature than the conditioned space, it will absorb heat from the conditioned space and transfer that heat to the condenser (or outdoor) coil where it will be discharged. As a result, the conditioned space is cooled.

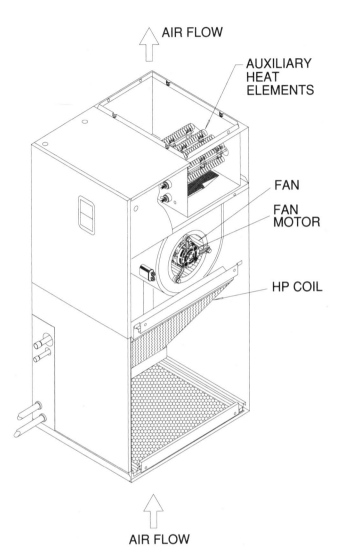

AIR FLOW

AUXILIARY
HEAT
ELEMENTS

FAN

FAN
MOTOR

HP COIL

AIR FLOW

Fig. 25–4   Air-to-air heat pump indoor unit. It is an electric furnace with a heat pump coil in it. Notice that the air flows through the heat pump coil and then through the heating elements.   *Courtesy Inter-City Products Corporation (U.S.A.)*

If the role of these two coils is reversed and the condenser (or outdoor) coil is kept at a cooler temperature than the ambient air outside of the conditioned space, then heat from the ambient could be absorbed and transferred to the evaporator (or indoor) coil where the heat would be released into the conditioned space for heating purposes. This is basically how a heat pump works. By means of a reversing valve and check valves, the refrigerant flow can be directed to provide for either a cooling or a heating cycle.

While the operation of a heat pump is essentially the same in all systems, there are many variations in system design, component parts, and sequence of operation by different manufacturers. As the service technician involved in heat pump installation and service, you must become familiar with the particular brand, application, specifications, and operating characteristics of the system you are servicing. The following sections discuss the operation of a typical heat pump.

Cooling Cycle

Referring to Figure 25–5, we can follow the **cooling cycle** of a typical air-to-air heat pump. The reversing valve will control the direction of flow of the discharge and suction gases in the system. Check valves are also installed at the outlets of both coils to control the refrigerant flow.

Discharge gas from the compressor enters the outdoor coil where it is condensed and flows through an open check valve number 1 to the indoor coil metering device. Check valve number 2 is closed at this time. The refrigerant enters the indoor coil where it absorbs heat, vaporizes, and is returned to the compressor through the reversing valve. This completes the cooling cycle.

Heating Cycle

In the **heating cycle** refrigerant flow in the system is reversed, as illustrated in Figure 25–6. Discharge gas from the compressor is directed to the indoor coil where it will release its heat and allow the vapor to condense to a liquid. Check valve number 2 is now open and check valve number 1 is closed. The condensed liquid is now directed to the outdoor coil metering device where it absorbs heat from the outdoor ambient, vaporizes, and returns to the compressor through the reversing valve. This completes the heating cycle.

Defrosting Cycle

When the heat pump is operating in the heating cycle and the temperature of the outdoor coil

COOLING MODE

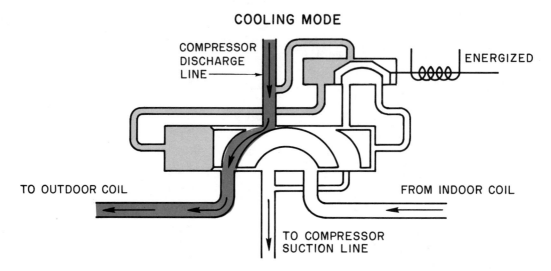

COMPRESSOR
DISCHARGE
LINE

ENERGIZED

TO OUTDOOR COIL

FROM INDOOR COIL

TO COMPRESSOR
SUCTION LINE

Fig. 25–5   Heat pump cooling cycle.

HEATING MODE

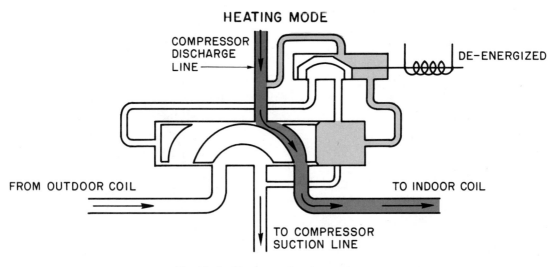

COMPRESSOR
DISCHARGE
LINE

DE-ENERGIZED

FROM OUTDOOR COIL

TO INDOOR COIL

TO COMPRESSOR
SUCTION LINE

Fig. 25–6   Heat pump heating cycle.

surface falls below 32°F, frost may accumulate on the coil surfaces. If allowed to continue, a gradual build-up of ice will form and the flow of air through the coils will eventually become restricted. This will decrease the heat transfer ability of the outdoor coil. When the heat transfer efficiency has decreased to the point where it will affect the overall system performance, the frost or ice must be removed.

In heat pump applications the process of frost or ice removal involves the following conditions:

1. The system must be able to detect the build up of frost in order to initiate a defrosting cycle.

2. The outdoor coil must go through the defrosting cycle in order to remove frost or ice accumulation.

3. The system must be able to terminate the defrosting cycle and return to the heating mode upon removal of frost or ice accumulation.

During the **defrosting cycle** of a heat pump the system returns to the cooling cycle to allow hot discharge gas from the compressor to defrost the outdoor coil. The indoor fan will continue to operate in order to evaporate the liquid refrigerant that is circulating through the indoor coil.

Auxiliary heat, usually in the form of electric heaters, will be energized to prevent cold air from circulating into the conditioned space. The most common methods used to initiate a defrosting cycle are by air pressure, time and temperature, or solid-state circuitry. In most applications the same method used to initiate a defrosting cycle is used to terminate it.

Air Pressure Method

This method measures the air pressure across the outdoor coil. As frost and ice start to accumulate on the coil, the flow of air across the coil will become restricted and a pressure drop across the coil will be experienced. At a predetermined control setting, the reversing valve will activate and initiate a defrosting cycle.

When the ice and frost have been removed from the surfaces of the outdoor coil, the pressure drop across the coil will be reduced. At a predetermined pressure differential, the defrost cycle will be terminated and the unit will return to the heating cycle.

Time and Temperature Method

In this method, a clock-timer is set to activate the reversing valve and initiate a defrosting cycle at predetermined intervals. If the temperature of the outdoor coil is above 32°F, however, the defrosting cycle will not start, and the timer will continue to run until the next scheduled defrosting cycle. When the temperature of the outdoor coil falls to a point where frost begins to accumulate on the coil surfaces, a defrosting cycle will be initiated.

If a timing mechanism is activated when the defrosting cycle is initiated, the system will remain in defrost for a preset period of time. At the end of the scheduled time period, the unit will return to the heating cycle.

Temperature is one of the more positive ways to terminate a defrosting cycle. During the defrosting cycle, the temperature of the outdoor coil remains fairly constant, usually around 32°F. As the ice and frost melt, the temperature of the coil will start to rise and the system will return to the heating cycle when a predetermined temperature is reached.

Solid-State Method

The use of solid-state circuitry is becoming more popular in controlling heat pump defrosting cycles. Thermistors, which sense the difference between the ambient air temperature and the refrigerant temperature, will initiate a defrosting cycle when the temperature difference reaches a predetermined point.

During the defrosting cycle, the pressure of the refrigerant remains fairly constant. As the ice and frost melt, the pressure of the refrigerant in the outdoor coil will increase and the system will return to the heating cycle when a predetermined pressure is reached.

COMPONENTS OF HEAT PUMPS
Suction Line Accumulator

One of the most important considerations in the operation of a heat pump is to protect the compressor from liquid refrigerant floodback during the changeover in cycles. Liquid refrigerant floodback may also occur during the heating cycle when, under colder operating temperatures, the outdoor coil may not be able to evaporate all of the refrigerant. For these reasons, a **suction line accumulator** should be installed in the suction line. The type and size of the accumulator will depend on the application of the system and the manufacturer's design.

Crankcase Heater

During the off cycle, liquid refrigerant will migrate to the oil in the compressor crankcase. The migration of refrigerant becomes more rapid during the heating cycle because the compressor is colder. This could result in starting problems, loss of lubrication, and damage to the compressor. To prevent this condition a **crankcase heater** is usually installed in all heat pump applications. Depending on the application and the design of the heat pump, the heater may be immersed directly in the oil in the compressor crankcase or mounted externally by being wrapped around the surface of the compressor crankcase. The heater may be energized all the time or only when the compressor is not running.

Reversing Valve

The **reversing valve** is the primary device that directs the direction of refrigerant flow for the heating and cooling cycles of a heat pump. In some applications two three-way valves are used. These valves have an opening to the compressor, the condenser, and the evaporator. The most common type of reversing valve is the four-way valve, which is shown in Figure 25–7. The valve operation is controlled by a single valve stem, which will open or close several ports in the valve body. The valve is activated by a solenoid that utilizes high-pressure compressor discharge vapor to move the valve piston to the cooling or the heating mode. Figures 25–8 and 25–9 illustrate the heating and cooling modes of operation of the four-way reversing valve.

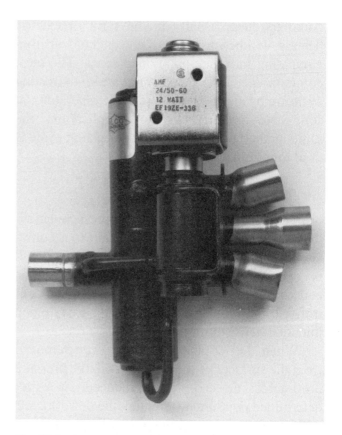

Fig. 25–8   The reversing valve has four refrigerant line connections. The single line connection on the valve body is always connected to the compressor discharge. The center line of the group of three refrigerant line connections is always suction pressured back to the accumulator. The other two refrigerant line connections will be connected, one to the indoor coil and the other to the outdoor coil.   *Courtesy ALCO Controls.*

Fig. 25–7   Four-way reversing valve.   *Courtesy Ranco Controls.*

Metering Device

In the operation of a heat pump, liquid refrigerant must be metered to the coil that is absorbing heat. During the cooling cycle, the refrigerant must be metered to the indoor coil, and during the heating cycle, the refrigerant must be metered to the outdoor coil.

In some applications the refrigerant flow to the indoor and outdoor coil is controlled by a single **metering device** that allows the refrigerant to flow in either direction. The device, which is located in the system liquid line, reverses the direction of refrigerant flow when the system changes cycle.

The most common method of controlling the flow of refrigerant to each coil is to use two sep-

Fig. 25–9 The reversing valve consists of a body that contains the slide valve and a solenoid operated pilot valve. *Courtesy ALCO Controls.*

arate metering devices and two check valves, as shown in Figure 25–10. The metering devices may be in the form of thermostatic expansion valves, capillary tubes, or a combination of the two. In this application, the check valves, which allow the refrigerant to flow in one direction only, bypass the unused metering device and direct the flow of refrigerant to the proper coil.

Filter-Drier

A bi-directional **filter-drier** is used in heat pump applications. This type of filter-drier, which is illustrated in Figure 25–11, utilizes check valves to control the flow of refrigerant so that any contamination is fully absorbed during the heating and cooling cycles. As a result, the reverse flushing of system contaminants through the system is not experienced.

FEATURES OF HEAT PUMPS

Heat Pump Efficiency

The **balance point** of a heat pump is reached when the heat pump cannot only supply the conditioned space with the same amount of heat that is lost but also maintain the desired temperature of the conditioned space. Below this balance point, the amount of heat lost from the conditioned space will be greater than the amount of heat that the heat pump is able to supply. Consequently, auxiliary heat will be required to maintain the desired temperature of the conditioned space.

In application, heat pumps are more economical than resistance heat. Two methods can be used

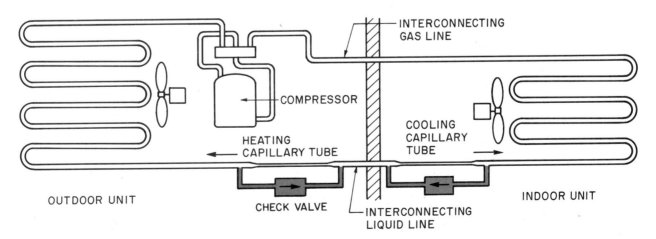

Fig. 25–10 Capillary tube metering device used on a heat pump. Notice that check valves are used and there are two different capillary tubes.

Fig. 25–11   Bi-directional filter-drier. *Courtesy Sporlan Valve Company.*

to measure the efficiency of a heat pump: the **energy efficiency ratio (EER)** and the **coefficient of performance (COP)**.

The energy efficiency ratio calculates how many Btu's are generated for every watt of PEER consumed. To determine the energy efficiency ratio of a heat pump, the following equation may be used:

$$EER = \frac{Btu/hr\ capacity}{unit\ wattage/hr}$$

For example, if a heat pump rated at 3 tons, or 36,000 Btu/hr, was operating at an outdoor ambient temperature of 40°F with a heating capacity of 39,000 Btu/hr, consuming 4,380 watts per hour, the energy efficiency ratio would be calculated as follows:

$$EER = \frac{39,000\ Btu/hr}{4,380\ watts/hr}$$

$$= 8.9\ Btu/watt$$

When the efficiency of a heat pump is compared to electric heat, we find that electric heat will generate 3.413 Btu's of heat for each watt with the maximum coefficient of electrical heat being 1.0. In order to determine the coefficient of performance of a heat pump, the following equation may be used:

$$COP = \frac{Btu/hr\ capacity}{unit\ wattage \times 3.413\ Btu/watt}$$

For example, if a system was rated at 39,000 Btu/hr and the unit wattage draw was 4,380 watts, the COP would be calculated as follows:

$$COP = \frac{39,000\ Btu/hr}{4,380\ watt/hr \times 3.413\ Btu/watt}$$

$$= 2.6\ Btu/watt$$

In this example the heat pump will deliver 2.6 times as much heat as electrical resistance heaters using the same wattage. In comparison, the heat pump will deliver more heat per watt of power used and is therefore more efficient.

Supplemental Heat

In application, the heat pump is the primary source of heat for the conditioned space.

However, when the outdoor temperature decreases to the extent that the system is operating below the balance point, **supplemental heat** must be utilized in order to maintain the desired temperature in the conditioned space. This may be in the form of electrical, oil, or gas heat. In most applications, electrical resistance heaters are used as a supplemental heat source due to their easy installation in the air delivery system.

Temperature Control

The **temperature control** of a heat pump system is usually in the form of a combination heating and cooling thermostat with a manual or automatic changeover from heating to cooling. In the automatic changeover, there will usually be a five degree differential setting between the heating and cooling cycles to prevent the unit from short cycling. Indoor fan operation may be set to cycle with the compressor or it may operate continuously.

The heating thermostat has a two-stage operation. The first stage of operation cycles the compressor on to initiate the heating cycle. If the heat is not sufficient to maintain the desired temperature in the conditioned space, the second stage of the thermostat will energize a supplemental heat source.

In systems utilizing electrical heat as a supplemental heat source, the second stage of the heating thermostat operation will energize an electrical resistance heater. In these applications, the second stage of the heating thermostat is wired in series with an outside thermostat, which senses the temperature of the outside air. The outside thermostat controls the operation of the supplemental electrical heaters, energizing them only when required for maximum efficiency. The outside thermostat is activated by the outside temperature during the normal heating cycle and can quickly sense any sudden decrease in the outside temperature. Thus, it will energize additional stages of heat before the temperature of the conditioned space becomes uncomfortable. The heat pump has a maximum outdoor temperature above which the heating cycle should not be energized. This is usually established at around 65°F. At these high outdoor temperatures, the system suction pressure and the density of the refrigerant vapor will increase.

The increased gas density will cause more pounds of refrigerant to circulate through the system, resulting in an increase of system capacity. Consequently, more power will be consumed by the system compressor, resulting in an overload condition that will create service problems and/or equipment damage.

Emergency Heat Setting

If a failure is experienced in the heat pump system, an **emergency heat setting** on the thermostat may be employed to energize supplementary heat for the conditioned space. In this mode of operation, the outdoor thermostat will be bypassed to allow for maximum heat output. The temperature of the conditioned space will then be controlled by the indoor thermostat. A red warning light on the thermostat will remain on until the system is returned to normal operation.

EVAPORATOR AIRFLOW IN A HEAT PUMP SYSTEM

One of the major problems concerning heat pump operation is the amount of air circulating across the indoor coil during the heating cycle. During the cooling cycle, when the indoor coil is being used as an evaporator, the reduction of airflow will result in a decrease of system capacity and in an increase of temperature in the conditioned space. This in turn will result in a customer complaint. During the heating cycle, however, when the indoor coil is being used as a condenser, a reduction of airflow across the coil will result in a higher discharge air temperature. There will be no indication of a service problem because the temperature of the conditioned space will not be affected. In fact, the supply air temperature will be hotter, and it will appear that the system is performing more efficiently.

During a typical cooling cycle of a heat pump, the average compression ratio of a compressor operating in the air-conditioning range is approximately 3.3. During the heating cycle, when the heat pump is operating in the refrigeration temperature range, the compression ratio could increase to 8.5. Most hermetic compressors are not designed to operate under this high of a compression ratio; even those that can operate at such a level will not tolerate operating above this point without experiencing service problems and eventual compressor failure.

If the compressor discharge pressure is increased due to a lack of airflow across the indoor coil during the heating cycle, the compression ratio will also be increased, subjecting the compressor to severe operating conditions. The high operating temperatures caused by the high discharge pressures and temperatures will eventually break down the lubricating qualities of the compressor oil and result in eventual compressor failure.

Because the heat pump compressor operates close to these maximum discharge temperature tolerances, it is extremely important that the quantity of airflow across the indoor coil during the heating cycle is maintained at the prescribed minimum, which is usually 450 cfm per ton of capacity.

The most common cause of insufficient airflow is dirty filters and coils. Other common causes of insufficient airflow include dirty fans, low fan speed, duct restrictions, and improperly sized duct systems. Consequently, it is important that a proper preventive maintenance program, including regular filter changes, coil inspection, and cleaning, be maintained in heat pump applications. Periodically inspecting and cleaning the fan blades is also recommended. If the fans are belt-driven, check the condition and tension of the belts.

Another source of inefficient airflow can be the result of blocked diffuse outlets. In the application of gas- or oil-fired furnaces, the temperature of the air delivered to the conditioned space will be 130°F to 140°F. In contrast, the temperature of the air delivered to the conditioned space by a heat pump is usually around 100°F. For those who are not familiar with heat pump operation, this lower temperature could indicate that there is too much air moving through the diffuser. If, in an attempt to correct the problem, the technician reduces the amount of air delivered to the conditioned space, serious service problems could result.

Measurement of Airflow

Two common methods employed to measure the quantity of air passing over the indoor coil are the pressure drop method and the temperature rise method.

In the **pressure drop method**, two ¼" pilot holes are drilled in the ductwork of the indoor coil—one hole upstream of the coil and one hole downstream of the coil, as shown in Figure 25–12. A piece of ¼" tubing is inserted approximately two inches into each hole and connected to a manometer. The resulting reading will be the static pressure drop across the coil in inches of water.

Once the static pressure drop has been determined, refer to the manufacturer's product guide for the pressure drop chart specifications of the specific coil being used to determine the cfm.

In the **temperature rise method**, the temperatures of the supply and return air across the indoor coil are recorded with a thermometer. In measuring the discharge air temperature when electric heat is used, the temperature reading should be taken at a point far enough downstream of the electrical heating elements to obtain an accurate reading.

After the air temperature readings are recorded, the voltage and current draw of the indoor section of the system must also be recorded. It is important to ensure that the total current draw of the indoor section, including the fan motor current draw, be measured. The temperature rise method is illustrated in Figures 25–13A and 25–13B.

To Measure Air Flow
Pressure Drop Method

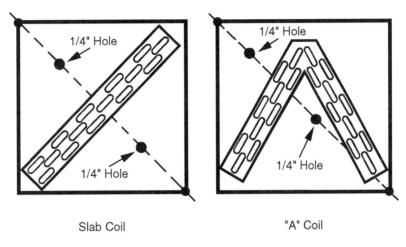

Slab Coil   "A" Coil

Fig. 25–12   Pressure drop method.

When the temperature current draw has been collected, the following formula may be used to determine the cfm:

$$cfm = \frac{volts \times amperes \times 3.413}{1.08 \times temperature\ difference}$$

For example, if the supply voltage to the indoor unit is 230 volts, the total current draw of the indoor section including the electrical heaters and the indoor fan motor is 45 amperes, and the temperature difference across the coil is 28°F, the cfm would be calculated as follows:

$$cfm = \frac{230 \times 45 \times 3.413}{1.08 \times 28}$$

$$cfm = 1,170$$

HEAT PUMP INSTALLATION

The installation procedures for a heat pump are similar to the installation procedures for a conventional air-conditioning system. There are, however, special considerations that must be taken to ensure proper system operation and to prevent unnecessary service problems.

Position the outdoor unit in a location where water or snow will not fall and collect on the out-door coil. Make sure that there is enough clearance around the unit to allow for the proper airflow and service access. Do not locate the unit where it will be directly exposed to the prevailing winds or where the airflow may be recirculated through the coil. In areas where there is a heavy accumulation of snow or prolonged subfreezing temperatures, the outdoor unit should be installed about 12 to 18 inches above the ground. A gravel apron should extend at least 12 inches around the base of the unit to provide proper drainage.

In most heat applications, the base pan of the outdoor unit contains drainage holes that allow the condensate created during the defrost cycle to drain, thus preventing a build up of ice that will restrict the flow of air across the outdoor coil.

Consult the manufacturer's installation instructions for the proper procedure to be followed for a particular application.

Installing Heat Pump Tubing

The tubing used in heat pump applications is basically the same type of tubing used in conventional air-conditioning systems. Precharged tubing with quick disconnect fittings is commonly used for liquid and suction lines. In some applica-

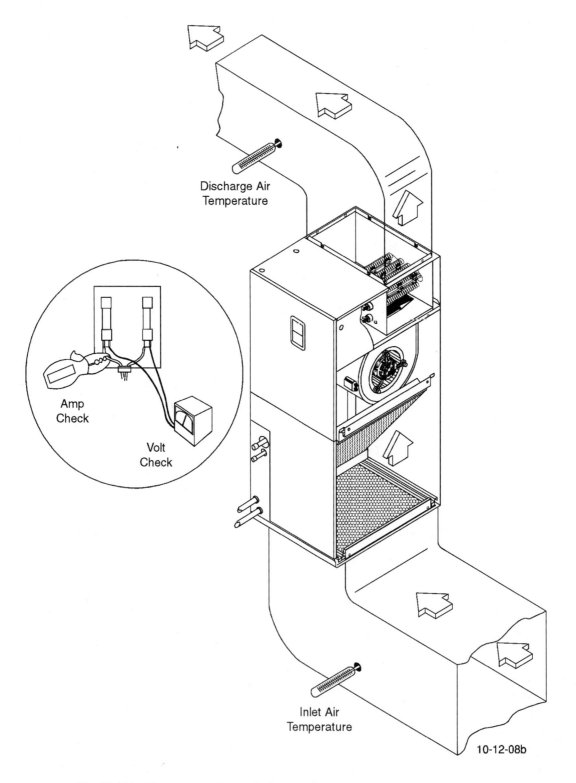

Discharge Air
Temperature

Amp
Check

Volt
Check

Inlet Air
Temperature

10-12-08b

Fig. 25–13A   Temperature rise method.   *Courtesy Inter-City Products Corporation (U.S.A.)*

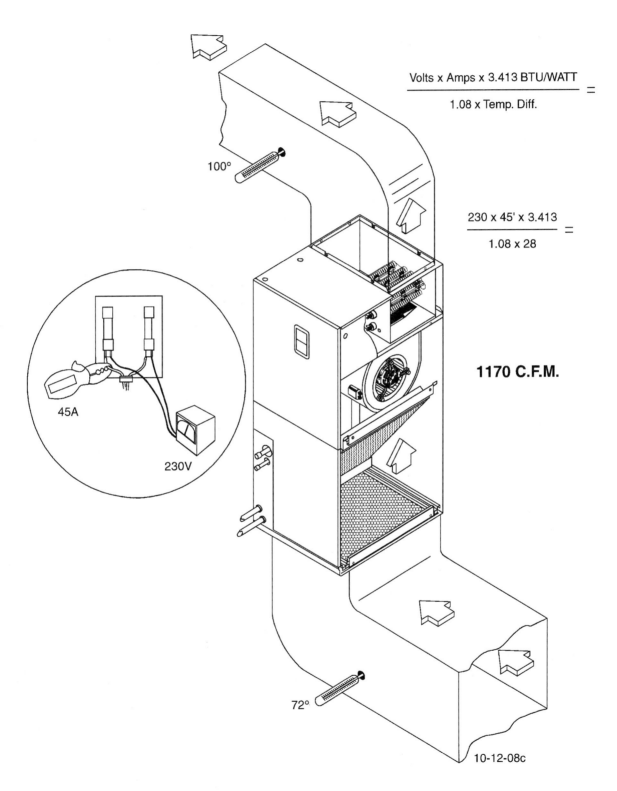

$$\frac{\text{Volts x Amps x 3.413 BTU/WATT}}{1.08 \times \text{Temp. Diff.}} =$$

$$\frac{230 \times 45' \times 3.413}{1.08 \times 28} =$$

100°

45A

230V

1170 C.F.M.

72°

10-12-08c

Fig. 25–13B   Temperature rise method.   *Courtesy Inter-City Products Corporation (U.S.A.)*

tions, the refrigerant lines may be charged with nitrogen to prevent contamination. The suction line in most heat pump applications is more heavily insulated than the suction line in conventional air-conditioning systems in order to contain the excessive amount of heat that may be experienced during the defrost cycle.

The sizes of the suction and liquid lines used depend upon the capacity of the unit and the length of the tubing run. Always refer to the manufacturer's installation instructions for the proper tubing size for a particular unit. The following list can serve as a guide for proper tubing installation:

1. Keep tubing bends to a minimum. An excessive amount of bends will result in an increased pressure drop, which will reduce system capacity.

2. Isolate the system piping from the structure to prevent vibration transmission.

3. Prevent the tubing from coming into contact with any object or structure that will eventually cause the tubing to wear or leak.

4. Do not use the liquid and suction lines as a heat exchanger. This will create an undesirable heat exchange, especially during the heating cycle.

Charging the Heat Pump System

The charging procedure for a heat pump is essentially the same as the charging procedure used in a conventional air-conditioning system. Cooling cycle charging charts, which will ensure that the correct amount of refrigerant is charged into a particular system under a variety of operating conditions, are usually provided by the manufacturer. These charts should be followed in order to check and adjust the proper refrigerant charge.

If the actual system suction pressure is lower than the pressure specified on the charging chart, additional refrigerant should be charged into the system. If the suction pressure is higher than the chart specifies, refrigerant should be removed from the system. When the system is properly charged, the operating suction pressure should match the pressure indicated on the charging chart. It is important to let the system operate at least 10 to 15 minutes after refrigerant has been added or removed to allow the system pressure to stabilize before making any further adjustments in the refrigerant charge. The refrigerant charge can be weighed into the system by means of a charging column or scale. The electronic sight glass can also be used.

Checking Heat Pump Operation

To check heat pump operation during the heating cycle, refer to the heating cycle check charts for that particular system. These charts will reflect the relationship between the system operating pressures and air temperatures entering the indoor and outdoor coils. If the system operating pressures do not correlate with the chart, the system charge should be checked. It is important to remember, however, that other system problems, such as an improper airflow across the indoor or outdoor coil, can also cause an improper pressure and temperature reading. Therefore, it is important to check the entire system for proper operation.

The heating cycle check chart should not be used to adjust the refrigerant charge in a system. If the amount of refrigerant charge is in question, the system should be evacuated of all refrigerant and the correct amount of refrigerant should be weighed in. Overcharging or undercharging a heat pump system will cause a decrease of system efficiency and could result in equipment damage.

Unit Twenty-Five Summary

- In a heat pump system, the coils are referred to as the indoor coil and the outdoor coil rather than the evaporator and the condenser coils. This is due to the fact that in the heating mode, the function of the coils is reversed and the evaporator technically becomes the condenser and the condenser technically becomes the evaporator.

- Heat pumps are classified according to the medium used to transfer heat. Some examples are water-to-air systems, ground-to-air systems, and air-to-air systems.

- Heat pump outdoor coils must be defrosted when ice and frost accumulate below 32°F.

- A suction line accumulator is used to prevent liquid refrigerant from reaching the compressor in a heat pump system. Supplementary heat in the form of resistance elements is used in a heat pump system as second stage heating and to temper the air during the defrost mode.

- Heat pumps should be charged by weight, an electronic sight glass, or through the use of a manufacturer's charging chart.

Unit Twenty-Five Key Terms

Heat Pump
Ground-to-Air Heat Pump
Water-to-Air Heat Pump
Air-to-Air Heat Pump
Cooling Cycle
Heating Cycle
Defrosting Cycle
Suction Line Accumulator
Crankcase Heater
Reversing Valve
Metering Device
Filter-Drier
Balance Point
Energy Efficiency Ratio (EER)
Coefficient of Performance (COP)
Supplemental Heat
Temperature Control
Emergency Heat Setting
Pressure Drop Method
Temperature Rise Method

UNIT TWENTY-FIVE REVIEW

1. What is a heat pump?

2. List the three mediums used to transfer heat to the outdoor coil.

3. What cycle does each statement describe?

    a. The indoor coil is kept at a lower temperature than the conditioned space and absorbs heat that is transferred to the outdoor coil.

    b. The outdoor coil is kept at a lower temperature than the ambient air outside and absorbs heat that is transferred to the indoor coil.

4. What cycle is the heat pump in for defrosting to begin and what causes frost to accumulate on the outdoor coil?

5. Describe the methods used to initiate a heat pump defrosting cycle.

6. What methods are used to terminate a heat pump defrosting cycle?

7. Match the following terms with their description: pressure drop method, auxiliary heat, solid-state method, time and temperature method, heat transfer coil, reversing valve, defrosting cycle, air pressure method, crankcase heater, suction line accumulator.

    a. Pressure drop across the coil will be reduced and at a predetermined pressure differential the defrosting cycle will stop.

    b. Protects the compressor from liquid refrigerant floodback.

   c. As the ice and frost melt, the temperature of the coil will start to rise and the system will return to the heating cycle.

   d. Used to defrost the outdoor coil.

   e. Hot discharge gas from the compressor flows to the outdoor coil.

   f. Prevents cold air from circulating into the conditioned space.

   g. Pressure of the refrigerant in the outdoor coil will increase and the system will return to the heating cycle when a predetermined pressure is reached.

   h. Prevents damage to the compressor when migration of refrigerant becomes rapid because the compressor is colder.

   i. Directs the flow of refrigerant.

   j. A manometer reading measures a static pressure drop across the indoor coil in inches of water.

8. What purpose do check valves serve in heat pump applications?

9. During the heating cycle the refrigerant must be metered to the ___ _____ coil. During the cooling cycle the refrigerant must be metered to the _____ coil.

10. Define the balance point of a heat pump.

11. What happens when the desired temperature of the conditioned space falls below the balance point?

12. What methods are used to measure the efficiency of a heat pump?

13. How does the efficiency of the heat pump compare to that of electric heat?

14. What is the purpose of supplemental heat in heat pump applications?

15. What is the average temperature differential between the heating and cooling cycles of a heat pump?

16. What is the maximum outdoor temperature at which the heating cycle of a heat pump should not be energized?

17. What is the purpose of the emergency heat setting in heat pump applications?

18. Why is it important to maintain minimum airflow across the indoor coil during the heating cycle of a heat pump?

19. What are the common causes of inefficient airflow across the indoor coil of a heat pump?

20. Give the two-stage operation of the heating thermostat.

21. What problem can a heat pump have with high outside temperatures?

22. What are the two common methods to measure the quantity of air passing over the indoor coil?

23. Electric heat will generate 3.413 Btu's for each watt.

   Electric heat's maximum coefficient of performance (COP) = 1.

$$\text{COP} = \frac{\text{Btu's per hour capacity}}{\text{watt/hr} \times 3.413 \text{ Btu's}}$$

    a. A 30,000 Btu heat pump draws 3,580 watts.

        A 24,000 Btu heat pump draws 2,950 watts.

        Which heat pump is more efficient? Show your work.

    b. What is its COP?

24.

$$cmp = \frac{volts \times amperes \times 3.413}{1.08 \times temperature\ difference}$$

A technician took airflow measurements on the same system but one year apart. Which measurement was taken at the time of installation and which measurement was taken recently?

What does the more recent measurement indicate about the airflow?

Supply voltage 230

76 amperes for electric heaters and fan motor.

    a. 45 degree temperature difference.

    b. 50 degree temperature difference.

25. Write a paragraph comparing the efficiency of electric heat to the heat pump.

26. Write a paragraph explaining your goal for this course.

UNIT TWENTY-SIX

Air-Conditioning System Controls

OBJECTIVES

After completing this unit, the student will be able to:

1. Explain the application of control systems in air-conditioning systems.

2. Explain the difference between bimetal and mercury bulb controls.

3. Explain the function of a heat anticipator.

4. Explain the function of a cooling anticipator.

5. Outline the basic procedures for troubleshooting control systems.

In residential air-conditioning systems, a variety of controls are employed to control system operation and protect it from abnormal operating conditions. As a service technician you'll find that approximately 85 percent of the problems experienced in HVAC/R work are the result of electrical failure, and many times these problems are centered in the control system and circuitry.

In this unit, we'll discuss the types of controls and protection devices used on comfort cooling systems and the procedures for evaluating and troubleshooting controls and control systems.

THERMOSTATS

The **thermostat** is an important control in a cooling and/or heating system because it provides the temperature control for a conditioned space as well as the operating control for the total system. The primary purpose of a thermostat is to cycle the cooling and/or heating system on and off in order to maintain a desired temperature in the conditioned space. The thermostat becomes a multi-function device when it:

1. Stops and starts the operation of a cooling and/or heating system.

2. Automatically or manually switches the system from the heating to the cooling mode and vice versa.

3. Automatically or manually controls fans and damper operation.

4. Indicates room temperature and, in some cases, the time of day.

5. Automatically anticipates changes in room temperature.

All residential air-conditioning systems as well as many commercial systems utilize a 24-volt control circuit and thermostat. When compared with other circuits, the low voltage control has the following advantages:

1. Provides more sensitivity.

2. Is more compact.

3. Is less costly.

4. Allows for the use of smaller wire sizes.

Types of Thermostats

The types and designs of thermostats that are available and in use today are as varied as the applications in which they are used. A bimetallic thermostat is the most popular thermostat in use. A remote bulb thermostat is used in special applications, and the solid-state programmable thermostat is increasing in popularity.

The **bimetallic thermostat** is the most common thermostat used in residential and commercial heating and cooling applications. The control utilizes a bimetallic element to create the movement of the switching contacts. The bimetallic element consists of two pieces of metal welded together. Each piece of metal has a different coefficient of heat transfer. At a given temperature both pieces of metal are the same length, as shown in Figure 26–1. As the temperature of the bimetallic element is increased, one metal will expand faster than the other. This will cause the bimetallic element to arch or flex, as shown in Figure 26–2. The other metal will move freely and, combined with a contact arrangement, will open or close a set of switch contacts.

Because of its sensitivity to slight temperature changes, the bimetallic thermostat has a tendency to react too quickly for stable temperature control. For this reason, a snap-action type of movement is used during the opening and closing of switch contacts. To execute this movement a magnet is located near the switch contacts. When the bimetallic element moves in response to a temperature change, the switch contacts will snap open as they overcome the magnetic attraction of the magnet. As the switch contacts move toward the magnet, they will snap closed when the bimetallic element is drawn close enough. This snap action during the opening and closing of the switch contacts will also prevent the contacts from arcing.

The **mercury bulb thermostat** utilizes a spiral wound bimetal to provide for its movement, as shown in Figure 26–3. A sealed globule of mercury that moves between sealed, fixed probes is attached to the spiral bimetal. As the temperature decreases, the bimetal will move in a counterclockwise direction, causing the mercury bulb to tilt in the opposite direction and, as a result, make or break mercury contact between the probes. The movement of the mercury in the sealed bulb provides the necessary snap action of the thermostat. Figure 26–4 illustrates the positions of a mercury bulb thermostat for two- and three-terminal applications.

The heart of a **remote bulb thermostat**, which is illustrated in Figure 26–5A, is a power element that is filled with a predetermined amount of refrigerant liquid and vapor. The bulb, which is connected to a bellows or diaphragm assembly by

Fig. 26–1   **Bimetal in closed position.**

Fig. 26–2   **Bimetal in open position.**

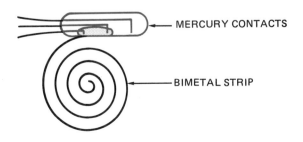

Fig. 26–3   **Mercury-type thermostat.**

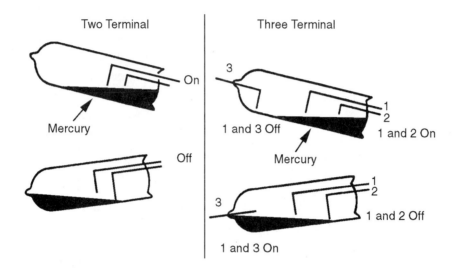

Fig. 26–4 Mercury switch operation.

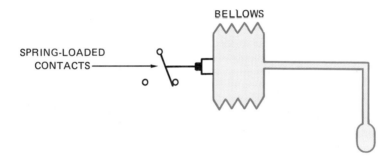

Fig. 26–5A Thermostat contacts are operated by pressure.

Fig. 26–5B Remote bulb thermostat.

means of a capillary tube, is sealed. When the temperature of the feeler bulb increases, the temperature and pressure of the confined refrigerant within the bulb will also increase. This will cause the diaphragm or bellows assembly of the thermostat to expand and exert pressure on the mechanical linkage that, in turn, will open or close a set of contact points, depending on the application. Conversely, as the temperature decreases, the temperature and the pressure within the bulb will also decrease, causing the bellows or diaphragm assembly to move in the opposite direction to open or close the switch contacts.

Remote bulb thermostats are most commonly used in commercial applications. They are usually adjustable, in that the cut-in and cut-out settings must be manually set upon installation. When selecting this type of thermostat, it is important to ensure that the control has the proper temperature range for the application.

Solid-state thermostats utilize microcomputer and solid-state sensors to provide precise temperature control. In most applications, these thermostats can provide several separate time/temperature settings in a 24-hour period. Features may include two sets of time/temperature schedules, one for weekdays and one for weekends.

Usually, any time/temperature combination can be scheduled for any period and for any five consecutive days of the week. Most use a 9-volt battery to maintain the stored program for approximately a week, in the event of power failure.

Features include LCD displays for set point, time, and room temperature. Heating thermostats have sensors to protect against temperature override; indicators are provided for "System Cycle"; and manual operation requires just the push of a button. Figure 26–6 illustrates a typical electronic thermostat.

Many heating and air-conditioning systems are designed to operate in stages in order to control system capacity in applications that experience fluctuating load conditions. A two-stage heating or cooling system operates at its first stage until the heating or cooling load increases to a point where the second stage of operation is required to maintain the desired temperature setting. Therefore, a **two-stage thermostat** would be utilized to control the operation of the two stages of heating or cooling. There is usually a temperature differential of two or three degrees

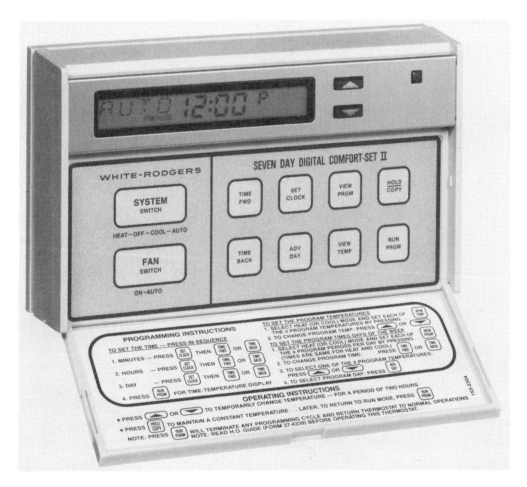

Fig. 26–6 A typical electronic thermostat that indicates present temperature, time, and heating or cooling mode. *Courtesy White-Rodgers Division, Emerson Electric Company.*

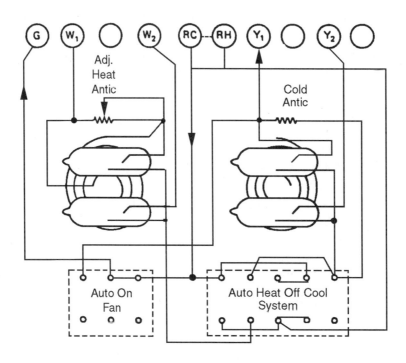

Fig. 26–7   Wiring diagram of simple two-stage heating and cooling thermostat.   *Courtesy Honeywell.*

between stages. For example, a first-stage cooling thermostat could be set to cut in at 70°F, and a second-stage cooling thermostat could be set to cut in at 73°F. Similarly, a two-stage heating thermostat could be set to cut in at 75°F, and a second-stage heating thermostat could be set to cut in at 72°F. This temperature differential will prevent the second stage of heating or cooling from short cycling and the first stage to operate until full system capacity is achieved. Figure 26–7 illustrates a typical wiring diagram of a two-stage heating and cooling thermostat. The thermostat provides for fan auto or on positions, and system control for auto, heat, or cool.

Clock thermostats or **chrono-therms** are used to reduce or set back conditioned space temperatures during low load conditions such as during the night or during periods when the conditioned space is not occupied. A clock thermostat is illustrated in Figures 26–8A and B. Numerous types and designs of clock thermostats are available to suit practically any application. The simplest type of clock thermostat has an

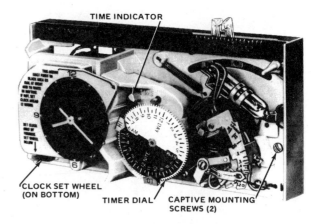

Fig. 26–8A   Single-setback clock thermostat with the cover removed.

adjustable program that sets back and picks up the conditioned space temperature every 24 hours. They usually utilize a 24-volt power supply. Multisetback thermostats allow the user to program various temperature settings throughout a period of 24 hours. Some of these thermostats utilize a 24-volt power supply while others are battery operated.

The most advanced versions of the clock thermostats are found in the form of solid-state units.

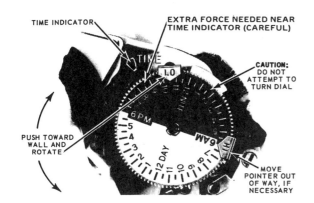

Fig. 26–8B   Setting procedure for a single-setback clock thermostat.

Fig. 26–8C   Multi-setback programmable clock thermostat.

Fig. 26–9A   Solid-state clock thermostat.

Fig. 26–9B   Solid-state clock thermostat with the cover removed.

These thermostats contain a variety of programming options including day and night setbacks, the omission of weekend setbacks, skip-a-day settings, and holiday settings. Figures 26–9A and B illustrate a solid-state programmable thermostat.

Thermostat Subbases

A thermostat is also used as a control center for total system operation. This is accomplished through the use of a **subbase** that contains manual switching circuits to provide for the complete remote control of the system. In general, the subbase will have one or more of the following switching control positions, and function as follows:

1. Fan control center

   a. Fan on—Indoor fan operates continuously.

   b. Fan auto—Indoor fan cycles with the compressor.

2. System control

   a. Heat—Controls heating cycle only.

   b. Cool—Controls cooling cycle only.

   c. Off—Total heating and cooling cycles are off.

Generally, the identification of thermostat subbases is standard within the industry. However, it is always wise to refer to the manufacturer's wiring instructions and installation procedures.

Cooling Anticipators

The problem encountered in cooling applications is that during the off cycle of a system, the temperature of the conditioned space may rise above the desired temperature setting before the thermostat cycles the cooling system back on. **Cooling anticipators** are used in cooling applications to prevent temperature fluctuations in the conditioned space. A

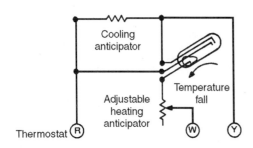

Fig. 26–10 Schematic diagram of cooling anticipator connected to a thermostat.

cooling anticipator is wired in parallel with the thermostat contacts, as shown in Figure 26–10.

During the system's off cycle, the heating resistor of the anticipator is energized to provide additional heat to the thermostat's bimetal. This will cause the thermostat contacts to close one or two degrees before the actual thermostat setting is reached. As room temperature continues to rise, the cooling system will cycle on to prevent the room

temperature from rising above the desired temperature setting. Cooling anticipators are nonadjustable. They are fixed by the manufacturer to provide maximum system performance at factory specifications.

Heating Anticipators

One of the problems with heating thermostats is that the room temperature rises above the temperature setting of the thermostat after the thermostat switch has opened and cycled the heating system off. This temperature override is caused by the residual heat that remains in the system. In order to control this condition, a small resistance heater called a **heating anticipator** is wired in series with the thermostat switch contacts so that the anticipator will remain energized when the thermostat switches are closed and the system is in operation. The wiring of a heating anticipator is illustrated in Figure 26–11.

Fig. 26–11 Wiring of a heat anticipator. *Courtesy Honeywell.*

During the on cycle, the heating anticipator will emit a small amount of heat, giving the thermostat bimetal a "false heat" and causing the thermostat contacts to open one or two degrees before the actual thermostat setting is reached. The room temperature will continue to rise after the thermostat cycles the unit off, but the temperature will not rise above the actual temperature setting of the thermostat. One factor that must be considered in dealing with heating thermostats is anticipator sizing. If the heating anticipator is too big, more resistance and more false heat will be created and the on cycle will be shorter. If the heating anticipator is too small, less false heat is created and the on cycle will be longer. This will result in an undesirable variation of room temperatures.

Thermostat Installation

A thermostat should always be mounted on an inside wall. If it is located on an outside wall, the thermostat will sense the infiltration of cold air and as a result will overheat the conditioned space. A thermostat should be mounted about five feet above the floor where it can best sense the average air temperature of the occupied space. It is important to avoid installation near false heat sources, such as lamps, warm air ducts, and direct sunlight. Installation near areas of vibration, such as sliding doors and windows, should also be avoided. In all installations, make sure that the thermostat is level and that all of the openings that were required in making the installation are sealed.

FAN RELAY CONTROL CENTERS

Fan relay control centers are used extensively in cooling and heating applications to provide simple, low voltage control over the heating and cooling system as well as over the fan operation and auxiliary circuit devices. The control consists of a low voltage transformer and a switching relay, which are designed to carry low voltage or line voltage to the load device being controlled. Manufacturers of fan relay control

centers have a variety of controls with various voltage and switch contact arrangements available to suit many system applications. Therefore, consider the following factors before selecting a fan relay control center:

1. Primary voltage of the transformer.
2. Secondary voltage of the transformer.
3. VA rating of the transformer.
4. Relay switching needs—double-pole, double-throw; single-pole, single-throw; or single-pole, double-throw.
5. Amperage ratings of the relay contact—full-load amperage or locked-rotor amperage.

Operation of a Fan Relay Control Center

A fan relay control center, Figure 26–12, operates as follows:

1. Power from the transformer, terminal R, is brought to terminal R of the thermostat. All

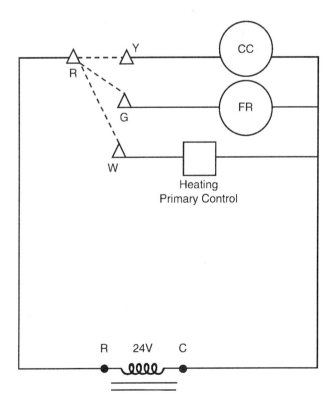

Fig. 26–12 Basic fan relay control center (CC) Contactor holding coil, (FR) Fan relay holding coil.

low voltage power will be distributed from this terminal to the various cooling, heating, and auxiliary circuits of the system, such as the compressor circuit, indoor blower motor, and dampers.

2. When the thermostat switch makes contact from terminal R to terminal G, current will flow to terminal G of the fan relay center, through the fan relay coil, and back to the source of supply at terminal C of the transformer. The fan relay coil will now energize and close the fan relay contacts, allowing the indoor blower motor to energize.

3. When the thermostat calls for cooling, thermostat contacts R to Y will make contact. Current will flow to terminal Y of the fan relay center, through the holding coil circuit of the compressor, and back to the source of supply at terminal C of the transformer. When the holding coil circuit of the compressor contactor is complete, the compressor will energize to start the cooling cycle.

4. When the thermostat calls for heat, thermostat terminals R to W will make contact. Current will flow to terminal W of the fan relay center, to the heating system primary control, and back to the source of supply at terminal C of the transformer. When the primary control of the heating system is energized, the system will operate in the heating mode.

Troubleshooting the Fan Relay Control Centers

After all other system components have been checked and eliminated as the possible source of a problem, the fan relay control center can be easily checked with an ohmmeter and a voltmeter.

If the indoor blower motor will not deenergize, the problem may be that the fan relay contacts are welded or stuck closed. To solve the problem, follow these steps:

1. Deenergize the fan relay coil by turning the thermostat selector to the off position.

2. Take a voltage reading at the fan relay coil. There should be no voltage present. If the blower motor continues to operate, the fan relay contacts are stuck in the closed position and the relay needs to be replaced.

Another check that can be made to confirm whether the fan relay contacts are stuck in the closed position is to disconnect all power to the unit and then disconnect one lead from the fan relay holding coil. Using an ohmmeter, take a continuity check across the fan relay contacts. If continuity exists, the fan relay contacts are stuck or welded closed and the relay needs to be replaced.

If the blower does not energize when the thermostat switch is placed in the appropriate position, the problem may be in the fan relay or the fan relay transformer. To check the control center, follow these steps:

1. Check for voltage at the fan relay holding coil. If voltage exists, disconnect all power and check the relay holding coil for proper resistance. If proper resistance is indicated, the relay coil is good and the relay contacts should be checked. If proper resistance is not indicated or if an open coil is indicated, the relay coil is defective and the relay needs to be replaced.

2. Check for proper voltage across the fan relay contacts. If proper voltage is present, the relay contacts are open and the relay needs to be replaced.

To test the transformer, follow these steps:

1. Using a voltmeter, check for power at the primary windings of the transformer. If an improper voltage condition exists, the supply of voltage needs to be corrected.

2. Check for proper voltage at the secondary windings of the transformer. If there is an improper voltage reading, disconnect all power to the system. Disconnect the primary and secondary electrical leads of the transformer and check the primary and secondary windings with an ohmmeter. If the primary

and secondary windings do not indicate proper resistance readings, the transformer is defective and needs to be replaced. Other sources of the problem may be due to loose, broken, or corroded electrical connections.

High-Pressure Control

The **high-pressure control (HPC)**, Figure 26–13, is installed in an air-conditioning system to cycle the compressor off before the system discharge pressures become dangerously high. Excessive discharge pressures place additional burdens on the compressor by causing higher than normal current draws and operating temperatures. These conditions can lead to the formation of carbon in the system and, in combination with other contaminants such as moisture, can result in the formation of an acid that will lead to component failure and hermetic compressor burnout. High

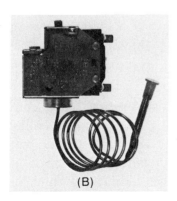

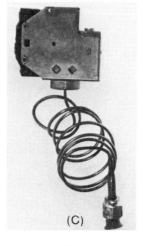

Fig. 26–13 High-pressure controls (a) enclosed high-pressure control, (b) manual reset, (c) automatic reset. *Photos by Bill Johnson.*

operating temperatures will also contribute to compressor discharge valve failure. Excessive discharge pressures can also result in the rupturing of refrigerant lines and fittings.

The high-pressure control is basically a single-pole, single-throw pressure-operated switch. The control has a built-in bellows or diaphragm assembly that is connected to the system where it can best sense the discharge pressure. When the system discharge pressure becomes too high, the bellows or diaphragm assembly will expand and move against a plunger-type assembly, opening the switch contacts. The high-pressure control is usually wired in series with the compressor contactor holding coil circuit. When the switch contacts of the pressure control open, power to the compressor contactor holding coil will be interrupted and the compressor will stop. The high-pressure control can be reset automatically or manually once the system pressure has been reduced to the cut-in point of the control. Most residential applications utilize a high-pressure control with an automatic resetting feature.

High-pressure controls are available in adjustable or nonadjustable designs. Usually adjustable controls should be set to cut out approximately 20 percent above the normal operating discharge pressures of the system. The maximum ambient temperatures under which the system will operate should also be taken into account. Generally, differential settings (the difference between the cut-out and cut-in settings) should be around 65 PSIG.

Low-Pressure Controls

The purpose of a **low-pressure control (LPC)**, Figure 26–14, is to cycle the compressor off once the suction pressure has decreased to a predetermined minimum. Low suction pressure will have a detrimental effect on system performance and compressor operation. The hermetic compressors employed in air-conditioning applications depend on cool suction vapor returning from the evaporator to pass over and around the motor windings to keep them cool. A system that

Fig. 26–14 Low-pressure control. *Photo by Bill Johnson.*

is short of refrigerant and allowed to operate will cause the compressor motor to overheat and eventually the equipment to fail. Another problem arises when there is a condition of reduced airflow across the evaporator coil. Suction pressures and temperatures will decrease, which may cause the evaporator coil to ice up and cause liquid refrigerant to flood to the compressor. The low-pressure control is a single-pole, single-throw pressure-operated switch that will cause the pressure to decrease. The control has a built-in bellows or diaphragm assembly that is connected to the system where it can best sense suction pressure. When the system low-side pressure decreases to a preset minimum, the bellows or diaphragm assembly will move and open the control contact switch. The low-pressure control is wired into the compressor contactor holding coil circuit and, when opened, will interrupt power to the control circuit and stop compressor operation.

The recommended settings of a low-pressure control vary from one manufacturer to another. Adjustable controls are usually set to cut out at approximately 25 PSIG and cut in at approximately 55 to 60 PSIG.

Servicing Pressure Controls

For the most part, high- and low-pressure controls require a minimal amount of service or attention. Under normal operating conditions, they will provide many years of trouble-free service. If a problem does occur, it could be attrib-

uted to corroded or burnt contact points, loose or corroded terminal connections, defective switching mechanisms within the control, or a leaking bellows assembly. Any of these problems can be easily checked and verified. The design of the particular control will govern whether it can be field repaired or should be replaced. In most applications the controls are factory sealed and cannot be disassembled for repair.

The best way to check and test pressure controls is to install manifold gauges on the system and observe the system pressures. This will verify whether the problem is within the sealed system or the control. If, for example, the low-pressure control is open and the system low-side pressure is above the cut-in setting of the control, the control is obviously defective and needs to be replaced. Similarly, if the high-pressure control is open and the system high-side pressure is below that of the cut-out setting of the control, the control is again defective and needs to be replaced.

Unit Twenty-Six Summary

- The thermostat automatically controls the operation of a heating and cooling system. The two basic types of thermostat designs are the bimetal design and the mercury bulb design.

- Two-stage thermostats are designed to control system capacity in applications that experience varying load conditions. One example of a two-stage thermostat would be a heat pump application in which the refrigeration system is energized by the first stage and the supplemental heat strips are energized by the second stage.

- The thermostat subbase is used as a total control center for heating and cooling. Cooling anticipators cycle the cooling system on prematurely to prevent temperature override of the conditioned space, and heating anticipators cycle the heating system off prematurely.

- A thermostat should be installed on an inside wall where it can best sense the temperature of the conditioned space and not be affected by the outside temperature.

- Fan relay centers are used to provide low voltage control of the heating and cooling system.

- The high-pressure control is installed on an air-conditioning system to protect it from excessive discharge pressure; the low-pressure control is installed on air-conditioning systems to protect the system in the event of low suction pressure.

Unit Twenty-Six Key Terms

Thermostat

Bimetallic Thermostat

Mercury Bulb Thermostat

Remote Bulb Thermostat

Solid-State Thermostat

Two-Stage Thermostat

Clock Thermostat

Chrono-Therm

Subbase

Cooling Anticipator

Heating Anticipator

Fan Relay Control Center

High-Pressure Control (HPC)

Low-Pressure Control (LPC)

UNIT TWENTY-SIX REVIEW

1. What causes a majority of the problems experienced by service technicians in the HVAC/R field?

2. What is the primary function of a thermostat?

3. How is the thermostat a multi-function device?

4. Why is low voltage used in control circuits?

5. A low-voltage control circuit provides _____ sensitivity, is _____ compact, _____ costly, and allows for the use of _____ wire sizes.

6. Match the following types of thermostats with the description that fits: bimetallic thermostat, remote bulb thermostat, mercury bulb thermostat, solid-state thermostat, staging thermostat, clock thermostat.

   a. Bellows assembly expands and exerts pressure on a mechanical linkage.

   b. Utilizes a spiral-wound bimetal to provide movement.

   c. Used to reduce or set back conditioned space temperature during low load conditions.

   d. The switch contacts will open as they overcome the magnetic attraction.

   e. Utilizes microcomputer and solid-state sensors.

   f. Will prevent the second stage of heating or cooling from short cycling.

7. What is the temperature differential between the first and second stage of a two-stage thermostat?

8. What is the purpose of the setback feature in a clock thermostat?

9. What is the function of a thermostat subbase?

10. Which small resistance heater, during a system's off cycle, provides additional heat to a thermostat bimetal, causing thermostat contacts to close one degree?

11. Which small resistance heater, during a system's on cycle, provides additional heat to a thermostat bimetal, causing thermostat contacts to open one degree?

12. What is the best way to check and test pressure controls?

13. What is the relationship between thermostat installation and lamps, warm air ducts, and direct sunlight?

14. What are two checks to determine why an indoor blower motor will not deenergize?

15. Why is the high-pressure control installed to cycle the compressor off before discharge pressures become dangerously high?

16. What happens to suction pressure to send the low-pressure control switch into operation?

17. How does the fan operate when the fan switch is in the auto position?

18. What are the designations for thermostat subbase terminals R, Y, G, and W?

19. What is the purpose of a cooling anticipator?

20. How is the cooling anticipator wired into the thermostat circuit?

21. What is the purpose of a heating anticipator?

22. How is the heating anticipator wired into the thermostat circuit?

23. Why should a thermostat be mounted on an inside wall?

24. What purpose does a fan relay center serve?

25. What considerations should be made when selecting a fan relay center for a given application?

26. What do broken or dashed lines represent in an electrical schematic?

27. Describe the procedure used to check the fan relay contacts.

28. Describe the procedure used to check the fan relay transformer.

29. What is the purpose of a low-pressure control?

30. How is the low-pressure control wired into the control circuit?

31. What are the approximate cut-in and cut-out settings of a low-pressure control in air-conditioning applications?

32. Describe the procedure used to check the high-pressure control.

33. Describe the procedure used to check the low-pressure control.

34. Percentage is useful for comparison between sets of numbers. For example, between 130° and 90° is a difference of 40°. Forty becomes useful, providing even more information when 90 ÷ 130 = 69%, or a 31% change in temperature.

    The following is a list of numbers for measurements of airflow taken from an air-conditioning system at different fan motor speeds. What is the percentage of drop in airflow between the different settings? Which change in settings recorded the greatest drop in airflow?

speed	cfm
High	1,750
Med-High	1,360
Med-Low	1,090
Low	930

35. Write an explanation in 30 words or less describing the function of the high-pressure control switch.

SECTION SIX

Commercial Systems

After completing a well-rounded HVAC/R training program, some entry-level technicians make the decision to specialize in the area of commercial refrigeration and HVAC. Specializing in this area would result in your spending the majority of your time in grocery stores, convenience stores, and equipment rooms of office buildings, hotels, and apartment complexes. A technician working in this area would also spend a great deal of time on preventive-maintenance programs.

In this section, we'll look at applications of commercial refrigeration systems as well as some of the service procedures that apply to this segment of the HVAC/R industry.

UNIT TWENTY-SEVEN

Commercial Refrigeration and HVAC

OBJECTIVES

After completing this unit, the student will be able to:

1. Describe the various applications of commercial refrigeration systems.

2. Explain the method of operation of EPR and CPR valves.

3. Describe the operation of a two-stage compressor system.

4. Explain the various defrost methods used in commercial refrigeration.

5. Explain the method of operation of an oil-pressure safety switch.

INTRODUCTION TO COMMERCIAL REFRIGERATION AND AIR-CONDITIONING

Commercial refrigeration represents the broadest scope of the refrigeration industry, along with the widest variations in storage boxes, hardware, piping, and electrical systems. The application of commercial refrigeration equipment can range from the storage of perishable merchandise to the temperature control of a manufacturing process involving nonperishable items. In general, the size, design, and installation of the commercial equipment is dependent upon the particular application for which it will be used and the individual needs of the user.

Commercial Cabinets

A wide variety of cabinet designs are available to meet the needs of the individual user. To promote easy maintenance, the finishes of com-

mercial cabinets are designed for easy cleaning. Porcelain and stainless steel cabinets are widely used; plastic liners are commonly used for the interior construction. Cabinet insulation may be in the form of urethane or polystyrene, either installed in pieces or foamed into place. Mullion heaters are installed in cabinets to prevent condensation around doors and other areas where moist, warm air may create condensate problems. Door systems are designed for heavy usage and magnetic vinyl door gaskets are commonly used to provide a heat seal against air leakage or infiltration.

Display Cases

A **display case**, as shown in Figure 27–1, is used for displaying and storing perishable merchandise and may be of an open or closed design. These cases are used to display or sell merchandise. They are designed with a full display and

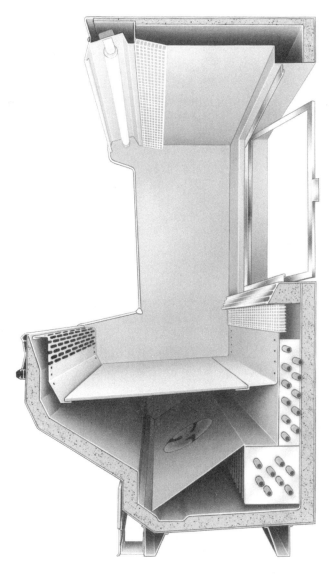

Fig. 27–1   Display case for perishable merchandise.
Courtesy Tyler Refrigeration Corporation.

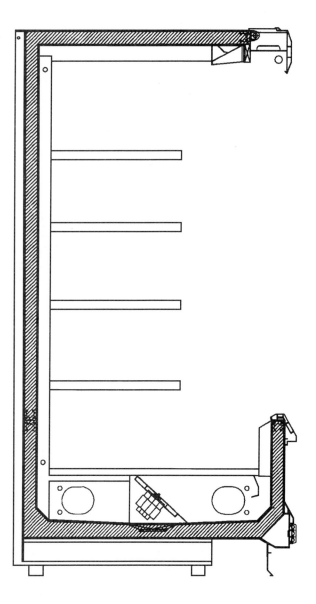

Fig. 27–2   Open display case.   *Courtesy Kysor/Warren Division of Kysor Industrial Co.*

full vision capacity, allowing the customer to pick and choose the merchandise desired.

Open Display Case

The **open display case**, which is illustrated in Figure 27–2, is among the most popular cabinet designs in use today. Merchandise stored in an open case is refrigerated by means of a cold blanket of conditioned air. In open display cases, the evaporator is usually located at the bottom of the fixture, and a fan circulates the conditioned air.

Moisture condensed from the warm air is removed from the case through a drain.

One of the primary problems with open display cases is in providing the additional refrigeration necessary to meet the high losses of conditioned air that result when the top layer of cool circulating air within the refrigerated cabinet comes into contact with the warm air above it. Even though the cold air within the refrigerated cabinet is heavier than the warm air above it, the circulating top layer of cool conditioned air will still come in contact with the warm air above the case, which is constantly rising and leaving the fixture to be replaced by new warm air from outside the fixture.

This results in an additional heat gain for the refrigeration system of the display case that requires additional refrigeration capacity to compensate for the loss. Not only must the warm air be reduced to the cabinet temperature, but the condensed moisture from this air must also be removed.

For these reasons, it is important that open display cases be located away from all external air sources, such as fans, entryways, air ducts, and other similar sources of external air movement. The merchandise stored in open cases should be placed within a certain zone within the cabinet to prevent exposure to warm external air sources. For this reason, many open display cases have a red line indicating where merchandise should be stored. When food is stored in open cases, it should be wrapped with a clear, protective material such as cellophane. This protects it from contamination due to customer handling and ambient air conditions that may introduce dust and germs.

Closed Display Case

A **closed display case**, or **reach-in case**, stores perishable merchandise such as dairy products, soda, beer, wine, and other food products. A closed display case may have a single-door or multiple-door design. Large glass doors display the merchandise to be sold.

In most applications closed display cases have self-contained condensing units. Because the merchandise is packaged, a forced convection evaporator is usually located on the top or in the rear wall of the cabinet. A closed display case is illustrated in Figure 27–3.

Double-Duty Display Case

The **double-duty display case** has two refrigerated storage cabinets. The upper compartment is used for display purposes and the lower cabi-

Fig. 27–3 A closed display case. *Courtesy Hill Refrigeration.*

net, usually located in the rear of the fixture, is used to store merchandise not yet ready for display. The evaporator coil is located in the bottom of the fixture and the conditioned air is circulated throughout the cabinet by means of a fan.

Floral Cabinet

The size, type, and design of a **floral cabinet** will vary with the individual application. These display cases are designed with large glass areas for displaying the merchandise they contain. Because humidity is the primary concern in most applications, large gravity evaporator coils are installed across the top of the fixture to prevent the stored merchandise from drying out. These cabinets usually maintain a storage temperature of 40°F to 55°F. A floral cabinet is illustrated in Figure 27–4.

Frozen Food Display Case

A **frozen-food display case** can be either opened or closed in design. The open type of dis-

Fig. 27–4 Floral refrigerator. *Courtesy Buchbinder, Chicago, IL.*

play case is more suitable for marketing the stored merchandise because the merchandise is openly displayed and the customer has easy access to the product desired. The disadvantage of an open frozen food case is that the lower storage temperatures induce a greater infiltration of warm air, resulting in a higher refrigeration load requirement. This infiltration of warm, moisture-laden air must be considered when the size of the condensing unit for the system is determined.

Opened or closed frozen-food display cases may have plate or blower evaporator coils, and their condensing units may be self-contained or remotely located. The average storage temperatures may be 0°F or lower. A frozen-food display case is illustrated in Figure 27–5.

Walk-In Units

A **walk-in unit** is illustrated in Figures 27–6A and B. A walk-in unit is used primarily for the short- or long-term storage of perishable merchan-

dise in applications such as restaurants, meat and vegetable markets, grocery stores, and taverns.

While older walk-ins were constructed of hardwood walls and floors, modern walk-ins are constructed of metal walls and hardwood floors. Insulation of the cabinet walls is usually in the form of polyurethane foam that is pumped, under pressure, between the inner and outer wall lining. The condensing unit of a walk-in may be installed on top of or along side of the cabinet, or it may be remotely installed outdoors. The door systems are light in weight, and magnetic gaskets are commonly used to seal the door, providing a positive seal against heat leakage.

An interior light is operated by a switch located outside of the refrigerated cabinet. In many applications the fan cycles off to prevent warm air from circulating inside the box when the user turns on the light and enters the cabinet. An indicator light installed in the light switch will alert the user that the interior light is on and that the fan is off.

Commercial Condensing Units

The condensing units in commercial installations will vary according to the application. Most applications utilize air-cooled condensers because of their reduced maintenance and operating costs. The condensing units may be located indoors or outdoors, and range in size from small, fractional horsepower condensing units to large, high horsepower units with several hundred tons of refrigeration capacity. Condensing units may be installed in self-contained refrigeration storage cabinets, such as in the case of the fractional horsepower units, or they may be installed in a location remote from the refrigerated space, usually in an equipment room or on the roof of the building. Because space is an important factor in many applications, the condensing units may even be remotely installed outdoors.

Multiplex Systems

It is common practice in many commercial refrigeration applications to operate several refrigerated fixtures with one compressor.

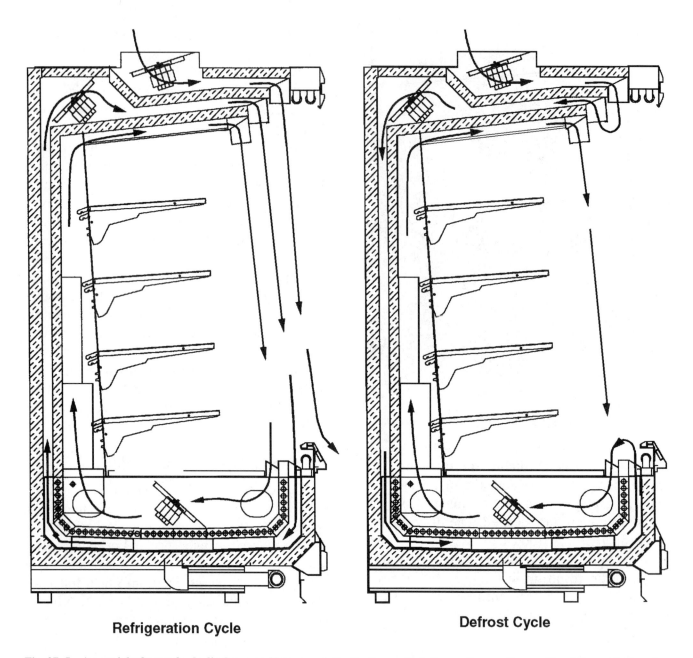

Refrigeration Cycle   **Defrost Cycle**

Fig. 27–5   An upright frozen-foods display case. Note, especially, the fans and airflow pattern.   *Courtesy Kysor/Warren Division of Kysor Industrial Co.*

Generally, the temperature control of each fixture is achieved by means of a thermostat that controls the opening and closing of a solenoid valve installed in the liquid line of the controlled evaporator or by means of an evaporator pressure regulator valve installed in the suction line of the controlled evaporator. In dealing with systems where several evaporators are operating at one temperature and one compressor is employed, the question of convenience versus economy of operation arises.

Whenever two or more vapors of dissimilar pressures are throttled so that they will enter the compressor at a common pressure (equal to that of the evaporator with the lowest temperature), an increase of superheat is experienced that, in turn, will increase the work requirement of the compressor. In small commercial applications, the

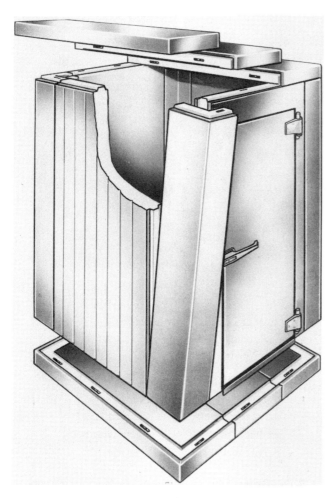

Fig. 27–6A Walk-in refrigerator. *Courtesy Bally Engineered Structures, Inc.*

convenience gained by multiplex systems overrides the thermal losses experienced by the compressor. In larger commercial applications, the amount of thermal loss sustained from common compression is so great that there is a tendency to favor alternate methods, such as using separate compressors for different evaporator temperatures. For the right application, a multiplex system can offer lower operating and installation costs as well as decreased maintenance.

Figures 27–7, 27–8, and 27–9 represent layouts of various dual- and multi-temperature systems that utilize one compressor.

Two-Stage Compression Systems

In order to overcome the problems encountered in using a single-stage compressor in low

Fig. 27–6B Walk-in refrigerator. *Courtesy Bally Engineered Structures, Inc.*

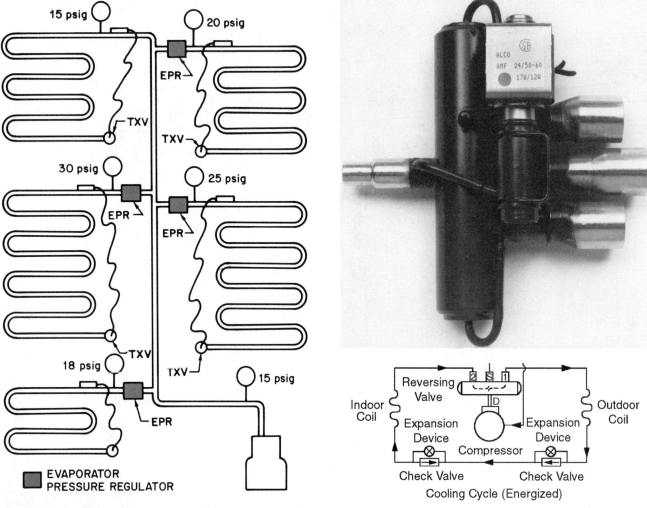

Fig. 27–7 Single compressor with multiple evaporator units.

15 psig

20 psig

EPR

TXV

TXV

30 psig

25 psig

EPR

EPR

TXV

18 psig

TXV

15 psig

EPR

■ EVAPORATOR
PRESSURE REGULATOR

Indoor
Coil

Reversing
Valve

2 S 1

D

Expansion
Device

Expansion
Device

Outdoor
Coil

Compressor

Check Valve

Check Valve

Cooling Cycle (Energized)

Fig. 27–9 Installation of solenoid valves. The solenoid controls the cabinet temperature of the cabinet. Valves and thermostats are using line voltages. *Courtesy ALCO Controls.*

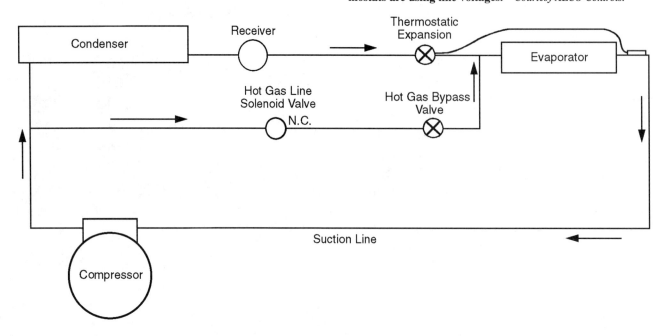

Condenser

Receiver

Thermostatic
Expansion

Evaporator

Hot Gas Line
Solenoid Valve

N.C.

Hot Gas Bypass
Valve

Suction Line

Compressor

Fig. 27–8 Schematic refrigerant piping diagram. *Courtesy Copeland Corporation.*

temperature applications, a two-stage compression system is employed. A two-stage compression system, also known as a compound system, can be accomplished through one of two methods:

1. The use of two compressors with the refrigerant discharge of one compressor pumping into the suction line of a second compressor.

2. The use of one compressor with multiple cylinders.

Because of the problems in maintaining the proper oil level in systems that utilize two compressors, it is more desirable to use one compressor with a multiple cylinder arrangement. In the application of a two-stage compression system, the total compression ratio is the sum of the ratio for each stage. While not exactly equal, the ratios will essentially be the same.

Figure 27–10 represents a typical compound system utilizing two compressors. The low-stage compressor discharges refrigerant into the suction line of the high-stage compressor, which will discharge the refrigerant into the condenser to be condensed. In order to prevent the high-stage compressor from overheating, the interstage refrigerant must be cooled. This is accomplished by a de-superheating expansion valve, which will inject liquid refrigerant into the interstage manifold cooling and de-superheat the hot refrigerant vapor before it enters the second stage compressor.

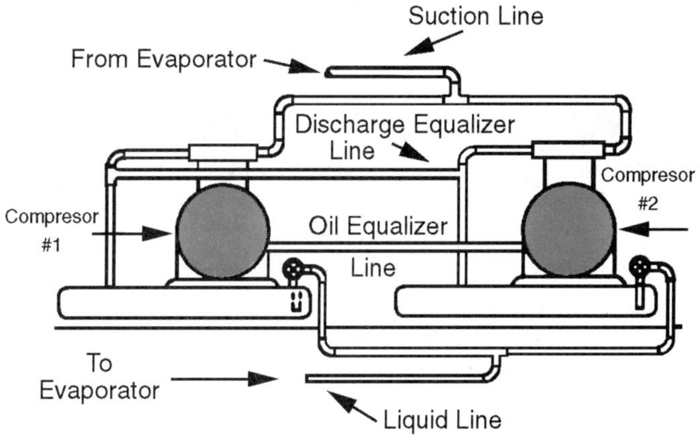

In Figure 27–11, suction vapor returning from the evaporator enters the four low-stage cylinders of a two-stage compressor. The hot discharge vapor leaving the first stage cylinders is cooled by a de-superheating expansion valve before it enters the compressor and the second stage cylinders, where it will be compressed and discharged into the condenser.

A two-stage compression system is used primarily in low temperature applications. Because of the high compression ratios experienced in low temperature applications, the capacity of a single-stage compressor decreases rapidly and is therefore not a good choice for use in these applications. Furthermore, a single-stage compressor should not be used because the weight of the refrigerant vapor returning to the compressor may be reduced to the point where motor overheating could result.

The weight decrease in the refrigerant vapor may also result in poor oil circulation throughout the system. Two major factors cause a loss of compressor efficiency when there is an increase in compression ratio, such as those encountered in low temperature systems. Figures 27–12 and 27–13 illustrate a typical six-cylinder, two-stage compressor and a three-cylinder, two-stage compressor, respectively.

1. When the compressor piston reaches top dead center on the discharge stroke, a certain amount of discharge vapor remains in the clearance pocket of the valve plate and cylinder. It reexpands on the suction stroke, reducing the amount of refrigerant that enters the cylinder from the suction line. As the discharge pressure increases, the density of the refrigerant also increases. This results in the reexpansion of more refrigerant vapor into the cylinder, further reducing compressor efficiency.

2. As the heat of compression increases, the temperature of the cylinder walls and the head of the compressor also increase. The refriger-

Fig. 27–10   Compound system using two compressors.

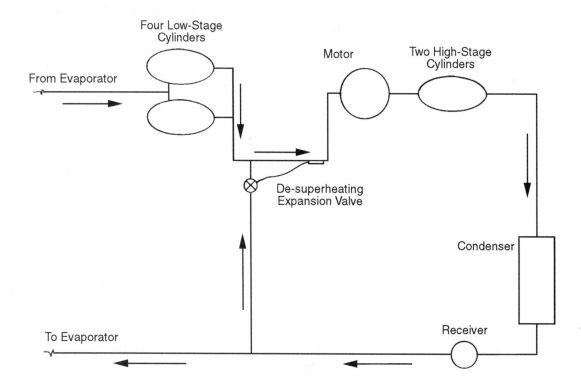

Fig. 27–11 Schematic two-stage system. *Courtesy Copeland Corporation.*

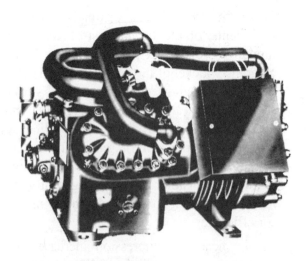

Fig. 27–12 Typical six-cylinder two-stage compressor.
Courtesy Copeland Corporation.

Fig. 27–13 Typical three-cylinder two-stage compressor.
Courtesy Copeland Corporation.

ant vapor entering the compressor will become heated and then expand, resulting in a reduced weight of refrigerant vapor entering the compressor.

Low Ambient Discharge Pressure Controls for Air-Cooled Condensers

While it is desirable to take advantage of lower condensing temperatures to increase system capacity, reduce discharge gas temperatures, and lower power requirements, a discharge pressure that is too low can result in serious system operating problems, such as these listed below:

1. Reduced capacity of the metering device, resulting in a decrease of evaporating pressure. (The capacity of an expansion valve or capillary tube is proportional to the pressure differential across the capillary tube or expansion valve.)

2. Starved or oil-logged evaporator coil.

3. Short cycling of the low-pressure control.

4. Reduction of system capacity.

5. Erratic expansion valve operation.

When an air-cooled condenser is operated in ambient temperatures below 50°F to 60°F, adequate minimum discharge pressures must be maintained to prevent system damage. This may be accomplished by one of several methods:

1. Air volume dampers that control the airflow across the condenser coil. (They are a simple and economical means of head pressure control and are usually controlled by the system discharge pressure. See Figures 27–14 and 27–15.)

2. A reverse acting, high-pressure sensing system that cycles on one or more condenser fans when the discharge pressure falls below the minimum setting of the control—Figures 27–16 and 27–17.

3. Flooding the condenser with liquid refrigerant, thus reducing the effective condenser surface area. (Several systems are utilized to control system discharge pressure by flooding the condenser and these systems vary from one manufacturer to the other. These applications require a considerable increase in the amount of the system refrigerant charge and adequate receiver capacity during cold and warm weather operation.)

4. Use of a pressure-regulating valve.

Figure 27–18 illustrates a simple method of discharge pressure control utilizing a pressure-regulating valve that is installed at the outlet of

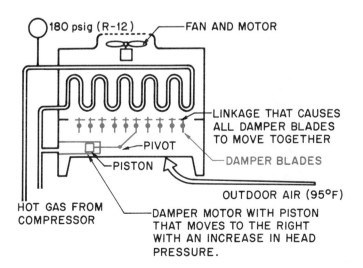

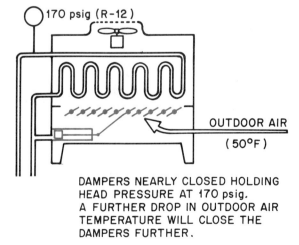

Fig. 27–14   An air-cooled condenser with pressure-operated damper. As condensing pressure decreases, damper will start to close, reducing condenser airflow.

Fig. 27–15 This 3 hp air-cooled condensing unit is equipped with an adjustable damper to vary airflow as ambient temperature changes. *Courtesy Tyler.*

the receiver. When the pressure at the outlet side of the valve is reduced below the valve setting, the valve will close, preventing liquid refrigerant from leaving the condenser until proper condensing pressure has been experienced. A check valve is installed to ensure one-way flow.

Another type of discharge pressure control system is illustrated in Figure 27–19. In this application, a pressure-regulating valve is designed to open as the pressure in the receiver decreases, allowing hot gas to bypass into the receiver. This will increase the pressure in the receiver and increase the flow of liquid refrigerant to the metering device. The valve has two openings. As one opening closes, the other will open, thereby controlling the flow of hot gas or liquid refrigerant to the receiver.

To prevent liquid refrigerant from collecting in the compressor during the off cycle, a trap is

Fig. 27–16 A fan speed control unit. *Courtesy Hoffman Controls Corporation.*

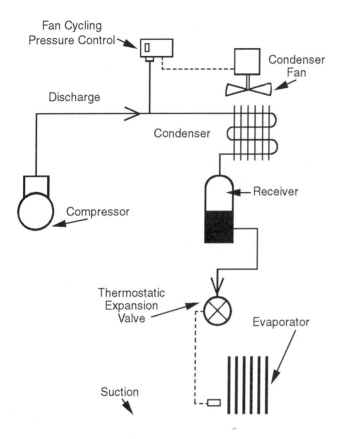

Fig. 27–17 This schematic shows refrigeration cycle with a fan cycling pressure control that senses condenser pressure. *Courtesy Ranco Controls.*

installed in the discharge line of the compressor and another inverted trap is installed at the condenser outlet. In many cold weather applications the receiver may be heavily insulated and warmed with electrical resistance heaters.

CRANKCASE PRESSURE REGULATOR VALVES

The purpose of a **crankcase pressure regulator (CPR) valve** is to protect the compressor motor against an overload condition due to the intermittent high suction pressures. Under normal operating conditions the current draw of the compressor motor will approach normal full load ratings. There are several operating conditions, however, when the suction pressure may rise above normal. This results in excessive current draw of the compressor motor and an overload condition.

On initial start-up, temperatures in the conditioned space are high and the rate at which the refrigerant evaporates is also high, resulting in higher than normal suction pressures. When the compressor starts, the high suction pressure will result in a high compressor motor current draw until the compressor is able to reduce the initial high load condition.

After a defrosting cycle is terminated, the evaporator is warm and the suction pressure is high. When the compressor is started, the high

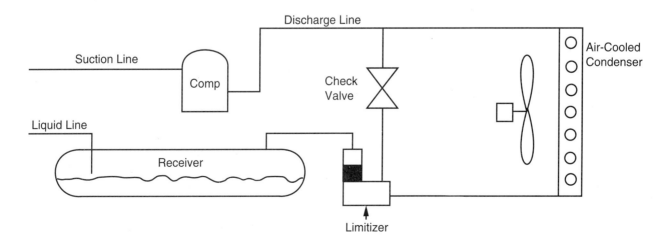

Fig. 27–18 Proper condensing pressure is accomplished in cold weather through the use of a pressure control system.

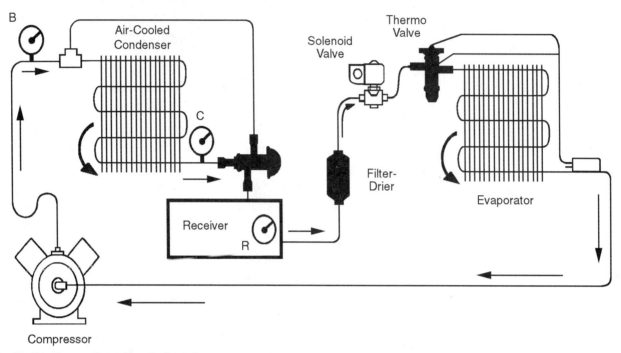

**Discharge Pressure
Above Valve Setting . . .
Flow Thru Condenser**

Fig. 27–19   Hot gas flows directly from the compressor into the receiver during cold weather when opening at "C" is closed and openings at "B" and "R" are open.   *Courtesy ALCO Controls.*

suction pressure will cause an overloading of the compressor motor and a higher than normal current draw.

In the case of hot gas defrost systems, hot gas is directed into the low side of the system during the defrosting cycle, resulting in a higher than normal suction pressure. In both cases, the suction pressure must be controlled or an overloading of the compressor motor, as well as equipment damage, may result.

If an additional heat load is placed on the conditioned space during normal operation, the suction pressure will increase and will be accompanied by an increase in compressor motor running current. The high current condition may cause the motor to cycle on overload several times before the load is reduced.

Valve Operation

The crankcase pressure-regulator valve, which is illustrated in Figure 27–20, is a spring-loaded

device designed to limit the suction pressure at the valve outlet or the compressor side of the control at a predetermined maximum pressure.

When the suction pressure is below the setting of the control during normal system operation, the crankcase pressure regulator valve is wide open and the system will function as if the valve were not there. When the suction pressure rises above the control setting, the valve will close, preventing the suction pressure from increasing at the compressor inlet. The valve will modulate from a fully opened position to a fully closed position in response to the pressure at the valve outlet.

Valve Installation

The crankcase pressure-regulator valve is installed in the suction line between the evaporator coil and the compressor, Figure 27–21. The valve is installed downstream of any other suction line accessories or controls. It is important to fol-

Fig. 27–20 **CPR valve to keep the compressor from running in an overloaded condition during a hot pull down.** *Courtesy Sporlan Valve Company.*

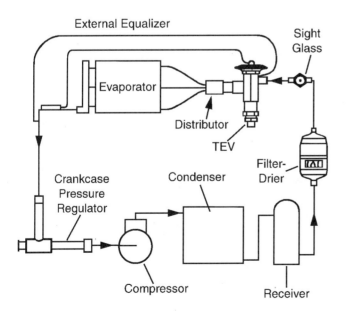

Fig. 27–21 **Installation of crankcase pressure-regulator valve.**

low good piping practices to prevent the valve from acting as an oil trap.

When installing the valve with solder connections, care should be taken to prevent the valve body from overheating. Always refer to the manufacturer's installation instructions. Wrapping the valve body in a damp cloth, using a proper sized torch tip, and directing the flame away from the valve body are factors that must be considered.

Valve Adjustment

The crankcase pressure-regulator valve should be adjusted when the suction pressure is above the desired maximum. The final setting of the valve should be below the maximum suction pressure recommended by the compressor manufacturer.

One of the best ways to check the valve setting is to check the compressor motor current draw on the initial start-up or after a defrosting cycle. If the current draw is high, which indicates an overload condition, the crankcase pressure-regulator valve may be set too high and the valve should be adjusted. With gauges installed, follow these steps to adjust the valve:

1. Keep the system off until the system pressures have equalized.

2. Start the system and observe the suction pressure. If the suction pressure is too high, slowly adjust the valve in a counterclockwise direction. For each one PSI of pressure change, turn the adjusting screw one-quarter of a turn.

3. Shut the system down after a few minutes of operation and allow the system pressures to equalize.

4. Repeat the process until the required adjustment is reached. Turning the adjusting screw clockwise will increase the valve setting; turning the screw counterclockwise will decrease the valve setting.

EVAPORATOR PRESSURE REGULATORS

The purpose of an **evaporator pressure regulator (EPR)** is to prevent evaporator pressure from decreasing below a predetermined point. The EPR is also referred to as a *two-temperature valve* because it is commonly used in multiple-evaporator systems where more than one temperature is maintained. In these applications the evaporators are connected to a single condensing unit.

An EPR is illustrated in Figure 27–22. It is installed at the outlet of the evaporator and responds to evaporator outlet pressure or valve inlet pressure. When the evaporator pressure is above the valve setting, the valve is wide open and the system operates as if the valve did not exist. As the evaporator pressure decreases, the valve will close to prevent the evaporator pressure from decreasing below the control setting. The EPR may be a modulating valve in which it opens or closes when the controlling pressure varies slightly, or it may be a snap-action valve in which it snaps opened or closed when the controlling pressure varies significantly. Figure 27–23 illustrates a typical EPR installation.

When EPR is used on multiple-evaporator systems, the controlled evaporator should not comprise more than 40 percent of the total system load or erratic cycling will result. In these

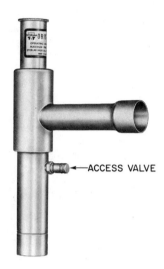

Fig. 27–22 Pressure tap provided on the evaporator side of the EPR valve so that the technician can observe the actual evaporator pressure. The true suction pressure can be obtained at the compressor suction service valve. *Courtesy Sporlan Valve Company.*

applications, the lower temperature evaporator must also be equipped with a check valve (installed at the evaporator outlet), to prevent the higher pressure vapor in the warmer evaporator from migrating to the lower temperature

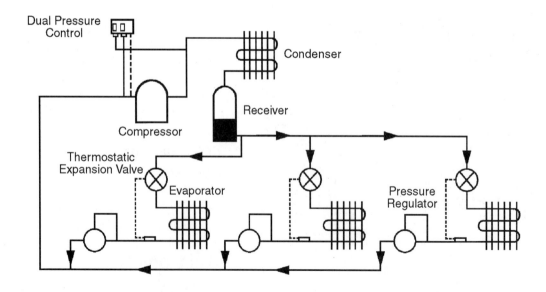

Fig. 27–23 Typical EPR installation.

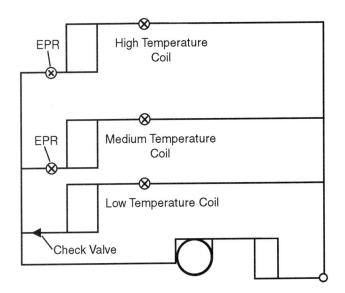

Fig. 27–24 Check valve installed on low-temperature evaporator suction line.

coil. An example of this check valve is shown in Figure 27–24.

Evaporator pressure regulator valves may have flared or soldered connections. An adjusting screw is provided with which to set the valve. A gauge port on the inlet side of the valve is also usually provided, giving the service technician access to monitor evaporator pressure and adjust the valve.

Surge Tanks

On multiple temperature systems regulated by a low-pressure control, the pressure fluctuation caused by the opening and closing of the two-temperature valve may cause the compressor to short cycle. An example of this is when the two-temperature valve closes after the temperature of the warm evaporator has been satisfied. The compressor will continue to operate until the temperature of the cold evaporator has been satisfied and the system is cycled off by the low-pressure control.

If a two-temperature valve controlling a warmer evaporator should open just after the compressor cycles off, the low-side pressure would rise rapidly and cause the low-pressure control to start the compressor, resulting in a short cycling that could cause the compressor to cycle on motor-overload protection.

To eliminate this problem, a **surge tank** is installed in the suction line ahead of the compressor. The surge tank is large enough to absorb and slow down the rapid pressure buildup in the suction line when a two-temperature valve opens. As a result, the short cycling problem is prevented. A line installed on the bottom of the surge tank allows oil to return to the compressor. A surge tank is illustrated in Figure 27–25.

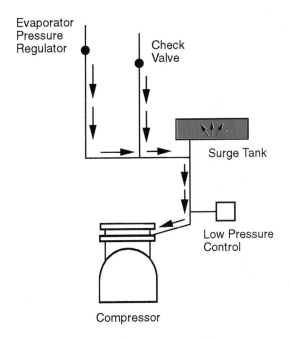

Fig. 27–25 Surge tank installed on two-temperature system with evaporator pressure-regulator valve.

Oil-Pressure Safety Switches

In positive-pressure lubricated compressors, it is important for proper oil pressure to be established by the time the compressor comes up to speed and for proper oil pressure to be maintained throughout the running cycle. Loss of oil pressure in a compressor will result in extreme damage. For this reason, the oil pressure must be monitored, and the compressor must be shut down if the oil pressure decreases below the minimum requirements.

In order to accomplish this, an **oil-pressure safety switch** is installed in the system to monitor compressor oil pressure. Figure 27–26 illustrates a typical oil-pressure safety control. If proper oil pressure is not established within a predetermined amount of time after the initial start-up of the compressor, or if the oil pressure should fail during the normal running cycle, the switch will remove the compressor from the circuit.

The operation of an oil-pressure safety switch is based on a pressure differential. The lubrication system of a compressor is contained within the compressor crankcase. In a positive lubrication system, the pressure at the discharge port of the compressor oil pump is the sum of the existing suction pressure experienced in the crankcase and the actual oil pump pressure. In order to determine the net oil pressure, the suction pressure existing in the compressor crankcase must be cancelled. For example, if the total pressure reading at the discharge port of the oil pump reads 50 PSIG and the suction pressure reads 30 PSIG, the net oil pressure will be 20 PSIG. Figure 27–27 illustrates this concept.

An oil-pressure safety switch will measure the difference between the oil pump discharge pressure and the suction pressure. The resulting net oil pressure will then act against a predetermined spring pressure in the switch, which must equal the minimum oil pressure allowed by the manufacturer. When a compressor starts, however, oil pressure is not fully obtained until the compressor comes up to full speed. To compensate for this condition, a time delay is built into the switch. This usually allows 60 to 120 seconds of operation for the oil pressure to build up before the compressor operation is terminated. This time delay may be a manufacturer's fixed setting or may be field adjustable.

Referring to Figure 27–28, the operation of an oil-pressure safety switch is as follows: Upon the initial start-up of the compressor, the oil-pressure safety switch and control timer switch contacts are in a closed position. If the compressor oil pressure does not overcome the opposing force of the suction pressure plus the differential spring setting of the control, the switch will remain closed and the time delay heater will remain energized.

If sufficient oil pressure has been established, the switch will open to remove the time delay heater from the circuit and the compressor will continue to operate. But if the compressor oil pressure fails to overcome the opposing forces

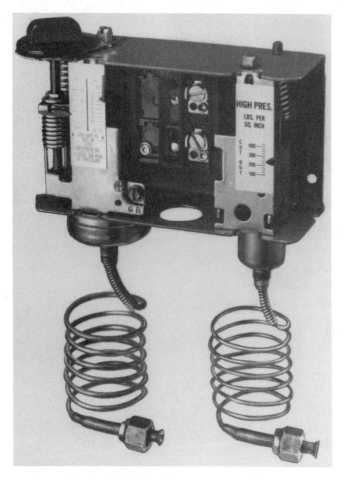

Fig. 27–26   Oil-pressure safety control. *Courtesy Johnson Controls, Inc.*

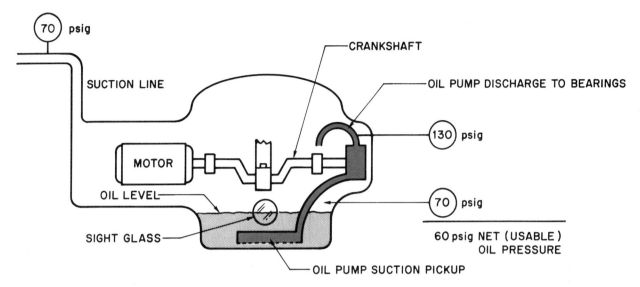

Fig. 27–27   The oil pump suction is actually the suction pressure of the compressor. This means that the true oil pump pressure is the oil pump discharge pressure less the compressor suction pressure. For example, if the oil pump discharge pressure is 130 PSIG and the compressor suction pressure is 70 PSIG, the net oil pressure is 60 PSIG. This is usable oil pressure.

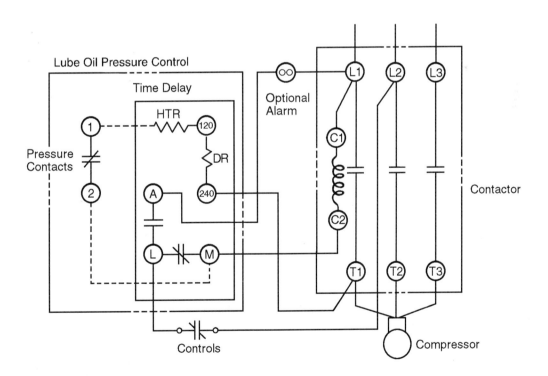

Fig. 27–28   Operation of oil-pressure safety switch.

within the prescribed time limit of the control, the heater will cause the bimetallic switch to warp, opening the time switch contacts and stopping compressor operation.

If the oil pressure decreases below the minimum set standard during normal operation of the compressor, the differential switch will close to energize the heater and, after the predetermined time delay, will open the timer switch contacts and stop compressor operation.

The switch will need to be reset manually after a shutdown from loss of oil pressure. A sufficient amount of time must be allowed before resetting the control to allow the heater to cool down. Figure 27–29 illustrates a typical installation of the oil-pressure safety control.

Low-Pressure Controls

In commercial applications, a **low-pressure control** is used as a safety control and as a temperature-control device.

The saturated temperature of the refrigerant in the evaporator is governed by the pressure at which it is maintained. Changes in the suction pressure will reflect changes in the temperature of the evaporator. The low-pressure control is activated by changes in the suction pressure and is utilized to control the temperature of the conditioned space by indirectly controlling the temperature of the evaporator. Figure 27–30 illustrates a typical low-pressure control.

Various methods are used to adjust the low-pressure control to obtain the desired fixture temperature. Figure 27–31 is an excellent guide to

Fig. 27–30   **Typical low-pressure control.**   *Courtesy Johnson Controls, Inc.*

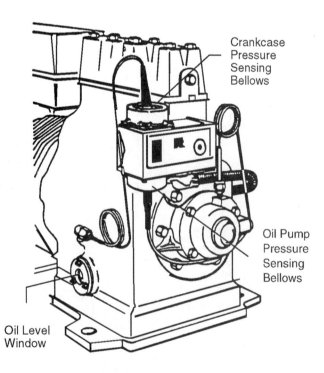

Crankcase Pressure Sensing Bellows

Oil Pump Pressure Sensing Bellows

Oil Level Window

Fig. 27–29   **Installation of oil-pressure safety switch.**
Courtesy Ranco Controls.

APPROXIMATE PRESSURE CONTROL SETTINGS

Vacuum – Inches of Mercury – ( )   Pressure – Pounds Per Square Inch Gauge

APPLICATION	REFRIGERANT					
	22		502		12	
	Out	In	Out	In	Out	In
Ice Cube Maker—Dry Type Coil	16	37	22	45	4	17
Sweet Water Bath—Soda Fountain	43	56	52	66	19	29
Beer, Water, Milk Cooler, Wet Type	40	56	48	66	19	29
Ice Cream Trucks, Hardening Rooms	13	34	18	41	2	15
Eutectic Plates, Ice Cream Truck	11	16	16	22	1	4
Walk In, Defrost Cycle	32	64	40	75	14	34
Reach In, Defrost Cycle	40	68	48	78	19	36
Vegetable Display, Defrost Cycle	30	66	38	77	13	35
Vegetable Display Case—Open Type	35	77	44	89	16	42
Beverage Cooler, Blower Dry Type.	34	64	42	75	15	34
Retail Florist—Blower Coil	55	77	65	89	28	42
Meat Display Case, Defrost Cycle	37	66	45	77	17	35
Meat Display Case—Open Type.	27	53	35	63	11	27
Dairy Case—Open Type.	26	66	33	77	10	35
Frozen Food—Open Type	4	17	8	24	(7)	5
Frozen Food—Open Type—Thermostat	—	—	—	—	2°F	10°F
Frozen Food—Closed Type	11	22	16	29	1	8

Fig. 27–31   Pressure-control settings for typical applications for temperature control. *Courtesy Sporlan Valve Company.*

follow for approximate pressure control settings for various equipment and refrigerants.

An accurate method used to adjust the low-pressure control is to calculate the temperature difference between the evaporator coil and the fixture's return air temperature. This method is called the *temperature difference method* and is determined as shown in the following example:

Assume that a medium temperature application is used with a desired fixture temperature of 35°F and an R-12 system. The permissible temperature fluctuation is 2°F; the highest allowable fixture temperature is 36°F; and the lowest allowable fixture temperature is 34°F. Determine the cut-in setting of the control.

During the system off cycle, the evaporator and fixture air temperature will rise together. When

the maximum allowable fixture temperature is reached, the suction temperature also will have risen to a corresponding saturated pressure. The low-pressure control is then adjusted to cut in at this pressure. In our example we've established that the highest allowable fixture temperature is 36°F. When the air temperature in the conditioned space reaches 36°F, the corresponding evaporator pressure will be 33.4 PSIG. This will be the cut-in setting of the pressure control.

To determine the operating temperature difference of the evaporator coil, follow these steps:

1. Allow the system to operate for at least 15 minutes to stabilize system pressures and temperatures.

2. Take a temperature reading of the return air that is entering the evaporator coil. This is

best done by placing a thermometer feeler bulb in a glycol solution and then placing it in the evaporator return airstream. For example, assume that the temperature of the evaporator return air is 45°F.

3. In order to determine the evaporator temperature, record the pressure at the compressor suction service valve and add 2 PSIG for suction line pressure loss. For example, if the suction pressure is 19 PSIG plus 2 PSIG for the suction pressure loss, then the evaporator pressure is 21 PSIG.

4. Convert this pressure to its corresponding saturation temperature to determine the evaporator coil temperature. For example, 21 PSIG converts to 20°F.

5. In order to determine the operating temperature difference between the evaporator and fixture return air temperature, subtract the evaporator coil temperature from the return air temperature. For example, if the return air temperature is 45°F and the evaporator temperature is 20°F, the operating temperature difference is 25°F.

To determine the cut-out setting of the control, subtract the evaporator operating temperature difference from the lowest allowable fixture temperature. During the running cycle of the system, evaporator temperature and suction pressure will decrease along with the fixture temperature. An example of this would be when the fixture temperature is reduced to its lowest allowable temperature of 34°F and the operating temperature difference is 25°F. The evaporator temperature will be 9°F.

To determine the cut-out setting of the low-pressure control, convert the 9°F to its corresponding saturation temperature. For example, 9°F converts to 14 PSIG.

Thus, the cut-in setting of the low-pressure control in this example is 33.4 PSIG, and the cut-out setting is 14 PSIG. Therefore, the differential setting is 19.4 PSIG.

After the initial control adjustment is made, allow the system to operate through several cycles, check the temperature, and make final adjustments to obtain the desired fixture temperature.

If time permits, another method commonly used to adjust the low-pressure control is as follows:

1. Determine the highest allowable fixture temperature and convert it to pressure for the cut-in setting of the control. For example, on an R-12 system, the highest allowable fixture temperature is 36°F, which converts to 33.4 PSIG. Thus, 33.4 PSIG is the cut-in setting of the low-pressure control.

2. Place a thermometer in a glycol solution and then in the conditioned space and start the system. When the thermometer indicates the desired fixture temperature, adjust the low-pressure control to cut out.

CAPACITY CONTROL

In many refrigeration and air-conditioning applications, the evaporator heat load may vary. This condition may be a result of varying product loads, usage of the occupied space, or varying ambient temperatures.

When a low evaporator heat load exists, the suction pressure of a system is reduced. This condition can result in a variety of service problems that could affect system performance or damage the compressor. For example:

1. Under low load conditions the compressor may short cycle.

2. Low refrigerant evaporating temperatures will result in reduced compressor capacity. Consequently, refrigerant density and velocity may be reduced to a point where oil return to the compressor may be affected.

3. The decreased volume of suction vapor returning to the compressor may result in the overheating of the compressor motor.

Under operating conditions where the evaporator heat load is relatively constant, an on and off cycle operation of the compressor is acceptable. In applications where fluctuating load conditions exist, some means must be provided to control compressor capacity in order to maintain the minimum suction pressures. This can be accomplished by multi-speed motors, hot gas bypass systems, cylinder unloading, and multiple compressors.

Hot Gas Bypass Systems

Hot gas bypass systems maintain minimum suction pressures by introducing hot discharge vapor directly into the low side of a system at a given controlled rate in order to prevent the suction pressure from falling below a predetermined point.

A hot gas bypass valve installed between the discharge side and the low-pressure side of the system will operate in response to system suction pressure. When the system suction pressure falls to a predetermined minimum, the valve will modulate to an opened position and allow discharge gas to be introduced into the low side of the system. As the suction pressure rises, the valve will modulate to a closed position.

A solenoid valve is normally installed upstream of the hot gas bypass valve to allow for manual or automatic system pump down and to

prevent leakage of hot gas into the low side of the system during the off cycle. Depending on the application, there are several methods in which hot gas may be introduced into the low side of a system.

Hot Gas Bypass into the Evaporator Inlet

For single evaporator systems, this is the most economical method of introducing hot gas into the low side of a system. System piping and controls are minimized. Hot discharge gas is introduced directly into the evaporator just after the metering device, as shown in Figure 27–32. The advantage of this method is that the hot gas will create a false heat load on the evaporator and the thermostatic expansion valve will modulate refrigerant flow as it normally would to maintain its superheat setting. As a result, suction vapor will return to the compressor at a normal operating temperature.

Hot Gas Bypass into the Suction Line

This arrangement, shown in Figure 27–33, is common when more than one evaporator is used with one compressor and where the compressor is

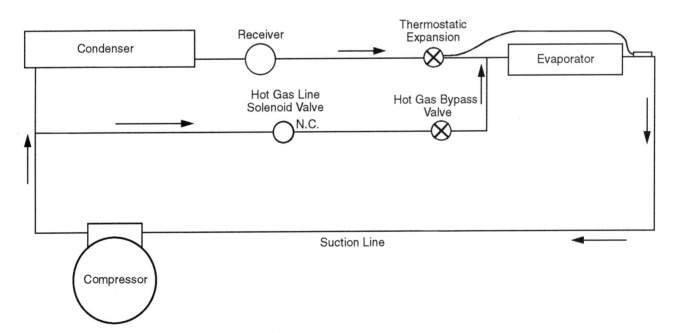

Fig. 27–32   Hot gas bypass into evaporator inlet. *Courtesy Copeland Corporation.*

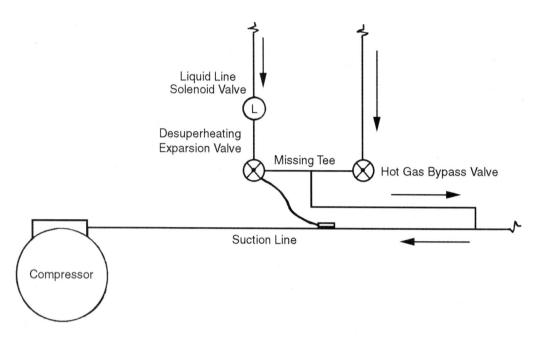

Fig. 27–33 System piping utilizing suction header to mix liquid and hot gas being piped into the suction line. *Courtesy Copeland Corporation.*

installed in a remote location from the evaporator. Several precautions should be taken when this method is used. For example, hot gas introduced into the suction line may return to the compressor at too high a temperature, resulting in the overheating of the compressor motor. For this reason, a de-superheating thermostatic expansion valve should be installed to meter the liquid refrigerant into the suction line at a controlled rate in order to cool the returning vapor.

Because there is a possibility of the liquid refrigerant returning to the compressor, a mixing chamber is required to thoroughly mix the hot gas and liquid refrigerant. A suction line accumulator may be installed for this purpose—at the same time, it will protect the compressor from liquid refrigerant floodback.

In installations where it is possible to introduce hot gas into the suction line at a distance from the compressor, a suction header may be used.

De-Superheating Expansion Valves

The purpose of a **de-superheating expansion valve** is to meter liquid refrigerant into the suction line in order to reduce the temperature of the

returning suction vapor under maximum bypass conditions. Suction vapor that enters the compressor should not exceed 65°F under low temperature load conditions and 90°F under high temperature load conditions. The feeler bulb of the de-superheating expansion valve should be located where it can best sense the temperature of the suction vapor returning to the compressor. The best location is downstream of where the refrigerant from the bypass line is introduced into the suction line.

Because the superheat setting of expansion valves used in hot gas bypass applications is not standard, special valves that have superheat settings over a wide range must be used. For this reason, the valve manufacturer should be consulted when selecting an expansion valve for hot gas bypass applications.

Because it is necessary to thoroughly mix the liquid refrigerant and the bypassed hot gas, the installation of a mixing chamber is recommended. A suction line accumulator, shown in Figures 27–34 and 27–35, can serve as an excellent mixing chamber. An alternate mixing method is to arrange the system piping so that a mixture of discharge gas and liquid refrigerant is intro-

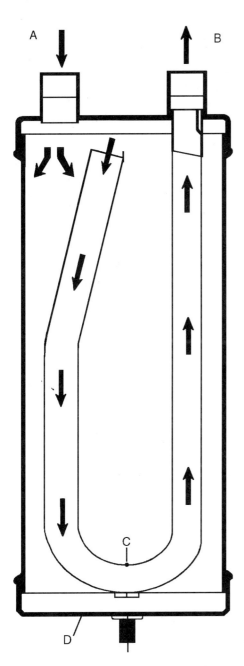

Fig. 27–34 An accumulator accomplishes its task when
suction gas flows into (a), while liquid is trapped at (d) and
vapor flows out at (b). (c) is the oil return line aspirator hole.

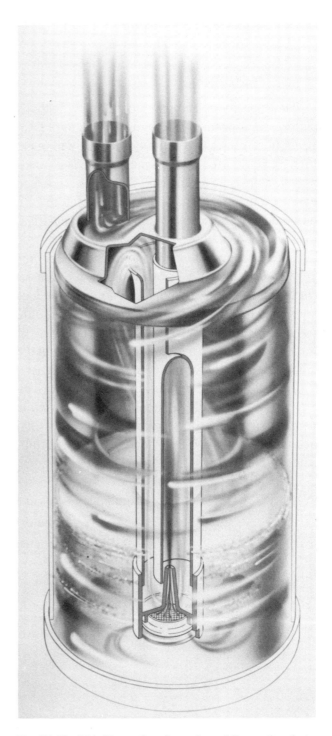

Fig. 27–35 This illustration shows the swirling action that
prevents liquid from getting to the compressor.

duced into the suction line some distance from the compressor.

Compressor Speed

The capacity of a compressor will vary with the compressor speed. An increase in compressor speed, in revolutions per minute (RPM), will result in an increase in compressor capacity. The maximum compressor capacity and RPM are established by the compressor manufacturer. Therefore, it's important to refer to the manufacturer's specifications when changing compressor speed. As the compressor RPM is increased, the drive motor current will also increase. For this reason, it is important to make sure that the motor current draw does not exceed the full current draw rating of the motor. Compressor RPM can be changed by an increase or decrease in the drive motor speed or a change in the pulley size of the drive motor or driven compressor.

This service procedure applies to belt-driven compressors only. In the application of hermetic compressors and other direct drive compressors, the compressor RPM may be controlled as established by the manufacturer's design specifications.

If a tachometer is not available to measure compressor RPM, the following formula may be used:

$$\text{Compressor RPM} = \text{motor RPM} \times \frac{\text{drive diameter}}{\text{driven diameter}}$$

Drive = motor pulley diameter
Driven = compressor pulley diameter

For example, assume that the motor RPM is 3,450, the drive diameter is 3 inches, and the driven diameter is 9 inches. The compressor RPM can be calculated as follows:

$$\text{Compressor RPM} = 3,450 \times \frac{3}{9}$$
$$= 3,450 \times .3$$
$$= 1,150 \text{ RPM}$$

LIQUID REFRIGERANT CONTROL

One of the major causes of compressor failure is due to liquid refrigerant entering the compressor crankcase during the running and off cycle periods of a system. An air-conditioning compressor is designed to move vapor. While a small amount of liquid refrigerant entering the compressor crankcase will do no harm, an excessive amount will cause damage.

Control Problems

Some common liquid refrigerant control problems in refrigeration and air-conditioning systems include migration, liquid slugging, and liquid flooding.

Migration

Liquid refrigerant is attracted to the oil in a sealed system. It will vaporize and move, or **migrate**, to the oil in the compressor crankcase during long off cycle periods. This condition is aggravated when the temperature of the compressor crankcase is colder than the evaporator. Consequently, a pressure differential is created to cause the refrigerant to flow into the crankcase. In some instances, when there are extremely long off cycle periods, migration will occur without the existence of a pressure or temperature differential but through the natural attraction of the refrigerant by the oil.

When a large amount of liquid refrigerant accumulates in the compressor crankcase, the following problems can result:

1. When the compressor starts, the pressure on the liquid and oil mixture in the crankcase is greatly reduced. This will cause a rapid boiling of the refrigerant and oil, resulting in a foaming of the oil and causing the compressor to pump excessive amounts of oil out of the crankcase into the system.

2. Liquid refrigerant present in the compressor crankcase will wash internal compressor components of lubricating oil, resulting in the failure of bearings, crankshafts, connecting rods, and the like.

Liquid Slugging

Liquid slugging is a condition that exists when liquid refrigerant passes through the compressor intake valves. This condition will cause compressor valve failure as well as damage to the internal components.

Depending on the system design and piping configuration, liquid refrigerant may condense in the suction line. On initial start-up, the liquid refrigerant could return to the compressor at a high enough velocity to damage the compressor valves. When this situation occurs, a clattering metallic sound will be heard in the compressor upon start-up.

Liquid Flooding

Liquid flooding is a condition that exists when liquid refrigerant returns to the compressor during the normal running cycle. This condition may be caused by a flow device, such as a thermostatic expansion valve, that is feeding too much liquid into the evaporator. It can also be attributed to a low load condition on the evaporator, such as when an air filter is restricted, an evaporator fan is inoperative, or an insufficient amount of air is being delivered across the evaporator coil. The liquid refrigerant will not vaporize completely as it passes through the evaporator, resulting in liquid entering the suction line and compressor.

Methods of Control

Depending upon the type and severity of the liquid refrigerant problem experienced and the application of the equipment, there are several alternatives that can be selected to resolve the problem.

Crankcase Heater

The primary purpose of a **crankcase heater** is to keep the temperature of the oil in the compressor crankcase high enough to prevent refrigerant vapor from entering and condensing in the compressor crankcase. This is the most common method of liquid refrigerant control employed in both residential and commercial applications.

The two basic types of crankcase heaters commonly used are the insert heater and the external wraparound heater. The insert crankcase heater, which is illustrated in Figure 27–36, is usually factory-installed by means of a threaded plug located in the compressor crankcase. The external wraparound crankcase heater, which is illustrated in Figures 27–37A and B, may be factory-or-field installed.

In some applications, the crankcase heater may be continuously energized when power is applied to the condensing unit. In other applications, its operation may be controlled by a thermostatic control. Another method of operation control is to use a switching relay to energize the heater during the off cycle of the compressor only.

One method of crankcase heating is accomplished through the use of a dual-run capacitor wired to provide a trickle of current flow through the compressor motor windings during the off cycle. This will keep the motor windings at a temperature high enough to prevent liquid migration, as well as vaporize any liquid refrigerant that has condensed in the compressor crankcase. Caution must be taken when replacing the run capacitor in this type of application. The replacement capacitor must have the same microfarad rating as the original capacitor, and the wiring configuration must be the same as well.

While the crankcase heater does a satisfactory job in preventing liquid refrigerant migration and vaporizing liquid that may have condensed in the compressor crankcase, it will not handle severe cases of migration or flooding.

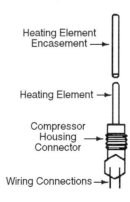

Heating Element Encasement →

Heating Element →

Compressor Housing Connector →

Wiring Connections →

Fig. 27–36 Insert-type crankcase heater.

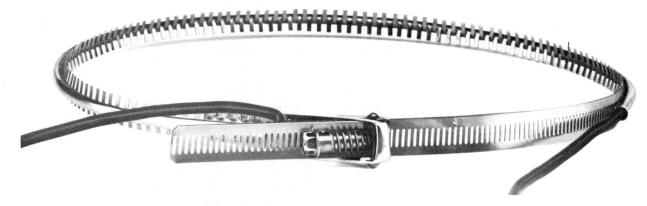

Fig. 27–37A   This type of crankcase heater is an accessory type that is fastened around the dome of the compressor.

Fig. 27–37B   The accessory type crankcase heater installed on a compressor.

Pump Down Cycle

The **pump down cycle** is the most effective method of controlling liquid refrigerant problems in a system. When this method is employed, the refrigerant is removed from the low side of the system and pumped into the condenser and/or receiver to be stored during the off cycle period of the system. Two types of pump down cycles used are recycling pump down and nonrecycling pump down.

Recycling Pump Down Cycle

Figure 27–38 illustrates a typical wiring schematic of a recycling pump down control circuit. The operation of a recycling pump down cycle is as follows:

1. When the thermostat calls for cooling, terminals R to Y of the thermostat will close to energize the liquid line solenoid valve.

2. When the liquid line solenoid valve opens, pressure from the high side of the system will bleed into the low side of the system.

3. When the low-side pressure rises to a predetermined level, the low-pressure control contacts will close.

4. Current will now flow across the compressor contactor holding coil circuit, energizing the compressor and placing the system into a normal refrigeration cycle.

5. When the thermostat is satisfied, terminals R to Y will open and the liquid line solenoid valve will close, restricting refrigerant flow into the low side of the system.

The compressor will continue to operate, pumping the refrigerant into the system condenser and/or receiver. When the low-side pressure is reduced to the cut-out setting of the low-pressure

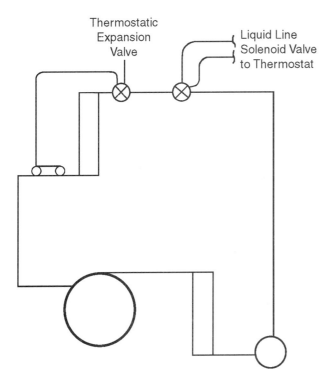

Fig. 27–38A   A typical piping diagram of a pump down system.

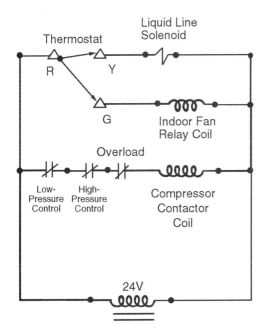

Fig. 27–38B   Cycle pump down system.

control, the control contacts will open, stopping compressor operation. The majority of the system refrigerant charge is now stored in the high side of the system.

The primary disadvantage of the recycling pump down cycle is that if the system low-side pressure increases to the cut-in point of the low-pressure control, the system will cycle back on, and in some cases, a short cycling condition may be experienced.

Nonrecycling Pump Down Cycle

The operation of a nonrecycling pump down cycle is identical to a recycling pump down cycle except that once the thermostat has been satisfied and the system has been pumped down, the compressor will not cycle back on until the thermostat calls for another cooling cycle. This holds true regardless of whether there is an increase in the low-side pressure that closes the low-pressure control contacts. The operation of the cycle is illustrated in Figure 27–39 and explained as follows:

1. When the thermostat calls for cooling, terminals R to Y will close, energizing the liquid line solenoid valve and the control relay coil and closing the control relay contacts in the compressor contactor holding coil circuit.

2. When the liquid line solenoid valve opens, system high-side pressure will bleed into the low side of the system, increasing the low-side pressure.

3. At a predetermined pressure, the low-pressure control contacts will close.

4. Current will now flow across the compressor contactor holding coil circuit, energizing the compressor and placing the system into a normal refrigeration cycle.

5. When the thermostat is satisfied, as shown in Figure 27–40, terminals R to Y of the thermostat will open and the liquid line solenoid valve will close, restricting refrigerant flow into the low side of the system. The control relay coil will deenergize, opening the control relay contacts in the compressor contactor holding coil circuit.

The compressor will continue to operate, pumping the refrigerant into the system con-

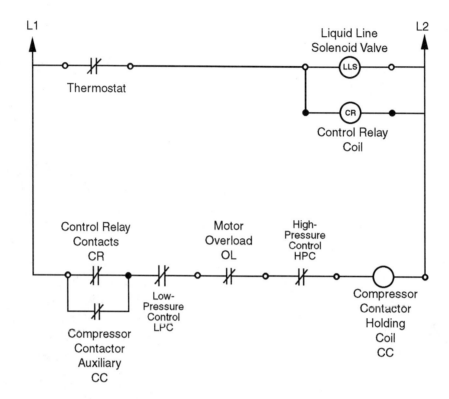

Fig. 27–39   Nonrecycling pump down system in cooling mode.

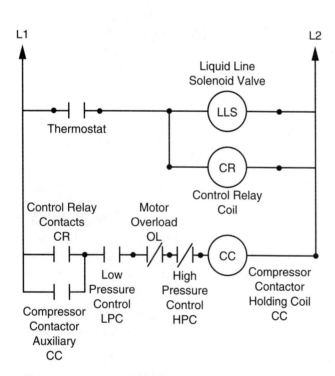

Fig. 27–40   Nonrecycling pump down system in pump down mode.

denser and/or receiver. (Notice that the compressor holding coil circuit is still complete through the closed contactor auxiliary contacts.) When the low-side pressure is reduced to the cut-out setting of the low-pressure control, the control contacts will open, stopping compressor operation. At this time, the compressor contactor auxiliary contacts will also open.

Suction Line Accumulator

The purpose of the **suction line accumulator** is to intercept any liquid refrigerant that may enter the suction line of the system before it reaches the compressor. The accumulator acts as a trap, holding the liquid refrigerant and allowing it to vaporize before it enters the suction line and compressor. Because the accumulator acts as a natural trap to collect oil, some applications utilize a separate oil-return line to meter the oil collected in the accumulator into the suction line at a controlled rate.

A suction line accumulator should be installed in the main suction line as close to the evaporator outlet as possible. In many commercial applications, the liquid line may be used as a heat transfer medium to help the vaporization process of the liquid refrigerant collected in the accumulator. The suction line accumulator should be large enough to hold at least 60 percent of the total refrigerant charge in the system. It should be designed for minimum pressure drop in the suction line and have adequate provisions for oil return.

REFRIGERANT PIPING

The piping design of a refrigeration or air-conditioning system is a critical element in the proper performance of the system. It involves an extremely complex relationship in the flow of refrigerant and refrigerant oil through the various parts of the system. The factors that must be considered in a piping design are the interrelationships between velocity, pressure, friction, and density, as well as the related variables required for proper fluid flow. These variables and relationships can be expressed in long mathematical calculations, performance charts, and pressure-drop tables for the fluid flow through the piping. The improper design and sizing of refrigerant piping in a sealed system may result in a loss of system efficiency that could lead to a loss of overall system capacity as well as eventual failure of components of the system.

The piping design and layout for a particular refrigeration or air-conditioning system is usually specified by an application design engineer. In order to design the refrigerant piping, the engineer must refer to pipe-sizing tables, charts, pressure-drop tables, and mathematical equations. In theory, the piping design and layout completed by the application engineer may be technically and diagrammatically correct. However, there may be no allowances for horizontal distances, vertical lifts, or other unforeseen installation problems. As a result, it is the responsibility of the installation and service technician to interpret the engineer's design and employ the proper modifications during the installation. Therefore, it is more important for the technician to become familiar with

some of the basic rules of proper system piping and layout than with the tables, charts, and equations dealing with system design.

Refrigerant piping serves two purposes:

1. It provides a passageway for the circulation of refrigerant liquid and vapor to the various component parts of the system.

2. It allows for the free travel of oil through the system and its return to the compressor.

A properly designed piping circuit should fulfill both of these requirements with a minimum amount of refrigerant pressure drop in the circuit.

Oil in the System

One of the major concerns in the design and layout of refrigerant piping is to ensure that any oil circulating in the system is returned to the compressor. In all compressors, a certain amount of oil will leave the compressor and circulate through the system piping and component parts. The presence of oil in evaporators and condensers can reduce the efficiency of these heat exchangers by as much as 20 percent. System piping that is not properly designed, sized, or installed can lead to at least one of the following problems:

1. Loss of oil in the compressor crankcase can lead to bearing failure and other internal compressor damage.

2. Liquid refrigerant and large quantities of oil returning to the compressor can result in internal component damage.

3. A loss of system capacity and efficiency that will result in complaints of higher-than-normal temperatures.

A properly designed, sized, and installed refrigerant piping circuit will allow for a minimum amount of refrigerant pressure drop through the various parts of the components and piping of the system, and a free return of oil circulating in the system to the compressor. Because there is no way to design a compressor in which none of the oil escapes into the system piping, the design of the refrigerant piping system must allow for the amount of oil leaving the compressor to always equal the amount of oil returning to the compressor.

Piping Guidelines

There are several basic rules that apply to all piping installations that will ensure safe and efficient system operation:

1. Properly size each section of the sealed system to ensure a minimal amount of refrigerant pressure drop as well as proper oil return.

2. Free all tubing, fittings, and parts of dirt, sludge, and moisture to prevent sealed system contamination.

3. Use as few fittings and return bends as possible to reduce not only the chance of refrigerant leaks but also the amount of pressure drop in the system.

4. Make all mechanical and solder connections in accordance with the manufacturer's specifications.

5. Pitch all refrigerant lines in the direction of the refrigerant flow. The amount of pitch should be one-half inch or more for each 10 feet of line run.

6. Avoid overheating of the tubing during soldering or brazing operations to prevent the interior of the tubing from oxidizing and building up carbon. To prevent this, purge the circuit with nitrogen during soldering and brazing operations.

7. Use a minimal amount of solder or brazing alloy to complete each connection.

8. Don't open the refrigerant tubing until it is ready to be installed. It should be capped when not in use.

9. Install caps at the ends of tubing that is installed through the walls.

Hot Gas Discharge Lines

Hot gas discharge lines on condensing units fabricated at the factory are preformed and connected and do not present any problem to the installation technicians. On field-installed systems with remote condensers, the size of discharge lines must allow for a minimum velocity of refrigerant flow to be maintained in order to ensure proper oil circulation and system performance. Pressure drop in the discharge line is perhaps the least critical of any other refrigerant circuit in the system.

When installing remote hot gas discharge lines, the line should be pitched toward the direction of flow, one-half inch for each 10 feet of tubing run. When the condenser is located less than eight feet above the compressor, the discharge line from the compressor should be pitched toward the direction of flow. When the condenser is located more than eight feet above the compressor, as shown in Figure 27–41, a "P" trap at the base of the rise may be used to prevent liquid refrigerant from migrating to the compressor and accumulating on the off cycle.

In applications where the compressor may be located in a cooler ambient temperature than the discharge line or condenser, there is a possibility that liquid refrigerant will migrate to the head of the compressor. To prevent this, a check valve may be installed in the discharge line near the condenser.

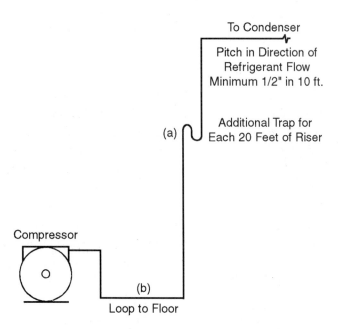

Fig. 27–41 Alternate methods of piping discharge line risers (a) loop to floor, (b) "P" trap. *Courtesy Copeland Corporation.*

In applications where the velocity of the hot gas in the discharge line is not sufficient enough to carry the oil up the riser, an oil separator may be installed to minimize the amount of oil circulating through the system.

Liquid Lines

The purpose of a liquid line is to deliver a solid column of liquid refrigerant to the system metering device. Because the liquid refrigerant and circulating oil will mix readily, velocity is not a primary concern for oil circulation in the liquid line. The primary concern in liquid line piping is to maintain a sufficient amount of pressure to prevent the liquid refrigerant from falling below its saturation temperature. If this occurs, some of the liquid in the line will flash off or vaporize, cooling the remaining liquid to its new saturation temperature. The liquid line will now contain a mixture of liquid refrigerant and vapor, which will increase pressure drop in the line due to friction, and reduce the capacity of the metering device as well as the system.

Pressure drop in the liquid line can be the result of long vertical risers and accessory devices, such as filter-driers, sight glasses, solenoid valves, and strainers. The total pressure drop in the liquid line is the sum of the pressure drop across each of these devices that may be installed in the line and the pressure drop due to friction and vertical lift. In most applications a pressure drop of 3 to 5 PSI should not be exceeded.

To prevent the occurrence of flash gas in the liquid line and to ensure proper system performance, it is essential that the liquid refrigerant entering the metering device be cooled slightly below its saturation temperature. This is normally accomplished in the lower tubes of the condenser and/or receiver as the liquid refrigerant enters the liquid line. In most applications a subcooling of 10°F is sufficient to provide for normal system pressure drops. In some instances, however, additional subcooling may be required. In these cases, heat exchangers or other special means should be used to provide the additional subcooling of the liquid refrigerant.

Suction Line

The sizing and piping of the suction line is the most critical element for proper and efficient system operation. The two most important elements of suction line piping are pressure drop and velocity.

Pressure Drop

Any pressure drop experienced in the suction line will result in a decrease in velocity and pressure of the refrigerant entering the compressor suction line. As the pressure of the suction vapor decreases, it will occupy a greater volume, resulting in a decrease in the weight of refrigerant pumped by the compressor and a loss of compressor capacity. Therefore, it is important that the suction line be sized for a minimal pressure drop.

The standard practice is to size the suction line for a pressure drop equivalent to a change in saturation temperature of 2°F or 1 PSIG. Generally, the maximum pressure drop in the suction line should not exceed 2 PSIG.

Velocity

The purpose of the suction line is to carry refrigerant vapor and oil from the evaporator to the compressor. Oil in the evaporator and suction line will return by gravity to the compressor by means of a vapor in motion. The movement of the oil in the suction line is dependent upon the mass of the refrigerant as well as its velocity. As the mass, or density, of the refrigerant in the suction line decreases, higher velocities of refrigerant vapor are required to move the oil. Minimum refrigerant velocities of 700 FPM for horizontal lines and 1,500 FPM for vertical suction lines have been the standards successfully employed by manufacturers over the years.

Suction Line Piping

When the compressor is located above the evaporator, there are no exceptional piping configurations that need to be considered. It is important, however, that the suction line be slightly

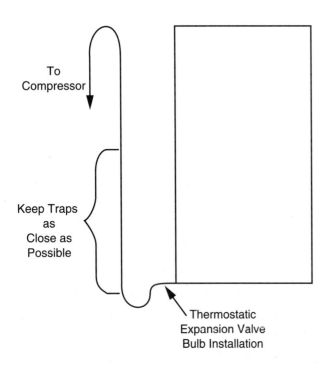

Fig. 27–42   Installation of "P" trap and vertical riser to help prevent liquid refrigerant migration.

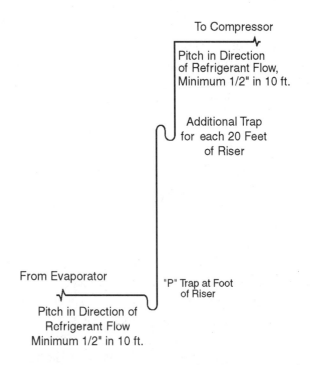

Fig. 27–43   Piping of suction line where vertical riser exceeds 20 feet.   *Courtesy Copeland Corporation.*

pitched toward the compressor to help ensure the return of oil by gravity.

When the compressor is located either at the same level or below the evaporator, as shown in Figure 27–42, a vertical riser should be installed at the outlet of the evaporator and piped to the top of the evaporator before returning to the compressor. This will help prevent the possibility of liquid refrigerant migration from the evaporator to the compressor. The "P" trap located at the bottom of the riser will facilitate the movement of oil up the riser.

In installations where the vertical riser exceeds 20 feet, as Figure 27–43 illustrates, a "P" trap should be installed every 20 feet to allow for oil return in smaller stages. The trap should be as short as possible to prevent excessive storage.

Figure 27–44 represents suction line piping with multiple evaporator installations.

DEFROSTING SYSTEMS

In commercial equipment applications, many evaporators operate at temperatures below freezing and, after prolonged periods of operation, an accumulation of frost or ice on the evaporator coil will likely develop. Frost and ice usually contain trapped air, which will produce the same effect as air trapped in an insulating material. In fact, one inch of frost can have the same insulating effect as up to one-half inch of cork. This condition can greatly reduce the heat transfer efficiency of the evaporator as well as overall system efficiency. Possible damage to the compressor may also result due to liquid refrigerant floodback. For these reasons, the evaporator coil should be defrosted periodically to remove accumulations of frost and ice. In low-temperature applications where the fin spacing of the evaporator is small, more frequent defrosting is required. In all applications the evaporator coil should be defrosted with a minimal rise in the fixture temperature.

Defrosting Methods

The methods used to **defrost** an evaporator coil include off cycle defrosting, hot gas defrosting, electric defrosting, water defrosting, warm

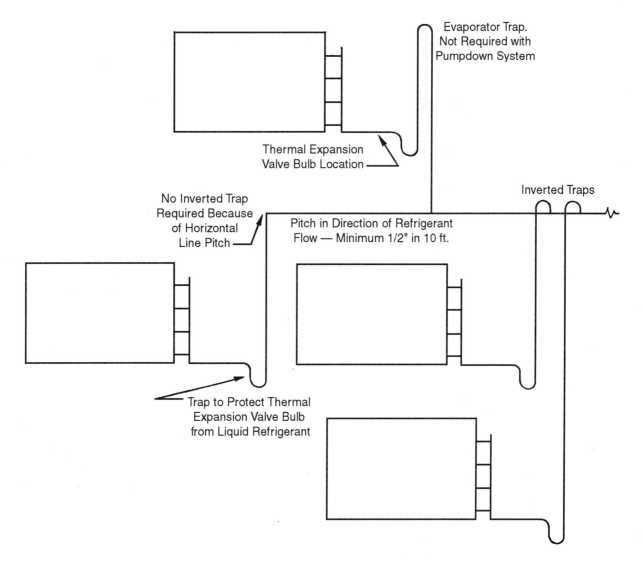

Fig. 27–44   Typical suction line piping with multiple evaporators.   *Courtesy Copeland Corporation.*

air defrosting, reverse cycle defrosting, and use of a nonfreezing solution.

Off Cycle Defrosting

In medium temperature applications, the evaporator coil will normally operate at below freezing temperatures and a certain amount of frost or ice will accumulate on the coil surfaces. In an **off cycle defrost** system, the temperature of the air in the conditioned space will rise above the freezing point during a compressor off cycle and can be used to defrost the evaporator coil. If fans are used in these applications, they will remain running to allow the air to circulate across the evaporator coil and facilitate defrosting.

Hot Gas Defrosting

The basic hot gas defrosting system, which is illustrated in Figure 27–45, utilizes hot gas from the compressor discharge line piped directly to the evaporator after the metering device to warm and defrost the evaporator coil. A solenoid valve in Figure 27–46, activated by a clock-timer, will open to allow the hot discharge refrigerant vapor to enter the evaporator through the bypass line. While this basic system may be acceptable in low-capacity refrigeration applications, hot gas defrosting in larger commercial applications is not satisfactory without the addition of accessory devices and specifically designed system piping arrangements.

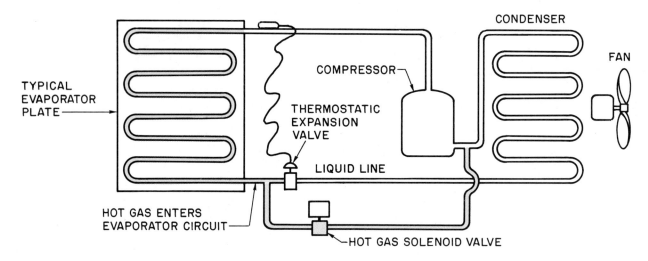

Fig. 27–45 Schematic drawing of simple hot gas defrosting system.

Fig. 27–46 Solenoid valve activated by a clock-timer opens to allow hot discharge refrigerant vapor to enter the evaporator through the bypass line. *Courtesy Parker-Hannifin Corporation.*

One of the problems with a hot gas defrosting system is that upon entering the cold evaporator, the hot gas from the compressor discharge line will condense and flood the evaporator with liquid refrigerant. This may result in liquid floodback to the compressor during the defrosting cycle or after the defrost cycle is terminated and the system has returned to normal operation.

Several methods can prevent liquid floodback in a hot gas defrosting system. In some applications, a suction line accumulator may be used to collect the liquid refrigerant as it leaves the evaporator and vaporize it before it reaches the compressor. Other applications may use a suction line, liquid line heat exchanger arrangement. Reevaporation of the condensed liquid refrigerant is another method used to prevent liquid floodback.

One method of reevaporation is the thermobank, shown in Figure 27–48, which is a combination of a heat bank and liquid reevaporator. During normal operation, hot gas from the compressor discharge valve will flow through a heat exchanger coil immersed in a tank of water. The hot gas will heat the water to form a heat bank. When the defrost cycle is initiated, the defrost solenoid valve will open, and the suction line solenoid valve will close. Liquid refrigerant leaving the evaporator will be diverted through the reevaporation coil in the heat bank. The warm water in the heat bank will evaporate the liquid refrigerant, allowing only the

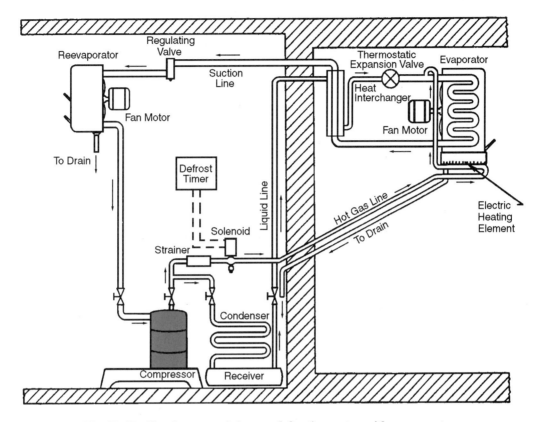

Fig. 27–47   Simple automatic hot gas defrosting system with reevaporator.

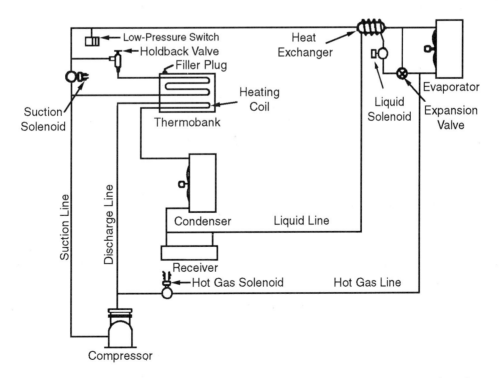

Fig. 27–48   Thermal bank system to reevaporate liquid refrigerant that may have been condensed during a hot gas defrost cycle.

vapor to return to the compressor. A crankcase pressure-regulator valve is installed in the suction line before the compressor to prevent an excessive rise of suction pressure. During the defrosting cycle some ice will accumulate on the reevaporator coil but will melt during the next running cycle of the system.

Electric Defrosting

Electric defrosting is a popular method used to defrost low-temperature coils. In this method, a resistance heating element is mounted in close thermal contact with the evaporator coil and will defrost the coil when energized. Figure 27–49 illustrates electric defrosting. While the control of

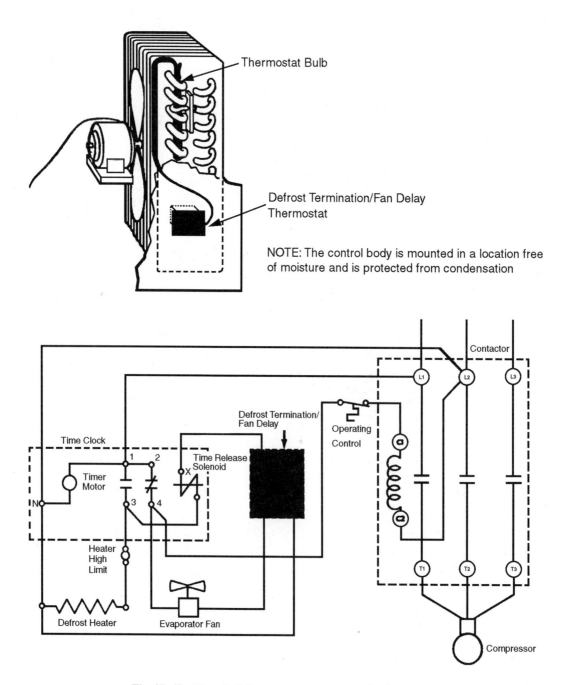

Fig. 27–49 Electric defrost system. *Courtesy Ranco Controls.*

electric defrosting systems is simple and sure, the operating costs are far greater than any other method.

Water Defrosting

A water defrosting system, Figure 27–50, utilizes tap water that is sprayed over the evaporator coil or is fed into a pan located over the evaporator. The pan has holes in its base that allow the water to flow evenly over the evaporator surface. The tap water is warm enough to melt any ice or frost accumulation on the evaporator coil. The melted ice or frost will drain into the evaporator drain pan. With this method, the evaporator is usually equipped with louvers that close during the defrosting cycle to prevent water from being introduced into the conditioned space. When the defrosting cycle is complete, a sufficient amount of time should be allowed for the water to drain completely from the evaporator and the drain pan to prevent the water from refreezing in the system. This is usually accomplished through a time delay programming of the defrost clock-timer.

Defrosting with a Nonfreezing Solution

The nonfreezing solution defrosting system, Figure 27–51, utilizes a brine solution that is stored in a container. During the normal refrigeration cycle, the solution is heated by hot discharge gas that is circulated through the brine solution container before it enters the condenser. The brine solution may also be heated by electrical heating elements. When the defrosting cycle is initiated, a liquid line solenoid valve will close to allow the system to pump down, and evaporator fans will normally be deenergized. The brine solution will be circulated through its own piping along the drain line and the evaporator coil and then returned to its storage container.

Warm Air Defrosting

Warm air defrosting is utilized where conditioned air is at a temperature high enough to allow the evaporator coil to defrost. In some installations, outside air introduced by means of a duct and fan system is used to circulate the warm air over the evaporator coil for defrosting purposes.

Reverse Cycle Defrosting

In the application of reverse cycle defrosting, Figures 27–52 and 27–53, the flow of refrigerant through the system is reversed by means of a reversing valve. As a result, the evaporator becomes the condenser and the condenser becomes the evaporator. When the evaporator functions as the condenser, the hot gas will melt any ice or frost that has accumulated on the evaporator coil. When the defrosting cycle is initiated, hot discharge gas from the compressor will be introduced into the suction line of the system where it will be directed to the evaporator. As the hot gas passes through the evaporator coil, the coil will defrost.

DEFROST CLOCK-TIMERS

In the application of automatic defrosting systems, an automatic **clock-timer** is utilized to ini-

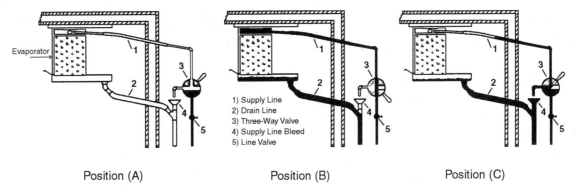

1) Supply Line
2) Drain Line
3) Three-Way Valve
4) Supply Line Bleed
5) Line Valve

Position (A)　　　　Position (B)　　　　Position (C)

Fig. 27–50 This system uses a water spray to accomplish the defrosting process.

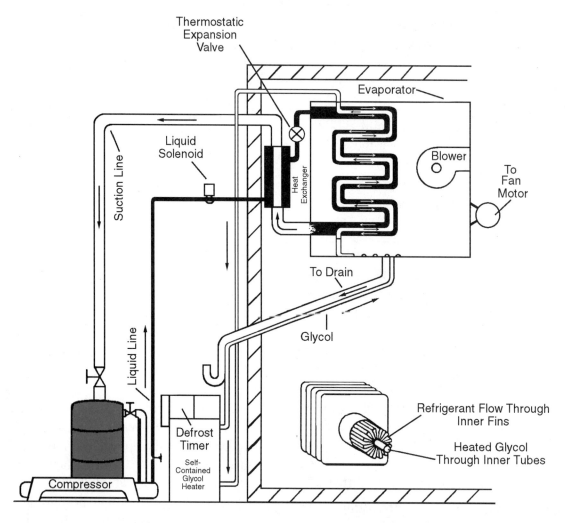

Fig. 27–51 A glycol solution is pumped through the inner tubing of the evaporator and along the drain system to prevent freezing.

tiate the defrosting cycle. The clock-timer is equipped with a cam assembly rotated by the clock mechanism. The cam assembly makes or breaks switches that control the operation of the compressor, fan motors, solenoid valves, defrost heaters, drain heaters, and other accessory devices that comprise the defrosting system.

Figure 27–54 illustrates a typical defrost timer. Depending on the application of the defrosting system, there are several variables that must be considered in the function of the system components. For example, in hot gas defrosting systems, the following events are controlled by the clock-timer:

1. Opening the hot gas solenoid valve

2. Stopping the evaporator fans

3. Continuous running of the compressor

4. Energizing the drain pan heaters

5. Programming the maximum defrosting time and termination of the cycle

6. Energizing the suction solenoid valves in a reevaporation

In some applications the condenser fans may be cycled off during the defrosting cycle to ensure that the hot gas entering the evaporator is at a maximum temperature.

In electric defrosting systems, the following events are controlled by the clock-timer:

1. Stopping the compressor (In some applications the system may be pumped down.)

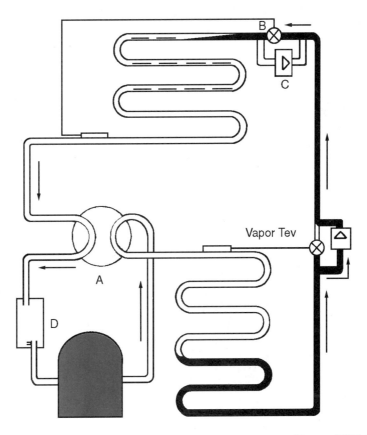

Fig. 27–52 The refrigeration mode of a reverse cycle defrost system. **(a) 4-way valve, (b) extra TEV, (c) check valves (two), (d) suction line accumulator.**

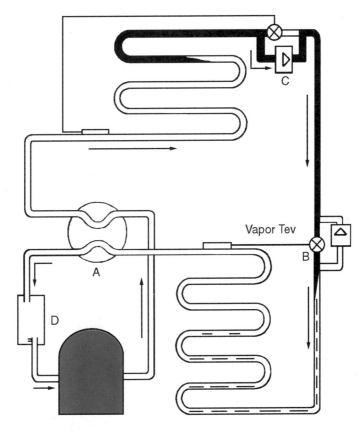

Fig. 27–53 The defrost mode of a reverse cycle defrost system. **(a) 4-way valve, (b) extra TEV, (c) check valves (two), (d) suction line accumulator.**

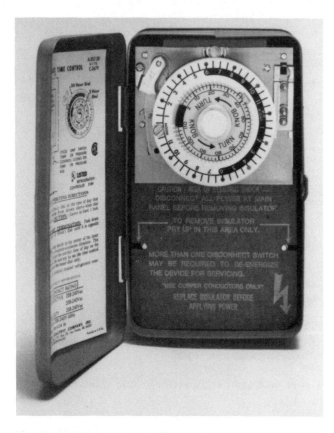

Fig. 27–54   Most defrost timers used in commercial HVAC/R equipment have wiring diagrams and installation instructions printed on the inside of the cover.

2. Stopping the evaporator fans

3. Energizing the electric heaters

4. Energizing the drain pan heaters

Defrost timers also include auxiliary switches that delay the operation of evaporator fans and drain pan heaters until after the defrosting cycle has been completed. This allows the evaporator temperature to decrease so that no warm air is circulated into the conditioned space. A time delay also allows moisture that has collected on the evaporator during the defrosting cycle to drain completely so that it isn't circulated into the conditioned space. The time delay allows all moisture to drain completely from the drain heaters, preventing a freeze-up of condensate water in the drain pan or drain tubes.

Solid-state programmable clock-timers are also in popular use. In many applications, the defrosting cycle is initiated by a thermistor that

senses the temperature difference of the evaporator airflow. In these applications the defrosting cycle is controlled by demand. As the temperature difference across the evaporator coil increases, frost and ice will accumulate on the coil. At a predetermined temperature difference, usually around 20°F to 30°F, the defrosting cycle will be initiated. The defrosting cycle may be terminated by temperature or pressure after the evaporator coil has warmed sufficiently to melt the frost and ice accumulation, usually at a temperature of around 40°F.

Figure 27–55 represents the various types of defrost wiring circuits.

Several methods are commonly used to control the initiation and termination of the defrosting cycle:

1. time initiated, time terminated method

2. time initiated, temperature terminated method

3. time initiated, pressure terminated method

4. airflow method

In applications where temperature or pressure termination is used, an overriding time termination feature is provided in case the defrosting cycle is prolonged for an abnormal length of time. Defrost clock-timers are available for 24-hour and 7-day cycles. The intervals between the initiation and termination of the cycle can be adjusted to suit the particular needs of a system.

Time Initiated, Time Terminated Method

The time initiated, time terminated method shuts down the compressor for a period of time to allow the evaporator to defrost. The evaporator fans usually operate to circulate air across the evaporator coil and provide a means of defrosting. After a predetermined period of time, the clock-timer will terminate the defrosting cycle and place the system back into normal operation.

Time Initiated, Temperature Terminated Method

In this defrosting application the clock-timer initiates the defrosting cycle. When the temperature of the evaporator coil (which is determined

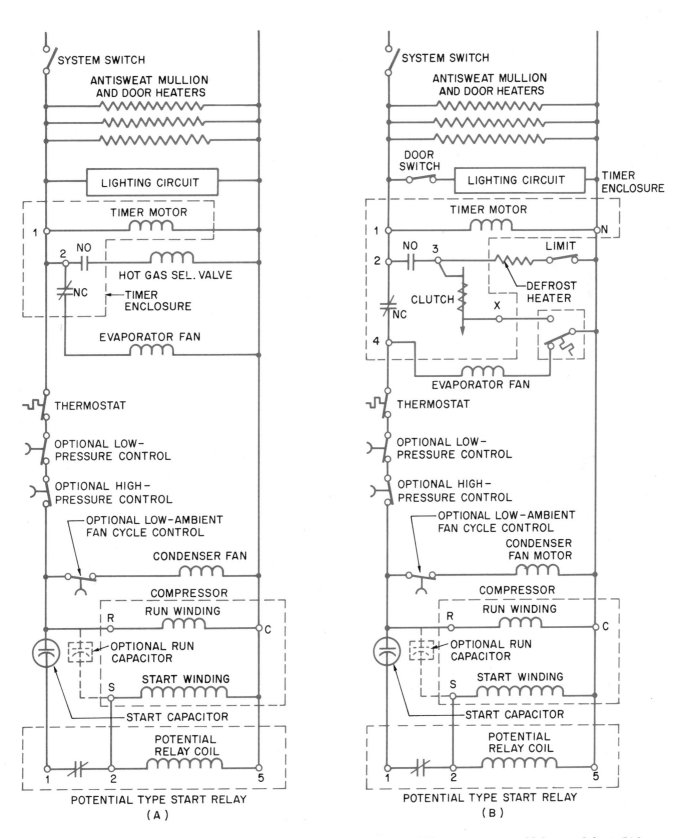

Fig. 27–55 Wiring diagrams for several types of defrost control arrangements (a) low temperature with hot gas defrost, (b) low temperature with electric defrost. *(continued)*

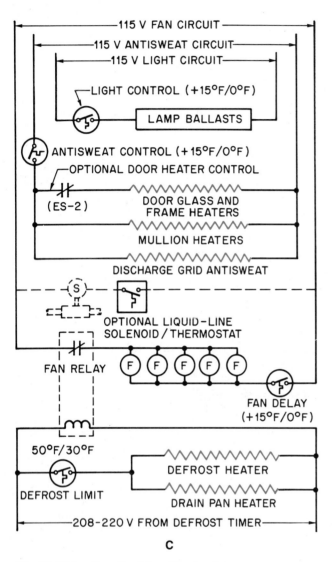

Fig. 27–55 *(continued)*   **(c) mullion heaters.**

by a sensing device) is above the freezing point, the defrosting cycle will be terminated.

Time Initiated, Pressure Terminated Method

The clock-timer initiates the defrosting cycle in this method. As the evaporator warms up during the defrosting period, the system low-side pressure will start to increase. When the pressure rises to a point where the temperature of the evaporator is above the freezing point, a low-pressure control will terminate the defrosting cycle.

Airflow Method

This method utilizes an air switch to initiate the defrosting cycle. As frost and ice accumulate on the evaporator coil, the amount of air passing across the coil as well as the pressure of the air will be reduced. At a predetermined point the air switch will initiate the defrosting cycle. The defrosting cycle can be terminated by pressure or temperature.

Unit Twenty-Seven Summary

- Commercial refrigeration cabinets are designed to meet a wide variety of needs and applications for use in restaurants, grocery stores, and taverns. Commercial refrigeration units range in size from fractional horsepower, self-contained units to high capacity systems of over 100 ton capability.

- Two-stage systems use either two compressors piped in series or a multiple-cylinder compressor.

- Low ambient operation has a dramatic effect on the capacity of air-cooled condensers and a variety of methods may be employed to maintain minimum required discharge pressures.

- Crankcase pressure-regulating valves (CPR) prevent the compressor suction pressure from going too high when a system starts up under a heavy load. Evaporator pressure regulators (EPR) prevent evaporator pressures from going too low.

- Check valves must be installed in low-temperature systems with multiple evaporator applications. Surge tanks prevent short cycling of the compressor on a two-temperature system.

- Oil-pressure safety switches protect the compressor in the event of a drop in oil pressure. Net oil pressure is the difference between suction pressure and oil pump pressure.

- Hot gas bypass systems prevent evaporator pressures from dropping too low and affecting the efficiency of a commercial refrigeration system. De-superheating expansion valves are used to cool hot gas introduced into the suction line.

- To prevent the migration of liquid refrigerant into the compressor crankcase, a pump down system is used on commercial refrigeration systems.

- Liquid slugging occurs when liquid refrigerant passes through the compressor intake valves. Liquid flooding occurs when liquid refrigerant enters the compressor crankcase.

- System piping on a commercial refrigeration system must provide for both refrigerant circulation and oil return to the compressor. Velocity of refrigerant and pressure drop are controlled by following proper piping procedures.

- Defrost systems are used on refrigeration systems to keep the evaporator clear and allow for maximum airflow and heat transfer, ensuring optimum performance of the equipment.

Unit Twenty-Seven Key Terms

Display Case

Open Display Case

Closed Display Case

Reach-in Case

Double-Duty Display Case

Floral Cabinet

Frozen-Food Display Case

Walk-In Unit

Crankcase Pressure-Regulator (CPR) Valves

Evaporator Pressure Regulator (EPR)

Surge Tank

Oil-Pressure Safety Switch

Low-Pressure Control

Hot Gas Bypass System

De-Superheating Expansion Valve

Migration

Liquid Slugging

Liquid Flooding

Crankcase Heater

Pump Down Cycle

Suction Line Accumulator

Defrost

Off Cycle Defrosting

Clock-Timer

UNIT TWENTY-SEVEN REVIEW

1. Which cabinet design has a lower cabinet to store merchandise not ready for display?

2. What is the primary problem with open display cases?

3. Describe the application of the closed display case.

4. Describe the application of the double-duty display case.

5. What is the primary disadvantage of an open frozen food case?

6. Match the following terms with their descriptions: mullion heaters, crankcase heater, evaporator pressure regulator, gauge port, suction line accumulator, surge tank, hot gas bypass valve, low-pressure controls, low evaporator heat load, hot gas bypass.

   a. Gives the service technician access to monitor evaporator pressure and adjust the valve.

   b. Slows down the rapid pressure buildup in the suction line.

    c. The suction pressure of a system is reduced.

    d. Prevents evaporator pressure from decreasing below a predetermined point.

    e. Prevents condensation around doors.

    f. Utilized to control the temperature of the conditioned space by indirectly controlling the temperature of the evaporator.

    g. Brings up the suction pressure.

    h. Brings hot gas into the low side of a system, creating a false heat load on the evaporator.

    i. Prevents refrigerant vapor from entering and condensing in the compressor crankcase.

    j. Intercepts any liquid refrigerant that may enter the suction line.

7. What is the overriding factor in the choice of multiplex systems?

8. Describe a multiplex refrigeration system.

9. In the application of a multiplex system, how is individual temperature control of each fixture maintained?

10. Starved or oil-logged evaporator coil, short cycling of the low-pressure control, and reduction of system capacity are operating problems associated with too _____ a discharge pressure.

11. What are the methods employed to maintain minimum discharge pressures?

12. What is the function of the de-superheating expansion valve in the two-stage compression systems?

13. How does a closed pressure-regulating valve installed on the outlet of the receiver affect the condensing process?

14. What is happening in the crankcase pressure regulator when suction pressure is below the control setting?

15. What is the purpose of a crankcase pressure-regulator valve?

16. Describe the procedure for adjusting a crankcase pressure-regulator valve.

17. How is a high current draw on the compressor motor during start-up or after the defrosting cycle related to the crankcase pressure regulator?

18. What is the relationship of speed to the compressor?

19. What is the difference between liquid slugging and migration, particularly when each occurs?

20. List four factors that must be considered in piping design.

21. Which of the following are not basic rules in piping installation?

    a. Use as few fittings and return bends as possible.

    b. Use a minimal amount of solder to complete each connection.

    c. To prevent oxidation and the buildup of carbon, purge with nitrogen during soldering and brazing operations.

22. Why are defrosting systems necessary?

23. In the application of multiple evaporators, what should be the maximum system load percentage of a controlled evaporator?

24. What purpose does a check valve serve in multiple evaporator applications that use an EPR?

25. What is the purpose of an oil safety switch?

26. How is net oil pressure determined?

27. Describe the procedure used to determine the operating temperature difference of an evaporator coil.

28. What factors may affect the evaporator heat load?

29. What system problems may result from a low evaporator heat load?

30. What methods may be used to control compressor capacity under low load conditions?

31. What methods are used to vary compressor speed?

32. What is liquid refrigerant migration?

33. What service problems can result from liquid refrigerant migration?

34. What are the causes of liquid flooding?

35. What methods can be used to prevent liquid refrigerant problems in a system?

36. What are the two types of crankcase heaters in common use?

37. What is the purpose of a pump down cycle?

38. Describe the difference between the recycling pump down system and the non-recycling pump down system.

39. What is the advantage of the nonrecycling pump down system over the recycling pump down system?

40. In the application of pump down systems, what device controls the cycling of the compressor?

41. What is the purpose of a "P" trap in the discharge line when the condenser is located above the compressor?

42. What is the accepted pressure drop in the sizing of liquid lines?

43. What is the maximum allowable pressure drop in the sizing of suction lines?

44. What methods are used to defrost an evaporator coil?

45. What is the purpose of reevaporation in a hot gas defrosting system?

46. What is the purpose of cycling the condenser fan motor off during a hot gas defrosting cycle?

47. Describe the operation of an electric defrosting system.

48. What methods are used to initiate and terminate a defrosting cycle?

49. With belt-driven compressors, the speed or RPM (revolutions per minute) can be changed. If the speed is changed, the capacity and RPM must stay within the manufacturer's recommendations.

The formula for determining a compressor's RPM:

$$\text{compressor RPM} = \text{motor RPM} \times \frac{\text{drive diameter (motor pulley)}}{\text{driven diameter (compressor pulley)}}$$

Find the compressor's RPM.

a. motor 3,450 RPM

motor pulley 2" diameter

compressor pulley 8" diameter

b. motor 1,725 RPM

motor pulley 5" diameter

compressor pulley 15" diameter

c. If the compressor's RPM must be reduced to 500 in part b, what size driven pulley could be used?

50. Write an explanation of the concerns about oil in the system.

UNIT TWENTY-EIGHT

Commercial HVAC/R Servicing

OBJECTIVES

After completing this unit, the student will be able to:

1. Describe basic service procedures related to commercial refrigeration systems.

2. Explain the processes of evacuating and dehydrating a system.

3. Describe the proper procedure to follow when servicing a contaminated refrigeration system.

4. Describe the proper procedure to follow when replacing a compressor in a burnout situation.

As an HVAC/R service technician it's important to understand the proper procedures to follow in servicing commercial refrigeration systems, both for personal safety and equipment protection. In this unit, we'll discuss the steps to take when charging commercial refrigeration systems, evaluate a system through acid tests, and then outline the proper procedures to follow when dealing with a contaminated system.

CHARGING THE SYSTEM

Every refrigeration and air-conditioning system is designed to produce a rated cooling effect under a given set of conditions and a specific refrigerant charge. If the system has an undercharge of refrigerant, the evaporator coil will be starved, resulting in excessively low suction pressures and a reduction in system capacity. Because modern hermetic compressors are dependent upon the return of cool suction vapor to cool motor windings, an undercharged system could cause the compressor motor windings to overheat, causing motor failure. Overcharging the system will result in higher than normal discharge pres-

sures and the possible floodback of liquid refrigerant to the compressor.

Checking the Refrigerant Charge

Most commercial refrigeration and air-conditioning systems allow for a slight variation in the amount of refrigerant charge and are not considered to be critically charged as are many small unitary systems such as those employing a capillary tube as the metering device.

In the application of commercial equipment, several methods may be employed to check the refrigerant charge. The method used depends on the application of the equipment and the facilities provided on the installation for checking the charge.

Sight Glass

The **sight glass** in Figure 28–1 is the most common method used to check the refrigerant charge of commercial equipment. When checking the refrigerant charge on a system equipped with a sight glass, start the system and allow it to operate until the system pressures have stabilized.

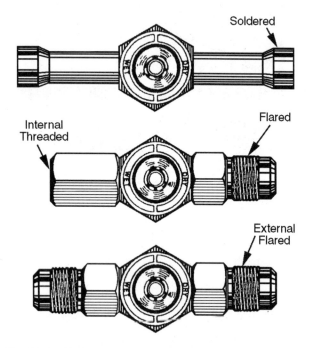

Fig. 28–1 Various types of sight glasses with different methods of installation. *Courtesy Virginia KMP Corp.*

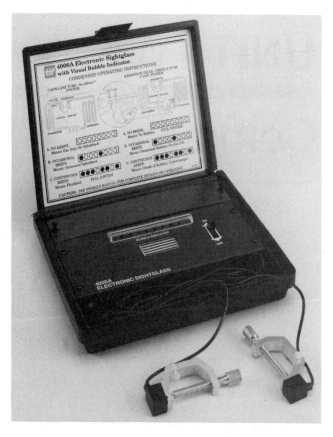

Fig. 28–2 When bubbles are present in an HVAC/R system liquid line, an electronic sight glass generates a tone. *Courtesy TIF Instruments, Inc.*

Observe the sight glass. A solid column of vapor-free liquid refrigerant should be flowing through the sight glass. If bubbles appear, this is an indicator of flash gas in the liquid line, which could be due to a refrigerant undercharge or some other problem in the system that is preventing the refrigerant from fully condensing.

The sight glass should be located just ahead of the metering device to allow the service technician to check and ensure that a solid column of liquid refrigerant is being introduced to the metering device. In remote condensing unit installations, however, the sight glass is usually installed at the condensing unit for convenience of service.

Electronic Sight Glass

An **electronic sight glass**, such as the one illustrated in Figure 28–2, is effective in checking the refrigerant charge of the capillary tube and thermostatic expansion valve as well as the accurator metering devices used on some systems. On systems equipped with expansion valves, the probes of the instrument are installed on the liquid line and an audible sound is emitted when the liquid line is

filled with liquid refrigerant, thus indicating that the system is properly charged. On accurator and capillary tube systems, the sensing probes are installed on the suction line and an audible sound is emitted when the system is properly charged.

Liquid Level Test Valve

Liquid level test valves, as illustrated in Figure 28–3, are installed on refrigerant lines, liquid receivers, or condensers. When the test valve is opened, a small amount of refrigerant is allowed to escape. If the refrigerant escapes as a vapor, the system is undercharged; if it escapes as a liquid, the system is overcharged. If the refrigerant escapes as both a liquid and a vapor, the refrigerant and the valve are at the same level, and the system is properly charged.

Always refer to the manufacturer's directions and specifications when servicing equipment with

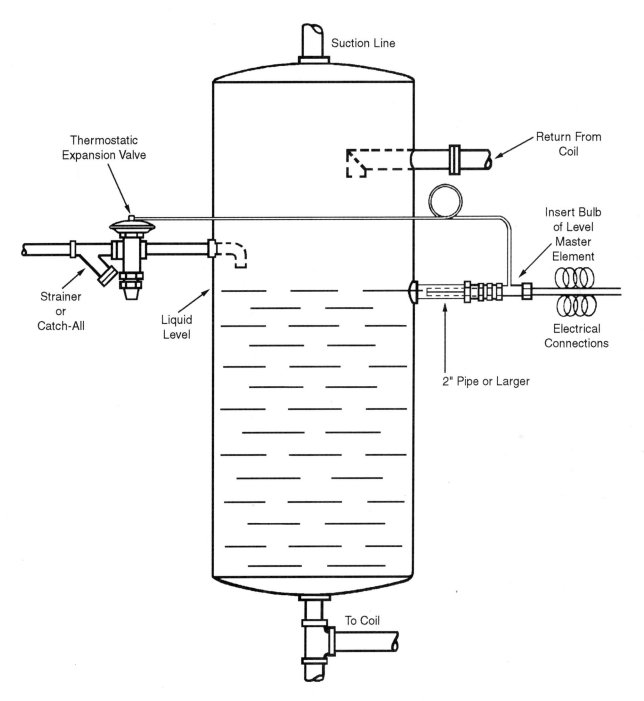

Suction Line

Thermostatic
Expansion Valve

Return From
Coil

Insert Bulb
of Level
Master
Element

Strainer
or
Catch-All

Liquid
Level

Electrical
Connections

2" Pipe or Larger

To Coil

Fig. 28–3   Liquid level test valve.   *Courtesy Sporlan Valve Company.*

liquid level test valves. On some large commercial equipment, the condenser or receiver may be equipped with a sight glass or float indicator that will indicate when the receiver or condenser is flooded with liquid refrigerant and when the system is properly charged.

Pressure and Temperature Charging Charts

Some manufacturers provide charging charts to determine the proper operating charge of a system. The chart shows in graph form the normal system suction and discharge operating pressures for various condenser and evaporator air temperatures under various ambient conditions. When using these charts, allow the system to operate until the system pressures have stabilized. Check the suction and discharge pressures, compare them to the charging chart, and add or remove refrigerant as required.

Charging by Weight

Charging a system by refrigerant weight is the most accurate method to ensure that the system is properly charged. This method is used when the entire system charge has to be replaced and when the exact amount of refrigerant required for the system is known. When this method is used, the exact amount of refrigerant required for the system is indicated on the unit's nameplate or in the manufacturer's installation guide. The refrigerant may be weighed on a scale or measured on a charging column.

Charging by Liquid

In most commercial applications where a large amount of refrigerant is required by the system, liquid charging is usually employed. This is accomplished through a charging or access valve installed in the liquid line, or a receiver outlet valve equipped with a charging port and service valve. While this procedure will vary from one application to the other depending on what means have been provided during the initial installation, the liquid refrigerant must never be introduced into the low side of the system. This could cause damage to the compressor or compressor valves.

On new installations, the system should be thoroughly evacuated and under a deep vacuum. To charge the system, refer to Figure 28–4 and the following steps:

1. Install the charging cylinder to the liquid line charging valve.

2. Purge the charging line of air.

3. Open the charging cylinder liquid valve and the liquid line charging valve. The vacuum in

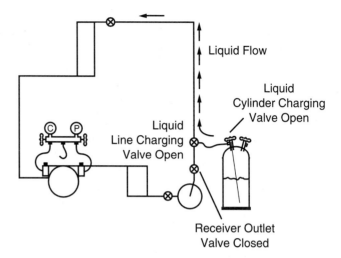

Fig. 28–4 Liquid charging utilizing system liquid line charging valve.

the system will draw the liquid refrigerant from the charging cylinder and introduce it into the system. Refrigerant will continue to flow from the charging cylinder into the system until the system and cylinder pressures have equalized.

4. After the system and cylinder pressures have equalized, close the liquid line charging valve.

5. Start the compressor and allow the system pressures to stabilize. Observe the suction and discharge pressures.

6. If the system requires more refrigerant, close the receiver outlet valve or the liquid line charging valve.

7. Open the liquid line charging valve. Liquid refrigerant will now flow from the charging cylinder, which is not acting as the system receiver, into the system. To determine if the system charge has reached its required amount, close the liquid charging valve and open the receiver outlet valve. Observe system pressures and repeat the charging process if necessary.

Charging by Vapor

Vapor charging is normally done when a small amount of refrigerant charge is required by the system. During this charging process, the refrigerant service cylinder is placed in an upright position and, if applicable, a vapor charging valve is used to ensure that only refrigerant vapor is introduced into the low side of the system. Manifold gauges are connected to the compressor or system service valves to monitor system suction and discharge pressures.

While this process is slower than the liquid charging method, a more precise control over the refrigerant being introduced into the system can be maintained. The rate of refrigerant flow into the system is governed by the gauge manifold. During the charging process, the suction pressure should not exceed the normal operating pressures of the compressor. Observing system operating pressures and controlling the rate of refrigerant flow into the system through the gauge manifold will prevent compressor overloading and system overcharge.

The vaporization of the liquid refrigerant in the refrigerant service cylinder will cool the remaining refrigerant in the cylinder and reduce the cylinder pressure. If this becomes a problem, fresh cylinders can be alternated or the service cylinder can be heated with warm water or a heat lamp. Under no conditions should the service cylinder be heated by a direct flame.

Evacuating the System

The purpose of evacuating the system is to reduce system pressure to a point where any moisture contamination will begin to boil or vaporize at normal atmospheric conditions and can be removed from the system with a vacuum pump. In order to accomplish this effectively, the system pressure must be reduced to an extremely deep vacuum.

Given the direct relationship between atmospheric pressure and a perfect vacuum, extremely low vacuum readings are read in microns. There are 25,400 microns to 1 inch of mercury. Figure 28–5 illustrates the comparisons of temperature, absolute pressure, inches of mercury, and the micron system of measurement.

Because standard refrigeration gauges are not capable of measuring deep vacuums accurately,

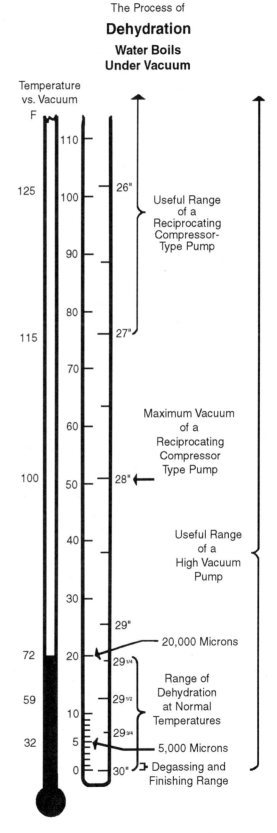

Fig. 28–5   Comparison of temperature, absolute pressure, inches of mercury, and micron system of measurement.
Courtesy Robinair Division, SPX Corporation.

electronic thermistor vacuum gauges are used. These instruments are capable of accurately monitoring and measuring system vacuums in the low micron ranges where moisture removal is most effective.

A vacuum pump should be capable of reducing system pressure to at least 500 microns. Because practical moisture removal occurs at 500 to 1,500 microns, the vacuum pump must be able to pull a deeper vacuum than the practical moisture removal rate. Consequently, a vacuum pump rate of 50 microns should be considered.

One of the critical elements of evacuating a system is the ability of the vacuum pump to hold a vacuum throughout the evacuation process and to pull a vacuum on the entire system. Single-stage vacuum pumps have the advantage of being lighter in weight, smaller, and less expensive, but they are not capable of pulling and holding a deep vacuum. The two-stage vacuum pump with its second pumping chamber has the ability to pull and maintain a deeper vacuum for prolonged periods of time and is best suited for system evacuation.

A gas ballast or vented exhaust is employed in two-stage vacuum pumps to allow relatively small amounts of dry atmospheric air to enter the second stage of the pump. This mixture of warm, dry air prevents the vaporized moisture from condensing in the pump, allowing it to be exhausted as a vapor.

In all vacuum pumps, even those with a gas ballast, some moisture will condense in the pump's oil. This will eventually result in corrosion, which will attack the internal parts of the pump. The presence of moisture in the vacuum pump oil will also increase the oil pressure vapor, resulting in a reduction of the capability of the vacuum pump to pull a deep vacuum. For these reasons, frequent oil changes are required for all vacuum pumps to prolong the service life of the pump and to maintain maximum efficiency. A special grade of vacuum pump oil is readily available. The manufacturer's specifications should also be followed when choosing an oil for a given pump application.

The service line connecting the vacuum pump to the system being serviced offers some resis-tance to the flow of the vapor being removed from the system. The amount of resistance is proportional to the length of the line and its inside diameter. Conventional neoprene hoses have too much resistance and leakage when subjected to deep vacuums. Therefore, seamless high vacuum metal hoses should be used. These hoses are available in standard lengths with a ⅜" inside diameter. The carrying capacity of these lines is 16 times greater than the conventional service hoses with a ¼" inside diameter.

Evacuating and Dehydrating the System

The presence of air and moisture in a sealed system is perhaps directly or indirectly the cause of more service problems and failures than all other factors combined. The purpose of dehydrating and evacuating a commercial refrigeration or air-conditioning system is to remove moisture and noncondensible gases that may have entered the system.

Moisture in the System

The presence of moisture in a system can cause a freeze-up at the system's metering device. This occurs when the circulating refrigerant picks up the water vapor and carries it through the system. When the moisture reaches the point of expansion at the metering device, ice crystals will start to form. The crystals will either reduce the amount of refrigerant entering the evaporator or they will restrict the amount of refrigerant entirely. In either case a loss of cooling results.

The moisture alone can cause corrosion of the internal parts of the system, but when it is combined with a refrigerant such as R-12 or R-22 and air, the air and water hydrolyze to form hydrochloric and hydrofluoric acids, which accelerate the rate of corrosion. As the metallic parts of the system are eaten away by the corrosion, sludge will start to form in the system.

Sludge deposits can take the form of a slimy liquid, fine powder, or granular or sticky solid. Because the sludge contains some of the corroding acids, the sludge will also corrode the surfaces of the system where it has accumulated.

Air in the System

Air is composed primarily of nitrogen and oxygen. When these elements enter a sealed system, they will remain in a gaseous state. Although these gases can be liquefied under low temperatures or high pressures, they are noncondensible under the normal operating pressures and temperatures of refrigeration and air-conditioning equipment. Reviewing the basic laws of physics, Dalton's law of partial pressures states, "When a combination of gases are contained in a sealed container, each gas will exert its own pressure independently." The total pressure of the container will be the combination of all gases present. Charles' law states, "If the space in which a gas is enclosed remains constant so that the gas cannot expand, the gas pressure will vary with the temperature." Applying these two laws, the air trapped in a sealed system will act additively to the existing system pressure. The pressure will increase as system temperature increases.

Because oxygen and nitrogen are noncondensible in the system, they will collect in the upper tubes of the condenser and reduce the capacity of the condenser. During system operation, the discharge pressure will be a combination of the refrigerant condensing pressure plus the oxygen and nitrogen pressures. This could result in discharge pressures as high as 35 to 45 PSIG above normal. For every one pound of air trapped in the system, there will be approximately a one pound increase in system pressure.

These high discharge pressures will create a loss of system cooling capacity as well as other service problems. High discharge pressures are a frequent cause of hermetic compressor motor burnouts. The increased pressure creates extremely high operating temperatures. As the refrigerant passes through the compressor, the chemical reaction that may exist in the system will increase, adding to or creating new corroding elements. High temperatures will also cause a

breakdown of the compressor lubricating oil, creating sludge and carbon deposits. Furthermore, as the discharge pressure increases, the compressor motor current draw will also increase, putting an additional strain on the compressor motor.

Along with the higher than normal operating discharge pressures, another indicator of air in a system is a higher than normal standing system pressure. To verify whether air is in the system, allow the system to remain idle for a period of time and observe the standing system pressure. The standing pressure should match the saturated pressure and temperature relationship of the refrigerant being used as indicated on a pressure/temperature chart. If the standing system pressure for a corresponding temperature is higher than normal, the system is contaminated with air.

Prevention of Moisture and Air Contamination

The best way to prevent moisture and air from contaminating a system is to follow the proper service procedures. The following list can serve as a general guide in preventing moisture and air contamination.

1. Properly purge all gauge manifold service lines prior to opening the system.

2. Do not use oil that has been exposed to the atmosphere.

3. Seal all refrigeration tubing after use to prevent contamination.

4. Acid test the oil of all defective compressors to ensure that acid contamination does not exist.

5. In cold weather, allow the replacement parts to warm up to room temperature before installing.

6. Install properly sized liquid and suction line filter-driers.

7. Properly evacuate the sealed system before placing the system back into operation.

Removal of Moisture and Air

Once it has been determined that air or moisture is in a sealed system, the method used to remove it will depend upon the system size and application. In applications where the refrigerant charge is small, a complete evacuation of the system is recommended. While several methods may be used to remove moisture and air from a system, the two most effective methods are to deep vacuum the system or to triple evacuate the system.

In triple evacuation, the system pressure is reduced to about 28 inches of vacuum or lower. The system is then charged with an anhydrous nitrogen or dry refrigerant. The dry gas will diffuse throughout the system, absorbing as much moisture as its saturation point will allow. The longer the dry gas is allowed to remain in the system, the more effective will be the moisture removal process. The moisture-laden gas is then released from the system, and the procedure is repeated.

When this process has been done three times and new filter-driers have been installed, the system should be charged and placed back into normal operation.

Servicing Hermetic Compressor Motor Burnout

Hermetic compressor motor burnout is a result of the compressor motor being exposed to contaminants, such as air and moisture, higher than normal operating discharge pressures, low voltage conditions, loss of refrigerant needed to cool the motor windings, and improper motor protection. System contaminants are absorbed by the oil in the system, which acts as a scavenger. This causes a breakdown in the lubricating qualities of the oil, resulting in higher than normal operating temperatures.

The increased temperatures will result in a chemical reaction among the oil, the refrigerant, and the contaminants that will create corroding elements in the system, usually in the form of hydrochloric and hydrofluoric acids. The oil will eventually break down and form sludge and carbon deposits that will circulate and collect in various parts of the system.

When a hermetic compressor motor experiences a burnout, the acid in the system will cause the motor windings to overheat and deteriorate. This will result in shorted motor windings, open motor windings, and grounded motor windings.

Acid Testing

Because the oil in a refrigeration or air-conditioning system acts as a scavenger in collecting system contaminants, the measurement of the acid concentration taken from the compressor's oil is a reliable method of determining the degree of contamination in a system. **Acid test kits** are readily available from a variety of manufacturers. These kits are quite accurate and should be used as part of a preventive maintenance program or whenever hermetic compressor failure is experienced.

Figure 28–6A and B illustrates a typical acid test kit. Follow the manufacturer's instructions for the test kit being used. Periodic checks will determine if the compressor oil should be changed and if other preventive maintenance procedures need to be performed.

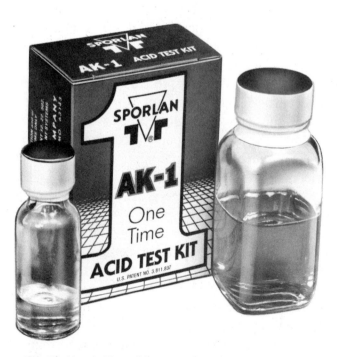

Fig. 28–6A Acid test kits. *Courtesy Sporlan Valve Company.*

Fig. 28–6B *(continued)*   **Acid test kits.** *Courtesy Sporlan Valve Company.*

HERMETIC COMPRESSOR MOTOR BURNOUT CLEANUP PROCEDURE

Compressor motor burnouts and the subsequent cleanup procedures have been a major topic of discussion among service technicians for many years. Flushing the system with R-11 or a cleaning agent was the original procedure developed to rid a system of contaminants. It will do a satisfactory job if applied properly. However, there are many disadvantages to the procedure, including the length of time required, the cost, and the risk of damage due to spillage. Experience on the part of the service technician is also required. Through the years, technology has evolved into the filter-drier method of burnout cleanup. This is a very practical method, which, if applied properly, will do an effective job in cleaning a system with burnout contamination.

Most cleanup procedures are essentially the same, but it is still wise to follow the manufacturer's instructions. The following discussion, however, can serve as a general guide.

Removal and Replacement of Existing Filter-Driers

All existing filter-driers in the system must be replaced. The filter-driers should be cut or uncoupled from the system. They should never be unsoldered because heat applied to a drier will drive the moisture trapped in the drier back into the system.

Installation of the New Filter-Driers

After the new compressor has been installed and the system has been leak tested, new filter-driers should be installed in the system liquid and suction lines. While the use of an oversized liquid line filter-drier is generally accepted, the suction line filter-drier must be properly sized for the system. A filter-drier should be equipped with a pressure test port on the inlet side so the pressure drop across the filter-drier can be checked easily. Once the new filter-driers have been installed, the system can be triple evacuated and charged with the correct amount of refrigerant.

After about two hours of operation, the pressure drop across the suction line filter-drier should be checked. If the pressure drop taken at the inlet side of the filter-drier and the compressor suction service port downstream of the filter-drier exceeds 2 PSIG, the filter-drier should be changed.

After about 48 hours of operation, an acid test of the system's oil should be performed, and the pressure drop across the suction line filter-drier should also be checked. If the oil sample shows any degree of contamination, it should be changed and new suction line and liquid line filter-driers should be installed. If the pressure drop across the suction line filter-drier exceeds 2 PSIG, the filter-drier should be changed. This procedure should be repeated in 48 hours and continued until the system is cleaned and the oil samples show no degree of contamination.

If a refrigeration or air-conditioning system experiences a hermetic compressor motor burnout and is not properly cleaned, a second burnout could be experienced within a year. Therefore, it is of the utmost importance to follow the proper repair procedure the first time a burnout is experienced.

Compressor Removal and Replacement

Once it has been established that a compressor has failed and needs to be replaced, an organized removal and replacement procedure will help ensure that the task is performed properly. Because the repair procedure depends on the application, the following can serve only as a general guide.

To avoid exposing the system to the atmosphere for an extended amount of time, it is advisable to have all replacement parts on hand before removing the defective parts from the existing system. The electrical characteristics and capacity rating of the replacement compressor should also be checked to ensure that they conform to the original compressor.

Follow these steps when replacing a compressor:

1. Disconnect all power to the system and lock the service disconnect switch or remove the fuses to prevent power from accidentally being applied.

2. Disconnect all wiring to the original compressor. It may be a good idea to tag each wire for easy identification during reinstallation.

3. Recover the refrigerant from the system in a liquid state whenever possible.

4. Evacuate the system to 0 PSIG before attempting to disconnect the refrigerant lines or remove accessory devices.

5. Remove the accessory devices after all of the refrigerant has been recovered from the system.

6. Disconnect the compressor discharge and suction lines by cutting or unsoldering. **Caution:** When unsoldering refrigerant lines, make sure that the system is properly vented to prevent a buildup of residual refrigerant pressure. Wear safety glasses and gloves to protect yourself and to avoid possible exposure to acids that may be present as a result of the hermetic compressor motor burnout.

7. Remove the compressor mounting bolts and the defective compressor from the system.

8. Take an oil sample from the compressor and perform an acid test. An acid test of the compressor oil should be performed on all defective compressors when they are removed from a system. There is the possibility of acid contamination even if the compressor failure was due to a mechanical failure. If the oil proves to be contaminated and a motor burnout has occurred, the burnout cleanup procedure should be followed. If no contamination is indicated, a liquid line filter-drier and suction line filter-drier should be installed after the new compressor is installed.

Unit Twenty-Eight Summary

- An undercharged refrigeration system operates at reduced efficiency and an overcharged system results in high discharge temperatures and liquid floodback. A sight glass is commonly used on commercial refrigeration systems to check the condition of the refrigerant in the liquid line, evaluating the refrigerant charge in the system. Liquid level test valves are often used to determine liquid levels in lines, receivers, and condensers.

- Capillary tube systems are referred to as critically charged systems.

- Charging by weight is one of the most efficient ways of charging a refrigeration system. Only vapor should be allowed to flow into the low side of a system during charging. Liquid refrigerant may be charged into a system high side through the use of a liquid line charging valve.

- NEVER heat a refrigerant cylinder with an open flame.

- Acid testing will evaluate the oil in a refrigeration system and indicate the presence of acid contamination. Triple evacuation helps to remove moisture from a system.

- Suction line and liquid line driers trap moisture and particles that can cause contamination in a refrigeration system. All systems that experience a hermetic compressor burnout must be properly cleaned to prevent early failure of the replacement compressor.

Unit Twenty-Eight Key Terms

Sight Glass

Electronic Sight Glass

Liquid Level Test Valve

Acid Test

UNIT TWENTY-EIGHT REVIEW

1. What are the possible consequences of an undercharge of refrigerant?

2. What do bubbles in the sight glass indicate?

3. Where in the system should a sight glass be located?

4. What is the purpose of an electronic sight glass?

5. Where are the instrument probes placed in a system to measure refrigerant charge with an expansion valve? with a capillary tube?

6. Using a liquid level test valve, what is the state of the refrigerant charge in the following situations?

    a. Small amount of liquid refrigerant escapes.

    b. Small amount of refrigerant vapor escapes.

    c. Small amount of refrigerant, both liquid and vapor, escapes.

7. What is the most accurate method to ensure a system is properly charged?

8. At what micron range does practical moisture removal begin?

9. Describe the procedures used to check for air in a system.

10. Describe the procedures used to remove air from a sealed system.

11. What is a hermetic motor burnout?

12. Why should liquid refrigerant not be added on the low side of a system?

13. When is vapor charging done?

14. What are two methods for warming a cool service cylinder?

15. Match the following terms with their descriptions: sludge, thermistor vacuum gauges, triple evacuation, moisture, vacuum pump, filter-drier, Dalton's law, Charles' law.

    a. Gas pressure will vary with the temperature.

    b. Can cause a freeze-up at system's metering device.

    c. Monitors and measures system vacuums in the low micron ranges.

    d. Contains corroding acids.

    e. Capable of reducing system pressure to at least 500 microns.

    f. Dry gas is used in moisture removal.

    g. Practical method for cleaning a system with burnout contamination.

16. How do oxygen and nitrogen affect discharge pressure in a system?

17. As the discharge pressure _____ , the compressor motor current draw will _____ .

18. Why must a filter-drier be cut out of the system and never unsoldered?

19. Describe the filter-drier method of cleaning up a burnout.

20. What can be done to prolong the service life of vacuum pumps?

21. 25,400 = 1 inch of mercury

    1 millimeter (mm) = .039 inch

    1,000 microns = ? mm   1/25,400 = .0000394

                                .0000394 × 1000 = .0393701 inch

                                .0393701/.039 = 1 mm

    A deep vacuum is about 50 to 200 microns.

    a. How many mm (millimeters) is 50?

    b. How many mm (millimeters) is 200?

22. Write a paragraph to explain what should be done to prevent moisture and air contamination in a system.

SECTION SEVEN

The HVAC/R Business

The intent of this section is to give you some insight into the "business end" of the HVAC/R business. A large percentage of technicians make the decision to operate independently rather than work as an employee of a firm and, unfortunately, many find themselves in a problem situation because they are unaware of some of the important considerations of owning and operating an HVAC/R business. Insurance, worker's compensation, taxes, and keeping office functions up-to-date are but a few of the things that must be considered by the independent operator.

If it's your desire to own and operate your own business in the HVAC/R area, take the time to thoroughly investigate before making the decision to go ahead.

UNIT TWENTY-NINE

The HVAC/R Business

Some surveys have stated that as many as 90 percent of the technicians in the HVAC/R field are either independently employed at the present time, have been independently employed at some time in the past, or have a desire to be independently employed in the future.

What this means, is that a large percentage of technicians make the decision to become business owners when, in fact, a person's technical ability is only one segment necessary for success as an independently employed service company owner. In this unit, we'll discuss and offer insight into the areas of regulation and licensing of HVAC/R businesses, insurance considerations, the factory authorized service center system, how computers fit into the HVAC/R business, and customer service.

REGULATING AND LICENSING AGENCIES

Regulation on a National Level

Until recently there was little regulation in regard to the HVAC/R industry on a national level. With the 1990 amendment to the Clean Air Act, however, that situation has changed dramatically. Technicians in the HVAC/R industry must now become certified in refrigerant-handling practices through testing.

Section 608 of the Clean Air Act defines technicians who must be certified as "any person who performs maintenance, service, or repair and could reasonably be expected to release Class 1 (CFC) and Class 2 (HCFC) substances into the atmosphere, including but not limited to installers, contractor employees, in-house service personnel, and in some cases, owners." The term technicians also extends to any person engaged in the disposal of major appliances, excluding small kitchen appliances.

To become certified in refrigerant handling practices, technicians are required to pass an EPA-sanctioned exam that consists of four parts.

Section One, the CORE section of the exam, must be completed by technicians wishing to become certified in any area. The CORE section of the exam consists of questions related to refrigeration fundamentals, refrigerants and their applications to various types of systems, and evacuation and dehydration requirements for refrigeration systems.

Also included in the CORE section of the exam is information on the ozone layer and the effects of chlorine-based refrigerants released into the atmosphere; the difference between recovery, recycle, and reclaim; information on lubricants used with replacement refrigerants; and the recovery practices and procedures that must be adhered to by technicians working in the HVAC/R industry.

For those wishing to be certified to service small appliances, Section Two of the exam must be completed along with the CORE (Section One). A person passing the CORE and Section Two of the test is referred to as "Type One" certified, meaning they are certified to work with equipment classified as Type One, small appliances.

The EPA defines small appliances as "any appliance that is fully manufactured, charged, and hermetically sealed in a factory with a charge of five pounds or less."

Section Two of the EPA exam focuses on leak-repair requirements in regard to small appliances, the types of access valves used in this equipment, and recovery requirements for small appliances. Safety, use of proper refrigerant containers, shipping, and disposal of refrigerants are also part of this section of the exam.

(NOTE: An on-site test is not the only method through which you can become Type-One certified. Some appliance service associations are offering a home study and mail-in testing system for those wishing to be certified only in the area of small appliances.)

For technicians working with comfort cooling systems that use high-pressure refrigerants, a certification known as "Type-Two" is required. In order to receive a Type-Two certification, you must pass Section Three of the EPA exam along with the CORE section. A Type-Two certification is the one most commonly needed by HVAC/R technicians since it encompasses R-22 systems.

Under EPA guidelines, Type-Two certification is required for servicing high and very high pressure refrigeration systems. High-pressure refrigerants with a boiling point between −58°F and 50°F at atmospheric pressure are R-12, R-22, R-134a, and R-502. Very high pressure refrigerants with a boiling point below 50°F at atmospheric pressure are R-13 and R-503.

Section Three of the EPA exam focuses on (1) the types of fittings required for high pressure equipment, (2) the types of repairs that are considered to be major, (3) leak-detection practices and recovery requirements, (4) the components commonly found in refrigeration systems of this type, (5) proper recovery procedures, (6) safety, and (7) shipping and disposal of refrigerants.

A "Type-Three" certification is required for those who work with low-pressure equipment. The EPA states that a Type-Three certification is required for servicing low-pressure refrigerants with boiling points above 50°F at atmospheric

pressure. Typical refrigerants in this classification are R-11, R-123, and R-113.

To become certified as a Type-Three technician, you must pass the CORE section and Section Four of the exam. Section Four focuses on components common to low-pressure equipment such as rupture disks and purge units. It also requires that you understand (1) equipment room requirements as outlined in ASHRAE Standards 15-1992, (2) leak-repair requirements for larger refrigeration systems, (3) service valves required with this type of equipment, (4) leak-detection procedures, (5) recovery device and evacuation level requirements, (6) safety, and (7) refrigerant transport and disposal.

Another certification available under Section 608 compliance testing is the "Type Four" technician certification. In order to receive this certification, you must pass all four sections of the EPA-sanctioned exam. This type of certification is also known as a "Universal" certification.

Information regarding trade schools, colleges, and associations offering training and certification testing is available from the EPA.

Regulation on a State Level

The terms "registered contractor" and "licensed contractor" relate to regulation in the HVAC/R industry on a state level. A state agency known as the Registrar of Contractors administers the required testing and grants registration for contractors.

Requirements vary from state to state, but most require that you have a minimum amount of field experience (in many cases, several years) as a technician before you're even allowed to take the contractor's exam in the area in which you wish to be registered and licensed.

In many cases, the Registrar of Contractors office makes a study guide available to those wishing to take an exam in their area of expertise. The study guide doesn't contain actual questions found on the exam, but outlines basic information about the exam. It may even outline the percentage of questions in a given area. As an example, a guide may tell you that 20 percent of the questions on the exam are related to refrigeration systems and 20 percent of the questions are related to business procedures.

Unlicensed independent contractors are a problem in the HVAC/R industry in almost every state. An important function of the Registrar Of Contractors office is to follow up on consumer complaints regarding shoddy workmanship and unfair business practices.

In many states you can operate as an independent service company without being registered or licensed by the state, but there is a limit on the type of work you can do and the amount of money you can earn on any given service call. Some states, for example, allow you to perform service up to a maximum of $600 on any given job.

What this type of situation means is that you could, for example, repair somebody's furnace (replace a gas valve, etc.) without being a licensed contractor, but if the furnace needed replacing, you would need to be a registered contractor to perform that service.

In some states, competency testing and certification is required not only to be a registered contractor, but also to work as a technician.

Regulation on a Local Level

In many cases, on the local or city level the only regulation required of you is to obtain a business license if you're planning on operating an independent service company. Some cities, however, may require that you demonstrate competency as a technician and become certified, whether you're operating as an independent or working as an employee.

INSURANCE FOR SERVICE COMPANIES

Operating as an independent contractor means that you'll have to provide insurance coverage for the protection of your company, yourself, and your customers. Two types of insurance policies you'll need are a commercial policy that covers your vehicle and a comprehensive liability policy that covers your business in general.

Commercial coverage for your service vehicle can be purchased in the same manner as you purchase standard insurance for your nonbusiness vehicle. Buying commercial liability insurance means that any damage caused by you in an accident will be covered (up to the limits of your policy), and buying commercial comprehensive insurance means that both your vehicle and any other vehicles or property damaged will be covered. If your service vehicle is being financed, the bank or credit union holding the loan will require that you have comprehensive coverage.

Aside from vehicle insurance, you'll also need a general insurance policy described as a comprehensive liability policy. A policy of this type covers the acts of anyone connected with your company, including your actions and those of your employees. In the event of damage to property or injury to customers, a comprehensive liability policy is designed to protect you and your business. The amount of coverage you can buy varies anywhere from a low of $100,000 worth of coverage to amounts over a million dollars.

When setting premiums for your comprehensive liability policy, an insurance agent will consult the commercial insurance manual to determine your cost. One general rule of thumb for service companies is that premiums are often based on the amount of payroll issued by the service company. The higher the payroll, the higher the premium.

When buying insurance for your business, always make sure the agent explains everything to your satisfaction. It can be an unpleasant experience to find that there are some limitations to your policy after the fact.

One example of this would be working with a subcontractor, such as a crane company. If, for example, you were using a crane company to place a roof unit, it may be your responsibility to confirm that they have their own insurance policy to cover any damage that may occur while they are lifting the equipment.

If something went wrong and a unit was dropped or a structure was damaged, your insurance company may determine that the responsibility for the damage lies with the crane company, and would therefore refuse to pay any claim.

In some cases, you may even find that the insurance company issuing your policy will ask you to list any subcontractors you use, and require that you provide a certificate of insurance that shows that your subcontractors are insured before a policy will be issued to your business. Some insurance companies may cancel your policy if they find that you're working with a sub that does not have insurance coverage of its own.

FACTORY-AUTHORIZED SERVICE CENTERS

If you're planning on operating an independent service company, one way you can boost your volume (and your credibility) is to become an authorized service center for a manufacturer.

Being an authorized service center means that you agree to perform warranty service for customers who have purchased HVAC/R equipment or appliances and find it necessary to call for service during the manufacturer's warranty period. Your customer expects you to perform the service at no charge to them, which means that you have to obtain the parts necessary and bill the manufacturer of the equipment for your labor charges.

Becoming an authorized service center requires an agreement in advance between you and the equipment manufacturer. It's more common to find independent service agents in the appliance industry than in the HVAC/R industry. In the HVAC/R industry, some manufacturers will only allow you to be an authorized service center if you are also selling and installing their equipment. In order to be selling and installing a manufacturer's equipment, you must be a registered, licensed contractor, as previously discussed in this unit.

In some cases, an authorized service center is paid in accordance with its standard rates for service calls. Some manufacturers, though, may use a flat-rate system, paying only a given amount for specific warranty services performed, regardless of the amount of time taken to complete the job.

When it comes to handling parts for warranty service calls, a service company will have to obtain the parts from a factory-authorized parts distributor. When parts for a warranty service call are obtained, they're billed to the service company's account (or paid for by the service company if they don't have a credit line with the parts distributor).

Once the service is completed, the parts purchase is reconciled. If the parts were billed to your account and the service was completed before payment was due, a credit for the returned parts will be issued and there won't be any out-of-pocket expense on your part. If the service wasn't completed before the billing date on your account, the parts distributor may require payment from the service company, then issue credit when the warranty parts are returned.

In some cases manufacturers require that the service company return failed parts, while in other cases, the service company is required to keep them on hand for a given period of time. If a representative of the factory doesn't visit the service company to check on the stored parts, they can be scrapped in accordance with the manufacturer's procedures.

When performing warranty service for a manufacturer, a service company may be required to use an invoice other than the standard one used for COD work. The manufacturer may supply them at no charge or the service company may be required to purchase the specialized manufacturer's invoice.

Technicians have to be aware of manufacturer's requirements when it comes to completing the paperwork. Model and serial numbers of equipment must be correct, part numbers must be listed correctly, and the customer's signature is required.

Manufacturers have a very detailed system for tracking service call records on specific equipment and recording repeated service calls for the same problem. Service companies have to guarantee their work to manufacturers just as they have to offer a guarantee to their COD customers.

COMPUTERS IN THE BUSINESS

As in all businesses and industry today, the computer has had an impact on the way HVAC/R technicians perform their jobs and the way service companies do business. Software programs are available for everything from payroll management and parts control to cost and equipment estimating.

EQUIPMENT AND BUSINESS RELATED SOFTWARE

Residential Load Estimating

One of the most popular software programs used in the HVAC/R industry is that designed for residential load estimating, Figure 29–1. Many programs are based on what is known as "Manual J," the manual of the Air Conditioning Contractors Association (ACCA).

When using this type of software, you have the ability to access design weather data for cities around the country and can calculate the peak heating and cooling loads for a given residence in a given climate.

You can input information about the amount of glass in a building, and about shading and ventilation. You can also input other important factors such as the type and amount of insulation you'll need, in order to calculate the necessary size of equipment and the amount of airflow in each room.

Commercial Load Estimating

Like the residential load-estimating software, commercial programs offer weather data for cities around the country, as shown in Figure 29–2. The maximum heating and cooling loads can be calculated for large and small office buildings. Calculation of heat loads created by lighting, equipment, and people can be accomplished. Software for commercial load-estimating also allows you to design a system with many different air handlers, factor in chilled water flow rates for hydronic systems, and consider factors such as seasonal infiltration and ventilation rates.

Duct-Sizing Programs

Duct-sizing software allows you to calculate the ductwork necessary for a given HVAC system, as illustrated in Figure 29–3. You can compute round, rectangular, and oval duct sizes to ensure that the heating and cooling system will have proper airflow to each room or zone. Air velocity, duct friction, and noise considerations are part of most programs. A common feature of duct-sizing programs is the ability to print a bill of materials that shows the material and labor costs for installing a duct system.

Fig. 29–1　Residential load-estimating software. *Courtesy Elite Software.*

Fig. 29–2　Commercial building load-estimating software. *Courtesy Elite Software.*

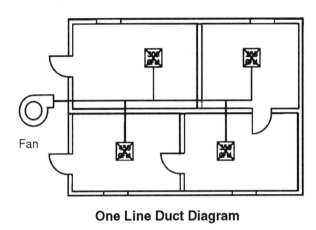

One Line Duct Diagram

Fig. 29–3 Duct design software. *Courtesy Elite Software.*

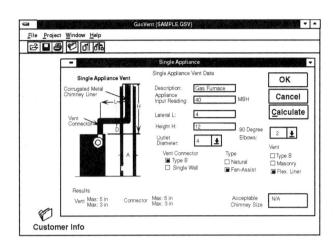

Fig. 29–4 Vent design software. *Courtesy Elite Software.*

Gas Venting Calculation Programs

In order to ensure that gas appliances operate with optimum efficiency and safety, the venting system must be correct. Software programs are available to calculate proper vent sizes in single- and multiple-story buildings and for single- and multiple-appliance applications, as in Figure 29–4. Most available software will allow you to calculate vent sizes for both single and double wall vent systems as well as calculate acceptable tile-lined masonry chimney sizes.

Software programs for vent calculating have become more common with the advent of higher efficiency furnaces that cause higher levels of condensation in vent piping that is not sized properly.

Industrial Ventilation Programs

Software designed for use in industrial applications allows you to design and to analyze industrial ventilation and exhaust duct systems, as shown in Figure 29–5. Often used in manufacturing situations, the software allows for the calculation of necessary duct sizes for both round and rectangular ductwork and the factoring in of elaborate filtering systems. The amount of makeup air required in high-velocity exhaust areas can also be determined.

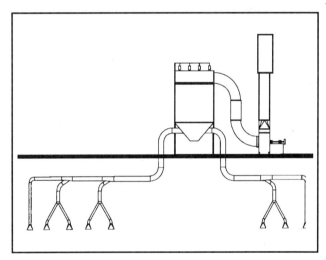

Fig. 29–5 Industrial ventilation design software. *Courtesy Elite Software.*

Food Preservation Systems Software

Software programs specifically designed for calculating refrigeration loads in grocery store equipment are available for those tasked with sizing and designing equipment for these applications, as shown in Figure 29–6. In addition to design weather data used in comfort cooling load calculations, software designed for this application also calculates the 24-hour refrigeration load in accordance with the amount of product, infiltration due to opening and closing of doors by shoppers, lighting, and defrost systems.

Many software programs of this type contain a library of product load data as well as specifications of coils and condensing units, which allows

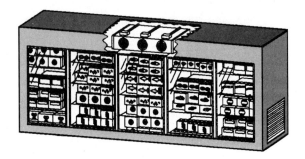

Refrigeration Loads

Fig. 29–6   Food preservation systems software. *Courtesy Elite Software.*

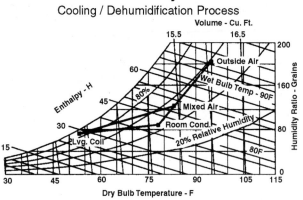

Fig. 29–7   Psychrometric chart software. *Courtesy Elite Software.*

you to quickly design a system or evaluate an existing system that isn't performing up to standards.

Psychrometric Chart Software

In the same way that a psychrometric chart is used in calculating heating and cooling loads, "psych" chart software allows you to determine the factors involved in the psychrometric processes of heating and humidification, cooling and dehumidification, and mixing. (see Figure 29–7) Programs such as this are often designed around the ASHRAE Handbook of Fundamentals.

Equipment Cost Analysis Software

Software that allows you to analyze the cost of operating HVAC/R equipment and calculate payback time frames as well as the cost effectiveness to replace or repair equipment is available. (refer to Figure 29–8) Generally, using the economic criteria from military sources, operational, maintenance, and other costs of operation can be calculated. The present worth of existing equipment can also be determined.

Service Call Scheduling Software

Computers can be used to enter service calls as they come into the office of the service com-

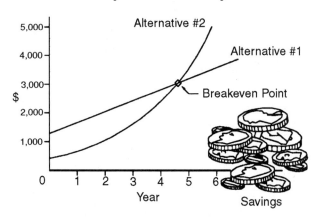

Fig. 29–8   Equipment cost analysis software. *Courtesy Elite Software.*

pany. A program such as this allows office personnel to determine if the customer calling is a new or a repeat customer, to schedule the service call, to select a technician to handle the work, and to print out the necessary service ticket. A program such as this allows you to track specific types of calls that have been coming in during a given season or time frame, group the calls by zip code, and print out mailing labels for follow-up after the service has been completed.

Inventory Control Programs

Inventory control programs designed for service companies keep track of parts on hand. The programs also work in conjunction with other programs related to service call scheduling to automatically delete parts used from the inventory. A program such as this can also keep track of truck stock and automatically reorder from the central parts storage when necessary.

Electronic Filing Systems

Software programs that enable you to control all aspects of a service company are available. You can maintain customer files, service history of their equipment, billing, and sales and service records of service contracts if a customer has purchased that option for his or her equipment. In many cases software packages such as this allow you to keep track of payroll for technicians (including accessing vacation time and sick time accrued and used), as well as paying utility, parts supply, and other expenses.

An electronic filing system such as this maintains tax records on a year-round basis, which makes things easier for the service company owner on April 15th.

CUSTOMER RELATIONS
Practical Communications

A big part of being a good communicator is common sense. Some suggestions for making yourself an effective communicator include:

Use the telephone. Call your customers when you are on the way to a service call. Always call if you are going to be late or early. You will save yourself a lot of time and trouble and have much happier customers.

Most of us do not have photographic memories. If it needs to be remembered, write it down. Carry small note pads or notebooks with you so you can store information, even when you are on a call.

Ask the customer about the problem. The customer often can help you solve the problem

and may save you time. They may also be able to explain earlier service done to the same unit that could have an effect on what you are working on.

Above all, listen. Listening to others before you jump to any conclusion is the key to being a good communicator and doing a good job.

Professional Appearance and Work Habits

You know that a person's abilities are often not represented by their physical appearance. As much as you may not like it, other people will most often judge you first by the way you look. Trying to look like a professional will make people think you are one. This is a key aspect of good customer relations.

Having a neat appearance makes other people believe that you believe in yourself. You can be confident and know your stuff, and still have customers be less trusting because you look sloppy. Ultimately, you make the choices that will affect others' first impressions of you. You control the way you dress, your cleanliness, and fitness. Choosing to look like a pro is an important choice for you to make.

Good grooming is a great place to start. While the HVAC/R field does not necessarily require you to conform to one set style, your cleanliness and neatness do make other people aware that you care about the way you look. This makes an important difference. For example, you may have the chance to wear your hair (facial hair too, if male) any way you choose. That's your choice—but whatever you do choose, choose to make it look good. It's one thing to express yourself with your appearance, and quite another to appear not to care about your appearance.

Wearing work clothes that identifies you as a technician will help give you a professional appearance. Whether you wear your own clothing or have uniforms issued by the employer, keeping those clothes neat, clean, and in good shape will make the public respect you more. Remember to always have coveralls with you for under-the-house work and also remember to remove them before going into a customer's house.

Personal health and fitness is another key to doing the best work you can. If you are in relatively good shape, and if you exercise regularly, you will do better work. You will also have more energy left for your personal life when the workday is done. Your health will have a direct bearing on your ability to do your job and keep the customers happy. Take care of yourself—many other things will fall into place. You will always be happy with the results if you follow what you know to be good living habits.

The same can be said of your service vehicle. Even though you will likely maintain the same vehicle for as long as is economical, that doesn't mean that it has to look a mess. Keeping tools and parts in an orderly fashion makes you work faster, more efficiently, and makes your customers feel that they are being assisted by a professional. Remember to be polite when driving and parking the vehicle, especially when near the customer's house. Obey the speed limit and do not block the customer's driveway or other entrances.

Once you arrive on the site you should concentrate on the job to be done. While having a sense of humor and being friendly are important, especially if you hope for repeat business, most customers appreciate a service technician whose diagnosis and repair are done promptly. This means neither rushing nor taking too much time— you will find your own pace at which you can do the job right the first time. What it does mean is staying focused on the task at hand and working steadily at it the entire time you are at the work site. You were called in to do a job so give the customer the impression that you are doing just that!

Keeping the work site clean and orderly is another key to being a professional. In commercial accounts this is critical. You may lose an important account if you leave a mess for your customer to clean up after you. The same is true in residential work. While you may be working on equipment in the dustiest, most cluttered corner of the basement, that doesn't give you the right to add to the mess. You must make it a rule to leave the place as you found it. This may mean grabbing a broom to sweep up, using drop cloths to catch dirt from falling when removing a unit, and so forth.

Remember, also, the little extras you can do for the customer that will take very little of your time, cost next to nothing, and can be shared with the customer and used to further build business. Examples include repairing frayed wires, cleaning or replacing air filters, replacing panels with correct fasteners, checking evaporator coils for frost or dirt buildup, checking gas lines for insulation, lubricating fan bearings, cleaning condensers, and fixing water leaks.

Always advise customers about the proper use of their equipment and the types of maintenance they can do themselves or those for which they should call you. As an example, many homeowners have no idea how inefficient a high-efficiency furnace can become if regular maintenance is not performed. Educating the customer can save them money in the long run and earn you money on a regular basis.

Don't Worry—Be Happy

Each of us has a personality all our own. We have our own beliefs, values, morals, and ethics. When working with the public, you will run into many types of people. Always remember that you are there to do a job. Be polite, be friendly, and stay focused on your work.

Do your job well, and enjoy the challenge each new job brings. You will build customers who will come to appreciate your efforts. If you work for a company, your employer will recognize you as a positive part of the team.

Work towards having a positive attitude and outlook. Everyone wants to be around happy people. If other things are going badly, use your work to relieve some of that stress. Work can be a great place to build up your own self-esteem, to have good relationships with other people, and to show the world that you are a capable person.

If you look for the worst in others, you will always find it. Others will also return the favor!

Your good attitude will make the job go better and will make you happier. Be proud of the trade you have learned. You will be called into a job to do things others *cannot do* themselves much more than *don't want to do* themselves. This is important work! Always do your best. Know that what you do makes a real difference.

UNIT TWENTY-NINE REVIEW

1. Describe the relationship between the following terms:

   a. HVAC/R technician/independence

   b. 1990 amendment to the Clean Air Act/certified

   c. ozone/chlorine

   d. Type-One certification/Type-Two certification

   e. Type-Three certification/Type-Four certification

2. Who is a certified technician under Section 608 of the Clean Air Act?

3. Match each section of the EPA exam with its description: CORE section, Section Two, Section Three, Section Four.

   a. Questions on the types of fittings for high-pressure equipment, major repairs, leak-detection practices, and recovery requirements.

   b. Questions on leak-repair requirements in regard to small appliances, types of access valves used in this equipment, and recovery requirements.

   c. Questions related to refrigeration fundamentals, refrigerants, and their applications to various types of systems.

   d. Questions on components common to low-pressure equipment, rupture disks, purge units, and equipment room requirements as outlined in ASHRAE Standards 15-1992.

   e. Questions on the difference between recovery, recycle, and reclaim, lubricants used with replacement refrigerants, and recovery practices.

4. How does the EPA define small appliances?

5. True or False. An on-site test is the only method through which you can become Type-One certified. Explain.

6. Which of the following areas is not a function of the Registrar of Contractors office?

   a. Administers the required testing and grants registration for contractors.

   b. Makes a study guide available to those wishing to take an exam in their area of expertise.

   c. Follows up on consumer complaints regarding shoddy workmanship and unfair business practices.

7. What is the purpose of a comprehensive liability insurance policy?

8. What does it mean to be an authorized service center?

9. How are computers and their software programs having an impact on HVAC/R technicians?

10. Name four things that can be done to improve communications with the customer.

11. Write a paragraph on the professional appearance of the technician you would want servicing your equipment or appliances.

12. Write an explanation of how health and fitness can affect your work as a service technician.

GLOSSARY

Note: English term appears boldface followed by Spanish term set in italic.

A

Absolute pressure Gauge pressure plus the pressure of the atmosphere, normally 14.696 at sea level at 68°F.

Presión absoluta. La presión del calibrador más la presión de la atmósfera, que generalmente es 14,696 al nivel del mar a 68°F (20°C).

Absolute zero temperature The lowest obtainable temperature where molecular motion stops, –460°F and –273°C.

Temperatura del cero absoluto. La temperatura más baja obtenible donde se detiene el movimiento molecular, –460°F y –273°C.

Absorbent (attractant) The salt solution used to attract water in an absorption chiller.

Hidrófilo. Solución salina utilizada para atraer el agua en un enfriador por absorción.

Absorber That part of an absorption chiller where the water is absorbed by the salt solution.

Absorbedor. El lugar en el enfriador por absorción donde la solución salina absorbe el agua.

Absorption air-conditioning chiller A system using a salt substance, water, and heat to provide cooling for an air-conditioning system.

Enfriador por absorción para acondicionamiento de aire. Sistema que utiliza una sustancia salina, agua y calor para proveer enfriamiento en un sistema de acondicionamiento de aire.

ABS pipe Acrylonitrilebutadiene styrene plastic pipe used for water, drains, waste, and venting.

Tubo de acronitrilo-butadieno-estireno. Tubo plástico de acronitrilo-butadieno-estireno utilizado para el agua, los drenajes, los desperdicios y la ventilación.

Accumulator A storage tank located in the suction line of a compressor. It allows small amounts of liquid refrigerant to boil away before entering the compressor. It is sometimes used to store excess refrigerant in heat pump systems during the winter cycle.

Acumulador. Tanque de almacenaje ubicado en el conducto de aspiración de un compresor. Permite que pequeñas cantidades de refrigerante líquido se evaporen antes de entrar al compresor. Algunas veces se utiliza para almacenar exceso de refrigerante en sistemas de bombas de calor durante el ciclo de invierno.

Acid-contaminated system A refrigeration system that contains acid due to contamination.

Sistema contaminado de ácido. Sistema de refrigeración que, debido a la contaminación, contiene ácido.

ACR tubing Air-conditioning and refrigeration tubing that is very clean, dry, and normally charged with dry nitrogen. The tubing is sealed at the ends to contain the nitrogen.

Tubería ACR. Tubería para el acondicionamiento de aire y la refrigeración que es muy limpia y seca, y que por lo general está cargada de nitrógeno seco. La tubería se sella en ambos extremos para contener el nitrógeno.

Activated alumina A chemical desiccant used in refrigerant driers.

Alúmina activada. Disecante químico utilizado en secadores de refrigerantes.

Active solar system A system that uses electrical and/or mechanical devices to help collect, store, and distribute the sun's energy.

Sistema solar activo. Sistema que utiliza dispositivos eléctricos y/o mecánicos para ayudar a acumular, almacenar y distribuir la energía del sol.

Air acetylene A mixture of air and acetylene gas that when ignited is used for soldering, brazing, and other applications.

Aire-acetilénico. Mezcla de aire y de gas acetileno que se utiliza en la soldadura, la broncesoldadura y otras aplicaciones al ser encendida.

Air conditioner Equipment that conditions air by either cleaning, cooling, heating, humidifying, or dehumidifying it. A term often applied to comfort cooling equipment.

Acondicionador de aire. Equipo que acondiciona el aire limpiándolo, enfriándolo, calentándolo, humidificándolo o deshumidificándolo. Término comúnmente aplicado al equipo de enfriamiento para comodidad.

Air-conditioning A process that maintains comfort conditions in a defined area.

Acondicionamiento de aire. Proceso que mantiene condiciones agradables en un área definida.

Air-cooled condenser One of the four main components of an air-cooled refrigeration system. It receives hot gas from the compressor and rejects it to a place where it makes no difference.

Condensador enfriado por aire. Uno de los cuatro componentes principales de un sistema de refrigeración enfriado por aire. Recibe el gas caliente del compresor y lo dirige a un lugar donde no afecte la temperatura.

Air gap The clearance between the rotating rotor and the stationary winding on an open motor. Known as a *vapor gap* in a hermetically sealed compressor motor.

Espacio de aire. Espacio libre entre el rotor giratorio y el devandado fijo en un motor abierto. Conocido como espacio de vapor en un motor de compresor sellado herméticamente.

Air handler The device that moves the air across the heat exchanger in a forced air system—normally considered to be the fan and its housing.

Tratante de aire. Dispositivo que dirige el aire a través del intercambiador de calor en un sistema de aire forzado—considerado generalmente como el abanico y su alojamiento.

Air heat exchanger A device used to exchange heat between air and another medium at different temperature levels, such as air-to-air, air-to-water, or air-to-refrigerant.

Intercambiador de aire y calor. Dispositivo utilizado para intercambiar el calor entre el aire y otro medio, como por ejemplo aire y aire, aire y agua o aire y refrigerante, a diferentes niveles de temperatura.

Air pressure control (switch) Used to detect air pressure drop across the coil in an outdoor heat pump unit due to ice buildup.

Regulador de la presión de aire (conmutador). Utilizado para detectar una caída en la presión del aire a través de la bobina en una unidad de bomba de calor para exteriores debido a la acumulación de hielo.

Air sensor A device that registers changes in air conditions such as pressure, velocity, temperature, or moisture content.

Sensor de aire. Dispositivo que registra los cambios en las condiciones del aire, como por ejemplo cambios en presión, velocidad, temperatura o contenido de humedad.

Air, standard Dry air at 70°F and 14.696 PSI at which it has a mass density of 0.075 lb/ft^3 and a specific volume of 13.33 ft^3/lb, ASHRAE 1986.

Aire, estándar. Aire seco a 70°F (21.11°C) y 14,696 PSI [libra por pulgada cuadrada]; a dicha temperatura tiene una densidad de masa de 0,075 pies/libras^3 y un volumen específico de 13,33 pies^3/libras, ASHRAE 1986.

Air vent A fitting used to vent air manually or automatically from a system.

Válvula de aire.   Accesorio utilizado para darle al aire salida manual o automática de un sistema.

Algae   A form of green or black slimy plant life that grows in water systems.

Alga.   Tipo de planta legamosa de color verde o negro que crece en sistemas acuáticos.

Allen head   A recessed hex head in a fastener.

Cabeza allen.   Cabeza de concavidad hexagonal en un asegurador.

Alternating current (AC)   An electric current that reverses its direction at regular intervals.

Corriente alterna.   Corriente eléctrica que invierte su dirección a intervalos regulares.

Altitude adjustment   An adjustment made to a refrigerator thermostat to account for a lower than normal atmospheric pressure such as may be found at a high altitude.

Ajuste para elevación.   Ajuste al termóstato de un refrigerador para regular una presión atmosférica más baja que la normal, como la que se encuentra en elevaciones altas.

Ambient temperature   The surrounding air temperature.

Temperatura ambiente.   Temperatura del aire circundante.

American standard pipe thread   Standard thread used on pipe to prevent leaks.

Rosca estándar estadounidense para tubos.   Rosca estándar utilizada en tubos para evitar fugas.

Ammeter   A meter used to measure current flow in an electrical circuit.

Amperímetro.   Instrumento utilizado para medir el flujo de corriente en un circuito eléctrico.

Amperage   Amount of electron or current flow (the number of electrons passing a point in a given time) in an electrical circuit.

Amperaje.   Cantidad de flujo de electrones o de corriente (el número de electrones que sobrepasa un punto específico en un tiempo fijo) en un circuito eléctrico.

Ampere   Unit of current flow.

Amperio.   Unidad de flujo de corriente.

Anemometer   An instrument used to measure the velocity of air.

Anemómetro.   Instrumento utilizado para medir la velocidad del aire.

Angle valve   Valve with one opening positioned at a 90° angle from the other opening.

Válvula en ángulo.   Válvula con una abertura a un ángulo de 90° con respecto a la otra abertura.

Anode   A terminal or connection point on a semiconductor.

Ánodo.   Punto de conexión o terminal en un semiconductor.

Approach temperature   The difference in temperature between the refrigerant and the leaving water in a chilled-water system.

Temperatura de acercamiento.   Diferencia en temperatura entre el refrigerante y el agua de salida en un sistema de agua enfriada.

ASA   Abbreviation for the American Standards Association, now known as the American National Standards Institute (ANSI).

A.S.A.   Abreviatura de Asociación Estadounidense de Normas, conocida ahora como Instituto Nacional Estadounidense de Normas (ANSI).

ASHRAE   Abbreviation for the American Society of Heating, Refrigerating, and Air-Conditioning Engineers.

ASHRAE.   Abreviatura de Sociedad Estadounidense de Ingenieros de Calefacción, Refrigeración y Acondicionamiento de Aire.

ASME   Abbreviation for the American Society of Mechanical Engineers.

ASME.   Abreviatura de Sociedad Estadounidense de Ingenieros Mecánicos.

Aspect ratio   The ratio of the length to width of a component.

Coeficiente de alargamiento.   Relación del largo al ancho de un componente.

Atmospheric pressure The weight of the atmosphere's gases pressing down on the earth. Equal to 14.696 PSI at sea level and 70°F.

Presión atmosférica. El peso de la presión ejercida por los gases de la atmósfera sobre la tierra, equivalente a 14,696 PSI al nivel del mar a 70°F.

Atom The smallest particle of an element.

Átomo. Partícula más pequeña de un elemento.

Atomize Using pressure to change liquid to small particles of vapor.

Atomizar. Utilizar la presión para cambiar un líquido a partículas pequeñas de vapor.

Automatic combination gas valve A valve for gas furnaces that incorporates a manual control, a gas supply for the pilot, adjustment and safety features for the pilot, a pressure regulator, the controls for and the main gas valve.

Válvula de gas de combinación automática. Válvula de gas para hornos de gas que incorpora un regulador manual, suministro de gas para la llama piloto, ajuste y dispositivos de seguridad, regulador de presión, la válvula de gas principal y los reguladores de la válvula.

Automatic control Controls that react to a change in condition, causing that condition to stabilize.

Regulador automático. Reguladores que reaccionan a un cambio en las condiciones para provocar la estabilidad de dicha condición.

Automatic defrost Using automatic means to remove ice from a refrigeration coil.

Desempañador automático. La utilización de medios automáticos para remover el hielo de una bobina de refrigeración.

Automatic expansion valve A refrigerant control valve that maintains a constant pressure in an evaporator.

Válvula de expansión automática. Válvula de regulación del refrigerante que mantiene una presión constante en un evaporador.

B

Back pressure The pressure on the low-pressure side of a refrigeration system (also known as *suction pressure*).

Contrapresión. La presión en el lado de baja presión de un sistema de refrigeración (conocido también como *presión de aspiración*).

Backseated The position of a refrigeration service valve when the stem is turned away from the valve body and seated.

Asiento trasero. Posición de una válvula de servicio de refrigeración cuando el vástago está orientado fuera del cuerpo de la válvula y aplicado sobre su asiento.

Baffle A plate used to keep fluids from moving back and forth freely in a container.

Deflector. Placa utilizada para evitar el libre movimiento de líquidos en un recipiente.

Balanced port (TXV) A valve that will meter refrigerant at the same rate as when the condenser head pressure is low.

Válvula electrónica de expansión con conducto equilibrado. Válvula que medirá el refrigerante a la misma proporción cuando la presión en la cabeza del condensador sea baja.

Ball check valve A valve with a ball-shaped internal assembly that only allows fluid flow in one direction.

Válvula de retención de bolas. Válvula con un conjunto interior en forma de bola que permite el flujo de fluido en una sola dirección.

Barometer A device used to measure atmospheric pressure that is commonly calibrated in inches or millimeters of mercury. There are two types: mercury column and aneroid.

Barómetro. Dispositivo comúnmente calibrado en pulgadas o en milímetros de mercurio que se utiliza para medir la presión atmosférica. Existen dos tipos: columna de mercurio y aneroide.

Base A terminal on a semiconductor.

Base. Punto terminal en un semiconductor.

Battery A device that produces electricity from the interaction of metals and acid.

Pila. Dispositivo que genera electricidad de la interacción entre metales y el ácido.

Bearing A device that surrounds a rotating shaft and provides a low-friction contact surface to reduce wear from the rotating shaft.

Cojinete. Dispositivo que rodea un árbol giratorio y provee una superficie de contacto de baja fricción para disminuir el desgaste de dicho árbol.

Bellows An accordion-like device that expands and contracts when internal pressure changes.

Fuelles. Dispositivo en forma de acordeón con pliegues que se expanden y contraen cuando la presión interna sufre cambios.

Bellows seal A method of sealing a rotating shaft or valve stem that allows rotary movement of the shaft or stem without leaking.

Cierre hermético de fuelles. Método de sellar un árbol giratorio o el vástago de una válvula que permite el movimiento giratorio del árbol o del vástago sin producir fugas.

Bending spring A coil spring that can be fitted inside or outside a piece of tubing to prevent its walls from collaPSIng when being formed.

Muelle de flexión. Muelle helicoidal que puede acomodarse dentro o fuera de una pieza de tubería para evitar que sus paredes se doblen al ser formadas.

Bimetal Two dissimilar metals fastened together to create a distortion of the assembly with temperature changes.

Bimetal. Dos metales distintos fijados entre sí para producir una distorción del conjunto al ocurrir cambios de temperatura.

Bimetal strip Two dissimilar metal strips fastened back to back.

Banda bimetálica. Dos bandas de metales distintos fijadas entre sí en su parte posterior.

Bleeding Allowing pressure to slowly move from one pressure level to another.

Sangradura. Proceso a través del cual se permite el movimiento de presión de un nivel a otro de manera muy lenta.

Bleed valve A valve with a small port usually used to bleed pressure from a vessel to the atmosphere.

Válvula de descarga. Válvula con un conducto pequeño utilizado normalmente para purgar la presión de un depósito a la atmósfera.

Blocked suction A method of cylinder unloading. The suction line passage to a cylinder in a reciprocating compressor is blocked, thus causing that cylinder to stop pumping.

Aspiración obturada. Método de descarga de un cilindro. El paso del conducto de aspiración a un cilindro en un compresor alternativo se obtura, provocando así que el cilindro deje de bombear.

Blowdown A system in a cooling tower whereby some of the circulating water is bled off and replaced with fresh water to dilute the sediment in the sump.

Vaciado. Sistema en una torre de refrigeración por medio del cual se purga parte del agua circulante y se reemplaza con agua fresca para diluir el sedimento en el sumidero.

Boiler A container in which a liquid may be heated using any heat source. When the liquid is heated to the point that vapor forms and is used as the circulating medium, it is called a *steam boiler.*

Caldera. Recipiente en el que se puede calentar un líquido utilizando cualquier fuente de calor. Cuando se calienta el líquido al punto en que se produce vapor y se utiliza éste como el medio para la circulación, se llama caldera de vapor.

Boiling point The temperature level of a liquid at which it begins to change to a vapor. The boiling temperature is controlled by the vapor pressure above the liquid.

Punto de ebullición. El nivel de temperatura de un líquido al que el líquido empieza a convertirse en vapor. La temperatura de ebullición se regula por medio de la presión del vapor sobre líquido.

Bore   The inside diameter of a cylinder.

Calibre.   Diámetro interior de un cilindro.

Bourdon tube   C-shaped tube manufactured of thin metal and closed on one end. When pressure is increased inside, it tends to straighten. It is used in a gauge to indicate pressure.

Tubo Bourdon.   Tubo en forma de C fabricado de metal delgado y cerrado en uno de los extremos. Al aumentarse la presión en su interior, el tubo tiende a enderezarse. Se utiliza dentro de un calibrador para indicar la presión.

Brazing   High-temperature (above 800°F) soldering of two metals.

Broncesoldadura.   Soldadura de dos metales a temperaturas altas (sobre los 800°F ó 430°C).

Breaker   A heat-activated electrical device used to open an electrical circuit to protect it from excessive current flow.

Interruptor.   Dispositivo eléctrico activado por el calor que se utiliza para abrir un circuito eléctrico a fin protegerlo de un flujo excesivo de corriente.

British thermal unit (Btu)   The amount (quantity) of heat required to raise the temperature of 1 lb of water 1°F.

Unidad térmica británica.   Cantidad de calor necesario para elevar en 1°F (–17.56°C) la temperatura de una libra inglesa de agua.

Btu   Abbreviation for British thermal unit.

Btu.   Abreviatura de unidad térmica británica.

Bulb, sensor   The part of a sealed automatic control used to sense temperature.

Bombilla sensora.   Pieza de un regulador automático sellado que se utiliza para advertir la temperatura.

Burner   A device used to prepare and burn fuel.

Quemador.   Dispositivo utilizado para la preparación y la quema de combustible.

Burr   Excess material squeezed into the end of tubing after a cut has been made. This burr must be removed.

Rebaba.   Exceso de material introducido por fuerza en el extremo de una tubería después de hacerse un corte. Esta rebaba debe removerse.

Butane gas   A liquefied petroleum gas burned for heat.

Gas butano.   Gas licuado derivado del petróleo que se quema para producir calor.

C

Cad cell   A device used to prove the flame in an oil burning furnace containing cadmium sulfide.

Celda de cadmio.   Dispositivo utilizado para probar la llama en un horno de aceite pesado que contiene sulfuro de cadmio.

Calibrate   To adjust instruments or gauges to the correct setting for conditions.

Calibrar.   Ajustar instrumentos o calibradores en posición correcta para su operación.

Capacitance   The term used to describe the electrical storage ability of a capacitor.

Capacitancia.   Término utilizado para describir la capacidad de almacenamiento eléctrico de un capacitador.

Capacitor   An electrical storage device used to start motors (start capacitor) and to improve the efficiency of motors (run capacitor).

Capacitador.   Dispositivo de almacenamiento eléctrico utilizado para arrancar motores (capacitador de arranque) y para mejorar el rendimiento de motores (capacitador de funcionamiento).

Capacity   The rating system of equipment used to heat or cool substances.

Capacidad.   Sistema de clasificación de equipo utilizado para calentar o enfriar sustancias.

Capillary attraction   The attraction of a liquid material between two pieces of material such as two pieces of copper or between copper and brass. For instance, in a joint made up of copper tubing and a brass fitting, the solder filler material has a greater attraction to the copper and brass than to itself and is drawn into the space between them.

Atracción capilar. Atracción de un material líquido entre dos piezas de material, como por ejemplo dos piezas de cobre o cobre y latón. Por ejemplo, en una junta fabricada de tubería de cobre y un accesorio de latón, el material de relleno de la soldadura tiene mayor atracción al cobre y al latón que a sí mismo y es arrastrado hacia el espacio entre éstos.

Capillary tube   A fixed-bore metering device. This is a small-diameter tube that can vary in length from a few inches to several feet. The amount of refrigerant flow needed is predetermined and the length and diameter of the capillary tube is sized accordingly.

Tubo capilar. Dispositivo de medición de calibre fijo. Este es un tubo de diámetro pequeño cuyo largo puede oscilar entre unas cuantas pulgadas a varios pies. La cantidad de flujo de refrigerante requerida es predeterminada y, de acuerdo a esto, se fijan el largo y el diámetro del tubo capilar.

Carbon dioxide   A by-product of natural gas combustion that is not harmful.

Bióxido de carbono. Subproducto de la combustión del gas natural que no es nocivo.

Carbon monoxide   A poisonous, colorless, odorless, tasteless gas generated by incomplete combustion.

Monóxido de carbono. Gas mortífero, inodoro, incoloro e insípido que se desprende en la combustión incompleta del carbono.

Catalytic combustor stove   A stove that contains a cell-like structure consisting of a substrate, a washcoat, and a catalyst, that produces a chemical reaction causing pollutants to be burned at much lower temperatures.

Estufa de combustor catalítico. Estufa con una estructura en forma de celda compuesta de una subestructura, una capa brochada y un catalizador que produce una reacción química. Esta reacción provoca la quema de contaminantes a temperaturas mucho más bajas.

Cathode   A terminal or connection point on a semiconductor.

Cátodo. Punto de conexión o terminal en un semiconductor.

Cavitation   A vapor formed due to a drop in pressure in a pumping system. The presence of air at a pump inlet may be caused at a cooling tower if the pressure is low and water is turned to vapor.

Cavitación. Vapor producido como consecuencia de una caída de presión en un sistema de bombeo. El aire a la entrada de una bomba puede ser producido en una torre de refrigeración si la presión es baja y el agua se convierte en vapor.

Celsius scale   A temperature scale with 1,200 graduations between water freezing (0°C) and water boiling (100°C).

Escala Celsio. Escala dividida en cien grados, con el cero marcado a la temperatura de fusión del hielo (0°C) y el cien a la de ebullición del agua (100°C).

Centigrade scale   See Celsius.

Centígrado. Véase escala Celsio.

Centrifugal compressor   A compressor used for large refrigeration systems. It is not positive displacement, but it is similar to a blower.

Compresor centrífugo. Compresor utilizado en sistemas grandes de refrigeración. No es desplazamiento positivo, pero es similar a un soplador.

Centrifugal switch   A switch that uses a centrifugal action to disconnect the start windings from the circuit.

Conmutador centrífugo. Conmutador que utiliza una acción centrífuga para desconectar los devanados de arranque del circuito.

Change of state The condition that occurs when a substance changes from one physical state to another, such as ice to water and water to steam.

Cambio de estado. Condición que ocurre cuando una sustancia cambia de un estado físico a otro, como por ejemplo el hielo a agua y el agua a vapor.

Charge The quantity of refrigerant in a system.

Carga. Cantidad de refrigerante en un sistema.

Charging cylinder A device that allows the technician to accurately charge a refrigeration system with refrigerant.

Cilindro cargador. Dispositivo que le permite al mecánico cargar correctamente un sistema de refrigeración con refrigerante.

Check valve A device that permits fluid flow in one direction only.

Válvula de retención. Dispositivo que permite el flujo de fluido en una sola dirección.

Chilled-water system An air-conditioning system that circulates refrigerated water to the area to be cooled. The refrigerated water picks up heat from the area, thus cooling the area.

Sistema de agua enfriada. Sistema de acondicionamiento de aire que hace circular agua refrigerada al área que será enfriada. El agua refrigerada atrapa el calor del área y la enfria.

Chiller purge unit A system that removes air from a low-pressure chiller.

Unidad enfriadora de purga. Sistema que remueve el aire de un enfriador de baja presión.

Chill factor A factor or number that is a combination of temperature, humidity, and wind velocity, which is used to compare a relative condition to a known condition.

Factor de frío. Factor o número que es una combinación de la temperatura, la humedad y la velocidad del viento utilizado para comparar una condición relativa a una condición conocida.

Chimney A vertical shaft used to convey flue gases above the rooftop.

Chimenea. Cañón vertical utilizado para conducir los gases de combustión por encima del techo.

Chimney effect A term used to describe air or gas that expands and rises when heated.

Efecto de chimenea. Término utilizado para describir el aire o el gas cuando se expande y sube al calentarse.

Chlorinated polyvinyl chloride (CPVC) Plastic pipe similar to PVC except that it can be used with temperatures up to 180°F at 100 PSIG.

Cloruro de polivinilo clorado (CPVC). Tubo plástico similar al PVC, pero que puede utilizarse a temperaturas de hasta 180°F (82°C) a 100 PSIG [indicador de libras por pulgada cuadrada].

Chlorofluocarbons (CFC) Those refrigerants thought to contribute to the depletion of the ozone layer.

Cloroflurocarburos. Líquidos refrigerantes que, según algunos, han contribuido a la reducción de la capa de ozono.

Circuit An electron or fluid flow path that makes a complete loop.

Circuito. Electrón o trayectoria del flujo de fluido que hace un ciclo completo.

Circuit breaker A device that opens an electric circuit when an overload occurs.

Interruptor para circuitos. Dispositivo que abre un circuito eléctrico cuando ocurre una sobrecarga.

Clamp-on ammeter An instrument that can be clamped around one conductor in an electrical circuit to measure the current.

Amperímetro fijado con abrazadera. Instrumento que puede fijarse con una abrazadera a un conductor en un circuito eléctrico y medir la corriente.

Clearance volume The volume at the top of the stroke in a compressor cylinder between the top of the piston and the valve plate.

Volumen de holgura. Volumen en la parte superior de una carrera en el cilindro de un compresor entre la parte superior del pistón y la placa de una válvula.

Closed circuit A complete path for electrons to flow on.

Circuito cerrado. Circuito de trayectoria ininterrumpida que permite un flujo continuo de electrones.

Closed loop Piping circuit that is complete and not open to the atmosphere.

Ciclo cerrado. Circuito de tubería completo y no abierto a la atmósfera.

Code The local, state, or national rules that govern safe installation and service of systems and equipment for the purpose of safety of the public and trade personnel.

Código. Reglamentos locales, estaduales o federales que rigen la instalación segura y el servicio de sistemas y equipo con el propósito de garantizar la seguridad del personal público y profesional.

Coefficient of performance (COP) The ratio of usable output energy divided by input energy.

Coeficiente de rendimiento. Relación de la de energía de salida utilizable dividida por la energía de entrada.

CO_2 indicator An instrument used to detect the quantity of carbon dioxide in flue gas for efficiency purposes.

Indicador del CO_2. Instrumento utilizado para detectar la cantidad de bióxido de carbono en el gas de combustión a fin de lograr un mejor rendimiento.

Cold The word used to describe heat at lower levels of intensity.

Frío. Término utilizado para describir el calor a niveles de intensidad más bajos.

Cold anticipator A device that anticipates a need for cooling and starts the cooling system early enough for it to reach capacity when it is needed.

Anticipador de frío. Dispositivo que anticipa la necesidad de enfriamiento y pone en marcha el sistema de enfriamiento con suficiente anticipación para que éste alcance su máxima capacidad cuando vaya a ser utilizado.

Cold junction The opposite junction to the hot junction in a thermocouple.

Empalme frío. El empalme opuesto al empalme caliente en un termopar.

Cold trap A device to help trap moisture in a refrigeration system.

Trampa del frío. Dispositivo utilizado para ayudar a atrapar la humedad en un sistema de refrigeración.

Cold wall The term used in comfort heating to describe a cold outside wall and its effect on human comfort.

Pared fría. Término utilizado en la calefacción para comodidad que describe una pared exterior fría y sus efectos en la comodidad de una persona.

Collector A terminal on a semiconductor.

Colector. Punto terminal en un semiconductor.

Combustion A reaction called rapid oxidation or burning produced with the right combination of a fuel, oxygen, and heat.

Combustión. Reacción conocida como oxidación rápida o quema producida con la combinación correcta de combustible, oxígeno y calor.

Comfort chart A chart used to compare the relative comfort of one temperature and humidity condition to another condition.

Esquema de comodidad. Esquema utilizado para comparar la comodidad relativa de una condición de temperatura y humedad a otra condición.

Compound gauge A gauge used to measure the pressure above and below the atmosphere's standard pressure. It is a Bourdon tube sensing device and can be found on all gauge manifolds used for air-conditioning and refrigeration service work.

Calibrador compuesto. Calibrador utilizado para medir la presión mayor y menor que la presión estándar de la atmósfera. Es un dispositivo sensor de tubo Bourdon que puede encontrarse en todos los distribuidores de calibrador utilizados para el servicio de sistemas de acondicionamiento de aire y de refrigeración.

Compression A term used to describe a vapor when pressure is applied and the molecules are compacted closer together.

Compresión. Término utilizado para describir un vapor cuando se aplica presión y se compactan las moléculas.

Compression ratio A term used with compressors to describe the actual difference in the low- and high-pressure sides of the compression cycle. It is absolute discharge pressure divided by absolute suction pressure.

Relación de compresión. Término utilizado con compresores para describir la diferencia real en los lados de baja y alta presión del ciclo de compresión. Es la presión absoluta de descarga dividida por la presión absoluta de aspiración.

Compressor A vapor pump that pumps vapor (refrigerant or air) from one pressure level to a higher pressure level.

Compresor. Bomba de vapor que bombea el vapor (refrigerante o aire) de un nivel de presión a un nivel de presión más alto.

Compressor displacement The internal volume of a compressor, used to calculate the pumping capacity of the compressor.

Desplazamiento del compresor. Volumen interno de un compresor, utilizado para calcular la capacidad de bombeo del mismo.

Compressor shaft seal The seal that prevents refrigerant inside the compressor from leaking around the rotating shaft.

Junta de estanqueidad del árbol del compresor. La junta de estanqueidad que evita la fuga, alrededor del árbol giratorio, del refrigerante en el interior del compresor.

Concentrator That part of an absorption chiller where the dilute salt solution is boiled to release the water.

Concentrador. El lugar en el enfriador por absorción donde se hierve la solución salina diluida para liberar el agua.

Condensate The moisture collected on an evaporator coil.

Condensado. Humedad acumulada en la bobina de un evaporador.

Condensate pump A small pump used to pump condensate to a higher level.

Bomba para condensado. Bomba pequeña utilizada para bombear el condensado a un nivel más alto.

Condensation Liquid formed when a vapor condenses.

Condensación. El líquido formado cuando se condensa un vapor.

Condense Changing a vapor to a liquid at a particular pressure.

Condensar. Convertir un vapor en líquido a una presión específica.

Condenser The component in a refrigeration system that transfers heat from the system by condensing refrigerant.

Condensador. Componente en un sistema de refrigeración que transmite el calor del sistema al condensar el refrigerante.

Condenser flooding A method of maintaining a correct head pressure by adding liquid refrigerant to the condenser from a receiver to increase the head pressure.

Inundación del condensador. Método de mantener una presión correcta en la cabeza agregando refrigerante líquido al condensador de un receptor para aumentar la presión en la cabeza.

Condensing gas furnace   A furnace with a condensing heat exchanger that condenses moisture from the flue gases resulting in greater efficiency.

Horno para condensación de gas.   Horno con un intercambiador de calor para condensación que condensa la humedad de los gases de combustión. El resultado será un mayor rendimiento.

Condensing pressure   The pressure that corresponds to the condensing temperature in a refrigeration system.

Presión para condensación.   La presión que corresponde a la temperatura de condensación en un sistema de refrigeración.

Condensing temperature   The temperature at which a vapor changes to a liquid.

Temperatura de condensación.   Temperatura a la que un vapor se convierte en líquido.

Condensing unit   A complete unit that includes the compressor and the condensing coil.

Conjunto del condensador.   Unidad completa que incluye el compresor y la bobina condensadora.

Conduction   Heat transfer from one molecule to another within a substance or from one substance to another.

Conducción.   Transmisión de calor de una molécula a otra dentro de una sustancia o de una sustancia a otra.

Conductivity   The ability of a substance to conduct electricity or heat.

Conductividad.   Capacidad de una sustancia de conducir electricidad o calor.

Conductor   A pathway through which electrical energy can flow.

Conductor.   Trayectoria que permite un flujo continuo de energía eléctrica.

Contactor   A larger version of the relay. It can be repaired or rebuilt and has movable and stationary contacts.

Contactador.   Versión más grande del relé. Puede ser reparado o reconstruido. Tiene contactos móviles y fijos.

Contaminant   Any substance in a refrigeration system that is foreign to the system, particularly if it causes damage.

Contaminante.   Cualquier sustancia en un sistema de refrigeración extraña a éste, principalmente si causa averías.

Control   A device to stop, start, or modulate flow of electricity or fluid to maintain a preset condition.

Regulador.   Dispositivo para detener, poner en marcha o modular el flujo de electricidad o de fluido a fin de mantener una condición establecida con anticipación.

Control system   A network of controls to maintain desired conditions in a system or space.

Sistema de regulación.   Red de reguladores que mantienen las condiciones deseadas en un sistema o un espacio.

Convection   Heat transfer from one place to another, using a fluid.

Convección.   Transmisión de calor de un lugar a otro por medio de un fluido.

Conversion factor   A number used to convert from one equivalent value to another.

Factor de conversión.   Número utilizado en la conversión de un valor equivalente a otro.

Cooler   A walk-in or reach-in refrigerated box.

Nevera.   Caja refrigerada donde se puede entrar o introducir la mano.

Cooling tower   The final device in many water-cooled systems that rejects heat from the system into the atmosphere by evaporation of water.

Torre de refrigeración.   Dispositivo final en muchos sistemas enfriados por agua, que dirige el calor del sistema a la atmósfera por medio de la evaporación de agua.

Copper plating Small amounts of copper are removed by electrolysis and deposited on the ferrous metal parts in a compressor.

Encobrado. Remoción de pequeñas cantidades de cobre por medio de electrólisis que luego se colocan en las piezas de metal férreo en un compresor.

Corrosion A chemical action that eats into or wears away material from a substance.

Corrosión. Acción química que carcome o desgasta el material de una sustancia.

Counter EMF Voltage generated or induced above the applied voltage in a single-phase motor.

Contra EMF. Tensión generada o inducida sobre la tensión aplicada en un motor unifásico.

Counterflow Two fluids flowing in opposite directions.

Contraflujo. Dos fluidos que fluyen en direcciones opuestas.

Coupling A device for joining two fluid flow lines. Also the device connecting a motor drive shaft to the driven shaft in a direct drive system.

Acoplamiento. Dispositivo utilizado para la conexión de dos conductos de flujo de fluido. Es también el dispositivo que conecta un árbol de mando del motor al árbol accionado en un sistema de mando directo.

Crackage Small spaces in a structure that allow air to infiltrate the structure.

Formación de grietas. Espacios pequeños en una estructura que permiten la infiltración del aire dentro de la misma.

Crankcase heat Heat provided to the compressor crankcase.

Calor para el cárter del cigüeñal. Calor suministrado al cárter del cigüeñal del compresor.

Crankcase pressure regulator (CPR) A valve installed in the suction line, usually close to the compressor. It is used to keep a low-temperature compressor from overloading on a hot pull-down.

Regulador de la presión del cárter del cigüeñal. Válvula instalada en el conducto de aspiración, normalmente cerca del compresor. Se utiliza para evitar la sobrecarga en un compresor de temperatura baja durante un arrastre caliente hacia abajo.

Crankshaft seal Same as the compressor shaft seal.

Junta de estanqueidad del árbol del cigüeñal. Exactamente igual que la junta de estanqueidad del árbol del compresor.

Crankshaft throw The off-center portion of a crankshaft that changes from rotating motion to reciprocating motion.

Excentricidad del cigüeñal. Porción descentrada de un cigüeñal que cambia el movimiento giratorio a un movimiento alternativo.

Creosote A mixture of unburned organic material found in the smoke from a wood-burning fire.

Creosota. Mezcla del material orgánico no quemado que se encuentra en el humo proveniente de un incendio de madera.

Crisper A refrigerated compartment that maintains a high humidity and a low temperature.

Encrespador. Compartimiento refrigerado que mantiene una humedad alta y una temperatura baja.

Cross charge A control with a sealed bulb that contains two different fluids that work together for a common specific condition.

Carga transversal. Regulador con una bombilla sellada compuesta de dos fluidos diferentes que pueden funcionar juntos para una condición común específica.

Cross liquid charge bulb A type of charge in the sensing bulb of the TXV that has different characteristics from the system refrigerant. This is designed to help prevent liquid refrigerant from flooding to the compressor at start-up.

Bombilla de carga del líquido transversal. Tipo de carga en la bombilla sensora de la válvula electrónica de expansión que tiene características diferentes a las del refrigerante del sistema. La carga está diseñada para ayudar a evitar que el refrigerante líquido se derrame dentro del compresor durante la puesta en marcha.

Cross vapor charge bulb Similar to the vapor charge bulb but contains a fluid different from the system refrigerant. This is a special type of charge that produces a different pressure/temperature relationship under different conditions.

Bombilla de carga del vapor transversal. Similar a la bombilla de carga del vapor pero contiene un fluido diferente al del refrigerante del sistema. Esta es una carga de tipo especial y produce una relación diferente entre la presión y la temperatura bajo condiciones diferentes.

Crystallization When a salt solution becomes too concentrated and part of the solution turns to salt.

Cristalización. Condición que ocurre cuando una solución salina se concentra demasiado y una parte de la solución se convierte en sal.

Current, electrical Electrons flowing along a conductor.

Corriente eléctrica. Electrones que fluyen a través de un conductor.

Current relay An electrical device activated by a change in current flow.

Relé para corriente. Dispositivo eléctrico accionado por un cambio en el flujo de corriente.

Cut-in and cut-out The two points at which a control opens or closes its contacts based on the condition it is supposed to maintain.

Puntos de conexión y desconexión. Los dos puntos en los que un regulador abre o cierra sus contactos según las condiciones que debe mantener.

Cycle A complete sequence of events (from start to finish) in a system.

Ciclo. Secuencia completa de eventos, de comienzo a fin, que ocurre en un sistema.

Cylinder A circular container with straight sides used to contain fluids or to contain the compression process (the piston movement) in a compressor.

Cilindro. Recipiente circular con lados rectos, utilizado para contener fluidos o el proceso de compresión (movimiento del pistón) en un compresor.

Cylinder, compressor The part of the compressor that contains the piston and its travel.

Cilindro del compresor. Pieza del compresor que contiene el pistón y su movimiento.

Cylinder head, compressor The top to the cylinder on the high-pressure side of the compressor.

Culata del cilindro del compresor. Tapa del cilindro en el lado de alta presión del compresor.

Cylinder, refrigerant The container that holds refrigerant.

Cilindro del refrigerante. El recipiente que contiene el refrigerante.

Cylinder unloading A method of providing capacity control by causing a cylinder in a reciprocating compressor to stop pumping.

Descarga del cilindro. Método de suministrar regulación de capacidad provocando que el cilindro en un compresor alternativo deje de bombear.

D

Damper A component in an air distribution system that restricts airflow for the purpose of air balance.

Desviador. Componente en un sistema de distribución de aire que limita el flujo de aire para mantener un equilibrio de aire.

Declination angle The angle of the tilt of the Earth on its axis.

Ángulo de declinación. Ángulo de inclinación de la Tierra en su eje.

Defrost Melting of ice.

Descongelar. Convertir hielo en líquido.

Defrost cycle The portion of the refrigeration cycle that melts the ice off the evaporator.

Ciclo de descongelación. Parte del ciclo de refrigeración que derrite el hielo del evaporador.

Defrost timer A timer used to start and stop the defrost cycle.

Temporizador de descongelación. Temporizador utilizado para poner en marcha y detener el ciclo de descongelación.

Degreaser A cleaning solution used to remove grease from parts and coils.

Desengrasador. Solución limpiadora utilizada para remover la grasa de piezas y bobinas.

Dehumidify To remove moisture from air.

Deshumidificar. Remover la humedad del aire.

Dehydrate To remove moisture from a sealed system or a product.

Deshidratar. Remover la humedad de un sistema sellado o un producto.

Density The weight per unit of volume of a substance.

Densidad. Relación entre el peso de una sustancia y su volumen.

Desiccant Substance in a refrigeration system drier that collects moisture.

Disecante. Sustancia en el secador de un sistema de refrigeración que acumula la humedad.

Design pressure The pressure at which the system is designed to operate under normal conditions.

Presión de diseño. Presión a la que el sistema ha sido diseñado para funcionar bajo condiciones normales.

De-superheating Removing heat from the superheated hot refrigerant gas down to the condensing temperature.

De sobrecalentamiento. Reducir el calor del gas caliente del refrigerante sobrecalentado hasta alcanzar la temperatura de condensación.

Detector A device to search and find.

Detector. Dispositivo de búsqueda y detección.

Dew Moisture droplets that form on a cool surface.

Rocío. Gotitas de humedad que se forman en una superficie fría.

Dew point The exact temperature at which moisture begins to form.

Punto de rocío. Temperatura exacta a la que la humedad comienza a formarse.

DIAC A semiconductor often used as a voltage-sensitive switching device.

DIAC. Semiconductor utilizado frecuentemente como dispositivo de conmutación sensible a la tensión.

Diaphragm A thin flexible material (metal, rubber, or plastic) that separates two pressure differences.

Diafragma. Material delgado y flexible, como por ejemplo el metal, el caucho o el plástico, que separa dos presiones diferentes.

Die A tool used to make an external thread such as on the end of a piece of pipe.

Troquel. Herramienta utilizada para formar un filete externo, como por ejemplo, en el extremo de un tubo.

Differential The difference in the cut-in and cut-out points of a control, pressure, time, temperature, or level.

Diferencial. Diferencia entre los puntos de conexión y desconexión de un regulador, una presión, un intervalo de tiempo, una temperatura o un nivel.

Diffuser The terminal or end device in an air distribution system that directs air in a specific direction using louvers.

Placa difusora. Punto o dispositivo terminal en un sistema de distribución de aire que dirige el aire a una dirección específica, utilizando aberturas tipo celosía.

Diode A solid-state device composed of both P-type and N-type material. Current connected in a circuit will only flow one way. When the diode is reversed, current will not flow.

Diodo. Dispositivo de estado sólido compuesto de material P y de material N. Cuando se conecta a un circuito de una manera, la corriente fluye. Cuando la dirección del diodo cambia, la corriente deja de fluir.

Direct current (DC) Electricity in which all electrons flow continuously in one direction.

Corriente continua. Electricidad en la que todos los electrones fluyen continuamente en una sola dirección.

Direct expansion The term used to describe an evaporator with an expansion device other than a low-side float type.

Expansión directa. Término utilizado para describir un evaporador con un dispositivo de expansión diferente al tipo de dispositivo flotador de lado bajo.

Direct spark ignition (DSI) A system that provides direct ignition to the main burner.

Encendido de chispa directa. Sistema que le provee un encendido directo al quemador principal.

Discus compressor A reciprocating compressor distinguished by its disc-type valve system.

Compresor de disco. Compresor alternativo caracterizado por su sistema de válvulas de tipo disco.

Discus valve A reciprocating compressor valve, designed with a low clearance volume and a larger bore.

Válvula de disco. Válvula de compresor alternativo diseñada con un volumen de holgura bajo y un calibre más grande.

Distributor A component installed at the outlet of the expansion valve that distributes the refrigerant to each evaporator circuit.

Distribuidor. Componente instalado a la salida de la váluva de expansión que distribuye el refrigerante a cada circuito del evaporador.

Doping Adding an impurity to a semiconductor to produce a desired charge.

Impurificación. La adición de una impureza para producir una carga deseada.

Double flare A connection used on copper, aluminum, or steel tubing that folds tubing wall to a double thickness.

Abocinado doble. Conexión utilizada en tuberías de cobre, aluminio, o acero que pliega la pared de la tubería y crea un espesor doble.

Dowel pin A pin, which may or may not be tapered, used to align and fasten two parts.

Pasador de espiga. Pasador, que puede o no ser cónico, utilizado para alinear y fijar dos piezas.

Draft gauge A gauge used to measure very small pressures (above and below atmospheric) and to compare them to the atmosphere's pressure. Used to determine the flow of flue gas in a chimney or vent.

Calibrador de tiro. Calibrador utilizado para medir presiones sumamente pequeñas, (mayores o menores que la atmosférica), y compararlas con la presión de la atmósfera. Utilizado para determinar el flujo de gas de combustión en una chimenea o válvula.

Drier A device used in a refrigerant line to remove moisture.

Secador. Dispositivo utilizado en un conducto de refrigerante para remover la humedad.

Drip pan A pan used to collect moisture condensing on an evaporator coil in an air-conditioning or refrigeration system.

Colector de goteo. Un colector formado para acumular la humedad que se condensa en la bobina de un evaporador en un sistema de acondicionamiento de aire o de refrigeración.

Dry bulb temperature The temperature measured using a plain thermometer.

Temperatura de bombilla seca. Temperatura que se mide con un termómetro sencillo.

Duct A sealed channel used to convey air from the system to and from the point of utilization.

Conducto. Canal sellado que se emplea para dirigir el aire del sistema hacia y desde el punto de utilización.

E

Eccentric An off-center device that rotates in a circular direction around a shaft.

Excéntrico. Dispositivo descentrado que gira en un círculo alrededor de un árbol.

Eddy current test A test with an instrument to find potential failures in evaporator or condenser tubes.

Prueba para la corriente de Foucault. Prueba que se realiza con un instrumento para detectar posibles fallas en los tubos del evaporador o del condensador.

Effective temperature Different combinations of temperature and humidity that provide the same comfort level.

Temperatura efectiva. Diferentes combinaciones de temperatura y humedad que proveen el mismo nivel de comodidad.

Electrical power Electrical power is measured in watts. One watt is equal to one ampere flowing with a potential of one volt. Watts = Volts × Amperes ($P = E \times I$)

Potencia eléctrica. La potencia eléctrica se mide en watios. Un watio equivale a un amperio que fluye con una potencia de un voltio. Watios = voltios × amperios $P = E \times I$)

Electrical shock An electrical current that travels through a human body.

Sacudida eléctrica. Paso brusco de una corriente eléctrica a través del cuerpo humano.

Electric heat The process of using resistance to convert electrical energy into heat.

Calor eléctrico. Proceso de convertir energía eléctrica en calor a través de la resistencia.

Electromagnet A coil of wire wrapped around a soft iron core that creates a magnet.

Electroimán. Bobina de alambre devanado alrededor de un núcleo de hierro blando que crea un imán.

Electron The smallest portion of an atom that carries a negative charge.

Electrón. Partícula más pequeña de un átomo que tiene carga negativa.

Electronic air filter A filter that charges dust particles using high-voltage direct current and then collects these particles on a plate of an opposite charge.

Filtro de aire electrónico. Filtro que carga partículas de polvo utilizando una corriente continua de alta tensión y luego las acumula en una placa de carga opuesta.

Electronic charging scale An electronically operated scale used to accurately charge refrigeration systems by weight.

Escala electrónica para carga. Escala accionada electrónicamente que se utiliza para cargar correctamente sistemas de refrigeración por peso.

Electronic expansion valve (EEV) A metering valve that uses a thermistor as a temperature-sensing element which varies the voltage to a heat motor-operated valve.

Válvula electrónica de expansión. Válvula de medición que utiliza un termistor como elemento sensor de temperatura para variar la tensión a una válvula de calor accionada por motor.

Electronic leak detector An instrument used to detect gases in very small portions by using electronic sensors and circuits.

Detector electrónico de fugas. Instrumento que se emplea para detectar cantidades de gases sumamente pequeñas utilizando sensores y circuitos electrónicos.

Electronics The use of electron flow in conductors, semiconductors, and other devices.

Electrónica. La utilización del flujo de electrones en conductores, semiconductores y otros dispositivos.

Emitter A terminal on a semiconductor.

Emisor. Punto terminal en un semiconductor.

End bell The end structure of an electric motor that normally contains the bearings and the lubrication system.

Extremo acampanado. Estructura terminal de un motor eléctrico que generalmente contiene los cojinetes y el sistema de lubrificación.

End play The amount of lateral travel in a motor or pump shaft.

Holgadura. Amplitud de movimiento lateral en un motor o en el árbol de una bomba.

Energy The capacity for doing work.

Energía. Capacidad para realizar un trabajo.

Energy efficiency ratio (EER) An equipment efficiency rating that is determined by dividing the output in Btu's by the input in watts. This does not take into account the start-up and shutdown for each cycle.

Relación del rendimiento de engería. Clasificación del rendimiento de un equipo que se determina al dividir la salida en Btuh por la entrada en watios. Esto no toma en cuenta la puesta en marcha y la parada de cada ciclo.

Enthalpy The amount of heat a substance contains determined from a predetermined base or point.

Entalpía. Cantidad de calor que contiene una sustancia, establecida desde una base o un punto predeterminado.

Environment Our surroundings, including the atmosphere.

Medio ambiente. Nuestros alrededores, incluyendo la atmósfera.

Ethane gas The fossil fuel, natural gas, used for heat.

Gas etano. Combustible fósil, gas natural, utilizados para generar calor.

Evacuation The removal of any gases not characteristic to a system or vessel.

Evacuación. Remoción de los gases no característicos de un sistema o depósito.

Evaporation The condition that occurs when heat is absorbed by liquid and is changed to vapor.

Evaporación. Condición que ocurre cuando un líquido absorbe calor y se convierte en vapor.

Evaporator The component in a refrigeration system that absorbs heat into the system and evaporates the liquid refrigerant.

Evaporador. El componente en un sistema de refrigeración que absorbe el calor hacia el sistema y evapora el refrigerante líquido.

Evaporator fan A forced convector used to improve the efficiency of an evaporator by air movement over the coil.

Abanico del evaporador. Convector forzado que se utiliza para mejorar el rendimiento de un evaporador por medio del movimiento de aire a través de la bobina.

Evaporator pressure regulator (EPR) A mechanical control installed in the suction line at the evaporator outlet that keeps the evaporator pressure from dropping below a certain point.

Regulador de presión del evaporador. Regulador mecánico instalado en el conducto de aspiración de la salida del evaporador; evita que la presión del evaporador caiga hasta alcanzar un nivel por debajo del nivel específico.

Evaporator types (1) Flooded—an evaporator where the liquid refrigerant level is maintained to the top of the heat exchange coil. (2) Dry type—an evaporator coil that achieves the heat exchange process with a minimum of refrigerant charge.

Clases de evaporadores. (1) Inundado—un evaporador en el que se mantiene el nivel del refrigerante líquido en la parte superior de la bobina de intercambio de calor. (2) Seco—una bobina de evaporador que logra el proceso de intercambio de calor con una mínima cantidad de carga de refrigerante.

Exhaust valve The movable component in a refrigeration compressor that allows hot gas to flow to the condenser and prevents it from refilling the cylinder on the downstroke.

Válvula de escape. Componente móvil en un compresor de refrigeración que permite el flujo de gas caliente al condensador y evita que este gas rellene el cilindro durante la carrera descendente.

Expansion (metering) device The component between the high-pressure liquid line and the evaporator that feeds the liquid refrigerant into the evaporator.

Dispositivo de (medición) de expansión. Componente entre el conducto de líquido de alta presión y el evaporador que alimenta el refrigerante líquido hacia el evaporador.

Expansion joint A flexible portion of a piping system or building structure that allows for expansion of the materials due to temperature changes.

Junta de expansión. Parte flexible de un sistema de tubería o de la estructura de un edificio que permite la expansión de los materiales debido a cambios de temperatura.

External drive An external type of compressor motor drive, as opposed to a hermetic compressor.

Motor externo. Motor tipo externo de un compresor, en comparación con un compresor hermético.

External equalizer The connection from the evaporator outlet to the bottom of the diaphragm on a thermostatic expansion valve.

Equilibrador externo. Conexión de la salida del evaporador a la parte inferior del diafragma en una válvula de expansión termostática.

F

Fahrenheit scale The temperature scale that places the boiling point of water at 212°F and the freezing point at 32°F.

Escala Fahrenheit. Escala de temperatura en la que el punto de ebullición del agua se encuentra a 212°F y el punto de fusión del hielo a 32°F.

Fan A device that produces a pressure difference in air, causing it to move.

Abanico. Dispositivo que produce una diferencia de presión en el aire para moverlo.

Fan cycling The use of a pressure control to turn a condenser fan on and off to maintain a correct pressure within the system.

Funcionamiento cíclico. La utilización de un regulador de presión para poner en marcha y detener el abanico de un condensador a fin de mantener una presión correcta dentro del sistema.

Fan relay coil A magnetic coil that controls the starting and stopping of a fan.

Bobina de relé del abanico. Bobina magnética que regula la puesta en marcha y la parada de un abanico.

Farad The unit of capacity of a capacitor. Capacitors in the HVAC/R industry are rated in microfarads.

Faradio. Unidad de capacidad de un capacitador. En nuestro medio, los capacitadores se clasifican en microfaradios.

Female thread The internal thread in a fitting.

Filete hembra. Filete interno en un accesorio.

Fill or wetted-surface method Water in a cooling tower is spread out over a wetted surface while air is passed over it to enhance evaporation.

Método de relleno o de superficie mojada. El agua en una torre de refrigeración se extiende sobre una superficie mojada mientras el aire se dirige por encima de la misma para facilitar la evaporación.

Film factor The relationship between the medium giving up heat and the heat exchange surface (evaporator). This relates to the velocity of the medium passing over the evaporator. When the velocity is too low, the film between the air and the evaporator becomes greater and becomes an insulator, which slows the heat exchange.

Factor de película. Relación entre el medio que emite calor y la superficie del intercambiador de calor (evaporador). Esto se refiere a la velocidad del medio que pasa sobre el evaporador. Cuando la velocidad es demasiado lenta, la película entre el aire y el evaporador se expande y se convierte en un aislador, disminuyendo así la velocidad del intercambio del calor.

Filter A fine mesh or porous material that removes particles from passing fluids.

Filtro. Malla fina o material poroso que remueve partículas de los fluidos que pasan por él.

Fin comb A hand tool used to straighten the fins on an air-cooled condenser.

Herramienta para aletas. Herramienta manual utilizada para enderezar las aletas en un condensador enfriado por aire.

Fixed-bore device An expansion device with a fixed diameter that does not adjust to varying load conditions.

Dispositivo de calibre fijo. Dispositivo de expansión con un diámetro fijo que no se ajusta a las condiciones de carga variables.

Fixed resistor A nonadjustable resistor. The resistance cannot be changed.

Resistor fijo. Resistor no ajustable. La resistencia no se puede cambiar.

Flapper valve See reed valve.

Chapaleta. Véase válvula de lámina.

Flare The angle that may be fashioned at the end of a piece of tubing to match a fitting and create a leak-free connection.

Abocinado. Ángulo que puede formarse en el extremo de una pieza de tubería para emparejar un accesorio y crear una conexión libre de fugas.

Flare nut A connector used in a flare assembly for tubing.

Tuerca abocinada. Conector utilizado en un conjunto abocinado para tuberías.

Flash gas A term used to describe the pressure drop in an expansion device when some of the liquid passing through the valve is changed quickly to a gas and thereby cools the remaining liquid to the corresponding temperature.

Gas instantáneo. Término utilizado para describir la caída de la presión en un dispositivo de expansión cuando una parte del líquido que pasa a través de la válvula se convierte rápidamente en gas y enfria el líquido restante a la temperatura correspondiente.

Float, valve or switch An assembly used to maintain or monitor a liquid level.

Válvula o conmutador de flotador. Conjunto utilizado para mantener o controlar el nivel de un líquido.

Flooded system A refrigeration system operated with the liquid refrigerant level very close to the outlet of the evaporator coil for improved heat exchange.

Sistema inundado. Sistema de refrigeración que funciona con el nivel del refrigerante líquido bastante próximo a la salida de la bobina del evaporador para mejorar el intercambio de calor.

Flooding The term applied to a refrigeration system when the liquid refrigerant reaches the compressor.

Inundación. Término aplicado a un sistema de refrigeración cuando el nivel del refrigerante líquido llega al compresor.

Flue The duct that carries the products of combustion out of a structure for a fossil or a solid fuel system.

Conducto de humo. Conducto que extrae los productos de combustión de una estructura en sistemas de combustible fósil o sólido.

Flue gas analysis instruments Instruments used to analyze the operation of fossil fuel burning equipment, such as oil and gas furnaces, by analyzing the flue gases.

Instrumentos para el análisis del gas de combustión. Instrumentos utilizados para llevar a cabo un análisis del funcionamiento de los quemadores de combustible fósil, como por ejemplo hornos de aceite pesado o gas, a través del estudio de los gases de combustión.

Fluid The state of matter of liquids and gases.

Fluido. Estado de la materia de líquidos y gases.

Fluid expansion device Using a bulb or sensor, tube, and diaphragm filled with fluid, this device will produce movement at the diaphragm when the fluid is heated or cooled. A bellows may be added to produce more movement. These devices may contain vapor and liquid.

Dispositivo para la expansión del fluido. Utilizando una bombilla o sensor, un tubo y un diafragma lleno de fluido, este dispositivo generará movimiento en el diafragma cuando se caliente o enfrie el fluido. Se le puede agregar un fuelle para generar aún más movimiento. Dichos dispositivos pueden contener vapor y líquido.

Flush The process of using a fluid to push contaminants from a system.

Descarga. Proceso de utilizar un fluido para remover los contaminantes de un sistema.

Flux A substance applied to soldered and brazed connections to prevent oxidation during the heating process.

Fundente. Sustancia aplicada a conexiones soldadas y broncesoldadas para evitar la oxidación durante el proceso de calentamiento.

Foaming A term used to describe oil when it has liquid refrigerant boiling out of it.

Espumación. Término utilizado para describir el aceite cuando el refrigerante líquido se derrama del mismo.

Foot-pound A unit of energy that describes the amount of work accomplished by lifting 1 lb of weight 1 ft.

Libra-pie. Medida de la cantidad de energía o fuerza que se requiere para levantar una libra a una distancia de un pie; unidad de energía.

Force Energy exerted.

Fuerza. Energía ejercida sobre un objeto.

Forced convection The movement of fluid by mechanical means.

Convección forzada. Movimiento de fluido por medios mecánicos.

Fossil fuels Natural gas, oil, and coal formed millions of years ago from dead plants and animals.

Combustibles fósiles. El gas natural, el petroleo y el carbón que se formaron hace millones de años de plantas y animales muertos.

Four-way valve The valve in a heat pump system that changes the direction of the refrigerant flow between the heating and cooling cycles.

Válvula con cuatro vías. Válvula en un sistema de bomba de calor que cambia la dirección del flujo de refrigerante entre los ciclos de calentamiento y enfriamiento.

Freezer burn The term applied to frozen food when it becomes dry and hard from dehydration due to poor packaging.

Quemadura del congelador. Término aplicado a la comida congelada cuando se seca y endurece debido a la deshidratación ocacionada por el empaque de calidad inferior.

Freeze up Excess ice or frost accumulation on an evaporator to the point that airflow may be affected.

Congelación. Acumulación excesiva de hielo o congelación en un evaporador a tal extremo que el flujo de aire puede ser afectado.

Freezing The change of state of water from a liquid to a solid.

Congelamiento. Cambio de estado del agua de líquido a sólido.

Freon The trade name for refrigerants manufactured by E.I. duPont de Nemours & Co., Inc.

Freón. Marca registrada para refrigerantes fabricados por la compañía E.I. duPont de Nemours, S.A.

Frequency The cycles per second (cps) of the electrical current supplied by the power company. This is normally 60 cps in the United States.

Frecuencia. Ciclos por segundo (cps), generalmente 60 cps en los Estados Unidos, de la corriente eléctrica suministrada por la empresa de fuerza motriz.

Frontseated A position on a valve that will not allow refrigerant flow in one direction.

Sentado delante. Posición en una válvula que no permite el flujo de refrigerante en una dirección.

Frost back A condition of frost on the suction line and even the compressor body usually due to liquid refrigerant in the suction line.

Obturación por congelación. Condición de congelación que ocurre en el conducto de aspiración e inclusive en el cuerpo del compresor, normalmente debido a la presencia de refrigerante líquido en el conducto de aspiración.

Frostbite When skin freezes.

Quemadura por frío. Congelación de la piel.

Frozen The term used to describe water in the solid state; also used to describe a rotating shaft that will not turn.

Congelado. Término utilizado para describir el agua en un estado sólido; utilizado también para describir un árbol giratorio que no gira.

Fuel oil The fossil fuel used for heating; a petroleum distillate.

Aceite pesado. Combustible fósil utilizado para calentar; un destilado de petróleo.

Full-load amperage (FLA) The current an electric motor draws while operating under a full-load condition. Also called the *run-load amperage.*

Amperaje de carga total. Corriente que un motor eléctrico consume mientras funciona en una condición de carga completa. Conocido también como *amperaje de carga de funcionamiento.*

Furnace Equipment used to convert heating energy, such as fuel oil, gas, or electricity, to usable heat. It usually contains a heat exchanger, a blower, and the controls to operate the system.

Horno. Equipo utilizado para la conversión de energía calórica, como por ejemplo el aceite pesado, el gas o la electricidad, en calor utilizable. Normalmente contiene un intercambiador de calor, un soplador y los reguladores para accionar el sistema.

Fuse A safety device used in electrical circuits for the protection of the circuit conductor and components.

Fusible. Dispositivo de segurdidad utilizado en circuitos eléctricos para la protección del conductor y de los componentes del circuito.

Fusible link An electrical safety device normally located in a furnace that burns and opens the circuit during an overheat situation.

Cartucho de fusible. Dispositivo eléctrico de seguridad ubicado por lo general en un horno, que quema y abre el circuito en caso de sobrecalentamiento.

Fusible plug A device (made of low-melting-temperature metal) used in pressure vessels that is sensitive to low temperatures and relieves the vessel contents in an overheating situation.

Tapón de fusible. Dispositivo utilizado en depósitos en presión, hecho de un metal que tiene una temperatura de fusión baja. Este dispositivo es sensible a temperaturas bajas y alivia el contenido del depósito en caso de sobrecalentamiento.

G

Gas The vapor state of matter.

Gas. Estado de vapor de una materia.

Gasket A thin piece of flexible material used between two metal plates to prevent leakage.

Guarnición. Pieza delgada de material flexible utilizada entre dos placas de metal para evitar fugas.

Gas pressure switch Used to detect gas pressure before gas burners are allowed to ignite.

Conmutador de presión del gas. Utilizado para detectar la presión del gas antes de que los quemadores de gas puedan encenderse.

Gas valve A valve used to stop, start, or modulate the flow of natural gas.

Válvula de gas. Válvula utilizada para detener, poner en marcha o modular el flujo de gas natural.

Gate A terminal on a semiconductor.

Compuerta. Punto terminal en un semiconductor.

Gauge An instrument used to detect pressure.

Calibrador. Instrumento utilizado para detectar presión.

Gauge manifold A tool that may have more than one gauge with a valve arrangement to control fluid flow.

Distribuidor de calibrador. Herramienta que puede tener más de un calibrador con las válvulas arregladas a fin de regular el flujo de fluido.

Gauge port The service port used to attach a gauge for service procedures.

Orificio de calibrador. Orificio de servicio utilizado con el propósito de fijar un calibrador para procedimientos de servicio.

Germanium A substance from which many semiconductors are made.

Germanio. Sustancia de la que se fabrican muchos semiconductores.

Glow coil A device that automatically reignites a pilot light if it goes out.

Bobina encendedora. Dispostivo que automáticamente vuelve a encender la llama piloto si ésta se apaga.

Graduated cylinder A cylinder with a visible column of liquid refrigerant used to measure the refrigerant charged into a system. Refrigerant temperatures can be dialed on the graduated cylinder.

Cilindro graduado. Cilindro con una columna visible de refrigerante líquido utilizado para medir el refrigerante inyectado al sistema. Las temperaturas del refrigerante pueden marcarse en el cilindro graduado.

Grain Unit of measure. One pound = 7,000 grains.

Grano. Unidad de medida. Una libra equivale a 7000 granos.

Gram Metric measurement term used to express weight.

Gramo. Término utilizado para referirse a la unidad básica de peso en el sistema métrico.

Grille A louvered, often decorative, component in an air system at the inlet or the outlet of the airflow.

Rejilla. Componente con celosías, comúnmente decorativo, en un sistema de aire que se encuentra a la entrada o a la salida del flujo de aire.

Grommet A rubber, plastic, or metal protector usually used where wire or pipe goes through a metal panel.

Guardaojal. Protector de caucho, plástico o metal normalmente utilizado donde un alambre o un tubo pasa a través de una base de metal.

Ground, electrical A circuit or path for electron flow to the earth ground.

Tierra eléctrica. Circuito o trayectoria para el flujo de electrones a la puesta a tierra.

Ground wire A wire from the frame of an electrical device to be wired to the earth ground.

Alambre a tierra. Alambre que va desde el armazón de un dispositivo eléctrico para ser conectado a la puesta a tierra.

Guide vanes Vanes used to produce capacity control in a centrifugal compressor. Also called *prerotation guide vanes.*

Paletas directrices. Paletas utilizadas para producir la regulación de capacidad en un compresor centrífugo. Conocidas también como *paletas directrices para prerotación.*

H

Halide refrigerants Refrigerants that contain halogen chemicals. Refrigerants include R-12, R-22, R-500, and R-502.

Refrigerantes de hálido. Refrigerantes que contienen productos químicos de halógeno; entre ellos se encuentran el R-12, R-22, R-500 y R-502.

Halide torch A torch-type leak detector used to detect halogen refrigerants.

Soplete de hálido. Detector de fugas de tipo soplete utilizado para detectar los refrigerantes de halógeno.

Halogens Chemical substances found in many refrigerants containing chlorine, bromine, iodine, and fluorine.

Halógenos. Sustancias químicas presentes en muchos refrigerantes que contienen cloro, bromo, yodo y flúor.

Hand truck A two-wheeled piece of equipment that can be used for moving heavy objects.

Vagoneta para mano. Equipo con dos ruedas que puede utilizarse para transportar objetos pesados.

Hanger A device used to support tubing, pipe, duct, or other components of a system.

Soporte. Dispositivo utilizado para apoyar tuberías, tubos, conductos u otros componentes de un sistema.

Head Another term for pressure, usually referring to gas or liquid.

Carga. Otro término para presión, refiriéndose normalmente a gas o líquido.

Header A pipe or containment to which other pipe lines are connected.

Conductor principal. Tubo o conducto al que se conectan otras conexiones.

Head pressure control A control that regulates the head pressure in a refrigeration or air-conditioning system.

Regulador de la presión de la carga. Regulador que controla la presión de la carga en un sistema de refrigeración o de acondicionamiento de aire.

Heat Energy that causes molecules to be in motion and to raise the temperature of a substance.

Calor. Energía que ocasiona el movimiento de las moléculas provocando un aumento de temperatura en una sustancia.

Heat anticipator A device that anticipates the need for cutting off the heating system prematurely so the fan can cool the furnace.

Anticipador de calor. Dispositivo que anticipa la necesidad de detener la marcha del sistema de calentamiento para que el abanico pueda enfriar el horno.

Heat coil A device, made of tubing or pipe, designed to transfer heat to a cooler substance by using fluids.

Bobina de calor. Dispositivo hecho de tubos, diseñado para transmitir calor a una sustancia más fría por medio de fluidos.

Heat exchanger A device that transfers heat from one substance to another.

Intercambiador de calor. Dispositivo que transmite calor de una sustancia a otra.

Heat of compression That part of the energy from the pressurization of a gas or a liquid converted to heat.

Calor de compresión. La parte de la energía generada de la presurización de un gas o un líquido que se ha convertido en calor.

Heat of fusion The heat released when a substance is changing from a liquid to a solid.

Calor de fusión. Calor liberado cuando una sustancia se convierte de líquido a sólido.

Heat of respiration When oxygen and carbon hydrates are taken in by a substance or when carbon dioxide and water are given off. Associated with fresh fruits and vegetables during their aging process while being stored.

Calor de respiración. Cuando se admiten oxígeno e hidratos de carbono en una sustancia o cuando se emiten bióxido de carbono y agua. Se asocia con el proceso de maduración de frutas y legumbres frescas durante su almacenamiento.

Heat pump A refrigeration system used to supply heating or cooling by using valves to reverse the refrigerant gas flow.

Bomba de calor. Sistema de refrigeración utilizado para suministrar calor o frío mediante válvulas que cambian la dirección del flujo de gas del refrigerante.

Heat reclaim Using heat from a condenser for purposes such as space and domestic water heating.

Reclamación de calor. La utilización del calor de un condensador para propósitos tales como la calefacción de espacio y el calentamiento doméstico de agua.

Heat sink A low-temperature surface to which heat can transfer.

Fuente fría. Superficie de temperatura baja a la que puede transmitírsele calor.

Heat transfer The transfer of heat from a warmer to a cooler substance.

Transmisión de calor. Cuando se transmite calor de una sustancia más caliente a una más fría.

Helix coil A bimetal formed into a helix-shaped coil that provides longer travel when heated.

Bobina en forma de hélice. Bimetal encofrado en una bobina en forma de hélice que provee mayor movimiento al ser calentado.

Hermetic system A totally enclosed refrigeration system where the motor and compressor are sealed within the same system with the refrigerant.

Sistema hermético. Sistema de refrigeración completamente cerrado donde el motor y el compresor se obturan dentro del mismo sistema con el refrigerante.

Hertz Cycles per second.

Hertz. Ciclos por segundo.

Hg Abbreviation for the element mercury.

Hg. Abreviatura del elemento mercurio.

High-pressure control A control that stops a boiler heating device or a compressor when the pressure becomes too high.

Regulador de alta presión. Regulador que detiene la marcha del dispositivo de calentamiento de una caldera o de un compresor cuando la presión alcanza un nivel demasiado alto.

High side A term used to indicate the high-pressure or condensing side of the refrigeration system.

Lado de alta presión. Término utilizado para indicar el lado de alta presión o de condensación del sistema de refrigeración.

High-temperature refrigeration A refrigeration temperature range starting with evaporator temperatures no lower than 35°F, a range usually used in air-conditioning (cooling).

Refrigeración a temperatura alta. Margen de la temperatura de refrigeración que comienza con temperaturas de evaporadores no menores de 35°F (2°C). Este margen se utiliza normalmente en el acondicionamiento de aire (enfriamiento).

High-vacuum pump A pump that can produce a vacuum in the low micron range.

Bomba de vacío alto. Bomba que puede generar un vacío dentro del margen de micrón bajo.

Horsepower A unit equal to 33,000 ft lb of work per minute.

Potencia en caballos. Unidad equivalente a 33.000 libras-pies de trabajo por minuto.

Hot gas The refrigerant vapor as it leaves the compressor. This is often used to defrost evaporators.

Gas caliente. El vapor del refrigerante al salir del compresor. Esto se utiliza con frecuencia para descongelar evaporadores.

Hot gas bypass Piping that allows hot refrigerant gas into the cooler low-pressure side of a refrigeration system usually for system capacity control.

Desviación de gas caliente. Tubería que permite la entrada de gas caliente del refrigerante en el lado más frío de baja presión de un sistema de refrigeración, normalmente para la regulación de la capacidad del sistema.

Hot gas defrost A system where the hot refrigerant gases are passed through the evaporator to defrost it.

Descongelación con gas caliente. Sistema en el que los gases calientes del refrigerante se pasan a través del evaporador para descongelarlo.

Hot gas line The tubing between the compressor and condenser.

Conducto de gas caliente. Tubería entre el compresor y el condensador.

Hot junction That part of a thermocouple or thermopile where heat is applied.

Empalme caliente. El lugar en un termopar o pila termoeléctrica donde se aplica el calor.

Hot pull-down The process of lowering the refrigerated space to the design temperature after it has been allowed to warm up considerably over this temperature.

Descenso caliente. Proceso de bajar la temperatura del espacio refrigerado a la temperatura de diseño luego de habérsele permitido calentarse a un punto sumamente superior a esta temperatura.

Hot-water heat A heating system using hot water to distribute the heat.

Calor de agua caliente. Sistema de calefacción que utiliza agua caliente para la distribución del calor.

Hot wire The wire in an electrical circuit that has a voltage potential between it and another electrical source or between it and ground.

Conductor electrizado. Conductor en un circuito eléctrico a través del cual fluye la tensión entre éste y otra fuente de electricidad o entre éste y la tierra.

Humidifier A device used to add moisture to the air.

Humedecedor. Dispositivo utilizado para agregarle humedad al aire.

Humidistat A control operated by a change in humidity.

Humidistato. Regulador activado por un cambio en la humedad.

Humidity Moisture in the air.

Humedad. Vapor de agua existente en el ambiente.

Hydraulics Producing mechanical motion by using liquids under pressure.

Hidráulico. Generación de movimiento mecánico por medio de líquidos bajo presión.

Hydrocarbons Organic compounds containing hydrogen and carbon and found in many heating fuels.

Hidrocarburos. Compuestos orgánicos que contienen el hidrógeno y el carbón presentes en muchos combustibles de calentamiento.

Hydrochlorofluorocarbons (HCFC) Refrigerants thought to contribute to the depletion of the ozone layer, although not to the extent of chlorofluorocarbons.

Hidroclorofluorocarburos. Líquidos refrigerantes que, según algunos, han contribuido a la reducción de la capa de ozono aunque no en tal grado como los clorofluorocarburos.

Hydrometer An instrument used to measure the specific gravity of a liquid.

Hidrómetro. Instrumento utilizado para medir la gravedad específica de un líquido.

Hydronic   Usually refers to a hot-water heating system.

Hidrónico.   Normalmente se refiere a un sistema de calefacción de agua caliente.

Hygrometer   An instrument used to measure the amount of moisture in the air.

Higrómetro.   Instrumento utilizado para medir la cantidad de humedad en el aire.

I

Idler   A pulley on which a belt rides. It does not transfer power but it is used to provide tension or reduce vibration.

Polea tensora.   Polea sobre la que se mueve una correa. No sirve para transmitir potencia, pero se utiliza para proveer tensión o disminuir la vibración.

Ignition transformer   Provides a high-voltage current, usually to produce a spark to ignite a furnace fuel, either gas or oil.

Transformador para encendido.   Provee una corriente de alta tensión, normalmente para generar una chispa a fin de encender el combustible de un horno, sea gas o aceite pesado.

Impedance   A form of resistance in an alternating current circuit.

Impedancia.   Forma de resistencia en un circuito de corriente alterna.

Impeller   The rotating part of a pump that causes the centrifugal force to develop fluid flow and pressure difference.

Impulsor.   Pieza giratoria de una bomba que hace que la fuerza centrífuga desarrolle flujo de fluido y una diferencia en presión.

Impingement   The condition in a gas or oil furnace when the flame strikes the sides of the combustion chamber, resulting in poor combustion efficiency.

Golpeo.   Condición que ocurre en un horno de gas o de aceite pesado cuando la llama golpea los lados de la cámara de combustión. Esta condición trae como resultado un rendimiento de combustión pobre.

Inclined water manometer   Indicates air pressures in very low pressure systems.

Manómetro de agua inclinada.   Señala las presiones de aire en sistemas de muy baja presión.

Induced magnetism   Magnetism produced, usually in a metal, from another magnetic field.

Magnetismo inducido.   Magnetismo generado, normalmente en un metal, desde otro campo magnético.

Inductance   An induced voltage, producing a resistance in an alternating current circuit.

Inductancia.   Tensión inducida que genera una resistencia en un circuito de corriente alterna.

Induction motor   An alternating current motor where the rotor turns from induced magnetism from the field windings.

Motor inductor.   Motor de corriente alterna donde el rotor gira debido al magnetismo inducido desde los devanados inductores.

Inductive reactance   A resistance to the flow of an alternating current produced by an electromagnetic induction.

Reactancia inductiva.   Resistencia al flujo de una corriente alterna generada por una inducción electromagnética.

Inert gas   A gas that will not support most chemical reactions, particularly oxidation.

Gas inerte.   Gas incapaz de resistir la mayoría de las reacciones químicas, especialmente la oxidación.

Infiltration   Air that leaks into a structure through cracks, windows, doors, or other openings due to less pressure inside the structure than outside the structure.

Infiltración.   Penetración de aire en una estructura a través de grietas, ventanas, puertas u otras aberturas debido a que la presión en el interior de la estructura es menor que en el exterior.

Infrared rays The rays that transfer heat by radiation.

Rayos infrarrojos. Rayos que transmiten calor por medio de la radiación.

In-phase When two or more alternating current circuits have the same polarity at all times.

En fase. Cuando dos o más circuitos de corriente alterna tienen siempre la misma polaridad.

Insulation, electric A substance that is a poor conductor of electricity.

Aislamiento eléctrico. Sustancia que es un conductor pobre de electricidad.

Insulation, thermal A substance that is a poor conductor of the flow of heat.

Aislamiento térmico. Sustancia que es un conductor pobre de flujo de calor.

Intermittent ignition Ignition system for a gas furnace that operates only when needed or when the furnace is operating.

Encendido interrumpido. Sistema de encendido para un horno de gas que funciona solamente cuando es necesario o cuando el horno está trabajando.

Isolation relays Components used to prevent stray unwanted electrical feedback that can cause erratic operation.

Relés de aislación. Componentes utilizados para evitar la realimentación eléctrica dispersa no deseada que puede ocasionar un funcionamiento errático.

J

Joule Metric measurement term used to express the quantity of heat.

Joule. Término utilizado para referirse a la unidad básica de cantidad de calor en el sistema métrico.

Junction box A metal or plastic box where electrical connections are made.

Caja de empalme. Caja metálica o plástica dentro de la cual se hacen conexiones eléctricas.

K

Kelvin A temperature scale where absolute 0 equals 0, or where molecular motion stops at 0. It has the same graduations per degree of change as the Celsius scale.

Escala absoluta. Escala de temperaturas donde el cero absoluto equivale a 0 ó donde el movimiento molecular se detiene en 0. Tiene las mismas graduaciones por grado de cambio que la escala Celsio.

Kilopascal A metric unit of measurement for pressure used in the air-conditioning, heating, and refrigeration field. There are 6.89 kilopascals in 1 PSI.

Kilopascal. Unidad métrica de medida de presión utilizada en el ramo del acondicionamiento de aire, calefacción y refrigeración. 6,89 kilopascales equivalen a 1 PSI.

Kilowatt A unit of electrical power equal to 1,000 watts.

Kilowatio. Unidad eléctrica de potencia equivalente a 1000 watios.

Kilowatt-hour One kilowatt (1,000 watts) of energy used for one hour.

Kilowatio hora. Unidad de energía equivalente a la que produce un kilowatio durante una hora.

King valve A service valve at the liquid receiver.

Válvula maestra. Válvula de servicio ubicada en el receptor del líquido.

L

Latent heat Heat energy absorbed or rejected when a substance is changing state but not changing in temperature.

Calor latente. Energía calórica absorbida o rechazada cuando un sustancia cambia de estado y no se experimentan cambios de temperatura.

Leak detector Any device used to detect leaks in a pressurized system.

Detector de fugas. Cualquier dispositivo utilizado para detectar fugas en un sistema presurizado.

Lever truck A long-handled, two-wheeled device that can be used to lift and assist in moving heavy objects.

Vagoneta con palanca. Dispositivo con dos ruedas y una manivela larga que puede utilizarse para levantar y ayudar a transportar objetos pesados.

Limit control A control used to make a change in a system—usually to stop it when predetermined limits of pressure or temperature are reached.

Regulador de límite. Regulador utilizado para realizar un cambio en un sistema, normalmente para detener su marcha cuando se alcanzan niveles predeterminados de presión o de temperatura.

Line set A term used for tubing sets furnished by the manufacturer.

Juego de conductos. Término utilizado para referise a los juegos de tubería suministrados por el fabricante.

Liquid A substance where molecules push outward and downward to seek a uniform level.

Líquido. Sustancia donde las moléculas empujan hacia afuera y hacia abajo y buscan un nivel uniforme.

Liquid charge bulb A type of charge in the sensing bulb of the thermostatic expansion valve. This charge is characteristic of the refrigerant in the system and contains enough liquid so that it will not totally boil away.

Bombilla de carga líquida. Tipo de carga en la bombilla sensora de la válvula de expansión termostática. Esta carga es característica del refrigerante en el sistema y contiene suficiente líquido para que el mismo no se evapore completamente.

Liquid line A term applied in the industry to refer to the tubing or piping from the condenser to the expansion device.

Conducto de líquido. Término aplicado en nuestro medio para referirse a la tubería que va del condensador al dispositivo de expansión.

Liquid nitrogen Nitrogen in liquid form.

Nitrógeno líquido. Nitrógeno en forma líquida.

Liquid receiver A container in the refrigeration system where liquid refrigerant is stored.

Receptor del líquido. Recipiente en el sistema de refrigeración donde se almacena el refrigerante líquido.

Liquid refrigerant charging The process of allowing liquid refrigerant to enter the refrigeration system through the liquid line to the condenser and evaporator.

Carga para refrigerante líquido. Proceso de permitir la entrada del refrigerante líquido al condensador y al evaporador en el sistema de refrigeración a través del conducto de líquido.

Liquid slugging A large amount of liquid refrigerant in the compressor cylinder, usually causing immediate damage.

Relleno de líquido. Acumulación de una gran cantidad de refrigerante líquido en el cilindro del compresor, que normalmente provoca una avería inmediata.

Liquified petroleum Liquified propane, butane, or a combination of these gases. The gas is kept as a liquid under pressure until ready to use.

Petróleo licuado. Propano o butano licuados, o una combinación de estos gases. El gas se mantiene en estado líquido bajo presión hasta que se encuentre listo para usar.

Lithium bromide A type of salt solution used in an absorption chiller.

Bromuro de litio. Tipo de solución salina utilizada en un enfriador por absorción.

Locked rotor amperage (LRA) The current an electric motor draws when it is first turned on. This is normally five times the full-load amperage.

Amperaje de rotor bloqueado. Corriente que un motor eléctrico consume al ser encendido, la cual generalmente es cinco veces mayor que el amperaje de carga completa.

Low-pressure control A pressure switch that can provide low-charge protection by shutting down the system on low pressure. It can also be used to control space temperature.

Regulador de baja presión. Conmutador de presión que puede proveer protección contra una carga baja al detener el sistema si éste alcanza una presión demasiado baja. Puede utilizarse también para regular la temperatura de un espacio.

Low side A term used to refer to that part of the refrigeration system that operates at the lowest pressure, between the expansion device and the compressor.

Lado bajo. Término utilizado para referirse a la parte del sistema de refrigeración que funciona a niveles de presión más baja, entre el dispositivo de expansión y el compresor.

Low-temperature refrigeration A refrigeration temperature range starting with evaporator temperatures no higher than 0°F for storing frozen food.

Refrigeración a temperatura baja. Margen de la temperatura de refrigeración que comienza con temperaturas de evaporadores no mayores de 0°F (–18°C) para almacenar comida congelada.

LP fuel Liquefied petroleum, a substance used as a gas for fuel. It is transported and stored in the liquid state.

Combustible PL. Petróleo licuado, sustancia utilizada como gas para combustible. El petróleo licuado se transporta y almacena en estado líquido.

M

Magnetic field A field or space where magnetic lines of force exist.

Campo magnético. Campo o espacio donde existen líneas de fuerza magnética.

Magnetism A force causing a magnetic field to attract ferrous metals, or where like poles of a magnet repel and unlike poles attract each other.

Magnetismo. Fuerza que hace que un campo magnético atraiga metales férreos, o cuando los polos iguales de un imán se rechazan y los opuestos se atraen.

Male thread The external thread of either a pipe, a fitting, or a cylinder.

Filete macho. Filete en la parte exterior de un tubo, accesorio o cilindro; filete externo.

Manometer An instrument used to check low vapor pressures. The pressures may be checked against a column of mercury or water.

Manómetro. Instrumento utilizado para revisar las presiones bajas de vapor. Las presiones pueden revisarse comparándolas con una columna de mercurio o de agua.

Mapp gas A composite gas similar to propane that may be used with air.

Gas Mapp. Gas compuesto similar al propano que puede utilizarse con aire.

Marine water box A water box with a removable cover.

Caja marina para agua. Caja para agua con un tapón desmontable.

Mass Matter held together to the extent that it is considered one body.

Masa. Materia compacta que se considera un solo cuerpo.

Mass spectrum analysis An absorption machine factory leak test performed using helium.

Análisis del límite de masa. Prueba para fugas y absorción llevada a cabo en la fábrica utilizando helio.

Matter A substance that takes up space and has mass.

Materia. Sustancia que ocupa espacio y tiene peso.

Medium temperature refrigeration Refrigeration where evaporator temperatures are 32°F or below, normally used for preserving fresh food.

Refrigeración a temperatura media. Refrigeración, donde las temperaturas del evaporador son 32°F (0°C) o menos, utilizada generalmente para preservar comida fresca.

Megger An instrument (megohmmeter) that can detect very high resistances, in millions of ohms. Megger relates to megohm or 1,000,000 ohms.

Megóhmetro. Instrumento que puede detectar resistencias sumamente altas, en millones de ohmios. Este término está relacionado al megohmio o 1.000.000 de ohmios.

Megohm A measure of electrical resistance equal to 1,000,000 ohms.

Megohmio. Medidad de resistencia eléctrica equivalente a 1.000.000 de ohmios.

Melting point The temperature at which a substance will change from a solid to a liquid.

Punto de fusión. Temperatura a la que una sustancia se convierte de sólido a líquido.

Mercury bulb A glass bulb containing a small amount of mercury and electrical contacts used to make and break the electrical circuit in a low-voltage thermostat.

Bombilla de mercurio. Bombilla de cristal que contiene una pequeña cantidad de mercurio y contactos eléctricos, utilizada para conectar y desconectar el circuito eléctrico en un termostato de baja tensión.

Metering device A valve or small fixed-size tubing or orifice that meters liquid refrigerant into the evaporator.

Dispositivo de medida. Válvula o tubería pequeña u orificio que mide la cantidad de refrigerante líquido que entra en el evaporador.

Methane Natural gas is composed of 90 percent to 95 percent methane, a combustible hydrocarbon.

Metano. El gas natural se compone de un 90 porcenta a un 95 porcenta de metano, un hidrocarburo combustible.

Metric system or **System International (SI)** System of measurement used by most countries in the world.

Sistema métrico o Sistema internacional. El sistema de medida utilizado por la mayoría de los países del mundo.

Micro A prefix meaning 1/1,000,000.

Micro. Prefijo que significa una parte de un millón.

Microfarad Capacitor capacity equal to 1/1,000,000 of a farad.

Microfaradio. Capacidad de un capacitador equivalente a 1/1.000.000 de un faradio.

Micrometer A precision measuring instrument.

Micrómetro. Instrumento de precisión utilizado para medir.

Micron A unit of length equal to 1/1000 of a millimeter, 1/1,000,000 of a meter.

Micrón. Unidad de largo equivalente a 1/1,000 de un milímetro, o 1/1.000.000 de un metro.

Micron gauge A gauge used when it is necessary to measure pressure close to a perfect vacuum.

Calibrador de micrón. Calibrador utilizado cuando es necesario medir la presión de un vacío casi perfecto.

Midseated (cracked) A position on a valve that allows refrigerant flow in all directions.

Sentado en el medio (agrietado). Posición en una válvula que permite el flujo de refrigerante en cualquier dirección.

Milli A prefix meaning 1/1,000.

Mili. Prefijo que significa una parte de mil.

Modulator A device that adjusts by small increments or changes.

Modulador. Dispositivo que se ajusta por medio de incrementos o cambios pequeños.

Moisture indicator A device for determining moisture in a refrigerant.

Indicador de humedad. Dispositivo utilizado para determinar la humedad en un refrigerante.

Molecular motion The movement of molecules within a substance.

Movimiento molecular. Movimiento de moléculas dentro de una sustancia.

Molecule The smallest particle that a substance can be broken into and still retain its chemical identity.

Molécula. La partícula más pequeña en la que una sustancia puede dividirse y aún conservar sus propias características.

Monochlorodifluoromethane The refrigerant R-22.

Monoclorodiflorometano. El refrigerante R-22.

Motor service factor A factor above an electric motor's normal operating design parameters, indicated on the nameplate, under which it can operate.

Factor de servicio del motor. Factor superior a los parámetros de diseño normales de funcionamiento de un motor eléctrico, indicados en el marbete; este factor indica su nivel de funcionamiento.

Motor starter Electromagnetic contactors that contain motor protection and are used for switching electric motors on and off.

Arrancador de motor. Contactadores electromagnéticos que contienen protección para el motor y se utilizan para arrancar y detener motores eléctricos.

Muffler compressor Sound absorber at the compressor.

Silenciador del compresor. Absorbedor de sonido ubicado en el compresor.

Mullion Stationary frame between two doors.

Parteluz. Armazón fijo entre dos puertas.

Mullion heater Heating element mounted in the mullion of a refrigerator to keep moisture from forming on it.

Calentador del parteluz. Elemento de calentamiento montado en el parteluz de un refrigerador para evitar la formación de humedad en el mismo.

Multimeter An instrument that will measure voltage, resistance, and milliamperes.

Multímetro. Instrumento que mide la tensión, la resistencia y los miliamperios.

Multiple evacuation A procedure for removing the refrigerant from a system. A vacuum is pulled, a small amount of refrigerant is allowed into the system, and the procedure is duplicated. This is often done three times.

Evacuación múltiple. Procedimiento para remover el refrigerante de un sistema. Se crea un vacío, se permite la entrada de una pequeña cantidad de refrigerante al sistema, y se repite el procedimiento. Con frecuencia esto se lleva a cabo tres veces.

N

National Electrical Code (NEC) A publication that sets the standards for all electrical installations, including motor overload protection.

Código estadounidense de electricidad. Publicación que establece las normas para todas las instalaciones eléctricas, incluyendo la protección contra la sobrecarga de un motor.

National pipe taper (NPT) The standard designation for a standard tapered pipe thread.

Cono estadounidense para tubos. Designación estándar para una rosca cónica para tubos estándar.

Natural convection The natural movement of a gas or fluid caused by differences in temperature.

Convección natural. Movimiento natural de un gas o fluido ocasionado por diferencias en temperatura.

Natural gas A fossil fuel formed over millions of years from dead vegetation and animals that were deposited or washed deep into the earth.

Gas natural. Combustible fósil formado a través de millones de años de la vegetación y los animales muertos que fueron depositados o arrastrados a una gran profundidad dentro la tierra.

Needlepoint valve   A device having a needle and a very small orifice for controlling the flow of a fluid.

Válvula de aguja.   Dispositivo que tiene una aguja y un orificio bastante pequeño para regular el flujo de un fluido.

Negative electrical charge   An atom or component that has an excess of electrons.

Carga eléctrica negativa.   Átomo o componente que tiene un exceso de electrones.

Neoprene   Synthetic flexible material used for gaskets and seals.

Neopreno.   Material sintético flexible utilizado en guarniciones y juntas de estanqueidad.

Net oil pressure   Difference in the suction pressure and the compressor oil pump outlet pressure.

Presión neta del aceite.   Diferencia en la presión de aspiración y la presión a la salida de la bomba de aceite del compresor.

Neutralizer   A substance used to counteract acids.

Neutralizador.   Sustancia utilizada para contrarrestar ácidos.

Newton/meter²   Metric unit of measurement for pressure. Also called a *pascal*.

Metro-Newton².   Unidad métrica de medida de presión. Conocido también como *pascal*.

Nichrome   A metal made of nickel chromium that when formed into a wire is used as a resistance heating element in electric heaters and furnaces.

Níquel-cromio.   Metal fabricado de níquel-cromio que al ser convertido en alambre, se utiliza como un elemento de calentamiento de resistencia en calentadores y hornos eléctricos.

Nitrogen   An inert gas often used to "sweep" a refrigeration system to help ensure that all refrigerant and contaminants have been removed.

Nitrógeno.   Gas inerte utilizado con frecuencia para purgar un sistema de refrigeración. Esta gas ayuda a asegurar la remoción de todo el refrigerante y los contaminantes del sistema.

Nominal   A rounded-off stated size. The nominal size is the closest rounded-off size.

Nominal.   Tamaño redondeado establecido. El tamaño nominal es el tamaño redondeado más cercano.

Noncondensable gas   A gas that does not change into a liquid under normal operating conditions.

Gas no condensable.   Gas que no se convierte en líquido bajo condiciones de funcionamiento normales.

Nonferrous   Metals containing no iron.

No férreos.   Metales que no contienen hierro.

North pole, magnetic   One end of a magnet.

Polo norte magnético.   El extremo de un imán.

Nut driver   These tools have a socket head used primarily to drive hex head screws on air-conditioning, heating, and refrigeration cabinets.

Extractor de tuercas.   Estas herramientas tienen una cabeza hueca utilizada principalmente para darles vueltas a tornillos de cabeza hexagonal en gabinetes de acondicionamiento de aire, de calefacción y de refrigeración.

O

Off cycle   A period when a system is not operating.

Ciclo de apagado.   Período de tiempo cuando un sistema no está en funcionamiento.

Ohm   A unit of measurement of electrical resistance.

Ohmio.   Unidad de medida de la resistencia eléctrica.

Ohmmeter   A meter that measures electrical resistance.

Ohmiómetro.   Instrumento que mide la resistencia eléctrica.

Ohm's law   A law involving electrical relationships discovered by Georg Ohm: $E = I \times R$.

Ley de ohm.   Ley que define las relaciones eléctricas, descubierta por Georg Ohm: $E = I \times R$.

Oil-pressure safety control (switch)   A control used to ensure that a compressor has adequate oil lubricating pressure.

Regulador de seguridad para la presión de aceite (conmutador).   Regulador utilizado para asegurar que un compresor tenga la presión de lubrificación de aceite adecuada.

Oil, refrigeration   Oil used in refrigeration systems.

Aceite de refrigeración.   Aceite utilizado en sistemas de refrigeración.

Oil separator   Apparatus that removes oil from a gaseous refrigerant.

Separador de aceite.   Aparato que remueve el aceite de un refrigerante gaseoso.

Open compressor   A compressor with an external drive.

Compresor abierto.   Compresor con un motor externo.

Operating pressure   The actual pressure under operating conditions.

Presión de funcionamiento.   La presión real bajo las condiciones de funcionamiento.

Organic   Materials formed from living organisms.

Orgánico.   Materiales formados de organismos vivos.

Orifice   A small opening through which fluid flows.

Orificio.   Pequeña abertura a través de la cual fluye un fluido.

Overload protection   A system or device that will shut down a system if an overcurrent condition exists.

Protección contra sobrecarga.   Sistema o dispositivo que detendrá la marcha de un sistema si existe una condición de sobreintensidad.

Oxidation   The combining of a material with oxygen to form a different substance. This results in the deterioration of the original substance.

Oxidación.   La combinación de un material con oxígeno para formar una sustancia diferente, lo que ocasiona el deterioro de la sustancia original.

Ozone   A form of oxygen (O_3). A layer of ozone located in the stratosphere that almost entirely protects the earth from radiation from the sun's ultraviolet wave lengths.

Ozono.   Forma de oxígeno (O_3). La capa de ozono en la estratósfera que protege la tierra de ciertos rayos ultravioletas del sol.

P

Package unit   A refrigerating system where all major components are located in one cabinet.

Unidad completa.   Sistema de refrigeración donde todos los componentes principales se encuentran en un solo gabinete.

Packing   A soft material that can be shaped and compressed to provide a seal. It is commonly applied around valve stems.

Empaquetadura.   Material blando que puede formarse y comprimirse para proveer una junta de estanqueidad. Comúnmente se aplica alrededor de los vástagos de válvulas.

Parallel circuit   An electrical or fluid circuit where the current or fluid takes more than one path at a junction.

Circuito paralelo.   Corriente eléctrica o fluida donde la corriente o el fluido siguen más de una trayectoria en un empalme.

Pascal   A metric unit of measurement of pressure.

Pascal.   Unidad métrica de medida de presión.

Passive solar design   The use of nonmoving parts of a building to provide heat or cooling, or to eliminate certain parts of a building that cause inefficient heating or cooling.

Diseño solar pasivo. La utilización de piezas fijas de un edificio para proveer calefacción o enfriamiento, o para eliminar ciertas piezas de un edificio que causan calefacción o enfriamiento ineficientes.

Permanent magnet An object that has its own permanent magnetic field.

Imán permanente. Objeto que tiene su propio campo magnético permanente.

Permanent split capacitor motor (PSC) A split phase motor with a run capacitor only, and a low starting torque.

Motor permanente de capacitador separado. Motor de fase separada que sólo tiene un capacitador de funcionamiento. Su par de arranque es sumamente bajo.

Phase One distinct part of a cycle.

Fase. Una parte específica de un ciclo.

Pilot light The flame that ignites the main burner on a gas furnace.

Llama piloto. Llama que enciende el quemador principal en un horno de gas.

Piston The part that moves up and down in a cylinder.

Pistón. La pieza que asciende y desciende dentro de un cilindro.

Piston displacement The volume within the cylinder that is displaced with the movement of the piston from top to bottom.

Desplazamiento del pistón. Volumen dentro del cilindro que se desplaza de arriba a abajo con el movimiento del pistón.

Pitot tube Part of an instrument for measuring air velocities.

Tubo Pitot. Pieza de un instrumento para medir velocidades de aire.

Planned defrost Shutting the compressor off with a timer so that the space temperature can provide the defrost.

Descongelación proyectada. Detención de la marcha de un compresor con un temporizador

para que la temperatura del espacio lleve a cabo la descongelación.

Plenum A sealed chamber at the inlet or outlet of an air handler. The duct attaches to the plenum.

Plenum. Cámara sellada a la entrada o a la salida de un tratante de aire. El conducto se fija al plenum.

Polycyclic organic matter By-products of wood combustion found in smoke; considered to be health hazards.

Materia orgánica policíclica. Subproductos de la combustión de madera presentes en el humo y considerados nocivos para la salud.

Polyethylene (PE) Plastic pipe used for water, gas, and irrigation systems.

Polietileno. Tubo plástico utilizado en sistemas de agua, de gas y de irrigación.

Polyphase Three or more phases.

Polifase. Tres o más fases.

Polyvinyl chloride (PVC) Plastic pipe used in pressure applications for water and gas as well as for sewage and certain industrial applications.

Cloruro de polivinilo (PVC). Tubo plástico utilizado tanto en aplicaciones de presión para agua y gas, como en ciertas aplicaciones industriales y de aguas negras.

Porcelain A ceramic material.

Porcelana. Material cerámico.

Portable dolly A small platform with four wheels on which heavy objects can be placed and moved.

Carretilla portátil. Plataforma pequeña con cuatro ruedas sobre la que pueden colocarse y transportarse objetos pesados.

Positive displacement A term used with a pumping device, such as a compressor, that is designed to move all matter from a volume, such as a cylinder, or the device will stall and possibly cause a part failure.

Desplazamiento positivo. Término utilizado con un dispositivo de bombeo, como por ejemplo un compresor, diseñado para mover toda la materia de un volumen, como un cilindro o se bloqueará, posiblemente causándole fallas a una pieza.

Positive electrical charge   An atom or component that has a shortage of electrons.

Carga eléctrica positiva.   Átomo o componente que tiene una insuficiencia de electrones.

Positive temperature coefficient start device   A thermistor used to provide start assistance to a permanent split capacitor motor.

Dispositivo de arranque de coeficiente de temperatura positiva.   Termistor utilizado para ayudar a arrancar un motor permanente de capacitador separado.

Potential relay   A switching device used with hermetic motors that breaks the circuit to the start windings after the motor has reached approximately 75 percent of its running speed.

Relé de potencial.   Dispositivo de conmutación utilizado con motores herméticos que interrupe el circuito de los devandos de arranque antes de que el motor haya alcanzado aproximadamente un 75 percenta de su velocidad de marcha.

Potentiometer   An instrument that controls electrical current.

Potenciómetro.   Instrumento que regula corriente eléctrica.

Power   The rate at which work is done.

Potencia.   Velocidad a la que se realiza un trabajo.

Pressure   Force per unit of area.

Presión.   Fuerza por unidad de área.

Pressure drop   The difference in pressure between two points.

Caída de presión.   Diferencia en presión entre dos puntos.

Pressure/enthalpy diagram   A chart indicating the pressure and heat content of a refrigerant and the extent to which the refrigerant is a liquid and vapor.

Diagrama de presión y entalpía.   Esquema que indica la presión y el contenido de calor de un refrigerante y el punto en que el refrigerante es líquido y vapor.

Pressure limiter   A device that opens when a certain pressure is reached.

Dispositivo limitador de presión.   Dispositivo que se abre cuando se alcanza una presión específica.

Pressure-limiting TXV   A valve designed to allow the evaporator to build only to a predetermined temperature before the valve will shut off the flow of refrigerant.

Válvula electrónica de expansión limitadora de presión.   Válvula diseñada para permitir que la temperatura del evaporador alcance un límite predeterminado cuando la válvula detenga el flujo de refrigerante.

Pressure regulator   A valve capable of maintaining a constant outlet pressure when a variable inlet pressure occurs. Used for regulating fluid flow such as natural gas, refrigerant, and water.

Regulador de presión.   Válvula capaz de mantener una presión constante a la salida cuando ocurre una presión variable a la entrada. Utilizado para regular el flujo de fluidos, como por ejemplo el gas natural, el refrigerante y el agua.

Pressure switch   A switch operated by a change in pressure.

Conmutador accionado por presión.   Conmutador accionado por un cambio de presión.

Pressure/temperature relationship   This refers to the pressure/temperature relationship of a liquid and vapor in a closed container. If the temperature increases, the pressure will also increase. If the temperature is lowered, the pressure will decrease.

Relación entre presión y temperatura.   Se refiere a la relación entre la presión y la temperatura de un líquido y un vapor en un recipiente cerrado. Si la temperatura aumenta, la presión también aumentará. Si la temperatura baja, habrá una caída de presión.

Pressure vessels and piping Piping, tubing, cylinders, drums, and other containers that have pressurized contents.

Depósitos y tubería con presión. Tubería, cilindros, tambores y otros recipientes que tienen un contenido presurizado.

Primary control A controlling device for an oil burner that ensures ignition within a specific time span, usually 90 seconds.

Regulador principal. Dispositivo de regulación para un quemador de aceite pesado. El regulador principal asegura el encendido dentro de un período de tiempo específico, normalmente 90 segundos.

Propane An LP gas used for heat.

Propano. Gas de petróleo licuado que se utiliza para producir calor.

Proton That part of an atom having a positive charge.

Protón. Parte de un átomo que tiene carga positiva.

PSI Abbreviation for pounds per square inch.

PSI. Abreviatura de libras por pulgada cuadrada.

PSIA Abbreviation for pounds per square inch absolute.

PSIA. Abreviatura de libras por pulgada cuadrada absoluta.

PSIG Abbreviation for pounds per square inch gauge.

PSIG. Abreviatura de indicador de libras por pulgada cuadrada.

Psychrometer An instrument for determining relative humidity.

Sicrómetro. Instrumento para medir la humedad relativa.

Psychrometric chart A chart that shows the relationship of temperature, pressure, and humidity in the air.

Esquema sicrométrico. Esquema que indica la relación entre la temperatura, la presión y la humedad en el aire.

Pump A device that forces fluids through a system.

Bomba. Dispositivo que introduce fluidos por fuerza a través de un sistema.

Pump down To use a compressor to pump the refrigerant charge into the condenser and/or receiver.

Extraer con bomba. Utilizar un compresor para bombear la carga del refrigerante dentro del condensador y/o receptor.

Purge To remove or release fluid from a system.

Purga. Remover o liberar el fluido de un sistema.

Q

Quench To submerge a hot object in a fluid for cooling.

Enfriamiento por inmersión. Sumersión de un objeto caliente en un fluido para enfriarlo.

Quick-connect coupling A device designed for easy connecting or disconnecting of fluid lines.

Acoplamiento de conexión rápida. Dispositivo diseñado para facilitar la conexión o desconexión de conductos de fluido.

R

R-12 Dichlorodifluoromethane, a popular refrigerant for refrigeration systems.

R-12. Diclorodiflorometano, refrigerante muy utilizado en sistemas de refrigeración.

R-22 Monochlorodifluoromethane, a popular refrigerant for air-conditioning systems.

R-22. Monoclorodiflorometano, refrigerante muy utilizado en sistemas de acondicionamiento de aire.

R-123 Dichlorotrifluoroethane, a refrigerant developed for low-pressure application.

R-123. Diclorotrifloroetano, refrigerante elaborado para aplicaciones de baja presión.

R-134a Tetrafluoroethane, a refrigerant developed for refrigeration systems and as a possible replacement for R-12.

R-134a. Tetrafloroetano, refrigerante elaborado para sistemas de refrigeración y como posible sustituto del R-12.

R-502 An azeotropic mixture of R-22 and R-115, a popular refrigerant for low-temperature refrigeration systems.

R-502. Mezcla azeotrópica de R-22 y R-115, refrigerante muy utilizado en sistemas de refrigeración de temperatura baja.

Radiant heat Heat that passes through air, heating solid objects that in turn heat the surrounding area.

Calor radiante. Calor que pasa a través del aire y calienta objetos sólidos que a su vez calientan el ambiente.

Radiation Heat transfer. See radiant heat.

Radiación. Transferencia de calor. Véase calor radiante.

Random or **off cycle defrost** Defrost provided by the space temperature during the normal off cycle.

Descongelación variable o de ciclo apagado. Descongelación llevada a cabo por la temperatura del espacio durante el ciclo normal de apagado.

Rankine The absolute Fahrenheit scale with 0 at the point where all molecular motion stops.

Rankine. Escala absoluta de Fahrenheit con el 0 al punto donde se detiene todo movimiento molecular.

Reactance A type of resistance in an alternating current circuit.

Reactancia. Tipo de resistencia en un circuito de corriente alterna.

Reamer A tool used to remove burrs from inside a pipe after the pipe has been cut.

Escariador. Herramienta utilizada para remover las rebabas de un tubo después de haber sido cortado.

Receiver-drier A component in a refrigeration system for storing and drying refrigerant.

Receptor-secador. Componente en un sistema de refrigeración que almacena y seca el refrigerante.

Reciprocating Back-and-forth motion.

Movimiento alternativo. Movimiento de atrás para adelante.

Reciprocating compressor A compressor that uses a piston in a cylinder and a back-and-forth motion to compress vapor.

Compresor alternativo. Compresor que utiliza un pistón en un cilindro y un movimiento de atrás para adelante a fin de comprir el vapor.

Rectifier A device for changing alternating current to direct current.

Rectificador. Dispositivo utilizado para convertir corriente alterna en corriente continua.

Reed valve A thin steel plate used as a valve in a compressor.

Válvula con lámina. Placa delgada de acero utilizada como una válvula en un compresor.

Refrigerant The fluid in a refrigeration system that changes from a liquid to a vapor and back to a liquid at practical pressures.

Refrigerante. Fluido en un sistema de refrigeración que se convierte de líquido en vapor y nuevamente en líquido a presiones prácticas.

Refrigerant reclaim (on site) Recovering the refrigerant and processing it so that it can be reused.

Recuperación del refrigerante. La recuperación del refrigerante y su procesamiento para que pueda ser utilizado de nuevo.

Refrigerant reclaim To process refrigerant to new product specifications by means that may include distillation. It will require chemical analysis of the refrigerant to determine that appropriate product specifications are met. This term usually implies the use of processes or procedures available only at a reprocessing or manufacturing facility.

Recuperación del refrigerante. Procesar refrigerante según nuevas especificaciones para productos a través de métodos que pueden incluir la destilación. Se requiere un análisis químico del refrigerante para asegurar el cumplimiento de las especificaciones para productos adecuadas. Por lo general este término supone la utilización de procesos o de procedimientos disponibles solamente en fábricas de reprocesamiento o manufactura.

Refrigerant recovery   To remove refrigerant in any condition from a system and store it in an external container without necessarily testing or processing it in any way.

Recobrar refrigerante líquido.   Remover refrigerante en cualquier estado de un sistema y almacenarlo en un recipiente externo sin ponerlo a prueba o elaborarlo de ninguna manera.

Refrigerant recycling   To clean the refrigerant by oil separation and single or multiple passes through devices, such as replaceable core filterdriers, which reduce moisture, acidity, and particulate matter. This term usually applies to procedures implemented at the job site or at a local service shop.

Recirculación de refrigerante.   Limpieza del refrigerante por medio de la separación del aceite y pasadas sencillas o múltiples a través de dispositivos, como por ejemplo secadores filtros con núcleos reemplazables que disminyen la humedad, la acidez y las partículas. Por lo general este término se aplica a los procedimientos utilizados en el lugar del trabajo o en un taller de servicio local.

Refrigeration   The process of removing heat from a place where it is not wanted and transferring that heat to a place where it makes little or no difference.

Refrigeración.   Proceso de remover el calor de un lugar donde no es deseado y transferirlo a un lugar donde no afecte la temperatura.

Register   A terminal device on an air distribution system that directs air but also has a damper that adjusts airflow.

Registro.   Dispositivo terminal en un sistema de distribución de aire que dirige el aire y además tiene un desviador para ajustar su flujo.

Relative humidity   The amount of moisture contained in the air as compared to the amount the air could hold at that temperature.

Humedad relativa.   Cantidad de humedad presente en el aire, comparada con la cantidad de humedad que el aire pueda contener a dicha temperatura.

Relay   A small electromagnetic device used to control a switch, motor, or valve.

Relé.   Pequeño dispositivo electromagnético utilizado para regular un conmutador, un motor o una válvula.

Relief valve   A valve designed to open and release liquids at a certain pressure.

Válvula para alivio.   Válvula diseñada para abrir y liberar líquidos a una presión específica.

Remote system   Often called a *split system*, where the condenser is located away from the evaporator and/or other parts of the system.

Sistema remoto.   Llamado muchas veces sistema separado donde el condensador se coloca lejos del evaporador y/u otras piezas del sistema.

Resistance   The opposition to the flow of an electrical current or a fluid.

Resistencia.   Oposición al flujo de una corriente eléctrica o de un fluido.

Resistor   An electrical or electronic component with a specific opposition to electron flow. It is used to create voltage drop or heat.

Resistor.   Componente eléctrico o eletrónico con una oposición específica al flujo de electrones; se utiliza para producir una caída de tensión o calor.

Restrictor   A device used to create a planned resistance to fluid flow.

Limitador.   Dispositivo utilizado para producir una resistencia proyectada al flujo de fluido.

Reverse cycle   The ability to direct the hot gas flow into the indoor or the outdoor coil in a heat pump to control the system for heating or cooling purposes.

Ciclo invertido.   Capacidad de dirigir el flujo de gas caliente dentro de la bobina interior o exterior en una bomba de calor a fin de regular el sistema para propósitos de calentamiento o enfriamiento.

Rod and tube The rod and tube are each made of a different metal. The tube has a high expansion rate and the rod has a low expansion rate.

Varilla y tubo. La varilla y el tubo se fabrican de un metal diferente. El tubo tiene una tasa de expansión alta y la varilla una tasa de expansión baja.

Rotary compressor A compressor that uses rotary motion to pump fluids. It is a positive displacement pump.

Compresor giratorio. Compresor que utiliza un movimiento giratorio para bombear fluidos. Es una bomba de desplazmiento positivo.

Rotor The rotating or moving component of a motor, including the shaft.

Rotor. Componente giratorio o en movimiento de un motor, incluyendo el árbol.

Running time The time a unit operates. Also called the *on time.*

Período de funcionamiento. El período de tiempo en que funciona una unidad. Conocido también como *período de conexión.*

Run winding The electrical winding in a motor that draws current during the entire running cycle.

Devanado de funcionamiento. Devanado eléctrico en un motor que consume corriente durante todo el ciclo de funcionamiento.

Rupture disk Safety device for a centrifugal low-pressure chiller.

Disco de ruptura. Dispositivo de seguridad para un enfriador centrífugo de baja presión.

S

Saddle valve A valve that straddles a fluid line and is fastened by solder or screws. It normally contains a device to puncture the line for pressure readings.

Válvula de silleta. Válvula que está sentada a horcajadas en un conducto de fluido y se fija por medio de la soldadura o tornillos. Por lo general contiene un dispositivo para agujerear el conducto a fin de que se puedan tomar lecturas de presión.

Safety control An electrical, mechanical, or electromechanical control to protect either the equipment or the public from harm.

Regulador de seguridad. Regulador eléctrico, mecánico o electromecánico para proteger al equipo de posibles averías o al público de sufrir alguna lesión.

Safety plug A fusible plug.

Tapón de seguridad. Tapón fusible.

Sail switch A safety switch with a lightweight sensitive sail that operates by sensing an airflow.

Conmutador con vela. Conmutador de seguridad con una vela liviana sensible que funciona al advertir el flujo de aire.

Saturated vapor The refrigerant when all of the liquid has changed to a vapor.

Vapor saturado. El refrigerante cuando todo el líquido se ha convertido en vapor.

Saturation A term used to describe a substance when it contains all of another substance it can hold.

Saturación. Término utilizado para describir una sustancia cuando contiene lo más que puede de otra sustancia.

Scavenger pump A pump used to remove the fluid from a sump.

Bomba de barrido. Bomba utilizada para remover el fluido de un sumidero.

Schraeder valve A valve similar to the valve on an auto tire that allows refrigerant to be charged or discharged from the system.

Válvula Schraeder. Válvula similar a la válvula del neumático de un automóvil que permite la entrada o la salida de refrigerante del sistema.

Scotch yoke A mechanism used to create reciprocating motion from the electric motor drive in very small compressors.

Yugo escocés. Mecanismo utilizado para producir movimiento alternativo del accionador del motor eléctrico en compresores bastante pequeños.

Screw compressor A form of positive displacement compressor that squeezes fluid from a low-pressure area to a high-pressure area, using screw-type mechanisms.

Compresor de tornillo. Forma de compresor de desplazamiento positivo que introduce por fuerza el fluido de un área de baja presión a un área de alta presión, a través de mecanismos de tipo de tornillo.

Scroll compressor A compressor that uses two scroll-type components to compress vapor.

Compresor espiral. Compresor que utiliza dos componentes de tipo espiral para comprimir el vapor.

Sealed unit The term used to describe a refrigeration system, including the compressor, that is completely welded closed. The pressures can be accessed by saddle valves.

Unidad sellada. Término utilizado para describir un sistema de refrigeración, incluyendo el compresor, que es soldado completamente cerrado. Las presiones son accesibles por medio de válvulas de silleta.

Seasonal energy efficiency ratio (SEER) An equipment efficiency rating that takes into account the start-up and shutdown for each cycle.

Relación del rendimiento de energía temporal. Clasificación del rendimiento de un equipo que toma en cuenta la puesta en marcha y la parada de cada ciclo.

Seat The stationary part of a valve that the moving part of the valve presses against for shutoff.

Asiento. Pieza fija de una válvula contra la que la pieza en movimiento de la válvula presiona para cerrarla.

Semiconductor A component in an electronic system that is considered neither an insulator nor a conductor, but rather a partial conductor.

Semiconductor. Componente en un sistema eléctrico que no se considera ni aislante ni conductor, sino conductor parcial.

Semihermetic compressor A motor compressor that can be opened or disassembled by removing bolts and flanges. Also known as a *serviceable hermetic.*

Compresor semihermético. Compresor de un motor que puede abrirse o desmontarse al removerle los pernos y bridas. Conocido también como *hermético utilizable.*

Sensible heat Heat that causes a change in the level of a thermometer.

Calor sensible. Calor que produce un cambio en el nivel de un termómetro.

Sensor A component for detection that changes shape, form, or resistance when a condition changes.

Sensor. Componente para la deteción que cambia de forma o de resistencia cuando cambia una condición.

Sequencer A control that causes a staging of events, such as a sequencer between stages of electric heat.

Regulador de secuencia. Regulador que produce una sucesión de acontecimientos, como por ejemplo etapas sucesivas de calor eléctrico.

Series circuit An electrical or piping circuit where all of the current or fluid flows through the entire circuit.

Circuito en serie. Circuito eléctrico o de tubería donde toda la corriente o todo el fluido fluye a través de todo el circuito.

Serviceable hermetic See semihermetic compressor.

Compresor hermético utilizable. Véase compresor semihermético.

Service valve A manually operated valve in a refrigeration system used for various service procedures.

Válvula de servicio. Válvula de un sistema de refrigeración accionada manualmente que se utiliza en varios procedimientos de servicio.

Shaded pole motor An alternating current motor used for very light loads.

Motor polar en sombra. Motor de corriente alterna utilizado en cargas sumamente livianas.

Shell and coil A vessel with a coil of tubing inside that is used as a heat exchanger.

Coraza y bobina. Depósito con una bobina de tubería en su interior que se utiliza como intercambiador de calor.

Shell and tube A heat exchanger with straight tubes in a shell that can normally be mechanically cleaned.

Coraza y tubo. Intercambiador de calor con tubos rectos en una coraza que por lo general puede limpiarse mecánicamente.

Short circuit A circuit that does not have the correct measurable resistance; too much current flows causing the conductors to overload.

Cortocircuito. Corriente que no tiene la resistencia medible correcta; un exceso de corriente fluye a través del circuito provocando una sobrecarga de los conductores.

Short cycle The term used to describe the running time (on time) of a unit when it is not running long enough.

Ciclo corto. Término utilizado para describir el período de funcionamiento (de encendido) de una unidad cuando no funciona por un período de tiempo suficiente.

Shroud A fan housing that ensures maximum airflow through the coil.

Bóveda. Alojamiento del abanico que asegura un flujo máximo de aire a través de la bobina.

Sight glass A clear window in a fluid line.

Mirilla para observación. Ventana clara en un conducto de fluido.

Silica gel A chemical compound often used in refrigerant driers to remove moisture from the refrigerant.

Gel silíceo. Compuesto químico utilizado a menudo en secadores de refrigerantes para remover la humedad del refrigerante.

Silicon A substance from which many semiconductors are made.

Silicio. Sustancia de la cual se fabrican muchos semiconductores.

Silicon-controlled rectifier (SCR) A semiconductor control device.

Rectificador controlado por silicio. Dispositivo para regular un semiconductor.

Silver brazing A high-temperature (above 800°F) brazing process for bonding metals.

Soldadura con plata. Soldadura a temperatura alta (sobre los 800°F ó 430°C) para unir metales.

Sine wave The graph or curve used to describe the characteristics of alternating current voltage.

Onda sinusoidal. Gráfica o curva utilizada para describir las características de tensión de corriente alterna.

Single-phase The electrical power supplied to equipment or small motors, normally under 7-1/2 hp.

Monofásico. Potencia eléctrica suministrada a equipos o motores pequeños, por lo general menor de 7-1/2 hp.

Single-phasing The condition in a three-phase motor when one phase of the power supply is open.

Fasaje sencillo. Condición en un motor trifásico cuando una fase de la fuente de alimentación está abierta.

Sling psychrometer A device with two thermometers, one a wet bulb and one a dry bulb, used for checking the condition of the air, the temperature, and the humidity.

Sicrómetro con eslinga.   Dispositivo con dos termómetros, uno con una bombilla húmeda y otro con una bombilla seca, utilizados para revisar las condiciones del aire, de la temperatura y de la humedad.

Slip   The difference in the rated RPM of a motor and the actual operating RPM.

Deslizamiento.   Diferencia entre las RPM nominales de un motor y las RPM de funcionamiento reales.

Slugging   A term used to describe the condition when large amounts of liquid enter a pumping compressor cylinder.

Relleno.   Término utilizado para describir la condición donde grandes cantidades de líquido entran en el cilindro de un compresor de bombeo.

Smoke test   A test performed to determine the amount of unburned fuel in an oil burner flue gas sample.

Prueba de humo.   Prueba llevada a cabo para determinar la cantidad de combustible no quemado en una muestra de gas de combustión que se obtiene de un quemador de aceite pesado.

Snap-disc   An application of the bimetal. Two different metals fastened together in the form of a disc that provide a warping condition when heated. This also provides a snap action that is beneficial in controls that start and stop current flow in electrical circuits.

Disco de acción rápida.   Aplicación del bimetal. Dos metales diferentes fijados entre sí en forma de un disco que provee una deformación al ser calentado. Esto provee también una acción rápida, ventajosa para reguladores que ponen en marcha y detienen el flujo de corriente en circuitos eléctricos.

Solar collectors   Components of a solar system designed to collect the heat from the sun, using air, a liquid, or a refrigerant as the medium.

Colectores solares.   Componentes de un sistema solar diseñados para acumular el calor emitido por el sol, utilizando el aire, un líquido o un refrigerante como el medio.

Solar heat   Heat from the sun's rays.

Calor solar.   Calor emitido por los rayos del sol.

Soldering   Fastening two base metals together by using a third filler metal that melts at a temperature below 800°F.

Soldadura.   La fijación entre sí de dos metales bases utilizando un tercer metal de relleno que se funde a una temperatura menor de 800°F (430°C).

Solder pot   A device using a low-melting solder and an overload heater sized for the amperage of the motor it is protecting. The solder will melt, opening the circuit when there is an overload.

Olla para soldadura.   Dispositivo que utiliza una soldadura con un punto de fusión bajo y un calentador de sobrecarga diseñado para el amperaje del motor al que provee protección. La soldadura se fundirá, abriendo así el circuito cuando ocurra una sobrecarga. Puede ser reconectado.

Solenoid   A coil of wire designed to carry an electrical current producing a magnetic field.

Solenoide.   Bobina de alambre diseñada para conducir una corriente eléctrica generando un campo magnético.

Solid   Molecules of a solid are highly attracted to each other and form a mass that exerts all of its weight downward.

Sólido.   Las moléculas de un sólido se atraen entre sí y forman una masa que ejerce todo su peso hacia abajo.

Specific gravity   The weight of a substance compared to the weight of an equal volume of water.

Gravedad específica.   El peso de una sustancia comparada con el peso de un volumen igual de agua.

Specific heat   The amount of heat required to raise the temperature of 1 pound of a substance 1°F.

Calor específico. La cantidad de calor requerido para elevar la temperatura de una libra de una sustancia 1°F (–17°C).

Specific volume The volume occupied by 1 pound of a fluid.

Volumen específico. Volumen que ocupa una libra de fluido.

Splash lubrication system A system of furnishing lubrication to a compressor by agitating the oil.

Sistema de lubrificación por salpicadura. Método de proveerle lubrificación a un compresor agitando el aceite.

Splash method A method for more efficient evaporation in which water, dropping from a higher level in a cooling tower, splashes on slots with air passing through.

Método de salpicaduras. Método de dejar caer agua desde un nivel más alto en una torre de refrigeración y salpicándola en ranuras, mientras el aire pasa a través de las mismas con el propósito de lograr una evaporación más eficaz.

Split phase motor A motor with run and start windings.

Motor de fase separada. Motor con devandos de funcionamiento y de arranque.

Split system A refrigeration or air-conditioning system that has the condensing unit remote from the indoor (evaporator) coil.

Sistema separado. Sistema de refrigeración o de acondicionamiento de aire cuya unidad de condensación se encuentra en un sitio alejado de la bobina interior del evaporador.

Spray pond A pond with spray heads used for cooling water in water-cooled air-conditioning or refrigeration systems.

Tanque de rociado. Tanque con una cabeza rociadora utilizada para enfriar el agua en sistemas de acondicionamiento de aire o de refrigeración enfriados por agua.

Squirrel cage fan A fan assembly used to move air.

Abanico con jaula de ardilla. Conjunto de abanico utilizado para mover el aire.

Standard atmosphere or **standard conditions** Air at sea level at 70°F when the atmosphere's pressure is 14.696 PSIA (29.92 in Hg). Air at this condition has a volume of 13.33 ft³/lb.

Atmósfera estándar o condiciones estándares. El aire al nivel del mar a una temperatura de 70°F (21°C) cuando la presión de la atmósfera es 14,696 PSIA (29,92 pulgadas Hg). Bajo esta condición, el aire tiene un volumen de 13,33 ft³lb (libras/pies).

Standing pilot Pilot flame that remains burning continuously.

Piloto constante. Llama piloto que se quema de manera continua.

Start capacitor A capacitor used to help start an electric motor.

Capacitador de arranque. Capacitador utilizado para ayudar en el arranque de un motor eléctrico.

Starting relay An electrical relay used to disconnect the start winding in a hermetic compressor.

Relé de arranque. Relé eléctrico utilizado para desconectar el devanado de arranque en un compresor hermético.

Start winding The winding in a motor used primarily to give the motor extra starting torque.

Devanado de arranque. Devanado en un motor utilizado principalmente para proveerle al motor mayor par de arranque.

Starved coil The condition in an evaporator when the metering device is not feeding enough refrigerant to the evaporator.

Bobina estrangulada. Condición que ocurre en un evaporador cuando el dispositivo de medida no le suministra suficiente refrigerante al evaporador.

Stator The component in a motor that contains the windings; it does not turn.

Estátor. Componente en un motor que contiene los devanados y que no gira.

Steam   The vapor state of water.

Vapor.   Estado de vapor del agua.

Strainer   A fine-mesh device that allows fluid flow and holds back solid particles.

Colador.   Dispositivo de malla fina que permite el flujo de fluido a través de él y atrapa partículas sólidas.

Stratification   The condition where a fluid appears in layers.

Estratificación.   Condición que ocurre cuando un fluido aparece en capas.

Stress crack   A crack in piping or in another component caused by age or abnormal conditions such as vibration.

Grieta por tensión.   Grieta que aparece en una tubería u otro componente ocasionada por envejecimiento o condiciones anormales, como por ejemplo vibración.

Subbase   The part of a space temperature thermostat that is wall-mounted and is attached to the interconnecting wiring.

Subbase.   Pieza de un termóstato que mide la temperatura de un espacio que se monta sobre la pared y a la que se fijan los conductores eléctricos interconectados.

Subcooling   The temperature of a liquid when it is cooled below its condensing temperature.

Subenfriamiento.   La temperatura de un líquido cuando se enfría a una temperatura menor que su temperatura de condensación.

Sublimation   When a substance changes from the solid state to the vapor state without going through the liquid state.

Sublimación.   Cuando una sustancia cambia de sólido a vapor sin convertirse primero en líquido.

Suction gas   The refrigerant vapor in an operating refrigeration system found in the tubing from the evaporator to the compressor and in the compressor shell.

Gas de aspiración.   El vapor del refrigerante en un sistema de refrigeración en funcionamiento presente en la tubería que va del evaporador al compresor y en la coraza del compresor.

Suction line   The pipe that carries the heat-laden refrigerant gas from the evaporator to the compressor.

Conducto de aspiración.   Tubo que conduce el gas de refrigerante lleno de calor del evaporador al compresor.

Suction service valve   A manually operated valve with front and back seats located at the compressor.

Válvula de aspiración para servicio.   Válvula accionada manualmente que tiene asientos delanteros y traseros ubicados en el compresor.

Suction valve lift unloading   When the suction valve in a reciprocating compressor cylinder is lifted, causing that cylinder to stop pumping.

Descarga por levantamiento de la válvula de aspiración.   La válvula de aspiración en el cilindro de un compresor alternativo se levanta, provocando que el cilindro deje de bombear.

Sump   A reservoir at the bottom of a cooling tower to collect the water that has passed through the tower.

Sumidero.   Tanque que se encuentra en el fondo de una torre de refrigeración para acumular el agua que ha pasado a través de la torre.

Superheat   The temperature of vapor refrigerant above its saturated change of state temperature.

Sobrecalor.   Temperatura del refrigerante de vapor mayor que su temperatura de cambio de estado de saturación.

Surge   When the head pressure becomes too great or the evaporator pressure too low, refrigerant will flow from the high- to the low-pressure side of a centrifugal compressor system, making a loud sound.

Movimiento repentino.   Cuando la presión en la cabeza aumenta demasiado o la presión en el evaporador es demasiado baja, el refrigerante fluye del lado de alta presión al lado de baja presión de un sistema de compresor centrífugo. Este movimiento produce un sonido fuerte.

Swaged joint The joining of two pieces of copper tubing by expanding or stretching the end of one piece of tubing to fit over the other piece.

Junta estampada. La conexión de dos piezas de tubería de cobre dilatando o alargando el extremo de una pieza de tubería para ajustarla sobre otra.

Swaging tool A tool used to enlarge a piece of tubing for a solder or braze connection.

Herramienta de estampado. Herramienta utilizada para agrandar una pieza de tubería a utilizarse en una conexión soldada o broncesoldada.

Swamp cooler A slang term used to describe an evaporative cooler.

Nevera pantanoso. Término del argot utilizado para describir una nevera de evaporación.

Sweating A word used to describe moisture collection on a line or coil that is operating below the dew point temperature of the air.

Exudación. Término utilizado para describir la acumulación de humedad en un conducto o una bobina que está funcionando a una temperatura menor que la del punto de rocío del aire.

T

Tank A closed vessel used to contain a fluid.

Tanque. Depósito cerrado utilizado para contener un fluido.

Tap A tool used to cut internal threads in a fastener or fitting.

Macho de roscar. Herramienta utilizada para cortar filetes internos en un aparto fijador o en un accesorio.

Temperature A word used to describe the level of heat or molecular activity, expressed in Fahrenheit, Rankine, Celsius, or Kelvin units.

Temperatura. Término utilizado para describir el nivel de calor o actividad molecular, expresado en unidades Fahrenheit, Rankine, Celsio o Kelvin.

Test light A lightbulb arrangement used to prove the presence of electrical power in a circuit.

Luz de prueba. Arreglo de bombillas utilizado para probar la presencia de fuerza eléctrica en un circuito.

Therm Quantity of heat, 100,000 Btu's.

Therm. Cantidad de calor, mil unidades térmicas inglesas.

Thermistor A semiconductor electronic device that changes resistance with a change in temperature.

Termistor. Dispositivo eléctrico semiconductor que cambia su resistencia cuando se produce un cambio en temperatura.

Thermocouple A device made of two unlike metals that generates electricity when there is a difference in temperature from one end to the other. Thermocouples have both a hot and a cold junction.

Thermopar. Dispositivo hecho de dos metales distintos que genera electricidad cuando hay una diferencia en temperatura de un extremo al otro. Los termopares tienen un empalme caliente y uno frío.

Thermometer An instrument used to detect differences in the level of heat.

Termómetro. Instrumento utilizado para detectar diferencias en el nivel de calor.

Thermopile A group of thermocouples connected in series to increase voltage output.

Pila termoeléctrica. Grupo de termopares conectados en serie para aumentar la salida de tensión.

Thermostat A device that senses temperature change and also changes some dimension or condition within to control an operating device.

Termostato. Dispositivo que advierte un cambio en temperatura y cambia alguna dimensión o condición dentro de sí para regular un dispositivo en funcionamiento.

Thermostatic expansion valve (TXV) A valve used in refrigeration systems to control the superheat in an evaporator by metering the correct refrigerant flow to the evaporator.

Válvula de gobierno termostático para expansión. Válvula utilizada en sistemas de refrigeración para regular el sobrecalor en un evaporador midiendo el flujo correcto de refrigerante al evaporador.

Three-phase power A type of power supply usually used for operating heavy loads. It consists of three sine waves that are out of phase with each other.

Potencia trifásica. Tipo de fuente de alimentación normalmente utilizada en el funcionamiento de cargas pesadas. Consiste de tres ondas sinusoidales que no están en fase la una con la otra.

Throttling Creating a restriction in a fluid line.

Estrangulamiento. Que ocasiona una restricción en un conducto de fluido.

Timers Clock-operated devices used to time various sequences of events in circuits.

Temporizadores. Dispositivos accionados por un reloj utilizados para medir el tiempo de varias secuencias de eventos en circuitos.

Ton of refrigeration The amount of heat required to melt a ton (2,000 lb) of ice at 32°F, 288,000 Btu/24 h, 12,000 Btu/h, or 200 Btu/min.

Tonelada de refrigeración. Cantidad de calor necesario para fundir una tonelada (2000 libras) de hielo a 32°F (0°C), 288.000 Btu/24 h, 12.000 Btu/h, o 200 Btu/min.

Torque The twisting force often applied to the starting power of a motor.

Par de torsión. Fuerza de torsión aplicada con frecuencia a la fuerza de arranque de un motor.

Torque wrench A wrench used to apply a prescribed amount of torque or tightening to a connector.

Llave de torsión. Llave utilizada para aplicar una cantidad específica de torsión o de apriete a un conector.

Total heat The total amount of sensible heat and latent heat contained in a substance from a reference point.

Calor total. Cantidad total de calor sensible o de calor latente presente en una sustancia desde un punto de referencia.

Transformer A coil of wire wrapped around an iron core that induces a current to another coil of wire wrapped around the same iron core. Note: A transformer can have an air core.

Transformador. Bobina de alambre devanado alrededor de un núcleo de hierro que induce una corriente a otra bobina de alambre devanado alrededor del mismo núcleo de hierro. Nota: Un transformador puede tener un núcleo de aire.

Transistor A semiconductor often used as a switch or amplifier.

Transistor. Semiconductor que suele utilizarse como conmutador o amplificador.

TRIAC A semiconductor switching device.

TRIAC. Dispositivo de conmutación para semiconductores.

Tube-within-a-tube coil A coil used for heat transfer that has a pipe in a pipe and is fastened together so that the outer tube becomes one circuit and the inner tube another.

Bobina de tubo dentro de un tubo. Bobina utilizada en la transferencia de calor que tiene un tubo dentro de otro y se sujeta de manera que el tubo exterior se convierte en un circuito y el tubo interior en otro circuito.

Tubing Pipe with a thin wall used to carry fluids.

Tubería. Tubo que tiene una pared delgada utilizado para conducir fluidos.

Two-temperature valve A valve used in systems with multiple evaporators to control the evaporator pressures and maintain different temperatures in each evaporator. Sometimes called a *hold-back valve.*

Válvula de dos temperaturas. Válvula utilizada en sistemas con evaporadores múltiples para regular las presiones de los evaporadores y mantener temperaturas diferentes en cada uno de ellos. Conocida también como *válvula de retención*.

U

Ultraviolet Light waves that can only be seen under a special lamp.

Ultravioleta. Ondas de luz que pueden observarse solamente utilizando una lámpara especial.

Urethane foam A foam that can be applied between two walls for insulation.

Espuma de uretano. Espuma que puede aplicarse entre dos paredes para crear un aislamiento.

U-Tube mercury manometer A U-tube containing mercury, which indicates the level of vacuum while evacuating a refrigeration system.

Manómetro de mercurio de tubo en U. Tubo en U que contiene mercurio y que indica el nivel del vacío mientras vacía un sistema de refrigeración.

U-Tube water manometer Indicates natural gas and propane gas pressures. It is usually calibrated in inches of water.

Manómetro de agua de tubo en U. Indica las presiones del gas natural y del propano. Se calibra normalmente en pulgadas de agua.

V

Vacuum The pressure range between the earth's atmosphere and no pressure, normally expressed in inches of mercury (in Hg) vacuum.

Vacío. Margen de presión entre la atmósfera de la Tierra y cero presión, por lo general expresado en pulgadas de mercurio (pulgadas Hg) en vacío.

Vacuum pump A pump used to remove some fluids such as air and moisture from a system at a pressure below the earth's atmosphere.

Bomba de vacío. Bomba utilizada para remover algunos fluidos, como por ejemplo aire y humedad de un sistema a una presión menor que la de la atmósfera de la Tierra.

Valve A device used to control fluid flow.

Válvula. Dispositivo utilizado para regular el flujo de fluido.

Valve plate A plate of steel bolted between the head and the body of a compressor that contains the suction and discharge reed or flapper valves.

Placa de válvula. Placa de acero empernado entre la cabeza y el cuerpo de un compresor que contiene la lámina de aspiración y de descarga o las chapaletas.

Valve seat That part of a valve that is usually stationary. The movable part comes in contact with the valve seat to stop the flow of fluids.

Asiento de la válvula. Pieza de una válvula que es normalmente fija. La pieza móvil entra en contacto con el asiento de la válvula para detener el flujo de fluidos.

Vapor The gaseous state of a substance.

Vapor. Estado gaseoso de una sustancia.

Vapor barrier A thin film used in construction to keep moisture from migrating through building materials.

Película impermeable. Película delgada utilizada en construcciones para evitar que la humeded penetre a través de los materiales de construcción.

Vapor charge valve A charge in a thermostatic expansion valve bulb that boils to a complete vapor. When this point is reached, an increase in temperature will not produce an increase in pressure.

Válvula para la carga de vapor. Carga en la bombilla de una válvula de expansión termostática que hierve a un vapor completo. Al llegar a este punto, un aumento en temperatura no produce un aumento en presión.

Vaporization The changing of a liquid to a gas or vapor.

Vaporización. Cuando un líquido se convierte en gas o vapor.

Vapor lock A condition where vapor is trapped in a liquid line and impedes liquid flow.

Bolsa de vapor. Condición que ocurre cuando el vapor queda atrapado en el conducto de líquido e impide el flujo de líquido.

Vapor pump Another term for compressor.

Bomba de vapor. Otro término para compresor.

Vapor refrigerant charging Adding refrigerant to a system by allowing vapor to move out of the vapor space of a refrigerant cylinder and into the low-pressure side of the refrigeration system.

Carga del refrigerante de vapor. Agregarle refrigerante a un sistema permitiendo que el vapor salga del espacio de vapor de un cilindro de refrigerante y que entre en el lado de baja presión del sistema de refrigeración.

Variable pitch pulley A pulley whose diameter can be adjusted.

Polea de paso variable. Polea cuyo diámetro puede ajustarse.

Variable resistor A type of resistor where the resistance can be varied.

Resistor variable. Tipo de resistor donde la resistencia puede variarse.

V-belt A belt that has a V-shaped contact surface and is used to drive compressors, fans, or pumps.

Correa en V. Correa que tiene una superficie de contacto en forma de V y se utiliza para accionar compresores, abanicos o bombas.

Velocity The speed at which a substance passes a point.

Velocidad. Rapidez a la que una sustancia sobrepasa un punto.

Velocity meter A meter used to detect the velocity of fluids, air, or water.

Velocímetro. Instrumento utilizado para medir la velocidad de fluidos, aire o agua.

Voltage The potential electrical difference for electron flow from one line to another in an electrical circuit.

Tensión. Diferencia de potencial eléctrico del flujo de electrones de un conducto a otro en un circuito eléctrico.

Voltmeter An instrument used for checking electrical potential.

Voltímetro. Instrumento utilizado para revisar la potencia eléctrica.

Volt-ohm-milliampmeter (VOM) A multimeter that measures voltage, resistance, and current in milliamperes.

Voltio-ohmio-miliamperímetro. Multímetro que mide tensión, resistencia y corriente en miliamperios.

Volumetric efficiency The pumping efficiency of a compressor or vacuum pump that describes the pumping capacity in relationship to the actual volume of the pump.

Rendimiento volumétrico. Rendimiento de bombeo de un compresor o de una bomba de vacío que describe la capacidad de bombeo con relación al volumen real de la bomba.

Vortexing A whirlpool action in the sump of a cooling tower.

Acción de vórtice. Torbellino en el sumidero de una torre de refrigeración.

W

Walk-in cooler A large refrigerated space used for storage of refrigerated products.

Nevera con acceso al interior. Espacio refrigerado grande utilizado para almacenar productos refrigerados.

Water box A container or reservoir at the end of a chiller where water is introduced and contained.

Caja de agua. Recipiente o depósito al extremo de un enfriador por donde entra y se retiene el agua.

Water column (WC) The pressure it takes to push a column of water up vertically. One inch of water column is the amount of pressure it would take to push a column of water in a tube up one inch.

Columna de agua. Presión necesaria para levantar una columna de agua verticalmente. Una pulgada de columna de agua es la cantidad de presión necesaria para levantar una columna de agua a una distancia de una pulgada en un tubo.

Water-cooled condenser A condenser used to reject heat from a refrigeration system into water.

Condensador enfriado por agua. Condensador utilizado para dirigir el calor de un sistema de refrigeración al agua.

Water regulating valve An operating control regulating the flow of water.

Válvula reguladora de agua. Regulador de mando que controla el flujo de agua.

Watt A unit of power applied to electron flow. One watt equals 3.414 Btu's.

Watio. Unidad de potencia eléctrica aplicada al flujo de electrones. Un watio equivale a 3,414 Btu.

Watt-hour The unit of power that takes into consideration the time of consumption. It is the equivalent of a 1-watt bulb burning for one hour.

Watio hora. Unidad de potencia eléctrica que toma en cuenta la duración de consumo. Es el equivalente de una bombilla de 1 watio encendida por espacio de una hora.

Wet bulb temperature A wet bulb temperature of air is used to evaluate the humidity in the air. It is obtained with a wet thermometer bulb to record the evaporation rate with an airstream passing over the bulb to help in evaporation.

Temperatura de una bombilla húmeda. La temperatura de una bombilla húmeda se utiliza para evaluar la humedad presente en el aire. Se obtiene con la bombilla húmeda de un termómetro para registrar el margen de evaporación con un flujo de aire circulando sobre la bombilla para ayudar en evaporar el agua.

Wet heat A heating system using steam or hot water as the heating medium.

Calor húmedo. Sistema de calentamiento que utiliza vapor o agua caliente como medio de calentamiento.

Window unit An air conditioner installed in a window that rejects the heat outside the structure.

Acondicionador de aire para la ventana. Acondicionador de aire instalado en una ventana que desvía el calor proveniente del exterior de la estructura.

Work A force moving an object in the direction of the force. Work = Force × Distance.

Trabajo. Fuerza que mueve un objeto en la dirección de la fuerza. Trabajo = Fuerza × Distancia.

INDEX